Lecture Notes in Physics

Volume 1011

The series Lecture Notes in Physics (LNP), founded in 1969, reports new developments in physics research and teaching - quickly and informally, but with a high quality and the explicit aim to summarize and communicate current knowledge in an accessible way. Books published in this series are conceived as bridging material between advanced graduate textbooks and the forefront of research and to serve three purposes:

- to be a compact and modern up-to-date source of reference on a well-defined topic;
- to serve as an accessible introduction to the field to postgraduate students and non-specialist researchers from related areas;
- to be a source of advanced teaching material for specialized seminars, courses and schools.

Both monographs and multi-author volumes will be considered for publication. Edited volumes should however consist of a very limited number of contributions only. Proceedings will not be considered for LNP.
Volumes published in LNP are disseminated both in print and in electronic formats, the electronic archive being available at springerlink.com. The series content is indexed, abstracted and referenced by many abstracting and information services, bibliographic networks, subscription agencies, library networks, and consortia.
Proposals should be sent to a member of the Editorial Board, or directly to the responsible editor at Springer:

Dr Lisa Scalone
Springer Nature
Physics
Tiergartenstrasse 17
69121 Heidelberg, Germany
lisa.scalone@springernature.com

Chon-Fai Kam • Wei-Min Zhang •
Da-Hsuan Feng

Coherent States

New Insights into Quantum Mechanics with Applications

Springer

Chon-Fai Kam
Physics Department
University at Buffalo
Buffalo, NY, USA

Wei-Min Zhang
Physics Department
National Cheng Kung University
Tainan, Taiwan

Da-Hsuan Feng
Formerly at Department of Physics
Drexel University
Philadelphia, PA, USA

ISSN 0075-8450 ISSN 1616-6361 (electronic)
Lecture Notes in Physics
ISBN 978-3-031-20765-5 ISBN 978-3-031-20766-2 (eBook)
https://doi.org/10.1007/978-3-031-20766-2

This Springer imprint is published by the registered company Springer Nature Switzerland AG
The registered company address is: Gewerbestrasse 11, 6330 Cham, Switzerland

This book is dedicated to our loving families who supported us unconditionally.

Preface and Acknowledgments

This book is a labor of love by the three of us, a love of the subject of coherent states and its applications in many areas of physics for nearly four decades. It is also unusual that the three of us belong to three intellectual generations!

The genesis of this book's idea came about when Wei-Min Zhang (WMZ) began his physics graduate studies with Da Hsuan Feng (DHF) at Drexel University in the mid-1980s. It was at a time when the research interests of DHF were in two major directions: the interacting Boson model (IBM) in nuclear structure physics and quantum optics.

The former was pioneered by Franco Iachello of Yale University and Akito Arima of the University of Tokyo. A fundamental aspect of the IBM is its generous deployment of Lie group and Lie algebra to classify the various collective motions of nuclei.

For the latter, DHF who was trained as a nuclear theorist was profoundly influenced by his colleague Professor Lorenzo Narducci at the Department of Physics, Drexel University, who was a leading theorist in the field of quantum optics. In our many initial scientific conversations to eventual intense collaborations, DHF learned from Narducci that in quantum optics, the coherent states that was proposed by Roy Glauber of Harvard University, who is also the first person to coin the term "Coherent States," is literally its foundation. From learning more about the technical details of coherent states, DHF came to the realization that Lie group and Lie algebra also play the underlying mathematical foundation of the field of quantum optics.

It was the juxtaposition of the IBM and quantum optics that allowed WMZ and DHF to collaborate on exploring the general mathematical structures of coherent states and with such a study to explore its many possible physics applications. In fact, in collaboration with another colleague at Drexel University, Robert Gilmore, who had made pioneering studies of the subject of coherent states, we were able to publish an extensive review article (Rev. Mod. Phys. 62, 867—Published 1 October 1990) with the title "Coherent states: Theory and some applications." The content of this review was based on the doctoral thesis of WMZ.

With the increasing importance of the utilization of coherent states in many areas of science, and soon after the review article was published, WMZ and DHF germinated the idea of expanding the review article into a book. In fact, we were even encouraged by some of the leading scientists to do so. Unfortunately, as it was often the case, while our deep interest in coherent states remains, the career paths of WMZ and DHF diverged into different areas and it became non-conducive to the necessary concentration that was required to write such a book.

Fast forward to about 6 years ago, when DHF became a senior administrator at the University of Macau. At that time, CFK, who was formally an undergraduate advisee of WMZ at the National Cheng Kung University in Taiwan, was a doctoral student in physics at the Chinese University of Hong Kong. Quite by serendipity, being a "native" of Macau, on weekends, CFK would travel back to Macau from Hong Kong to visit his family. DHF thus exploited this opportunity by inviting WMZ to visit Macau so we could discuss physics together.

As things progressed, it was not surprising in hindsight that our discussions quickly centered on the dormant interest of WMZ and DHF to write a book about coherent states and applications. It was entirely obvious that CFK by then was already a highly sophisticated and enthusiastic theoretical physicist, and that the idea of us collaborating together to write a book on such a profoundly important subject, however arduous it would be, appealed to his intellectual taste. That was how our collaboration began.

One of us (DHF) was extremely honored to have met Roy Glauber only once, and it was at a conference on quantum optics which was held in 1995 at Jilin University, China. During our extensive discussions on many areas of physics at the conference, especially about his pioneering work on coherent states and some areas that are not in physics, DHF already felt so much wiser having had that conversation with Glauber (Fig. 1).

It is worth underscoring that when DHF met him, which was way before he became a Nobel laureate (which he received in 2005), DHF remembers distinctly

Fig. 1 In the middle of the front row of the photo is Roy Glauber. DHF had a green shirt on

telling him that he definitely should be bestowed the Nobel prize for his work on coherent states. DHF remembers the only answer he got from him was a faint smile!

Furthermore, for the more than three decades of our love of coherent states, we had benefitted enormously from having in-depth discussions, in person and/or in communication with many experts in this vast field. Discussions with pioneers such as Lorenzo Narducci, John Klauder, Robert Gilmore, and Elliott Lieb have certainly greatly shaped our understanding of coherent states and its applications in immeasurable ways, for which we are forever indebted.

We are also deeply grateful to receive the opportunities to have many decades of scientific interaction with one of the founders of the interacting Boson model (IBM), Franco Iachello. His influence on us with his deep knowledge about Lie algebra and Lie group is truly innumerable.

Last but certainly not least, we must thank Dr. Wei Ge, the Vice Rector of Research at the University of Macau. Without his sustainable support of our collaboration, completing this book could have been impossible.

For the readers who are interested in getting hints or solutions (we suggest hints) to the exercises, please contact any one of the three authors (CFK's email address: dubussygauss@gmail.com).

Buffalo, NY, USA — Chon-Fai Kam
Tainan, Taiwan — Wei-Min Zhang
Philadelphia, PA, USA — Da Hsuan Feng
August 15, 2022

Contents

1 Introduction

Over the past half-century, the developments of coherent states have been breathtaking, and their applications have been explored in a wide range of fields. Yet, it should be underscored that the idea of constructing coherent states for a quantum system can be dated back to the genesis of quantum theory in the 1920s. In fact, Erwin Schrödinger himself was the first person to propose the existence of what is now called "coherent states." By 1926, in the same year that Schrödinger formulated his wave equation that accurately calculated the energy levels of electrons in atoms, he published a paper which made the attempt to connect coherent states with the classical mechanics of a quantum harmonic oscillator [1]. In this regard, one could credit Schrödinger with the invention of coherent states immediately after the birth of quantum mechanics.

However, research on coherent states remained dormant in the years between 1926 and 1963. It was not until three and a half decades after Schrödinger's pioneering work that the first important and specific application of coherent states in quantum optics was made by Roy J. Glauber [2–4] and E. C. George Sudarshan [5]. In his two papers, in which the term "coherent states" was first coined, Glauber constructed the eigenstates of the annihilation operator of an electromagnetic field in order to study the electromagnetic correlation functions, which leads to a more comprehensive understanding of optical coherence. Glauber thereby won the 2005 Nobel Prize in Physics for his development of the theory of quantum optics. At the same time as Glauber and Sudarshan, John R. Klauder also developed a method to generate a set of continuous states in which the basic ideas of coherent states for arbitrary Lie groups are introduced [6, 7]. Since then, the field of coherent states has literally become an integral part of modern physics.

Roughly a decade later, by 1972, after the pioneering works of Glauber, Sudarshan, and Klauder, the explicit construction of coherent states for arbitrary Lie groups was successfully demonstrated by Askold M. Perelomov [8] and Robert Gilmore [9, 10]. It should also be underscored that around the same period, in order to study quantization on arbitrary Kähler manifolds, Felix A. Berezin proposed the

C.-F. Kam et al., *Coherent States*, Lecture Notes in Physics 1011,
https://doi.org/10.1007/978-3-031-20766-2_1

usage of a family of over-complete vectors which are mathematically equivalent to the Perelomov-Gilmore coherent states [11–14]. The central idea of Perelomov and Gilmore's constructions was to construct coherent states directly from the dynamical symmetry group for each physical system. For example, a quantized single-mode electromagnetic field possess the Heisenberg-Weyl group H_4 which contains the creation, annihilation, identity, and number operators as generators [15]. Hence, one may construct the Glauber coherent states as an element in the coset space $H_4/U(1) \otimes U(1)$ by applying the displacement operator on the vacuum state [16, 17]. From this perspective, as most physical systems possess dynamical symmetries beyond H_4, the concept of coherent states should not merely be restricted to quantum harmonic oscillator. Indeed, one should be able to generalize it to a broad range of physical systems.

A successful application of the generalized coherent states was in path integrals. As is inherently the main reason, the Feynman original path integral [18,19] required a phase-space structure. As such, it certainly hinders its application in spin systems which do not possess a simple phase-space structure [20]. Also, as a key ingredient of the standard path integral, the resolution of identity which is expressed in terms of the coordinate or momentum states does not directly apply to spin systems. Such difficulties were elegantly resolved through the usage of coherent states. As the coherent states always possess a resolution of identity and a phase-space structure in the geometrical coset space, the coherent state formalism of the path integral can be applied to all physical systems in principle in the sense that the Hilbert space of any quantum mechanical system is given by a unitary representation space of some Lie groups. In other words, one can always find a continuous representation for arbitrary quantum system in terms of the coherent states of the associate Lie group. The coherent state formalism of path integrals expressed in terms of arbitrary continuous representations was first recognized by Klauder in 1960 [21, 22] and was later applied to many-body systems [23–30], single-molecule magnet [31,32], quantum gravity [33–35], and quantum entanglement [36,37].

Another successful application of the generalized coherent states was in the coherent state representation of thermodynamics. In 1973, Elliott H. Lieb derived an ingenious thermodynamics inequality of the partition function for quantum spin systems using atomic coherent states [38]. Prior to this, in 1972, Felix A. Berezin derived the same inequality through the usage of the covariant and contravariant symbols related to an over-complete family of coherent states [39]. The Lieb-Berezin inequality gives both the upper and lower bounds of the quantum free energy. Using this inequality, one may readily construct approximation descriptions of a quantum statistics in terms of the coherent state representation. In particular, in the zero-temperature limit, the free energy can be identified as the ground state energy. Hence, the Lieb-Berezin inequality can be used to construct the upper and lower bounds of the ground state energy in the zero-temperature limit, where the ground state energies are determined by minimizing the Q and P representations of the Hamiltonian, respectively. Having such a theoretical foundation, the coherent approach to the approximation of ground state energies was among the earliest studies of quantum phase transition [40].

In the post 1980s era, a new branch of physics known as quantum computation emerged. It begins with Richard Feynman's realization that sufficiently well-controlled quantum mechanical systems can be used to simulating quantum dynamics in a more efficient way. One key concept in quantum computation is entanglement, which is a unique resource in quantum computation processing. The idea of entanglement was implicitly discussed in Schrödinger's ingenious 1935 seminal paper, in which he proposed that a cat can be simultaneously both alive and dead as a result of a random atomic event such as atomic decay that may or may not occur. Interestingly, Schrödinger's hypothetical cat becomes real when a mesoscopic cat-like states can be prepared through a superposition of coherent states of light. As such, one may even use Schrödinger's cats which composed of entangled coherent states to perform quantum information task. Experimental generating and measuring Schrödinger's cat states and its decoherence evolution is one of the major contributions archived by Serge Haroche and David J. Wineland for winning the 2012 Nobel Prize in Physics.

At the end, readers may find that through the coherent states theory of quantum mechanics, a solution to the long-standing problem about the quantum-classical correspondence, namely, to drive the classical and statistical mechanics solely from the quantum principles, has been presented unambiguously in this book. Simply speaking, the Glauber's coherent states make the direct connection of classical electric field with quantum state, sec Eq. (2.29) of Chap. 1. On the other hand, the path integral formulation given in Chap. 4 makes a clear connection of quantum mechanics with classical mechanics through Feynman's path integrals, where the Lagrangian formulation of classical mechanics can be obtained from the stationary paths of quantum evolution. The coherent state path integrals generalize the phase space structure of classical mechanics to arbitrary quantum systems. Last but not least, the solution of open quantum system by completely integrating out the environmental degrees of freedom, i.e., Eq. 12.48 in Chap. 12, results in the reduced density matrix of the system becoming the standard Gibbs state. From which, conventional statistical mechanics and thermodynamics are recovered from quantum mechanics.

Coherent States of Harmonic Oscillator

2

2.1 Schrödinger's Wave Packet

The coherent states [1, 17, 41] are built on the foundation of the simplest object in quantum mechanics, the harmonic oscillator. As such, it is not surprising that it has been invented many times and for different reasons and perspectives [1, 21, 42–46]. In a series of papers published in 1926 [47], Erwin Schrödinger built the wave formalism of quantum mechanics. Schrödinger is unquestionably one of the founders of quantum mechanics, a body of knowledge that not only reinvented physics but redefined humanity's existence. For example, he was the one who proposed the seemingly illogical gedanken experiment of a "live" and "death" cat to illustrate a fundamental quantum principle, which is also one of the central themes in today's research in quantum information [48, 49]. As such, Schrödinger must have intuitively recognized that there is a fundamentally irreconcilable discrepancy between quantum waves and classical trajectories, a discrepancy which, remarkably after more than 90 years since quantum mechanics was fully established in the 1926 Solvay conference, remains today a challenge to our basic understanding of our physical environment.

Throughout his scientific career, one of the central problems Schrödinger attempted to answer was how to reproduce classical trajectories from quantum wave functions. In his seminal paper titled "*The Continuous Transition From Micro- to Macro-Mechanics*" [1], Schrödinger demonstrated quite stunningly that the discrepancy between the quantum and classical mechanics, while not entirely resolved, can be partially reconciled. This remarkable conclusion of Schrödinger is based on a demonstration that a particular wave packet of proper vibrations with large quantum number can completely represent a particle that follows precisely the classical mechanical path. With this, Schrödinger's wave packet, built entirely on the wave functions of the harmonic oscillator, was how the appearance of coherent states first appeared in the scientific literature.

C.-F. Kam et al., *Coherent States*, Lecture Notes in Physics 1011,
https://doi.org/10.1007/978-3-031-20766-2_2

Since the work of Schrödinger is so fundamental to the discussion of coherent states, we shall present in the following a detail discussion of how Schrödinger constructed such wave packets.

It is of course well known now that the harmonic oscillator Hamiltonian is

$$\hat{H} = \frac{\hat{p}^2}{2m} + \frac{1}{2} m\omega_0^2 q^2,\ \hat{p} = -i\hbar \frac{\partial}{\partial q}, \tag{2.1}$$

where m is the oscillator mass, ω_0 is the oscillation angular frequency of the oscillation, and $\hbar$ is the Planck constant. The wave packet is determined by the Schrödinger equation $i\hbar\partial_t \psi = \hat{H}\psi$, where ψ is the wave function, which is the superposition of the proper vibrations

$$\begin{cases} \psi(q,t) = \sum_{n=0}^{\infty} c_n \psi_n(q,t), \\ \psi_n(q,t) = \phi_n(q) e^{-\frac{i}{\hbar} E_n t}. \end{cases} \tag{2.2}$$

Utilizing the length in units of $\sqrt{\hbar/m\omega_0}$ and introduce a dimensionless variable $x = q\sqrt{m\omega_0/\hbar}$, the proper vibrations can be defined by the Hermite polynomials

$$\begin{cases} \phi_n(x) = \dfrac{1}{\sqrt{2^n n! \sqrt{\pi}}} e^{-\frac{1}{2}x^2} H_n(x), \\ E_n = \left(n + \dfrac{1}{2}\right) \hbar\omega_0, \end{cases} \tag{2.3}$$

where n is a positive integer, $\phi_n(x)$ is a set of complete orthonormal functions satisfying

$$\int_{-\infty}^{\infty} \phi_m^*(x)\phi_n(x)dx = \delta_{mn}, \tag{2.4}$$

and $H_n(x)$ are the Hermite polynomials defined by the generating function

$$e^{-s^2+2sx} = \sum_{n=0}^{\infty} \frac{H_n(x)}{n!} s^n. \tag{2.5}$$

With the above, it immediately yields that the n-th Hermite polynomial is $(-1)^n e^{x^2}$ times the n-th derivative of the function e^{-x^2}

$$H_n(x) = (-1)^n e^{x^2} \frac{d^n}{dx^n} e^{-x^2}. \tag{2.6}$$

For a given real number A, a harmonic oscillator wave packet can be defined as the superposition of the normalized proper vibrations with coefficients $A^n/\sqrt{2^n n!}$ (Fig. 2.1)

$$\psi(x,t) = \sum_{n=0}^{\infty} \frac{A^n}{\sqrt{2^n n!}} \psi_n(x,t) = \frac{1}{\pi^{1/4}} e^{-\frac{i}{2}\omega_0 t} e^{-\frac{1}{2}x^2} \sum_{n=0}^{\infty} \frac{H_n(x)}{n!} \left(\frac{A}{2} e^{-i\omega_0 t}\right)^n .$$

Using Eq. (2.5), the wave packet can be brought into a closed form

$$\psi(x,t) = \frac{1}{\pi^{1/4}} \exp\left(-\frac{i}{2}\omega_0 t - \frac{A^2}{4} e^{-2i\omega_0 t} + A e^{-i\omega_0 t} x - \frac{x^2}{2}\right). \tag{2.7}$$

With additional derivation, we can obtain the final result

$$\psi(x,t) = \frac{1}{\pi^{1/4}} \exp\left[\frac{A^2}{4} - \frac{1}{2}(x - A\cos\omega_0 t)^2\right] e^{-i[\frac{1}{2}\omega_0 t + A\sin\omega_0 t(x - \frac{A}{2}\cos\omega_0 t)]}.$$

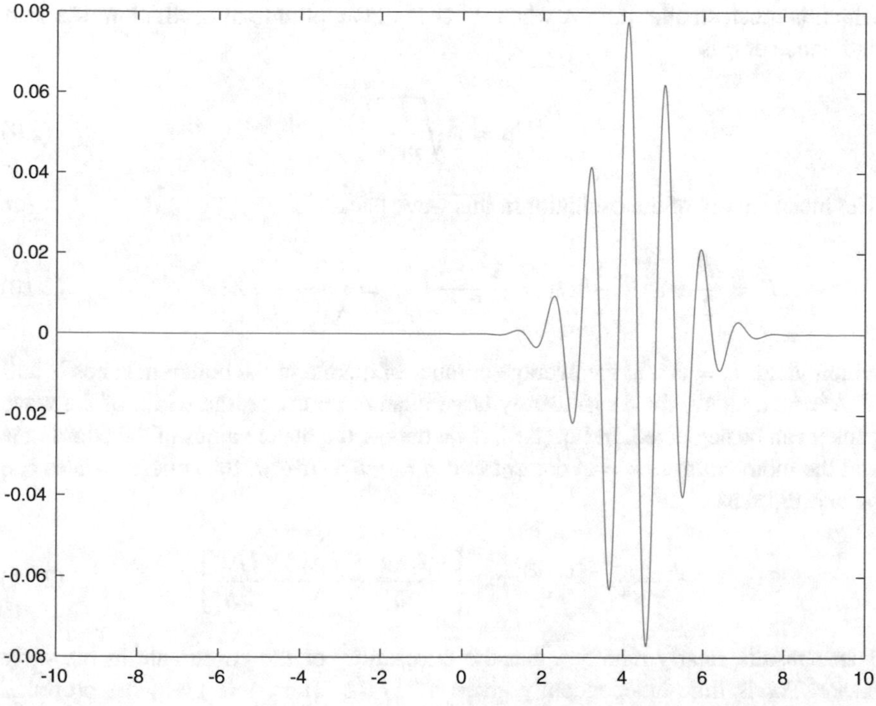

Fig. 2.1 Schrödinger's wave packet $\psi(x,t)$, Eq. (2.8), expressed as a function of x with $A = 8$ and $\omega_0 t = 1$, will contain a time-dependent phase factor and a Gaussian envelope. In the profile of the Gaussian envelope, the phase factor varies rapidly with x and ploughs through many deep and narrow furrows

It is noted that when $A = 0$, the wave packet becomes $\psi = \pi^{-1/4} e^{-x^2 - i\omega_0 t/2}$, which is the wave function of the ground state of the oscillator. The requirement that the result of integrating the probability over the whole space to be equal to unity yields a normalized wave packet

$$\psi(x,t) = e^{-A^2/4} \sum_{n=0}^{\infty} \frac{A^n}{\sqrt{2^n n!}} \psi_n(x,t) \tag{2.8}$$

$$= \frac{1}{\pi^{1/4}} \exp\left[-\frac{1}{2}(x - A\cos\omega_0 t)^2\right] e^{-i[\frac{1}{2}\omega_0 t + (x - A/2\cos\omega_0 t) A \sin\omega_0 t]}.$$

The above gives the earliest version of coherent states. It contains two different parts: a time-dependent phase factor and a Gaussian envelope. At any given instant of time, the Gaussian function has a peak at $x = A\cos\omega_0 t$. It has significant values only in the region of unity on both sides of the peak and diminishes rapidly beyond of this region. By confining our consideration in a restricted region, it shows that the wave packet can execute a periodic motion, which is entirely similar to a pendulum in classical mechanics. The width of the Gaussian envelope has the order of unity, which is much smaller than A when $A \gg 1$; the amplitude of oscillation of x is A and hence of q is

$$a = A\sqrt{\frac{\hbar}{m\omega}}. \tag{2.9}$$

The mean energy of the oscillator in this wave packet is

$$\bar{E} = \frac{a^2}{2} m\omega_0^2 + \frac{1}{2}\hbar\omega_0 = \frac{A^2 + 1}{2}\hbar\omega_0 = \left(\bar{n} + \frac{1}{2}\right)\hbar\omega_0, \tag{2.10}$$

which yields $\bar{n} = A^2/2$; the average number of quanta in the coherent states is half of A^2. As a result, for a sufficiently large quantum number, the width of the wave packet can be neglected. In Eq. (2.8), if we denote the mean values of the coordinate and the momentum as $\bar{x} = A\cos\omega_0 t$ and $\bar{p} = -\hbar A \sin\omega_0 t$, the coherent states can be expressed as

$$\psi(x,t) = \frac{1}{\pi^{1/4}} e^{-\frac{1}{2}(x-\bar{x})^2} \exp\left[-\frac{i\omega_0 t}{2} + \frac{i\bar{p}x}{\hbar} - \frac{i\bar{p}\bar{x}}{2\hbar}\right]. \tag{2.11}$$

These results clearly manifest that the uncertainty of the coordinate in the wave packet Δx is time-independently given by $1/\sqrt{2}$. Thus, this gives the profound result that the wave packet shall always remain compact, i.e., will not diffuse in time. Furthermore, the uncertainty of the momentum in the coherent states Δp is always given by $\hbar/\sqrt{2}$, which immediately yields $\Delta x \Delta p = \hbar/2$. The fact that the wave packet minimizes the uncertainty relation shows that it *de facto* is a classical particle located at the peak of the wave packet, and is performing a periodic motion with the amplitude A and the angular frequency ω_0.

The above result for the harmonic oscillator is absolutely "spectacular," namely, the complete agreement between classical and quantum description of the harmonic oscillator. Unfortunately, it simply cannot be generalized. Indeed, in contrast, one can demonstrate that a wave packet for a free particle shall always diffuse. Such a wave packet is the superposition of a series of monochromatic plane waves, which can be written as

$$\psi(x,t) = \int_{-\infty}^{\infty} g(k) e^{i(kx-\omega_k t)} dk. \tag{2.12}$$

where ω_k is the angular frequency of the plane wave with the propagation constant k. For a free particle, the dispersion relation is $\omega_k = \hbar k^2/2m$. Thus, if the wave packet at $t = 0$ is described by a Gaussian function $\psi(x,0) = \pi^{-1/4} e^{-x^2/2} e^{ikx}$, the wave packet at time t can be expressed as

$$\psi(x,t) = \int_{-\infty}^{\infty} g(k') \exp\left[i \left(k'x - \frac{\hbar k'^2}{2m} t \right) \right] dk', \tag{2.13}$$

where the amplitude $g(k')$ has the form

$$g(k') = \frac{\pi^{-1/4}}{\sqrt{2\pi}} \exp\left[-\frac{1}{2} (k'-k)^2 \right]. \tag{2.14}$$

The amplitude $g(k')$ has a peak at $k' = k$ and has significant values only in the region of unity on both sides of the peak. One can easily demonstrate that the wave packet at time t can be brought into a closed form

$$\psi(x,t) = \frac{\pi^{-1/4}}{\sqrt{1 + i\hbar t/m}} \exp\left[\frac{-x^2 + 2ik(x - \hbar k t/2m)}{2(1 + i\hbar t/m)} \right]. \tag{2.15}$$

Hence, at all times, the wave packet remains Gaussian, and the Gaussian envelope and the phase factor are given by

$$\begin{cases} \psi(x,t) = \pi^{-1/4} \left(1 + \dfrac{\hbar^2 t^2}{m^2} \right)^{-\frac{1}{4}} \exp\left\{ -\dfrac{1}{2} \left(1 + \dfrac{\hbar^2 t^2}{m^2} \right)^{-1} \left[x - \dfrac{\hbar k}{m} t \right]^2 \right\} e^{i\phi}, \\ \phi = -\dfrac{1}{2} \arctan\left(\dfrac{\hbar t}{m} \right) + \left(1 + \dfrac{\hbar^2 t^2}{m^2} \right)^{-1} \left[k \left(x - \dfrac{\hbar k}{2m} t \right) + \dfrac{\hbar x^2}{2m} t \right]. \end{cases}$$

From the above, one sees that the Gaussian function peaks at $\bar{x} = \hbar k t/m$ and the wave packet always moves with a constant velocity $\hbar k/m$. In other words, the center of the wave packet can be identified as a free particle in classical mechanics. However, as the width of the Gaussian function increases according to $\sqrt{1 + \hbar^2 t^2/m^2}$ and the height of the peak diminishes according to $(1 + \hbar^2 t^2/m^2)^{-1/4}$, the wave packet will gradually spread out. As time progresses, the wave packet will cease to be a particle.

2.2 Glauber's Coherent States

In the last section, we have discussed in detail as to how to construct the wave packet from the quantum harmonic oscillator eigenstates. In view of the fact that there exists similarity between the harmonic oscillator and the modal amplitude of the electromagnetic field, in principle such a wave packet should also be applicable to quantum optics. Yet, between 1920s and 1960s, the results of Schrödinger's wave packet remained at best a historical anecdote in the scientific literature. In fact, the term "coherent states" was first deployed in 1963 by Roy J. Glauber in his seminal paper on the quantum theory of optical coherence [3], almost 40 years later. At roughly the same time, E. C. G. Sudarshan independently discovered the coherent states and deployed them to construct his diagonal coherent state representation of quantum operators [5].

Reflecting on Schrödinger's ideas on the harmonic oscillator nonspreading wave packet, Glauber established the mathematical foundations of quantum optics by leveraging coherent states as a set of basic states in the Hilbert space. In his seminal paper titled "*Coherent and Incoherent States of the Radiation Field*" published in 1963 [4], Glauber discussed in terms of the coherent states the correlation and coherence properties of the optical field. Glauber's quantum theory of light was soon demonstrated to be the explanation of the experimental results regarding correlation effects and quantitatively the distribution of photon numbers. In 2005, for his coherent states work, Glauber received the Nobel Prize in Physics where the prize motivation is "**for his contribution to quantum theory of optical coherence**."

In the following, we shall discuss in detail Glauber's coherent states. Before this, we first discuss some necessary background materials of quantization of a free electromagnetic field. In a region without any sources, it is well known that the classical electromagnetic field satisfies the Maxwell equations

$$\begin{aligned} \nabla \cdot \mathbf{E} = 0, \nabla \times \mathbf{E} &= -\frac{1}{c}\frac{\partial \mathbf{B}}{\partial t}, \\ \nabla \cdot \mathbf{B} = 0, \nabla \times \mathbf{B} &= \frac{1}{c}\frac{\partial \mathbf{E}}{\partial t}, \end{aligned} \tag{2.16}$$

where one can derive the electric and magnetic field, $\mathbf{E}(\mathbf{r}, t)$ and $\mathbf{B}(\mathbf{r}, t)$, from the vector potential $\mathbf{A}(\mathbf{r}, t)$ via the relations

$$\mathbf{E} = -\frac{1}{c}\frac{\partial \mathbf{A}}{\partial t} \text{ and } \mathbf{B} = \nabla \times \mathbf{A}. \tag{2.17}$$

For our present purpose, we shall regard the electromagnetic field as being confined within a spatial volume of finite size and satisfying certain boundary conditions at the edge. By imposing these conditions, one can expand the vector potential by a set of discrete mode functions labeled by a number k, also known as the mode index

$$\mathbf{A}(\mathbf{r}, t) = c \sum_k \sqrt{\frac{\hbar}{2\omega_k}} \left(a_k \mathbf{u}_k(\mathbf{r}) e^{-i\omega_k t} + a_k^\dagger \mathbf{u}_k^*(\mathbf{r}) e^{i\omega_k t} \right). \tag{2.18}$$

where a_k and $a_k^\dagger$ are a pair of complex mode amplitudes of the field and $\mathbf{u}_k$ is the vector mode function that corresponds to the frequency ω_k, satisfying the wave equation

$$\left(\nabla^2 + \frac{\omega_k^2}{c^2} \right) \mathbf{u}_k = 0, \tag{2.19}$$

with the transversality condition $\nabla \cdot \mathbf{u}_k = 0$. All the mode functions should be normalized and orthogonal to each other

$$\int \mathbf{u}_k^*(\mathbf{r}) \cdot \mathbf{u}_{k'}(\mathbf{r}) d\mathbf{r} = \delta_{kk'}. \tag{2.20}$$

One can show that the energy function of the electromagnetic field takes on the form

$$H = \frac{1}{2} \sum_k \hbar\omega_k (a_k^\dagger a_k + a_k a_k^\dagger). \tag{2.21}$$

The above can be interpreted as the energy of an assembly of independent linear harmonic oscillators with vibrational frequencies ω_k. The analogy between the mode amplitudes and the harmonic oscillators suggests the following canonical commutation relations

$$[a_k, a_{k'}] = [a_k^\dagger, a_{k'}^\dagger] = 0, [a_k, a_{k'}^\dagger] = \delta_{kk'}. \tag{2.22}$$

The states of the quantized electromagnetic field are vectors in the product space of the Hilbert space for all of the modes, and the number of photons in each mode can vary from zero to infinity. A set of basic states $\{|n_k\rangle\}$, which is labeled by the number of photons for each discrete mode, can be obtained by multiplying the integral powers of the operator $a_k^\dagger$ to the ground state $|0_k\rangle$

$$|n_k\rangle = \frac{(a_k^\dagger)^{n_k}}{\sqrt{n_k!}} |0_k\rangle, \tag{2.23}$$

where $|0_k\rangle$ is the ground state defined by $a_k|0_k\rangle = 0$ and the basis states $|n_k\rangle$ are normalized and orthogonal in the usual sense, $\langle n_k|n_{k'}\rangle = \delta_{kk'}$. The operators a_k and $a_k^\dagger$ acting on the basis states will lower or raise the number of photons by one, thus yielding

$$\begin{aligned} a_k|n_k\rangle &= \sqrt{n_k}|n_k - 1\rangle, \\ a_k^\dagger|n_k\rangle &= \sqrt{n_k + 1}|n_k + 1\rangle, \\ a_k^\dagger a_k|n_k\rangle &= n_k|n_k\rangle. \end{aligned} \tag{2.24}$$

Hence, a_k and $a_k^\dagger$ can be interpreted as operators which annihilate or create a photon in the kth mode, respectively.

For quantum optics, the above background materials are sufficient to introduce the coherent states. In the quantum theory of light, what distinguishes a quantum electromagnetic field from its classical counterpart is the existence of the vacuum fluctuations. For example, the expectation value of the square of the electric field operator, $\langle \text{vac}|\mathbf{E}^2(\mathbf{r}, t)|\text{vac}\rangle$, is larger than zero in the vacuum state and is therefore irrelevant for the detection of photons. The measurement of the optical field actually involves only photon annihilation processes, and the relevant quantity is the positive frequency part of the electric field operator, $\mathbf{E}^{(+)}(\mathbf{r}, t)$, defined by $\mathbf{E}^{(+)}(\mathbf{r}, t)|\text{vac}\rangle = 0$. For an ideal photodetector placed at point $\mathbf{r}$ and measured at time t, the outcome of the measurement is

$$\langle \text{vac}||\mathbf{E}^{(+)}(\mathbf{r}, t)|^2|\text{vac}\rangle = \langle \text{vac}|\mathbf{E}^{(-)}(\mathbf{r}, t)\mathbf{E}^{(+)}(\mathbf{r}, t)|\text{vac}\rangle,$$

which vanishes identically in the vacuum state. Here, $\mathbf{E}^{(-)}(\mathbf{r}, t)$ is the negative frequency part of the electric field operator, $\mathbf{E}^{(-)}(\mathbf{r}, t) = \{\mathbf{E}^{(+)}(\mathbf{r}, t)\}^\dagger$. In other words, no photon is detected in the vacuum state by an ideal photodetector. For a general light beam, the counting rate of the ideal photodetector will be proportional to $G^{(1)}(\mathbf{r}t, \mathbf{r}t)$, where $G^{(1)}(\mathbf{r}t, \mathbf{r}'t')$ is the quantum mechanical correlation function defined by

$$G^{(1)}(\mathbf{r}t, \mathbf{r}'t') = \text{Tr}\left\{\rho\mathbf{E}^{(-)}(\mathbf{r}, t)\mathbf{E}^{(+)}(\mathbf{r}', t')\right\}. \tag{2.25}$$

Here ρ is the density operator specifying the state of the field. It can be shown that the quantum mechanical correlation function will obey

$$|G^{(1)}(\mathbf{r}t, \mathbf{r}'t')|^2 \leq G^{(1)}(\mathbf{r}t, \mathbf{r}t)G^{(1)}(\mathbf{r}'t', \mathbf{r}'t'). \tag{2.26}$$

The above equality sign is satisfied only if one can factorize the correlation function into the form $G^{(1)}(\mathbf{r}t, \mathbf{r}'t') = \mathcal{E}^*(\mathbf{r}, t)\mathcal{E}(\mathbf{r}', t')$. In this case, the complex function $\mathcal{E}(\mathbf{r}, t)$ is nearly completely determined by the coherent properties of the optical field to within a phase factor. According to Eqs. (2.17) and (2.18), one can express the positive frequency part of the electric field operator as

$$\mathbf{E}^{(+)}(\mathbf{r}, t) = i \sum_k \sqrt{\frac{\hbar\omega_k}{2}} a_k \mathbf{u}_k(\mathbf{r}) e^{-i\omega_k t}. \tag{2.27}$$

By introducing the coherent states of the individual modes as the eigenstates of the annihilation operators via $a_k|\alpha_k\rangle = \alpha_k|\alpha_k\rangle$, it yields

$$\mathcal{E}(\mathbf{r}, t) = i \sum_k \sqrt{\frac{\hbar\omega_k}{2}} \alpha_k \mathbf{u}_k(\mathbf{r}) e^{-i\omega_k t}, \tag{2.28}$$

where $\Pi_k|\alpha_k\rangle$ is the coherent state of the entire field satisfying

$$\mathbf{E}^{(+)}(\mathbf{r}, t) \prod_k |\alpha_k\rangle = \mathcal{E}(\mathbf{r}, t) \prod_k |\alpha_k\rangle. \tag{2.29}$$

With the above as background, we can now discuss in more detail the coherent states of a single mode optical field by dropping the mode index k. The **coherent state** of a single mode harmonic oscillator $|\alpha\rangle$ is defined as the *eigenstate of the annihilation operator a* via

$$a|\alpha\rangle = \alpha|\alpha\rangle, \tag{2.30}$$

where α is a complex number. In hindsight, defining the coherent state as the eigenstate of the annihilation operator may superficially appear to be a simple assumption. But one has to realize that historically, before Glauber's groundbreaking work, the scientific communities were only interested in the eigenstates of Hermitian operators, such as the position or momentum operators, which are experimentally measurable. It was exactly Glauber who took the first step to bring in an eigenstate of a non-Hermitian operator, which corresponds to the quantum electric field. With such a brave assumption, a great new era of quantum theory of light was to unveiled.

Applying the nth excited state of the harmonic oscillator $\langle n|$ to the both sides of Eq. (2.30), one obtains the recursion relation for the scalar product $\langle n|\alpha\rangle$, which is

$$\sqrt{n+1}\langle n+1|\alpha\rangle = \alpha\langle n|\alpha\rangle. \tag{2.31}$$

Solving the recursion relation by repeated substitution, one arrives at

$$\langle n|\alpha\rangle = \frac{\alpha^n}{\sqrt{n!}}\langle 0|\alpha\rangle. \tag{2.32}$$

Utilizing the completeness relation, the coherent states can be expanded in terms of the complete orthogonal basis $\{|n\rangle\}$ via

$$|\alpha\rangle = \sum_n |n\rangle\langle n|\alpha\rangle = \langle 0|\alpha\rangle \sum_n \frac{\alpha^n}{\sqrt{n!}}|n\rangle. \tag{2.33}$$

By imposing the state vector $|\alpha\rangle$ to have unit length, $\langle\alpha|\alpha\rangle = |\langle 0|\alpha\rangle|^2 e^{|\alpha|^2} = 1$, the coherent states up to within a phase factor can be expressed as

$$|\alpha\rangle = e^{-\frac{1}{2}|\alpha|^2} \sum_n \frac{\alpha^n}{\sqrt{n!}}|n\rangle. \tag{2.34}$$

The definition of coherent states gives the probability p_n of finding n photons in the coherent states as

$$p_n \equiv |\langle n|\alpha\rangle|^2 = e^{-|\alpha|^2}\frac{|\alpha|^{2n}}{n!}, \tag{2.35}$$

which shows that a coherent state has a photon statistics described by a **Poisson distribution** centered at $|\alpha|^2$ and with standard derivation $|\alpha|$, as illustrated in Fig. 2.2.

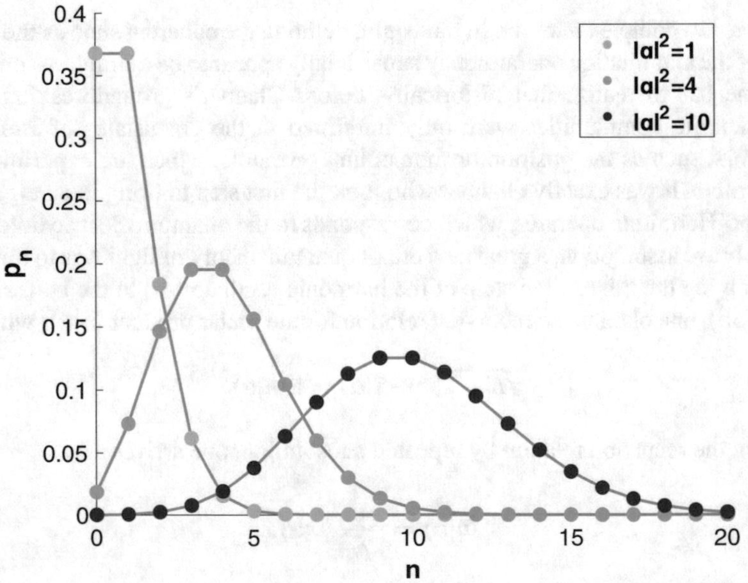

Fig. 2.2 Schematic of the Poisson distributions for coherent states $|\alpha\rangle$ with different mean photon numbers $\langle n\rangle = |\alpha|^2$

The coherent state which corresponds to $\alpha = 0$ is the ground state of the harmonic oscillator. An alternative way which provides a more intuitive understanding is to define the coherent state as the displaced form of the ground state

$$|\alpha\rangle = D(\alpha)|0\rangle, \tag{2.36}$$

where $D(\alpha)$ is a unitary operator acting on the annihilation operator a, according to the scheme

$$D^{-1}(\alpha)aD(\alpha) = a + \alpha. \tag{2.37}$$

It follows that since $D(\alpha)$ is a unitary operator, $|\alpha\rangle$ is already normalized. By multiplying $D^{-1}(\alpha)$ to the right side of Eq. (2.37), when applied to the state $|\alpha\rangle$, one yields

$$D^{-1}(\alpha)a|\alpha\rangle = aD^{-1}(\alpha)|\alpha\rangle + \alpha D^{-1}(\alpha)|\alpha\rangle. \tag{2.38}$$

According to the new definition, $D^{-1}(\alpha)|\alpha\rangle = |0\rangle$ is the ground state of the oscillator, and thus $aD^{-1}(\alpha)|\alpha\rangle$ vanishes identically. Hence, we recover the original definition of coherent states, $a|\alpha\rangle = \alpha|\alpha\rangle$. One can now explicitly obtain the expression of the displacement operator $D(\alpha)$, if the phase of the coherent states is chosen by selecting $D(0) = 1$. Employing equation (2.37) for an infinitesimal displacement $d\alpha$, $D(d\alpha)$ can be uniquely written as

$$D(d\alpha) = 1 + a^\dagger d\alpha - a d\alpha^*. \tag{2.39}$$

One can show rather easily that the unitary operator $D(\alpha)$ has the form for a finite displacement

$$D(\alpha) = e^{\alpha a^\dagger - \alpha^* a}. \tag{2.40}$$

As a result, the coherent states $|\alpha\rangle$ can be written as

$$|\alpha\rangle = e^{\alpha a^\dagger - \alpha^* a}|0\rangle. \tag{2.41}$$

In this definition, the coherent states can also be expanded in terms of the complete orthogonal basis $\{|n\rangle\}$. For this purpose, one can write the displacement operator $D(\alpha)$ as a product of exponential operators

$$D(\alpha) = e^{\alpha a^\dagger - \alpha^* a} = e^{-\frac{1}{2}|\alpha|^2} e^{\alpha a^\dagger} e^{-\alpha^* a} \tag{2.42}$$

by using the Baker-Campbell-Hausdorff (BCH) formula: if A and B are two arbitrary operators such that the commutator $[A, B]$ commutes with each of them, $[A, [A, B]] = [B, [A, B]] = 0$, then $e^A e^B = e^{A+B+\frac{1}{2}[A,B]}$. If applying the BCH formula to the case $A = \alpha a^\dagger - \alpha^* a$ and $B = \beta a^\dagger - \beta^* a$, one immediately finds that the displacement operators obey the multiplication law

$$D(\alpha)D(\beta) = D(\alpha+\beta)e^{\frac{1}{2}(\alpha\beta^*-\alpha^*\beta)}. \tag{2.43}$$

It can be shown that, besides a phase factor $e^{\alpha\beta^*-\alpha^*\beta}$, the displacement operators form a commutative group under multiplication, in which the result of applying the multiplication operation to two group elements does not depend on the order in which they are written. Utilizing the multiplication law, Eq. (2.43), one can derive the commutative law

$$D(\alpha)D(\beta) = D(\beta)D(\alpha)e^{\alpha\beta^*-\alpha^*\beta}. \tag{2.44}$$

The commutative law of the displacement operators can be used to calculate the matrix element $\langle\alpha|D(\beta)|\alpha\rangle$ which connects the two coherent states $|\alpha\rangle$ and $\langle\alpha|$ as

$$\begin{aligned}\langle\alpha|D(\beta)|\alpha\rangle &= \langle 0|D^{-1}(\alpha)D(\beta)D(\alpha)|0\rangle \qquad (2.45)\\ &= \langle 0|D^{-1}(\alpha)D(\alpha)D(\beta)e^{\beta\alpha^*-\beta^*\alpha}|0\rangle\\ &= e^{\beta\alpha^*-\beta^*\alpha}\langle 0|\beta\rangle = e^{\beta\alpha^*-\beta^*\alpha-\frac{1}{2}|\beta|^2}.\end{aligned}$$

Analogously, one can show that the matrix element $\langle\beta|D(\xi)|\alpha\rangle$ which connects the two unlike coherent states $|\alpha\rangle$ and $\langle\beta|$ is

$$\langle\beta|D(\xi)|\alpha\rangle = e^{\beta^*\alpha-\frac{1}{2}|\alpha|^2-\frac{1}{2}|\beta|^2}e^{\xi\beta^*-\xi^*\alpha-\frac{1}{2}|\xi|^2}. \tag{2.46}$$

Using $e^{-\alpha^* a}|0\rangle = |0\rangle$, one can express the coherent states as

$$|\alpha\rangle = e^{-\frac{1}{2}|\alpha|^2}e^{\alpha a^\dagger}|0\rangle = e^{-\frac{1}{2}|\alpha|^2}\sum_n \frac{(\alpha a^\dagger)^n}{n!}|0\rangle. \tag{2.47}$$

Hence, using the definition of the nth excited state, $|n\rangle = 1/\sqrt{n!}(a^\dagger)^n|0\rangle$, one can obtain the expected relation

$$|\alpha\rangle = e^{-\frac{1}{2}|\alpha|^2}\sum_n \frac{\alpha^n}{\sqrt{n!}}|n\rangle. \tag{2.48}$$

2.3 Mathematical Properties

One of the fundamental mathematical principles discussed in any standard quantum mechanics textbook [50–52] is the concept of a "complete Hilbert space." In this concept, any quantum mechanical action "can be expanded" in terms of an orthogonal set which is "complete." To this end, as coherent states have proven to be a most useful tool in the study of many quantum mechanical objects, it is worthy to profoundly understand their mathematical structures.

A basic property of the coherent states $|\alpha\rangle$ is that any two such states are **not orthogonal**. For values of α' close to α, we expect that the coherent states are similar in form with substantial overlap. In contrast, for values of α' very different from α, one would expect that the two coherent states are well separated with small overlap. To show this, the scalar product $\langle\alpha|\beta\rangle$ is

$$\langle\alpha|\beta\rangle = e^{\alpha^*\beta-\frac{1}{2}|\alpha|^2-\frac{1}{2}|\beta|^2} = e^{-\frac{1}{2}|\alpha-\beta|^2+i\Im(\alpha^*\beta)}. \tag{2.49}$$

It follows from above, $|\langle\alpha|\beta\rangle| = e^{-\frac{1}{2}|\alpha-\beta|^2}$ which is the overlap of any two well-separated coherent states can be shown to be exponentially small, and they tend to become approximately orthogonal.

As was previously mentioned, the states $|\alpha'\rangle$ and $|\alpha\rangle$ should be approximately the same for two nearby points in the complex α plane. This continuity property can be proved by using the vector norm $\||\alpha\rangle\|^2 = \langle\alpha|\alpha\rangle$ which measures the distance between two states. To establish the validity of this argument, we evaluate the vector norm $\||\alpha\rangle - |\beta\rangle\|^2$ as

$$\||\alpha\rangle - |\beta\rangle\|^2 = 2(1 - \Re\langle\alpha|\beta\rangle) \leq 2(|\alpha| + |\beta|)|\alpha - \beta|. \tag{2.50}$$

The vector norm $\||\alpha\rangle - |\beta\rangle\|$ can be arbitrarily small, if the α and β are sufficiently close. It should be noted that although the coherent states $|\alpha\rangle$ for a single mode oscillator do not form an orthogonal set, it can be shown that they do form a complete set. The key formula is the resolution of unity into projection operators, which establishes the correspondence between the unit operator and the integral over the complex α plane of the projection operator $|\alpha\rangle\langle\alpha|$

$$I = \sum_n |n\rangle\langle n| = \frac{1}{\pi}\int |\alpha\rangle\langle\alpha| d^2\alpha, \tag{2.51}$$

where $d^2\alpha = d(\Re\alpha)d(\Im\alpha)$ is the differential element of area in the complex α plane and the integration extends over the entire complex α plane. The precise meaning of the resolution of unity is embodied in the formula for the scalar product

$$\langle\phi|\psi\rangle = \sum_n \langle\phi|n\rangle\langle n|\psi\rangle = \frac{1}{\pi}\int \langle\phi|\alpha\rangle\langle\alpha|\psi\rangle d^2\alpha, \tag{2.52}$$

where $|\phi\rangle$ and $|\psi\rangle$ are two arbitrary states. A direct elementary proof of Eq. (2.52) is the following

$$\frac{1}{\pi}\int \langle\phi|\alpha\rangle\langle\alpha|\psi\rangle d^2\alpha = \frac{1}{\pi}\int \sum_{n,m=0}^{\infty} \frac{\alpha^{*n}\alpha^m}{\sqrt{m!n!}}\langle\phi|m\rangle\langle n|\psi\rangle e^{-|\alpha|^2} d^2\alpha \tag{2.53}$$
$$= \frac{1}{\pi}\sum_{n,m=0}^{\infty} \frac{\langle\phi|m\rangle\langle n|\psi\rangle}{\sqrt{m!n!}}\int \alpha^{*n}\alpha^m e^{-|\alpha|^2} d^2\alpha,$$

where the order of summation and integration in the last step was interchanged. If by introducing the polar coordinates such that $\alpha = re^{i\theta}$ and $d^2\alpha = rdrd\theta$, the integral can be shown as

$$\int \alpha^{*n}\alpha^m e^{-|\alpha|^2} d^2\alpha = \int_0^{\infty} r^{n+m+1} e^{-r^2} dr \int_0^{2\pi} e^{i(m-n)\theta} d\theta = \pi n!\delta_{mn}. \tag{2.54}$$

Substituting Eq. (2.54) into Eq. (2.53), it yields

$$\frac{1}{\pi}\int \langle\phi|\alpha\rangle\langle\alpha|\psi\rangle d^2\alpha = \sum_{n=0}^{\infty}\langle\phi|n\rangle\langle n|\psi\rangle = \langle\phi|\psi\rangle, \tag{2.55}$$

where the sum of the projection operators $|n\rangle\langle n|$ over n is the unit operator in the last step. With the resolution of unity, one obtains the expansion rule for an arbitrary state $|\psi\rangle$

$$|\psi\rangle = \frac{1}{\pi}\int |\alpha\rangle\langle\alpha|\psi\rangle d^2\alpha. \tag{2.56}$$

The above demonstrates the completeness relation for the coherent states: the vanishing of the function $\psi(\alpha^*) \equiv \langle\alpha|\psi\rangle$ for all values of α^* implies the vanishing of the state vector $|\psi\rangle$ itself and vice versa. It establishes the important one-to-one correspondence between the states $|\psi\rangle$ of an oscillator and the functions $\psi(\alpha^*)$ in the complex α plane.

It is worth noting that the interchange of summation and integration in Eq. (2.53) can be rigorously justified by the **dominated convergence theorem**: let $f_1, f_2, \cdots, f_N$ and R: $\mathbb{Z} \to \mathbb{R}$ be functions such that $|f_N(n)| \le g(n)$, $\sum_{n=0}^{\infty} g(n) < \infty$. Then one can prove that

$$\lim_{N\to\infty}\sum_{n=-\infty}^{\infty} f_N(n) = \sum_{n=-\infty}^{\infty}\lim_{N\to\infty} f_N(n). \tag{2.57}$$

To establish the completeness relation rigorously, one may first evaluate the special case when $|\phi\rangle$ and $|\psi\rangle$ are the same states, and notice that

$$\frac{1}{\pi}\int |\langle\alpha|\psi\rangle|^2 d^2\alpha = \lim_{N\to\infty}\sum_{n=0}^{\infty} |\langle n|\psi\rangle|^2 \gamma_N(n), \tag{2.58a}$$

$$\gamma_N(n) \equiv \frac{2}{n!}\int_0^N r^{2n+1} e^{-r^2} dr, \tag{2.58b}$$

where $\gamma_N(n)$ is a non-negative function satisfying the conditions $0 < \gamma_N(n) < 1$ and $\lim_{N\to\infty}\gamma_N(n) = 1$ for all n. Hence, the norm of the state $|\psi\rangle$ can always be expressed as

$$\langle\psi|\psi\rangle = \sum_{n=0}^{\infty} |\langle n|\psi\rangle|^2 = \sum_{n=0}^{\infty} \lim_{N\to\infty} |\langle n|\psi\rangle|^2 \gamma_N(n). \tag{2.59}$$

Defining $f_N(n) \equiv \gamma_N(n) g(n)$ and $g(n) \equiv |\langle n|\psi\rangle|^2$, one immediately obtains

$$|f_N(n)| = |\gamma_N(n) g(n)| \le g(n), \tag{2.60a}$$

$$\sum_{n=0}^{\infty} g(n) = \sum_{n=0}^{\infty} |\langle n|\psi\rangle|^2 = \langle\psi|\psi\rangle < \infty, \tag{2.60b}$$

where the last step in Eq. (2.60b) comes from the completeness of the Hilbert space. Hence the convergence of the sequences can be readily established by using the dominated convergence theorem

$$\lim_{N\to\infty}\sum_{n=0}^{\infty} |\langle n|\psi\rangle|^2 \gamma_N(n) = \sum_{n=0}^{\infty} \lim_{N\to\infty} |\langle n|\psi\rangle|^2 \gamma_N(n). \tag{2.61}$$

It immediately yields

$$\frac{1}{\pi}\int \langle\psi|\alpha\rangle\langle\alpha|\psi\rangle d^2\alpha = \sum_{n=0}^{\infty} |\langle n|\psi\rangle|^2 = \langle\psi|\psi\rangle. \tag{2.62}$$

For the general case, the formula for the scalar product $\langle\phi|\psi\rangle$ is obtained from linearity.

Let us now consider the continuity properties for the function $\psi(\alpha^*) = \langle\alpha|\psi\rangle$, which can be expanded as

$$\psi(\alpha^*) = e^{-\frac{1}{2}|\alpha|^2}\sum_{n=0}^{\infty} \frac{\alpha^{*n}}{\sqrt{n!}}\langle n|\psi\rangle. \tag{2.63}$$

Utilizing the Cauchy-Schwarz inequality, it immediately follows that every such $\psi(\alpha^*)$ is bounded, $|\langle\alpha|\psi\rangle| \leq \langle\psi|\psi\rangle^{1/2} = \||\psi\rangle\| < \infty$. In addition, the function $\psi(\alpha)$ is continuous on the entire complex plane,

$$|\psi(\alpha^*) - \psi(\beta^*)| \leq \||\alpha\rangle - |\beta\rangle\| \cdot \||\psi\rangle\| \leq \sqrt{2(|\alpha| + |\beta|)|\alpha - \beta|} \cdot \||\psi\rangle\|,$$

where we have used the inequality equation (2.50) in the last step. While continuity is a convenient property for complex-valued functions, it is not the strongest one. In fact, $\psi(\alpha^*)$ possesses stronger properties than continuity: it can be expanded as a power series that converges on the entire complex plane. To show the validity of this argument, one can define $f(\alpha^*) = e^{\frac{1}{2}|\alpha|^2}\psi(\alpha^*)$ and evaluate the radius of convergence of $f(\alpha^*)$. It can be shown that the series in Eq. (2.63) is absolutely convergent, since we have

$$\sum_{n=0}^{\infty} \frac{|\alpha^*|^n}{\sqrt{n!}} |\langle n|\psi\rangle| \leq \||\psi\rangle\| \cdot \sum_{n=0}^{\infty} \frac{|\alpha^*|^n}{\sqrt{n!}},$$

which converges for all α^* as the ratio of the nth term to the $(n-1)$th term tends to zero as n tends to infinity. This convergent series thus represents a function which is analytic throughout the entire complex plane. Namely, the function $f(\alpha^*) = e^{\frac{1}{2}|\alpha|^2}\langle\alpha|\psi\rangle$ defines an entire function of α^* for each $|\psi\rangle$. Applying Schwarz's inequality, it immediately leads to $|f(\alpha^*)| \leq e^{\frac{1}{2}|\alpha|^2}\||\psi\rangle\|$, which shows that $f(\alpha^*)$ is an entire function that grows no faster than $e^{\frac{1}{2}|\alpha|^2}$. Substituting Eq. (2.63) in Eq. (2.56), we immediately obtain the expansion of arbitrary states in terms of coherent states

$$|\psi\rangle = \frac{1}{\pi}\int |\alpha\rangle f(\alpha^*) e^{-\frac{1}{2}|\alpha|^2} d^2\alpha. \tag{2.64}$$

Now by taking the scalar product of both sites of Eq. (2.64) with the state $\langle\beta|$, and using the relation $\langle\beta|\alpha\rangle = e^{\beta^*\alpha - \frac{1}{2}|\beta|^2 - \frac{1}{2}|\alpha|^2}$, one can easily derive the general equation

$$f(\beta^*) = e^{\frac{1}{2}|\beta|^2}\langle\beta|\psi\rangle = \frac{1}{\pi}\int e^{\beta^*\alpha - |\alpha|^2} f(\alpha^*) d^2\alpha, \tag{2.65}$$

which reveals that the function $f(\alpha^*)$ fulfills an integral equation

$$f(\beta^*) = \int K(\beta, \alpha) f(\alpha^*) d\mu(\alpha), \; d\mu(\alpha) = \frac{1}{\pi} e^{-|\alpha|^2} d^2\alpha, \tag{2.66}$$

where $d\mu(\alpha)$ is the element of measure and $K(\beta, \alpha) = e^{\beta^*\alpha}$ is the reproducing kernel of the associated integral transform which satisfies $K(\beta, \alpha) = K^*(\alpha, \beta)$. It is worth nothing that the expansion of arbitrary adjoint states is similar to that in

Eq. (2.64). Too see this, we define $g(\beta) = e^{\frac{1}{2}|\beta|^2}\langle\phi|\beta\rangle$ as an entire function of β, then the state $\langle\phi|$ can be expanded as

$$\langle\phi| = \frac{1}{\pi}\int \langle\phi|\beta\rangle\langle\beta|d^2\beta = \frac{1}{\pi}\int \langle\beta|g(\beta)e^{-\frac{1}{2}|\beta|^2}d^2\beta. \tag{2.67}$$

Employing the relation $\langle\beta|\alpha\rangle = e^{\beta^*\alpha-\frac{1}{2}|\alpha|^2-\frac{1}{2}|\beta|^2}$ again, the scalar product $\langle\phi|\psi\rangle$ can be readily expressed as

$$\begin{aligned}\langle\phi|\psi\rangle =& \frac{1}{\pi^2}\iint g(\beta)f(\alpha^*)e^{\beta^*\alpha-|\alpha|^2-|\beta|^2}d^2\alpha d^2\beta \\ =& \iint g(\beta)f(\alpha^*)K(\beta,\alpha)d\mu(\alpha)d\mu(\beta).\end{aligned} \tag{2.68}$$

We can use the integral identity (2.65) to carry out the integration over the variable α to find

$$\langle\phi|\psi\rangle = \int g(\beta)f(\beta^*)d\mu(\beta) = \frac{1}{\pi}\int \langle\phi|\beta\rangle\langle\beta|\psi\rangle d^2\beta. \tag{2.69}$$

The above is exactly our basic starting relation, Eq. (2.52), the resolution of unity. It should be underscored that the coherent states are not linearly independent of one another, since any two such states are not in general orthogonal to each other. A concrete example is that any given coherent states $|\alpha\rangle$ can be expressed linearly in terms of all other states such that

$$|\alpha\rangle = \frac{1}{\pi}\int |\beta\rangle e^{\beta^*\alpha-\frac{1}{2}|\alpha|^2-\frac{1}{2}|\beta|^2}d^2\beta. \tag{2.70}$$

Such a multiplicity of decompositions is exactly the consequence of the over-completeness of the coherent states. In other words, there should be many linear dependencies among the coherent states. For example, we may have the following identity

$$\frac{1}{\pi}\int |\alpha\rangle\alpha^n e^{-\frac{1}{2}|\alpha|^2}d^2\alpha = 0, \tag{2.71}$$

which holds for all integer $n > 0$. Therefore, the expansion amplitudes $f(\alpha^*)$ in Eq. (2.64) cannot be replaced by more general functions like $F(\alpha,\alpha^*)$. Otherwise, there would be many additional ways to express the arbitrary states in terms of the coherent states. The uniqueness of the expansion equation (2.64) requires that the expansion amplitude should depend analytically upon the variable α^*. Such a restriction is crucial to calculate the expansion amplitudes since one can construct an explicit solution for them regardless of what initial representation is used.

As an application of the completeness relation for the coherent states, the matrix element of the displacement operator $\langle m|D(\xi)|n\rangle$ which connect the two different number states $\langle m|$ and $|n\rangle$ can be evaluated as

$$\langle m|D(\xi)|n\rangle = \frac{1}{\pi^2}\iint \langle m|\beta\rangle\langle\beta|D(\xi)|\alpha\rangle\langle\alpha|n\rangle d^2\alpha d^2\beta \tag{2.72}$$

$$= \frac{e^{-\frac{1}{2}|\xi|^2}}{\pi^2\sqrt{m!n!}}\iint e^{\beta^*\alpha+\xi\beta^*-\xi^*\alpha-|\alpha|^2-|\beta|^2}\alpha^{*n}\beta^m d^2\alpha d^2\beta.$$

where we have used the identity $\langle\beta|D(\xi)|\alpha\rangle = e^{\beta^*\alpha-\frac{1}{2}|\alpha|^2-\frac{1}{2}|\beta|^2}e^{\xi\beta^*-\xi^*\alpha-\frac{1}{2}|\xi|^2}$ in the last step. Applying the integral identity equation (2.65) to the case $f(\alpha^*) = \alpha^{*n}$, we obtain

$$\frac{1}{\pi}\int e^{(\beta^*-\xi^*)\alpha-|\alpha|^2}\alpha^{*n}d^2\alpha = (\beta^*-\xi^*)^n. \tag{2.73}$$

Substituting the above formula into the double integral, one obtains the following integral representation for the matrix element of the displacement operator

$$\langle m|D(\xi)|n\rangle = \frac{e^{-\frac{1}{2}|\xi|^2}}{\pi\sqrt{m!n!}}\int e^{\xi\beta^*-|\beta|^2}(\beta^*-\xi^*)^n\beta^m d^2\beta. \tag{2.74}$$

It is of interest to see how this integral representation is expressed in terms of the special functions. For the case $m \geq n$, the matrix elements can be summed in a closed form by the introduction of an auxiliary parameter t as

$$\sum_n t^n e^{\frac{1}{2}|\xi|^2}\sqrt{\frac{m!}{n!}}\langle m|D(\xi)|n\rangle = \frac{1}{\pi}\sum_n\int e^{\xi\beta^*-|\beta|^2}\frac{(t|\beta|^2-t\xi^*\beta)^n}{n!}\beta^k d^2\beta$$

$$= \frac{1}{\pi}\int e^{\xi\beta^*-t\xi^*\beta-(1-t)|\beta|^2}\beta^k d^2\beta$$

$$= \sum_l \frac{(-t\xi^*)^l}{l!}\frac{1}{\pi}\int e^{\xi\beta^*-(1-t)|\beta|^2}\beta^{k+l}d^2\beta,$$

where $0 < t < 1$ is an auxiliary parameter and $k = m - n$ is a positive integer. Applying the integral identity

$$\frac{1}{\pi}\int e^{\xi\beta^*-\lambda|\beta|^2}f(\beta)d^2\beta = \frac{1}{\lambda}f\left(\frac{\xi}{\lambda}\right) \tag{2.75}$$

to the case $\lambda = 1 - t$ and $f(\beta) = \beta^{k+l}$ and substituting the above formula into the integration over β, one gets a simple formula for the infinite sum

$$\sum_n t^n e^{\frac{1}{2}|\xi|^2} \sqrt{\frac{m!}{n!}} \langle m|D(\xi)|n\rangle = \frac{\xi^k}{(1-t)^{k+1}} \exp\left(\frac{-t|\xi|^2}{1-t}\right). \tag{2.76}$$

It is worth noting that what emerged in the above formula is precisely the generating function for the associated Laguerre polynomials $L_n^k(x)$

$$\sum_n t^n L_n^k(x) = \frac{1}{(1-t)^{k+1}} \exp\left(\frac{-tx}{1-t}\right). \tag{2.77}$$

Hence, the matrix element for the displacement operator can be expressed in a closed form

$$\langle m|D(\xi)|n\rangle = e^{-\frac{1}{2}|\xi|^2} \sqrt{\frac{n!}{m!}} \xi^{n-m} L_n^{m-n}(|\xi|^2). \tag{2.78}$$

2.4 Quantum Coherence and Distribution

From the above, we discussed the general properties of the coherent states, especially its usage as a continuous basis of the Hilbert space. In the following, we shall discuss the expansion of certain important operators, namely, the density operators, in terms of the coherent states.

The density operator for a pure coherent state $|\alpha\rangle$ is simply the projection operator $\rho = |\alpha\rangle\langle\alpha|$. Besides the pure coherent states $|\alpha\rangle$, another type of oscillator states, which is also important for quantum optics, is the **statistical mixtures** of the coherent states. The density operator for such a state is a well-known superposition of the projection operator $|\alpha\rangle\langle\alpha|$

$$\rho = \int d^2\alpha P(\alpha)|\alpha\rangle\langle\alpha|. \tag{2.79}$$

This kind of operator ensures that the oscillator is a coherent state associated with an unknown eigenvalue α. Here the function $P(\alpha)$, known as the Glauber-Sudarshan P-representation for the density operator, can be regarded as a probability distribution function of α over the entire complex plane. Since the density operator is Hermitian and normalized, i.e., $\rho^\dagger = \rho$ and $\text{Tr}\rho = 1$, it implies that $P(\alpha)$ is a real function and obey the normalization condition

$$\text{Tr}\rho = \int d^2\alpha P(\alpha) = 1, \tag{2.80}$$

where the trace of an outer product $|\alpha\rangle\langle\beta|$ is the inner product $\langle\beta|\alpha\rangle$.

In quantum optics, the primary goal is to understand the correlation and coherence properties of the optical field. To this end, it is natural to describes the field by the $2n$-points correlation functions, which are defined as the expectation value

$$\begin{aligned}&G^{(n)}(x_1, x_2, \cdots x_n; y_1, y_2, \cdots y_n)\\&= \mathrm{Tr}\{\rho \mathbf{E}^{(-)}(x_1)\cdots \mathbf{E}^{(-)}(x_n)\mathbf{E}^{(+)}(y_1)\cdots \mathbf{E}^{(+)}(y_n)\},\end{aligned} \tag{2.81}$$

where $\mathbf{E}^{(-)}(x_l)$ and $\mathbf{E}^{(+)}(y_l)$ are the negative and positive frequency parts of the electric field operator at the space-time points x_l and y_l, respectively. Since the positive frequency part of the electric field operator, $\mathbf{E}^{(+)}(\mathbf{r}, t)$, can be expanded by a complete set of mode functions $\mathbf{u}_k$

$$\mathbf{E}^{(+)}(\mathbf{r}, t) = i\sum_k \sqrt{\frac{\hbar\omega_k}{2}} a_k \mathbf{u}_k(\mathbf{r}) e^{-i\omega_k t}, \tag{2.82}$$

the first-order correlation function for a single mode electric field between two distinct space-time points, i.e., $G^{(1)}(x, x')$, simply reduces to the form

$$\begin{aligned}G^{(1)}(x, x') &= \mathrm{Tr}\{\rho \mathbf{E}^{(-)}(x)\mathbf{E}^{(+)}(x')\}\\&= \frac{\hbar\omega}{2} e^{i\omega(t-t')} \mathbf{u}^*(\mathbf{r})\mathbf{u}(\mathbf{r}')\mathrm{Tr}(\rho a^\dagger a).\end{aligned} \tag{2.83}$$

The above relation clearly demonstrated that the first-order correlation function is not sensitive to the photon statistics, as it depends only on the mean photon number. In order words, a laser beam and a thermal light can both have the same first-order correlation properties, regardless of their photon statistics. As such, one needs the second- and higher-order correlation function to distinguish the nature of the light source. Similar to the condition for first-order coherence, an optical system is said to exhibit nth-order coherence if all of its mth-order correlations for $m \leq n$ factorized, i.e.,

$$G^{(n)}(x_1, \cdots, x_m; y_1, \cdots, y_m) = \mathcal{E}^*(x_1)\cdots \mathcal{E}^*(x_m)\mathcal{E}(y_1)\cdots \mathcal{E}(y_m), \tag{2.84}$$

where $\mathcal{E}(x_l)$ is a complex function, and all other cases with non-factorizable correlations characterize partial coherence. In particular, the second-order correlation function for a single mode electric field simply takes on the form

$$\begin{aligned}&G^{(2)}(x_1, x_2; x_1', x_2') = \mathrm{Tr}\{\rho \mathbf{E}^{(-)}(x_1)\mathbf{E}^{(-)}(x_2)\mathbf{E}^{(+)}(x_1')\mathbf{E}^{(+)}(x_2')\}\\&= \left(\frac{\hbar\omega}{2}\right)^2 e^{i\omega(t_1+t_2-t_1'-t_2')} \mathbf{u}^*(\mathbf{r}_1)\mathbf{u}^*(\mathbf{r}_2)\mathbf{u}(\mathbf{r}_1')\mathbf{u}(\mathbf{r}_2')\mathrm{Tr}(\rho a^{\dagger 2} a^2).\end{aligned} \tag{2.85}$$

To illustrate the physical meaning of the second-order correlation function, one may turn to the **Hanbury Brown-Twiss experiment** [53, 54], in which a beam of light from a mercury lamp is split into two beams by a half-silvered mirror, and fell on the cathodes of two photo-multipliers, whose outputs were sent through band-limited amplifiers to a correlator. This experiment is significantly different from the Young double-slit experiment in that intensities of incident light, rather than the amplitudes, are compared. Two absorption measurements are performed on the same optical field, one at time t and one at $t + \tau$. One can show that this procedure measures the quantity $|\mathbf{E}^{(+)}(\mathbf{r}, t + \tau)\mathbf{E}^{(+)}(\mathbf{r}, t)|^2$. After averaging, this quantity is precisely the second-order correlation function of light at the same space point $\mathbf{r}$ and two different time points at t and $t + \tau$

$$\begin{aligned} G^{(2)}(x_1, x_2; x_2, x_1) &= \mathrm{Tr}\{\rho \mathbf{E}^{(-)}(x_1)\mathbf{E}^{(-)}(x_2)\mathbf{E}^{(+)}(x_2)\mathbf{E}^{(+)}(x_1)\} \\ &= \left(\frac{\hbar\omega}{2}\right)^2 |\mathbf{u}(\mathbf{r})|^4 \mathrm{Tr}(\rho a^{\dagger 2} a^2), \end{aligned} \tag{2.86}$$

where $x_1 \equiv (\mathbf{r}, t)$ and $x_2 \equiv (\mathbf{r}, t + \tau)$. To better display the influence of the photon statistics on the coherence properties of the optical field, one may compute the normalized second-order correlation function given by

$$g^{(2)}(0) \equiv \frac{G^{(2)}(x_1, x_2; x_2, x_1)}{|G^{(1)}(x_1, x_1)G^{(1)}(x_2, x_2)|} = \frac{\mathrm{Tr}(\rho a^{\dagger 2} a^2)}{[\mathrm{Tr}(\rho a^\dagger a)]^2}. \tag{2.87}$$

Clearly, $g^{(2)}(0) = 1$ for coherent states $|\alpha\rangle$ of the field, which implies that the coherent states exhibit second-order coherence. In fact, as can be seen from Eq. (2.81), the coherent state is the only pure state which factorizes the $2n$-points correlation function and thus exhibits nth-order coherence for arbitrary n. As another example, $g^{(2)}(0) = 1 - 1/n < 1$ for the n-photon state $|n\rangle$ of the field.

In general, the calculation of high-order correlation functions involve the statistical expectation of normally ordered product of the annihilation and creation operators in the form $a^{\dagger n} a^m$. Employing equation (2.79), the P-representation for the density operator, the expectation of $a^{\dagger n} a^m$ reduces to a simple mean of $\alpha^{*n}\alpha^m$ with respect to the function $P(\alpha)$, that is

$$\mathrm{Tr}(\rho a^{\dagger n} a^m) = \int d^2\alpha P(\alpha)\langle\alpha| a^{\dagger n} a^m |\alpha\rangle = \int d^2\alpha P(\alpha)\alpha^{*n}\alpha^m. \tag{2.88}$$

Hence, when an arbitrary operator expressed in terms of normally ordered product of the annihilation and creation operators, i.e., $\mathcal{O} \equiv \sum_{m,n} C_{nm} a^{\dagger n} a^m$, there exists the following **optical equivalence theorem**:

$$\langle \mathcal{O} \rangle \equiv \mathrm{Tr}(\mathcal{O}\rho) = \int d^2\alpha P(\alpha)\mathcal{O}(\alpha, \alpha^*), \tag{2.89}$$

where $\mathcal{O}(\alpha, \alpha^*) \equiv \langle\alpha|\mathcal{O}|\alpha\rangle = \sum_{m,n} C_{nm}\alpha^{*n}\alpha^m$. As an example, the average number of photons in a single mode, defined by the statistical average of $a^\dagger a$, is just the mean squared absolute value of α

$$\langle n\rangle \equiv \mathrm{Tr}(\rho a^\dagger a) = \int d^2\alpha\, P(\alpha)|\alpha|^2. \tag{2.90}$$

Applying the resolution of unity, $I = \sum_n |n\rangle\langle n|$, we can expand the density operator in terms of the complete orthogonal basis $\{|n\rangle\}$

$$\rho = \sum_{n,m} |n\rangle\langle n|\rho|m\rangle\langle m| = \sum_{n,m} \rho_{nm}|n\rangle\langle m|. \tag{2.91}$$

Hence, once the P-representation for the density operator is specified, the matrix elements ρ_{nm} can be easily described by the function $P(\alpha)$

$$\rho_{nm} = \int d^2\alpha\, P(\alpha)\langle n|\alpha\rangle\langle\alpha|m\rangle = \frac{1}{\sqrt{n!m!}}\int d^2\alpha\, P(\alpha)\alpha^n\alpha^{*m}e^{-|\alpha|^2}. \tag{2.92}$$

It follows then that the off-diagonal matrix elements vanish when $P(\alpha)$ has the spherical symmetry, i.e., $P(\alpha)$ depends only on $|\alpha|$. Using the Glauber-Sudarshan P-representation for normally ordered product of the annihilation and creation operators, the normalized second-order correlation function for a single mode optical field becomes

$$g^{(2)}(0) = \frac{\langle a^{\dagger 2}a^2\rangle}{\langle a^\dagger a\rangle^2} = \frac{\int d^2\alpha\, P(\alpha)|\alpha|^4}{(\int d^2\alpha\, P(\alpha)|\alpha|^2)^2}. \tag{2.93}$$

Hence, one immediately obtains the following non-classical inequality

$$g^{(2)}(0) = 1 + \frac{\int d^2\alpha\, P(\alpha)(|\alpha|^2 - \langle a^\dagger a\rangle)^2}{\langle a^\dagger a\rangle^2} < 1, \tag{2.94}$$

provided that the Glauber-Sudarshan P-representation of the field is not non-negative definite, i.e., $P(\alpha)$ is negative for at least some values of α. In this regard, $P(\alpha)$ ceases to be considered as a classical probability distribution if the non-classical inequality $g^{(2)}(0) < 1$ is fulfilled, i.e., the associated photon distribution is narrower than the Poisson distribution, which is referred to as a **sub-Poissonian** distribution. In contrast, when the Glauber-Sudarshan P-representation of the field is non-negative definite, $P(\alpha)$ can be regarded as a classical probability distribution, and the associated photon distribution is wider than the Poisson distribution, which is referred to as a **super-Poissonian** distribution.

In order to compute the Glauber-Sudarshan P-representation $P(\alpha)$ from a given density operator ρ, one needs two distinct coherent states $|\beta\rangle$ and $|-\beta\rangle$. Then from

Eqs. (2.79) and (2.49), it can be shown that

$$\begin{aligned}\langle -\beta|\rho|\beta\rangle &= \int d^2\alpha P(\alpha)\langle -\beta|\alpha\rangle\langle\alpha|\beta\rangle \\ &= e^{-|\beta|^2}\int d^2\alpha P(\alpha) e^{-|\alpha|^2} e^{\beta\alpha^*-\beta^*\alpha}.\end{aligned} \tag{2.95}$$

To better display this relation, one may let $\alpha \equiv x_\alpha + iy_\alpha$ and $\beta \equiv x_\beta + iy_\beta$. Then Eq. (2.95) immediately leads to

$$\langle -\beta|\rho|\beta\rangle e^{|\beta|^2} = \iint dx_\alpha dy_\alpha P(\alpha) e^{-|\alpha|^2} e^{2i(y_\beta x_\alpha - x_\beta y_\alpha)}. \tag{2.96}$$

Hence, the function $\langle -\beta|\rho|\beta\rangle e^{|\beta|^2}$ is nothing but the two-dimensional Fourier transform of the function $P(\alpha)e^{-|\alpha|^2}$. Taking the inverse Fourier transform of Eq. (2.96), one obtains

$$\begin{aligned}P(\alpha) &= \frac{e^{|\alpha|^2}}{\pi^2}\iint dx_\beta dy_\beta \langle \ \beta|\rho|\beta\rangle e^{|\beta|^2} e^{2i(y_\alpha x_\beta - x_\alpha y_\beta)} \\ &= \frac{e^{|\alpha|^2}}{\pi^2}\int d^2\beta \langle -\beta|\rho|\beta\rangle e^{|\beta|^2} e^{\beta^*\alpha-\alpha^*\beta}.\end{aligned} \tag{2.97}$$

As an example, one may compute the Glauber-Sudarshan P-representation for a thermal state. The thermal state, which is generated by a source in thermal equilibrium at temperature T, is described by a canonical ensemble

$$\rho = \frac{\exp(-H/k_BT)}{\mathrm{Tr}[\exp(-H/k_BT)]}, \tag{2.98}$$

where k_B is the Boltzmann constant and $H = \hbar\omega(a^\dagger a + 1/2)$ is the free-field Hamiltonian. For simplicity, we consider here only the thermal state for a single-mode optical field. Applying the resolution of unity, $I = \sum_n |n\rangle\langle n|$, one can readily expand the density operator for the thermal state as

$$\rho = \left[1 - \exp\left(\frac{-\hbar\omega}{k_BT}\right)\right]\sum_n \exp\left(\frac{-n\hbar\omega}{k_BT}\right)|n\rangle\langle n|. \tag{2.99}$$

From the above relation, one can compute the photon statistics, i.e., the probability $p_n \equiv \rho_{nn}$ that the field contains n photons, which is given by the **Boltzmann distribution**, as illustrated in Fig. 2.3

$$p_n = \left[1 - \exp\left(\frac{-\hbar\omega}{k_BT}\right)\right]\exp\left(\frac{-n\hbar\omega}{k_BT}\right), \tag{2.100}$$

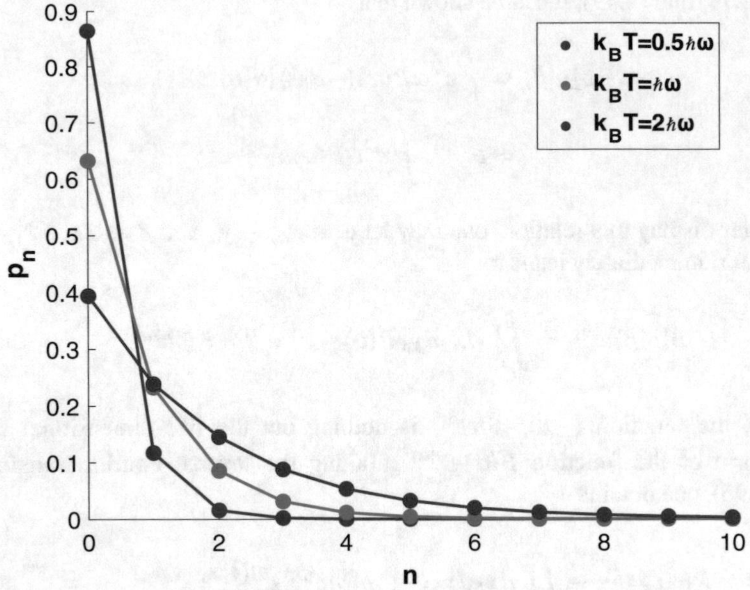

Fig. 2.3 Schematic of the photon statistics for thermal states with different temperatures, where $k_B T = 0.5\,\hbar\omega$, $\hbar\omega$, and $2\,\hbar\omega$ for the blue, green, and red solid curves, respectively

As the average number of photons in the thermal state is given by

$$\langle n\rangle = \mathrm{Tr}(\rho a^\dagger a) = \frac{1}{e^{\hbar\omega/k_B T} - 1}, \tag{2.101}$$

one can rewrite the density operator for the thermal state in terms of $\langle n\rangle$ as

$$\rho = \sum_n \frac{\langle n\rangle^n}{(1+\langle n\rangle)^{n+1}} |n\rangle\langle n|. \tag{2.102}$$

A direction computation yields

$$\begin{aligned}
\langle -\beta|\rho|\beta\rangle &= \sum_n \frac{\langle n\rangle^n}{(1+\langle n\rangle)^{n+1}} \langle -\beta|n\rangle\langle n|\beta\rangle \qquad (2.103)\\
&= \frac{e^{-|\beta|^2}}{1+\langle n\rangle} \sum_{n=0}^{\infty} \frac{(-|\beta|^2)^n}{n!} \left(\frac{\langle n\rangle}{1+\langle n\rangle}\right)^n \\
&= \frac{e^{-|\beta|^2}}{1+\langle n\rangle} \exp\left(\frac{-|\beta|^2}{1+\langle n\rangle^{-1}}\right).
\end{aligned}$$

Hence, using Eq. (2.97), one immediately obtains

$$P(\alpha) = \frac{1}{\pi \langle n \rangle} e^{-|\alpha|^2/\langle n \rangle}. \tag{2.104}$$

In other words, the Glauber-Sudarshan P-representation for a thermal state is given by a Gaussian distribution, which is a non-negative definite function on the complex α-plane. As a direct application of the Glauber-Sudarshan P-representation, one may compute the normalized second-order correlation function $g^{(2)}(0)$ for a thermal state, which is given by

$$g^{(2)}(0) = \frac{\frac{1}{\pi \langle n \rangle} \int d^2\alpha e^{-|\alpha|^2/\langle n \rangle} |\alpha|^4}{\langle n \rangle^2} = \frac{2\langle n \rangle^2}{\langle n \rangle^2} = 2, \tag{2.105}$$

which satisfies the classical inequality $g^{(2)}(0) > 1$. As another example of the Glauber-Sudarshan P-representation, one may compute the P-representation for a coherent state $|\alpha_0\rangle$, which is described by the density operator $\rho = |\alpha_0\rangle\langle\alpha_0|$. A direct computation yields

$$\langle -\beta|\rho|\beta\rangle = \exp(-|\alpha_0|^2 - |\beta|^2 + \beta\alpha_0^* - \beta^*\alpha_0). \tag{2.106}$$

Hence, from Eq. (2.97), the P-representation for a coherent state is given by

$$P(\alpha) = \delta^{(2)}(\alpha - \alpha_0), \tag{2.107}$$

which shows that the Glauber-Sudarshan P-representation for a coherent state is a two-dimensional delta function. As a final example, one can evaluate the Glauber-Sudarshan P-representation for a number state $|n\rangle$, of which the density operator is described by $\rho \equiv |n\rangle\langle n|$. A direct computation yields

$$\langle -\beta|\rho|\beta\rangle = \langle -\beta|n\rangle\langle n|\beta\rangle = e^{-|\beta|^2} \frac{(-|\beta|^2)^n}{n!}. \tag{2.108}$$

Hence, from Eq. (2.97), the P-representation for the number state is

$$\begin{aligned} P(\alpha) &= \frac{e^{|\alpha|^2}}{\pi^2 n!} \int d^2\beta (-|\beta|^2)^n e^{\beta^*\alpha - \beta\alpha^*} \\ &= \frac{e^{|\alpha|^2}}{n!} \frac{\partial^{2n}}{\partial\alpha^n \partial\alpha^{*n}} \delta^{(2)}(n). \end{aligned} \tag{2.109}$$

Hence, the Glauber-Sudarshan P-representation for a number state with $n > 0$ is not a non-negative definite function on the complex α-plane, which is a consequence of the quantum nature of the number state.

Similar to the Glauber-Sudarshan P-representation, which is useful in evaluating the normally ordered correlation functions of the creation and annihilation operators, one may define another distribution on the complex α-plane, namely, the Husimi Q-presentation, which is useful in evaluating anti-normally ordered correlation functions of the annihilation and creations. The Husimi Q-representation for the density operator ρ is defined as $1/\pi$ times the diagonal matrix element of ρ with respect to the coherent states

$$Q(\alpha) \equiv \frac{1}{\pi}\langle\alpha|\rho|\alpha\rangle = \frac{1}{\pi}e^{-|\alpha|^2}\sum_{n,m}\rho_{nm}\frac{\alpha^{*n}\alpha^m}{\sqrt{n!m!}}. \tag{2.110}$$

The function $Q(\alpha)$ can be interpreted as a probability distribution function so long as the statistical expectation value of the operator involved is expressed in anti-normal order product of the annihilation and creation operators

$$\mathrm{Tr}(\rho a^m a^{\dagger n}) = \frac{1}{\pi}\int d^2\alpha\langle\alpha|\rho|\alpha\rangle\alpha^m\alpha^{*n} = \int d^2\alpha\, Q(\alpha)\alpha^m\alpha^{*n}. \tag{2.111}$$

Hermiticity and normalization of the density operator require that $Q(\alpha)$ is a real function that obeys the normalization condition

$$\int d^2\alpha\, Q(\alpha) = 1. \tag{2.112}$$

Unlike the function $P(\alpha)$, the function $Q(\alpha)$ is finite in the complex α plane and positive everywhere, which satisfies the inequality $0 \le Q(\alpha) \le 1/\pi$. To establish the validity of this inequality, we shall express the density operator in the diagonal form

$$\rho = \sum_n \lambda_n|\psi_n\rangle\langle\psi_n|, \tag{2.113}$$

where $|\psi_n\rangle$ are the eigenstates of ρ and λ_n are the associated non-negative eigenvalues that satisfy $0 \le \lambda_n \le 1$, which leads to

$$Q(\alpha) = \frac{1}{\pi}\sum_n \lambda_n|\langle\alpha|\psi_n\rangle|^2 \ge 0, \tag{2.114a}$$

$$\le \frac{1}{\pi}\sum_n \langle\alpha|\psi_n\rangle\langle\psi_n|\alpha\rangle \le \frac{1}{\pi}. \tag{2.114b}$$

Using the expression of the scalar product $\langle\beta|\alpha\rangle = \exp\{-|\alpha-\beta|^2/2 + i\Im(\beta^*\alpha)\}$, we immediately obtain the relation between the P- and Q-representations

$$Q(\beta) = \frac{1}{\pi}\int d^2\alpha\, P(\alpha)e^{-|\alpha-\beta|^2}. \tag{2.115}$$

As an example, the Q-representation for a number state $|n\rangle$ is given by

$$Q(\alpha) = \frac{e^{-|\alpha|^2}|\alpha|^{2n}}{\pi n!}, \tag{2.116}$$

which is a non-negative and bounded function on the complex α-plane. As another example, the Q-representation for a thermal state, whose density matrix is described by Eq. (2.102), has the form

$$\begin{aligned} Q(\alpha) &= \frac{1}{\pi}\sum_n \frac{\langle n\rangle^n}{(1+\langle n\rangle)^{n+1}}\langle\alpha|n\rangle\langle n|\alpha\rangle \\ &= \frac{1}{\pi(1+\langle n\rangle)}e^{-\frac{|\alpha|^2}{1+\langle n\rangle}}. \end{aligned} \tag{2.117}$$

Comparing Eqs. (2.104) and (2.117), one notices that in the limit of large mean photon numbers, the expressions of the P- and Q-representations coincide. This is because in the limit of large mean photon numbers, i.e., $\langle n\rangle \to \infty$, the distinctions between normally and anti-normally ordered correlation functions of the creation and annihilation operators vanish.

Using Eq. (2.91), the expansion of a density operator in terms of the basis $\{|n\rangle\}$, any density operator ρ can be represented in a unique way by a function $R(\alpha^*, \beta)$ of two complex variables defined by

$$R(\alpha^*, \beta) = \frac{1}{\pi}\langle\alpha|\rho|\beta\rangle \exp\left[\frac{1}{2}|\alpha|^2 + \frac{1}{2}|\beta|^2\right] = \frac{1}{\pi}\sum_{n,m}\rho_{nm}\frac{\alpha^{*n}\beta^m}{\sqrt{n!m!}}. \tag{2.118}$$

The series on the right-hand side is absolutely convergent for all finite α^* and β, so that the function $R(\alpha^*, \beta)$ is analytic in the entire complex α^* and β plane. By definition, the functions $Q(\alpha)$ and $R(\alpha^*, \beta)$ are simply related by

$$Q(\alpha) = R(\alpha^*, \alpha)e^{-|\alpha^2|}. \tag{2.119}$$

Hence $Q(\alpha)$ is just a boundary value of the analytic function $R(\alpha^*, \beta)$. The matrix elements of the density operator can easily found by multiplying $R(\alpha^*, \beta)$ by $\alpha^i\beta^{*j}e^{-|\alpha|^2-|\beta|^2}$ and integrating over the complex α^* and β plane

$$\begin{aligned} &\int \alpha^i\beta^{*j}e^{-|\alpha|^2-|\beta|^2}R(\alpha^*, \beta)d^2\alpha d^2\beta \\ &= \sum_{n,m}\frac{\rho_{nm}}{\pi\sqrt{n!m!}}\int \alpha^{*n}\alpha^i e^{-|\alpha|^2}d^2\alpha \int \beta^{*j}\beta^m e^{-|\beta|^2}d^2\beta \\ &= \sum_{n,m}\frac{\rho_{nm}}{\pi\sqrt{n!m!}}\pi^2 n!m!\delta_{ni}\delta_{jm} = \pi\sqrt{i!j!}\rho_{ij}. \end{aligned} \tag{2.120}$$

where we have used the identity equation (2.54) in the last step, so that all the terms except $n = i$ and $m = j$ vanish in the sum and we obtain

$$\rho_{ij} \equiv \langle i|\rho|j\rangle = \frac{1}{\pi}\int \frac{\alpha^i \beta^{*j}}{\sqrt{i!j!}} e^{-|\alpha|^2-|\beta|^2} R(\alpha^*, \beta) d^2\alpha d^2\beta. \tag{2.121}$$

Up to now, we only discussed the Glauber-Sudarshan P-representation and the Husimi Q-presentation associated with normally and anti-normally ordered correlation functions of the creation and annihilation operators. But there exists one other possibility, that is, a distribution function associated with a symmetric ordering of the creation and annihilation operators in a classical fashion. It is well known that in classical mechanics, the state of a system is always specified by the values of the coordinate q and the momentum p in the phase space. For a given initial condition, the system evolves deterministically under the equations of motion. It can be visualized by considering the momentum and coordinate as the coordinates of a point in the two-dimensional phase space. The P-representation, or the Q-representation for the density operator, if interpreted correctly, could be regarded as a probability distribution function in the phase space, where the complex α plane plays the role of the two-dimensional phase space. However, due to the Heisenberg uncertainty principle, the coordinate q and the momentum p are not simultaneously measurable, and hence the concept of a **joint probability distribution** for the coordinate and momentum, or even the concept of a phase space, is at best subtle in quantum mechanics.

Since the birth of quantum mechanics, there have been significant efforts to restore the phase space concept for the description of quantum mechanical uncertainties. Among these attempts, the Wigner distribution, introduced by Eugene Wigner in his 1932 seminal work titled "*On the Quantum Correction for Thermodynamics Equilibrium*" [55], can be regarded as the prototype of all those phase space probability distribution functions. For a single particle specified by a wave function $\psi(q)$, the Wigner distribution $W(p, q)$ is defined as

$$W(p, q) \equiv \frac{1}{\pi\hbar}\int \psi^*(q + y)e^{2ipy/\hbar}\psi(q - y)dy, \tag{2.122}$$

where the integration with respect to y is carried out from $-\infty$ to ∞. The Wigner distribution has many properties similar to a joint probability distribution for the coordinate and momentum. For example, the probability of finding the particle at a given coordinate q or momentum p can be obtained by integrating over the other variable. If we integrate the Wigner function $W(p, q)$ with respect to p, we immediately obtain

$$\int W(p, q)dp = |\psi(q)|^2, \tag{2.123}$$

which is exactly the probability of finding the particle at coordinate q. Likewise, if we integrate the Wigner function $W(p, q)$ with respect to q, we have

$$\int W(p,q)dq = \frac{1}{\pi\hbar}\iint \psi^*(q+y)e^{2ipy/\hbar}\psi(q-y)dqdy \tag{2.124}$$
$$= \left|\frac{1}{\sqrt{2\pi\hbar}}\int \psi(u)e^{-ipu/\hbar}du\right|^2,$$

which also gives the correct probability of finding the particle with momentum p. Unfortunately, the Wigner distribution $W(p, q)$ cannot be interpreted as the simultaneous probability for the coordinate and momentum, since it may take on negative values for some regions of p and q, as indicated by Wigner himself. Hence, the difference between classical probabilistic mechanics and quantum mechanics is that quantum probabilities, in some way, assume negative values.

As a first example, we shall calculate the Wigner function $W(p, q)$ for a one-dimensional harmonic oscillator. We can now use the eigenstates of the quantum harmonic oscillator

$$\phi_n(q) = \frac{1}{\sqrt{2^n n!\xi\sqrt{\pi}}}e^{-\frac{q^2}{2\xi^2}}H_n\left(\frac{q}{\xi}\right), \tag{2.125}$$

to express the Wigner function in terms of an integral

$$W_n(p,q) = \frac{1}{\pi\hbar}\frac{e^{-q^2/\xi^2}}{2^n n!\xi\sqrt{\pi}}\int e^{-\frac{y^2}{\xi^2}+\frac{2ipy}{\hbar}}H_n\left(\frac{q+y}{\xi}\right)H_n\left(\frac{q-y}{\xi}\right)dy \tag{2.126}$$
$$= \frac{1}{\pi\hbar}\frac{e^{-x^2+z_0^2}}{2^n n!\sqrt{\pi}}\int e^{-z^2}H_n(x+z+z_0)H_n(x-z-z_0)dz,$$

where $\xi = \sqrt{\hbar/m\omega_0}$ is the characteristic length for the harmonic oscillator, $z = y/\xi - z_0$, $z_0 = i\xi p/\hbar$, $x = q/\xi$ are all dimensionless parameters, and $H_n(z)$ is the n-th Hermite polynomial. If we employ the identity $H_n(-z) = (-1)^n H_n(z)$, the Wigner function is given by

$$W_n(p,q) = \frac{(-1)^n}{\pi\hbar}\frac{e^{-\alpha\beta}}{2^n n!\sqrt{\pi}}\int e^{-z^2}H_n(z+\alpha)H_n(z-\beta)dz, \tag{2.127}$$

where $\alpha = x + z_0$ and $\beta = x - z_0$. We can reduce this integral to the known special functions by using the orthogonality of the Hermite polynomials and the relation

$H_n'(z) = 2nH_{n-1}(z)$, which leads to

$$\int e^{-z^2} H_n(z+\alpha)H_n(z-\beta)dz \tag{2.128}$$
$$= \sum_{l,m} \frac{2^l n!}{(n-l)!} \frac{2^m n!}{(n-m)!} \frac{\alpha^l}{l!} \frac{(-\beta)^m}{m!} \int e^{-z^2} H_{n-l}(z)H_{n-m}(z)dz.$$

By applying the orthogonality relations

$$\int e^{-z^2} H_n(z)H_m(z)dz = 2^n n! \sqrt{\pi}\,\delta_{nm}, \tag{2.129}$$

we obtain the following integral identity,

$$\int e^{-z^2} H_n(z+\alpha)H_n(z-\beta)dz \tag{2.130}$$
$$= 2^n n! \sqrt{\pi} \sum_l \binom{n}{l} \frac{(-1)^l}{l!} (2\alpha\beta)^l = 2^n n! \sqrt{\pi} L_n(2\alpha\beta),$$

where $L_n(z)$ is the n-th order Laguerre polynomial. Substituting this integral identity into Eq. (2.127), we obtain

$$W_n(p,q) = \frac{(-1)^n}{\pi\hbar} e^{-\frac{q^2}{\xi^2} - \frac{\xi^2 p^2}{\hbar^2}} L_n\left(\frac{2q^2}{\xi^2} + \frac{2\xi^2 p^2}{\hbar^2}\right) \tag{2.131}$$
$$= \frac{(-1)^n}{\pi\hbar} e^{-\frac{2H}{\hbar\omega_0}} L_n\left(\frac{4H}{\hbar\omega_0}\right),$$

where $H = (p^2/m + m\omega_0^2 q^2)/2$ is the Hamiltonian of the harmonic oscillator. As a result of the oscillatory structure in the Laguerre polynomials, the Wigner function is positive only for the ground state of the harmonic oscillator, and it may take negative values for all the higher excited states. Specifically, the first three Laguerre polynomials are:

$$L_0(z) = 1,\ L_1(z) = 1 - z,\ L_2(z) = 1 - 2z + z^2/2. \tag{2.132}$$

To better display the Wigner distribution for a one-dimensional harmonic oscillator, one may rescale the momentum and coordinate according to $P \to \xi p/\hbar$ and $Q \to q/\xi$, where $\xi = \sqrt{\hbar/m\omega_0}$ is the characteristic length for the harmonic oscillator. Hence, in the dimensionless unit, the Wigner distribution $W(P, Q)$ for the ground state, first excited state, and the second excited state is explicitly given by (Figs. 2.4, 2.5, and 2.6)

$$W_0(P, Q) = \frac{1}{\pi\hbar} e^{-(P^2+Q^2)}, \tag{2.133}$$

$$W_1(P, Q) = \frac{-1}{\pi\hbar} e^{-(P^2+Q^2)} \left[1 - 2\left(P^2 + Q^2\right)\right],$$

$$W_2(P, Q) = \frac{1}{\pi\hbar} e^{-(P^2+Q^2)} \left[1 - 4\left(P^2 + Q^2\right) + 2\left(P^2 + Q^2\right)^2\right].$$

For an arbitrary density matrix ρ, one may define the associated Wigner distribution $W(p, q)$ as

$$W(p, q) \equiv \frac{1}{(2\pi)^2} \int dudv e^{i(uq+vp)} \mathrm{Tr}[e^{-i(u\hat{q}+v\hat{p})} \rho], \tag{2.134}$$

where $[\hat{q}, \hat{p}] = i\hbar$. To show that the above definition coincides with that of Wigner's original distribution when the density operator is a pure state, one may apply the Baker-Campbell-Hausdorff formula $e^{A+B} = e^A e^B e^{-[A,B]/2}$ to the case $A = -iv\hat{p}$

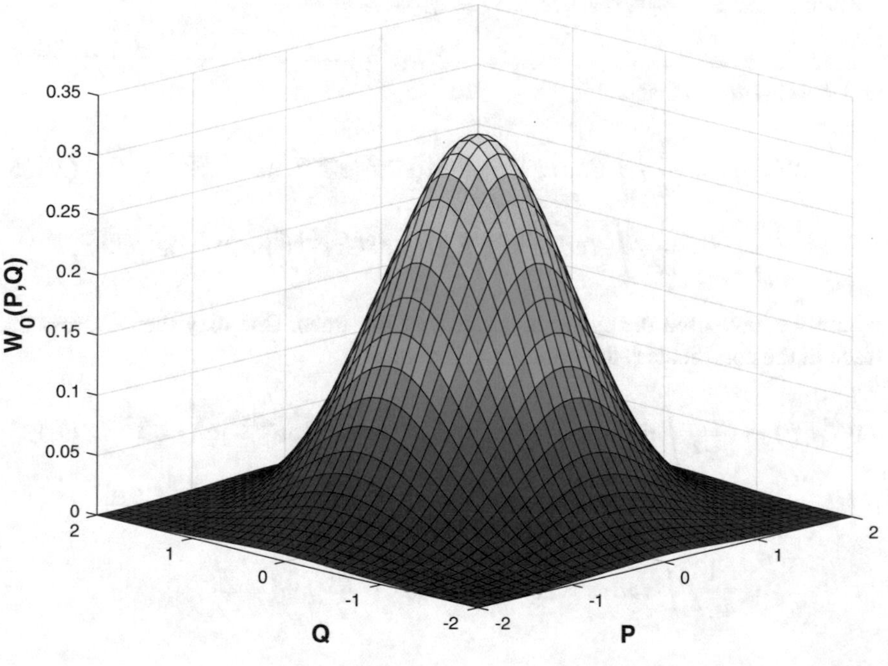

Fig. 2.4 The Wigner distribution function $W_0(P, Q)$ for the ground state of the harmonic oscillator, where $P \equiv \xi p/\hbar$, $Q \equiv q/\xi$, and $\xi = \sqrt{\hbar/m\omega_0}$

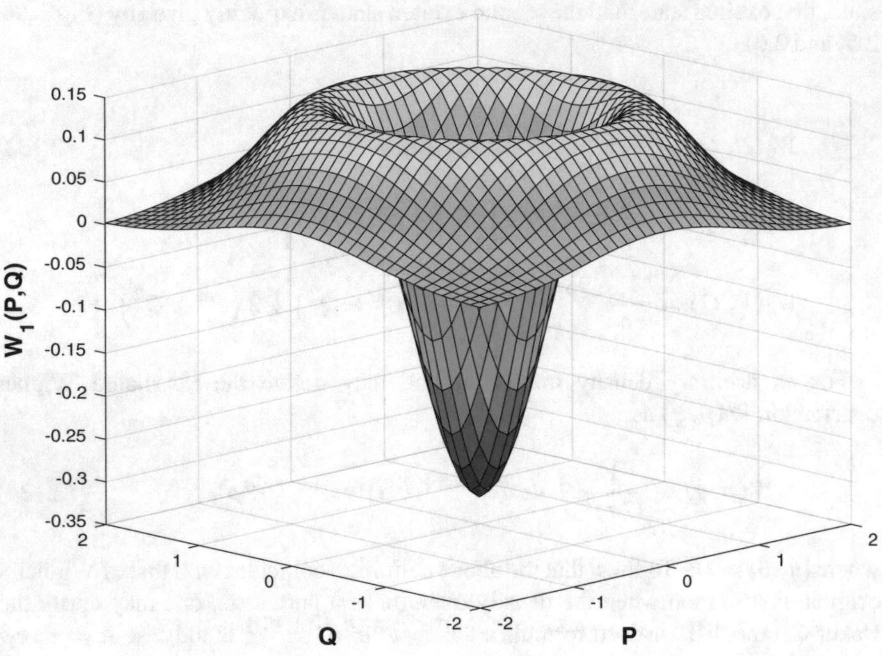

Fig. 2.5 The Wigner distribution function $W_1(P, Q)$ for the first excited state of the harmonic oscillator, where $P \equiv \xi p/\hbar$, $Q \equiv q/\xi$, and $\xi = \sqrt{\hbar/m\omega_0}$

and $B = -iu\hat{q}$, so that

$$W(p,q) = \frac{1}{4\pi^2}\int dudve^{i(vp+uq)}\mathrm{Tr}(e^{-iv\hat{p}}e^{-iu\hat{q}}\rho)e^{-iuv\hbar/2} \tag{2.135}$$
$$= \frac{1}{4\pi^2}\int dudve^{i(vp+uq)}\mathrm{Tr}(e^{-iv\hat{p}/2}e^{-iu\hat{q}}\rho e^{-iv\hat{p}/2})e^{-iuv\hbar/2},$$

where we have used the cyclic invariance of the trace. One may then express the trace in the coordinate representation and obtains

$$W(p,q) = \frac{1}{4\pi^2}\int dudve^{i(vp+uq)}dq'\langle q'|e^{-\frac{iv\hat{p}}{2}}e^{-iu\hat{q}}\rho e^{-\frac{iv\hat{p}}{2}}|q'\rangle e^{-\frac{iuv\hbar}{2}} \tag{2.136}$$
$$= \frac{1}{4\pi^2}\int dudve^{i(vp+uq)}dq'\langle q'+\frac{v\hbar}{2}|e^{-iu\hat{q}}\rho|q'-\frac{v\hbar}{2}\rangle e^{-\frac{iuv\hbar}{2}}$$
$$= \frac{1}{4\pi^2}\int dudve^{ivp}dq'e^{iu(q-q')}\langle q'+\frac{v\hbar}{2}|\rho|q'-\frac{v\hbar}{2}\rangle,$$

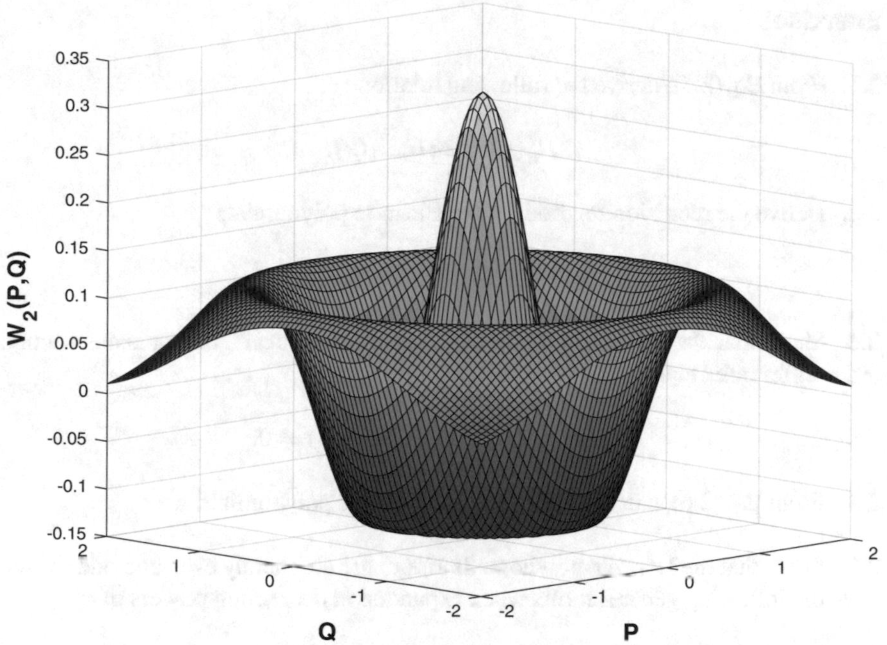

Fig. 2.6 The Wigner distribution function $W_2(P, Q)$ for the second excited state of the harmonic oscillator, where $P \equiv \xi p/\hbar$, $Q \equiv q/\xi$, and $\xi = \sqrt{\hbar/m\omega_0}$

where we have used the relation $e^{-iv\hat{p}/2}|q'\rangle = |q' - v\hbar/2\rangle$. After evaluating the integral with respect to u, one immediately obtains

$$
\begin{aligned}
W(p,q) &= \frac{1}{4\pi^2}\int dudve^{ivp}dq'e^{iu(q-q')}\langle q' + \frac{v\hbar}{2}|\rho|q' - \frac{v\hbar}{2}\rangle, \qquad (2.137) \\
&= \frac{1}{2\pi}\int dve^{ivp}dq'\delta(q-q')\langle q' + \frac{v\hbar}{2}|\rho|q' - \frac{v\hbar}{2}\rangle, \\
&= \frac{1}{2\pi}\int dve^{ivp}\langle q + \frac{v\hbar}{2}|\rho|q - \frac{v\hbar}{2}\rangle,
\end{aligned}
$$

After the change of coordinate $y = v\hbar/2$, one readily see that Eq. (2.137) is equivalent to Wigner's original definition, Eq. (2.122), when the density operator $\rho \equiv |\psi\rangle\langle\psi|$ is a pure state. Finally, the definition equation (2.137) of the Wigner distribution can be recast into a more convenient form

$$
W(\alpha) \equiv \frac{1}{\pi^2}\int d^2\eta e^{\alpha\eta^* - \alpha^*\eta}\mathrm{Tr}[D(\eta)\rho], \qquad (2.138)
$$

where $D(\eta) \equiv e^{\eta a^\dagger - \eta^* a}$ is the displacement operator and $\eta \equiv (v - iu)/\sqrt{2}$.

Exercises

2.1. From Eq. (2.5), derive the following relation

$$H_n'(x) = 2nH_{n-1}(x).$$

2.2. Derive the recursion relation for the Hermite polynomials

$$H_{n+1}(x) - 2xH_n(x) + 2nH_{n-1}(x) = 0.$$

2.3. Show that the Hermite polynomials satisfy the linear homogeneous second order differential equation

$$H_n''(x) - 2xH_n'(x) + 2nH_n(x) = 0.$$

2.4. From Eq. (2.6), calculate the first five Hermite polynomials.

2.5. Prove that the Hermite polynomials $H_n(x)$ are alternately even and odd, which have the following general forms when expanded in decreasing powers of x

$$(2x)^n - \frac{n(n-1)}{1!}(2x)^{n-2} + \frac{n(n-1)(n-2)(n-3)}{2!}(2x)^{n-4} - \ldots.$$

2.6. Using Eq. (2.8), show that the mean values of the coordinate and the momentum of the coherent states are $\bar{x} = A\cos\omega_0 t$, $\bar{p} = -\hbar A\sin\omega_0 t$.

2.7. Using Eq. (2.11), show that the coherent state is a minimum uncertainty state: the uncertainties of the coordinate and the momentum are $\Delta x = 1/\sqrt{2}$ and $\Delta p = \hbar/\sqrt{2}$.

2.8. Using Eqs. (2.17) to (2.20), derive equation (2.21) from the expression of the electromagnetic energy in the rationalized units

$$H = \frac{1}{2}\int (\mathbf{E}^2 + \mathbf{B}^2)d\mathbf{r}.$$

2.9. Using Eqs. (2.22) and (2.23), prove that the basis states $|n_k\rangle$ are normalized and orthogonal, $\langle n_k|n_{k'}\rangle = \delta_{kk'}$.

2.10. Show that the probability of having n photons in a coherent state $|\alpha\rangle$ is a Poisson distribution with mean value $|\alpha|^2$

$$|\langle n|\alpha\rangle|^2 = \frac{|\alpha|^{2n}}{n!}e^{-|\alpha|^2}.$$

2.11. Calculate the average photon number $\langle n\rangle = \langle\alpha|a^\dagger a|\alpha\rangle$ and the variance $(\Delta n)^2 = \langle\alpha|a^\dagger a a^\dagger a|\alpha\rangle - \langle\alpha|a^\dagger a|\alpha\rangle^2$. Show that the average photon number and the variance are both equal to $|\alpha|^2$

$$(\Delta n)^2 = \langle n\rangle = |\alpha|^2.$$

2.12. Let A and B be two arbitrary operators, and let $f(s) = e^{sA}Be^{-sA}$. Derive the operator expansion theorem

$$e^{sA}Be^{-sA} = B + s[A, B] + \frac{s^2}{2!}[A, [A, B]] + \ldots.$$

For any operator function $f(B)$ having a power series in B, derive the similarity transformation theorem

$$e^{sA}f(B)e^{-sA} = f(e^{sA}Be^{-sA}).$$

Specifically, for any pair of operators A and B whose commutator $[A, B]$ is an ordinary number c, prove that

$$e^{sA}f(B)e^{-sA} = f(B + cs).$$

2.13. Using the operator expansion theorem, derive the similarity transformations for the creation and annihilation operators

$$e^{sa^\dagger a}ae^{-sa^\dagger a} = ae^{-s},\, e^{s^\dagger a}a^\dagger e^{-sa^\dagger a} = a^\dagger e^s.$$

$$e^{-\alpha a^\dagger+\alpha^* a}ae^{\alpha a^\dagger-\alpha^* a} = a + \alpha,\, e^{-\alpha a^\dagger+\alpha^* a}a^\dagger e^{\alpha a^\dagger-\alpha^* a} = a^\dagger + \alpha^*.$$

Use them to prove the following identities.

$$e^{sa^\dagger a}f(a, a^\dagger)e^{-sa^\dagger a} = f(ae^{-s}, a^\dagger e^s),$$

2.14. Let $f(a, a^\dagger)$ be a function of the creation and annihilation operators; prove the following derivative theorems

$$\frac{\partial f(a, a^\dagger)}{\partial a} = -[a^\dagger, f(a, a^\dagger)], \quad \frac{\partial f(a, a^\dagger)}{\partial a^\dagger} = [a, f(a, a^\dagger)].$$

2.15. If A and B are two operators such that the commutator $[A, B]$ commutes with each of them, $[A, [A, B]] = [B, [A, B]] = 0$, prove the Baker-Campbell-Hausdorff formula

$$e^A e^B = e^{A+B+\frac{1}{2}[A,B]}.$$

2.16. Using the multiplication law of the displacement operators, Eq. (2.43), derive equation (2.46).

2.17. From the definition of coherent states, derive equation (2.49).

2.18. Using Eq. (2.49), the expression of the scalar product $\langle\alpha|\beta\rangle$, prove the inequality equation (2.50).

2.19. Prove the **dominated convergence theorem**: let $f_1, f_2, \cdots, f_N$ and R: $\mathbb{Z} \to \mathbb{R}$ be functions such that $|f_N(n)| \leq g(n)$, $\sum_{n=0}^{\infty} g(n) < \infty$. Prove that

$$\lim_{N\to\infty} \sum_{n=-\infty}^{\infty} f_N(n) = \sum_{n=-\infty}^{\infty} \lim_{N\to\infty} f_N(n).$$

2.20. Using Eq. (2.62) and the following identity

$$\langle\phi|\psi\rangle = \frac{1}{4}(\langle\phi+\psi|\phi+\psi\rangle - \langle\phi-\psi|\phi-\psi\rangle + i\langle\phi+i\psi|\phi+i\psi\rangle - i\langle\phi-i\psi|\phi-i\psi\rangle)$$

to derive the general case for $\langle\phi|\psi\rangle$ from linearity.

2.21. By direct calculation, prove the following integral identity

$$\frac{1}{\pi}\int e^{\beta^*\alpha-|\alpha|^2}\alpha^{*n} d^2\alpha = \beta^{*n},$$

which is a special case of Eq. (2.65).

2.22. By direct calculation, prove the following integral identity

$$\frac{1}{\pi}\int e^{\beta^*\alpha-|\alpha|^2}\alpha^n d^2\alpha = \delta_{n0},$$

and use it to derive equation (2.71).

2.23. From the integral representation for the matrix element of the displacement, Eq. (2.74), prove directly that $\langle m|D(0)|n\rangle = \delta_{mn}$ and

$$\langle m|D(\xi)|0\rangle = e^{-\frac{1}{2}|\xi|^2}\frac{\xi^m}{\sqrt{m!}},\ \langle 0|D(\xi)|n\rangle = e^{-\frac{1}{2}|\xi|^2}\frac{(-\xi^*)^n}{\sqrt{n!}}.$$

2.24. Show that the normalized second-order correlation function $g^{(2)}(0)$ for the n-photon state $|n\rangle$ of the field has the form

$$g^{(2)}(0) = 1 - 1/n.$$

2.25. Using Eq. (2.92), examine the validity of $\mathrm{Tr}\rho = \sum \rho_{nn} = 1$.

2.26. Verify Eq. (2.97).

2.27. Prove that the series $\sum_{n=1}^{\infty} nx^n$ converges and equals to $x/(1-x)^2$ when $|x| < 1$. Use the result to verity equation (2.101).

2.28. Verify Eq. (2.104).

2.29. Using $\rho^\dagger = \rho$ and $\mathrm{Tr}\rho = 1$, prove that all the diagonal matrix elements of the density operator are real and non-negative; using the positive definiteness of the density operator, prove the inequalities $|\rho_{nm}|^2 \le \rho_{nn}\rho_{mm}$.

2.30. Prove that if for any bounded operator $\hat{A}$, $A(\alpha) = \langle\alpha|\hat{A}|\alpha\rangle = 0$ in any finite domain over the complex α plane, then $A(\alpha) \equiv 0$ over the entire complex α plane, that is, the operator $\hat{A}$ itself is identity zero.

End of Chapter Problems

1. In this problem, we will show that a classical prescribed source of radiation always gives rise to a coherent state of the electromagnetic field. This fact was first indicated by Glauber in 1963. To simplify the discussion, we consider a single-electron atom interacting with a quantized electromagnetic field. In the electric dipole approximation, the interaction between atom and electromagnetic field is described by the Hamiltonian $H = H_A + H_F - e\mathbf{r}\cdot\mathbf{E}$, where $\mathbf{r}$ is the position vector of the atom, $H_F = \sum_k \hbar\omega_k(a_k^\dagger a_k + 1/2)$ is the energy of the free-field expressed via the creation and annihilation operators, and $H_A = \sum_i E_i|i\rangle\langle i| = \sum_i E_i\sigma_{ii}$ is the energy of the atom expressed via the atomic

transition operators $\sigma_{ij} \equiv |i\rangle\langle j|$. In terms of the electric-dipole transition matrix elements $d_{ij} \equiv e\langle i|\mathbf{r}|j\rangle$, we have $e\mathbf{r} = \sum_{ij} e|i\rangle\langle i|\mathbf{r}|j\rangle\langle j| = \sum_{ij} \mathbf{d}_{ij}\sigma_{ij}$. We now proceed to a two-level atom with $\mathbf{d}_{eg} = \mathbf{d}_{ge}$, which yields $e\mathbf{r} = \mathbf{d}_{ge}(\sigma_+ + \sigma_-)$ and $H_A = \hbar\Omega\sigma_z + \frac{1}{2}(E_e + E_g)$, where $\sigma_- \equiv |g\rangle\langle e|$, $\hbar\Omega \equiv \frac{1}{2}(E_e - E_g)$ and $\sigma_z \equiv \frac{1}{2}(\sigma_{ee} - \sigma_{gg})$. According to Eqs. (2.17) and (2.18), we obtain

$$
\begin{aligned}
-e\mathbf{r}\cdot\mathbf{E} &= -i\sum_k \sqrt{\frac{\hbar\omega_k}{2}}(\sigma_+ + \sigma_-)\mathbf{d}_{ge}\cdot(a_k\mathbf{u}_k(\mathbf{r})e^{-i\omega_k t} + a_k^\dagger\mathbf{u}_k^*(\mathbf{r})e^{i\omega_k t}) \\
&\equiv \hbar\sum_k(\sigma_+ + \sigma_-)(g_k a_k + g_k^* a_k^\dagger).
\end{aligned}
$$

In the rotating-wave approximation, we keep terms like $\sigma_- a_k^\dagger$ and $\sigma_+ a_k$, which describes the process of creating a photon of mode k by moving the electron from the upper to the lower state, and the process of moving the electron from the lower to the upper state by absorbing a photon of mode k. Thus, the Hamiltonian which describes the interaction between a two-level atom and an electromagnetic field can be written as $H = \hbar\Omega\sigma_z + \hbar\sum_k \omega_k a_k^\dagger a_k + \hbar\sum_k(g_k\sigma_+ a_k + g_k^*\sigma_- a_k^\dagger)$, where we have omitted the constant energy term $\frac{1}{2}(E_e + E_g)$ in H_A, and the zero-point energy $\sum_k \frac{1}{2}\hbar\omega_k$ in H_F. In general, the Hamiltonian in quantum optics which describes the interaction between a collection of two-level atoms and a quantized electromagnetic field can be written as

$$
H = \hbar\sum_i \Omega_i\sigma_z^i + \hbar\sum_k \omega_k a_k^\dagger a_k + \hbar\sum_{i,k}(g_{ik}\sigma_+^i a_k + g_{ik}^*\sigma_-^i a_k^\dagger).
$$

If we regard the atomic system as a classical source, and treat the Pauli operators as c-numbers, we obtain

$$
\begin{aligned}
\mathcal{H}_F &= \hbar\sum_i\langle\Omega_i\sigma_z^i\rangle + \hbar\sum_k \omega_k a_k^\dagger a_k + \hbar\sum_{i,k}(g_{ik}\langle\sigma_+^i\rangle a_k + g_{ik}^*\langle\sigma_-^i\rangle a_k^\dagger) \\
&\equiv \hbar\sum_k \omega_k a_k^\dagger a_k + \hbar\sum_k[\lambda_k(t)a_k^\dagger + \lambda_k^*(t)a_k] + \text{constant},
\end{aligned}
$$

where the constant energy term $\hbar\sum_i\langle\Omega_i\sigma_z^i\rangle$ is omitted in $\mathcal{H}_F$. Without loss of generality, we may only consider a single-mode optical field, so that the equation of motion for the field becomes $i\partial_t|\varphi(t)\rangle = [\omega a^\dagger a + \lambda(t)a^\dagger + \lambda^*(t)a]|\varphi(t)\rangle$. Prove that if the state of the field is initially prepared in the vacuum state $|0\rangle$, it is a coherent state at any time t up to a time-dependent phase factor.

Schrödinger's Cat States

3

3.1 EPR Paradox, Cat States and Entanglement

As one of the founders of quantum mechanics, Einstein never ceased to raise doubts as to whether the quantum theory is a complete theory of nature. In the May 15, 1935 issue of *Physical Review*, Einstein co-authored a landmark paper with his two assistants at the Institute for Advanced Study, Boris Podolsky and Nathan Rosen [56], in which he initiated a profound and insightful reflection on what was generally accepted quantum mechanical orthodoxy. In this paper, entitled "*Can Quantum-Mechanical Description of Physical Reality Be Considered Complete?*", Einstein, Podolsky, and Rosen (a paper now known as the initials of the three authors "EPR") demonstrated that there is an incompatibility between locality, separability, and completeness in the description of physical systems by means of wave functions which may lead to a dilemma—one can agree on any one of these premises but never all three. Although quantum nonlocality and nonseparability were not validated until decades after Einstein's death, the EPR paper is still among the top cited papers ever published in *physical review* journals, as it was the first to introduce a special property of composite quantum systems now known as **entanglement**. Quite remarkable, this concept becomes the cornerstone of the contemporary quantum information science.

The EPR incompleteness argument is based on two principles: (1) the separability principle asserts that any two spatially separated systems possess their own separable real states, and (2) the locality principle asserts that the real state of a system cannot be altered immediately as a consequence of measurements made on another system at a spatially separated location, i.e., the real state of a system can only be altered by local effects propagating with finite, sub-luminal velocities. In the EPR paper, Einstein and co-authors wrote that "at the time of measurement the two systems no longer interact, no real change can take place in the second system in consequence of anything that may be done to the first system." Both of these

C.-F. Kam et al., *Coherent States*, Lecture Notes in Physics 1011,
https://doi.org/10.1007/978-3-031-20766-2_3

principles are fundamentally rooted in everyday thought and are fundamental to the theory of relativity and other field theories.

The logical flow of the EPR paper is as follows: Einstein and co-authors first established a completeness condition and then aimed to show that at least in one special case involving previously interacting systems, the quantum theory failed to satisfy this necessary condition. The completeness condition in the original EPR paper is a one-sentence assertion: "every element of the physical reality must have a counterpart in the physical theory." But the existence of elements of physical reality requires a criterion of physical reality: "If, without in any way disturbing a system, we can predict with certainty (i.e. with probability equal to unity) the value of a physical quantity, then there exists an element of physical reality corresponding to this physical quantity." In other words, the sufficient condition for the existence of elements of physical reality is the existence of predetermined values for physical quantities.

With these premises, Einstein and co-authors proposed a gedanken experiment which involves a *non-factorizable* wave function describing two particles moving away from a source into spatially separated regions, and yet always having maximally correlated position, and anti-correlated momenta. Here, *non-factorizable* means that a wave function cannot be factorized as a simple product of wave functions of its local constituents. In particular, Einstein and co-authors considered the wave function defined in a position representation

$$\Psi(x_1, x_2) = \int_{-\infty}^{\infty} e^{\frac{ip}{\hbar}(x_1 - x_2 + x_0)} dp, \tag{3.1}$$

where x_0 is a constant separation between the two particles. A key issue indicated by Einstein and co-authors is that one may expand the wave function $\Psi(x_1, x_2)$ in more than one basis, which corresponds to different experimental settings. For example, one may expand the wave function $\Psi(x_1, x_2)$ in terms of the eigenfunction $u_p(x_1)$ of the momentum of the first particle

$$\Psi(x_1, x_2) = \int_{-\infty}^{\infty} \psi_p(x_2) u_p(x_1) dp, \tag{3.2}$$

where $u_p(x_1) \equiv e^{\frac{i}{\hbar}px_1}$ and $\psi_p(x_2) \equiv e^{\frac{i}{\hbar}p(x_0 - x_2)}$ satisfy $\hat{p}_1 u_p(x_1) = p u_p(x_1)$ and $\hat{p}_2 \psi_p(x_2) = -p\psi_p(x_2)$. Then, according to quantum mechanics, one may make an instant prediction with certainty that a measurement of the momentum of the second particle is $-p$, if the momentum of the first particle is measured to be p. Similarly, one may expand the wave function $\Psi(x_1, x_2)$ in terms of the eigenfunction $v_x(x_1)$ of the position of the first particle

$$\Psi(x_1, x_2) = \int_{-\infty}^{\infty} \varphi_x(x_2) v_x(x_1) dx, \tag{3.3}$$

where $v_x(x_1) \equiv \delta(x_1 - x)$ and $\varphi_x(x_2) \equiv h\delta(x - x_2 + x_0)$ satisfy $\hat{q}_1 v_x(x_1) = x v_x(x_1)$ and $\hat{q}_2 \varphi_x(x_2) = (x + x_0)\varphi_x(x_2)$. Then, one may make an instant prediction with certainty that a measurement of the position of the second particle is $x + x_0$, if the position of the first particle is measured to be x. As long as the two particles are sufficiently far apart, these predictions of experiment outcomes are made without disturbing the second particle, based on the locality principle, and so, based upon the criterion of physical reality mentioned above, Einstein and co-authors deduced that both the measurement outcomes for the second particle preexist, in the form of "elements of reality"—the perfect correlation in positions and anti-correlation in momenta between the two spatially separated particles implies the existence of two "elements of reality" that are simultaneously predetermined with absolute definiteness for both the measurement outcomes for the second particle. However, according to the uncertainty principle in quantum mechanics, simultaneous determinacy for both the position and momentum is not allowed for any quantum state. Hence, Einstein and co-authors arrived at the conclusion that "the quantum-mechanical description of physical reality given by wave functions is not complete." In short, one can only avoid this conclusion either by assuming that the separate real state of a system is changed by what happens in a causally separated region, or spatially separated systems do not possess separate states at all. But later, as was validated by various experiments over the last half-century, nature indeed prefers quantum-mechanical description of reality without the separability, assigning non-decomposable joint state even to largely separated previously connecting systems.

Unlike the EPR incompleteness argument that is entirely based on the separability and locality principles, Schrödinger focuses on a putative incompleteness in the quantum-mechanical description of macroscopic observables through his famous cat paradox, which first appeared in his seminal paper titled "*The Present Situation in Quantum Mechanics*" published in 1935 [57]. In this mind-penetrating paper, Schrödinger designed a gedanken experiment to highlight a seemingly absurdity in human's normal physical mindset of the quantum superposition principle—a cat which is hidden in a box could be simultaneously alive and dead. In the original text, Schrödinger wrote: "a cat is penned up in a steel chamber, along with the following device (which must be secured against direct interference by the cat): in a Geiger counter there is a tiny bit of radioactive substance, so small, that perhaps in the course of the hour one of the atoms decays, but also, with equal probability, perhaps none; if it happens, the counter tube discharges and through a relay releases a hammer which shatters a small flask of hydrocyanic acid. If one has left this entire system to itself for an hour, one would say that the cat still lives if meanwhile no atom has decayed. The psi-function of the entire system would express this by having in it the living and dead cat (pardon the expression) mixed or smeared out in equal parts." A measurement apparatus that is capable of measuring the atom in a superposition of not decayed and decayed which projects the cat into a superposition of alive and dead defines what is the "Schrödinger's cat paradox," a proposal which has profoundly challenged our common sense that a cat is, or can be, either alive or dead in the human-scale world. With the vast amount of research on quantum information processing in the last three decades, we came to realize that mesoscopic Schrödinger cat states, such as the superpositions of coherent states, are extremely sensitive to quantum decoherence, or the environmentally induced reduction of quantum superpositions into statistical mixtures and classical behavior [58–60]. Thus, in the macroscopic scale where such decoherence is palpable, it makes interference pattern observation exceedingly difficult, if not impossible.

The Schrödinger's cat paradox has still baffled the physicists and laymen alike after more than eight decades. It is still under intense debate regarding its implications in the measurement problem and the quantum-classical boundary. Although Schrödinger's hypothetical cat states seem not to appear in the macroscopic world, employing the groundbreaking experimental techniques developed in recent years, researchers are able to measure and manipulate individual quantum systems and have succeeded in producing Schrödinger cat-like states in mesoscopic systems. For example, one may prepare Schrödinger cat-like states through coherent states of light, which may be represented by the entangled superposition state

$$|\psi\rangle = \frac{1}{\sqrt{2}}(|\alpha\rangle|\uparrow\rangle + |-\alpha\rangle|\downarrow\rangle), \tag{3.4}$$

where $|\alpha\rangle$ and $|-\alpha\rangle$ are a pair of coherent states with opposite amplitudes which refer to the states of a live and dead cat and $|\uparrow\rangle$ and $|\downarrow\rangle$ are the spin-1/2 eigenstates for $\hat{J}_z$ which refer to the internal states of an atom which has and has not radioactively decayed.

In contrast to the entangled Schrödinger cat states for the whole atom-cat system, the coherent superposition of quasi-classical states such as coherent states of light also resembles the essential features of Schrödinger's superposition of dead and alive cat states, which may be represented by

$$|\psi\rangle = \frac{1}{\sqrt{2(1 \pm e^{-2|\alpha|^2})}}(|\alpha\rangle \pm |-\alpha\rangle), \tag{3.5}$$

where the plus and minus signs are for the *even-* and *odd-cat* states. Another example of the coherent superposition of dead and alive cat is the multi-atom Greenberger-Horne-Zeilinger (GHZ) state which consists of N spin-1/2 particles being in an equal superposition of all spins up or all spins down

$$|\psi\rangle = \frac{1}{\sqrt{2}}(|\uparrow\rangle^{\otimes N} + |\downarrow\rangle^{\otimes N}), \tag{3.6}$$

which has already been demonstrated experimentally up to $N = 20$ [48, 49]. In the next section, we will discuss experimental realizations of the entangled light-matter Schrödinger cat state and the optical cat states consisting of a superposition of two coherent states of light.

3.2 Experimental Realization

In the past decades, research in connect to coherent states have been experiencing a steady but significant growth due to their increasing and fundamental connection in quantum information processing and communication. By being non-spreading wave packets, which are centered along the classical particle trajectories, coherent states utilized in quantum optics become key to the understanding of how in a large-scale quantum-to-classical transition can occur. With this aim, it is quite natural that the preparation, maintenance, and calibration of superposition of multiple coherent states had become the central theme of quantum information science.

The superpositions of distinct coherent states are of special interest, since they could mimic Schrödinger's famous cat states. In 1986, Yurke and Stoler made the earliest attempt to construct superpositions of coherent states. They presented an ingenious proposal which was based on the physics of nonlinear fiber optics. In their paper titled "*Generating Quantum Mechanical Superpositions of Macroscopically Distinguishable States via Amplitude Dispersion*" [61], the authors demonstrated that a coherent state of light propagating along an optical fiber with Kerr nonlinearity could split into a superposition of two distinct coherent states with opposite amplitudes. To be precise, the Hamiltonian which describes the transmission of light along an optical fiber with Kerr nonlinearity can be written as

$$\hat{H} = \omega\hat{n} + \Omega\hat{n}^2, \tag{3.7}$$

where ω is the energy-level splitting for the linear Hamiltonian and Ω is the Kerr frequency associated with the optical Kerr effect. Without nonlinearity ($\Omega = 0$), an initial coherent state $|\alpha\rangle$ evolves according to

$$e^{-i\omega\hat{n}}|\alpha\rangle = e^{-\frac{1}{2}|\alpha|^2}\sum_n \frac{(\alpha e^{-i\omega t})^n}{\sqrt{n!}}|n\rangle = |\alpha e^{-i\omega t}\rangle, \tag{3.8}$$

which causes a global phase shift on the coherent states $|\alpha\rangle$. With nonlinearity, in the interaction picture where $\Omega\hat{n}^2$ is regarded as the interaction part of the Hamiltonian 3.7, an initial coherent state $|\alpha\rangle$ evolves according to

$$|\alpha, t\rangle = e^{-i\Omega\hat{n}^2}|\alpha\rangle = e^{-\frac{1}{2}|\alpha|^2}\sum_n \alpha^n \frac{e^{-i\Omega n^2 t}}{\sqrt{n!}}|n\rangle, \tag{3.9}$$

which causes a nonlinear phase distortion in the initial coherent states $|\alpha\rangle$. Hence, the state $|\alpha, t\rangle$ is non-coherent in general. For $t = 2\pi/\Omega \equiv T$, we have $|\alpha, T\rangle = |\alpha\rangle$, as n^2 is an integer. It implies that under the influence of the Kerr nonlinearity, the state $|\alpha, t\rangle$ experiences periodic revivals of the initial coherent states. At the intermediate time $t = T/2$, we have

$$
\begin{aligned}
|\alpha, T/2\rangle &= e^{-\frac{1}{2}|\alpha|^2} \sum_n \alpha^n \frac{e^{-i\pi n^2}}{\sqrt{n!}} |n\rangle \\
&= e^{-\frac{1}{2}|\alpha|^2} \sum_n \alpha^n \frac{(-1)^{n^2}}{\sqrt{n!}} |n\rangle = |-\alpha\rangle,
\end{aligned}
\tag{3.10}
$$

where in the last step we have used the relation $(-1)^{n^2} = (-1)^n$. Equation (3.10) shows that in an optical fiber with Kerr nonlinearity, the initial coherent state $|\alpha\rangle$ evolves into another coherent state $|-\alpha\rangle$ with an opposite amplitude at an intermediate time $T/2$. More interestingly, at one quarter of the period T, we have $e^{-i\Omega n^2 T/4} = e^{-i\pi n^2/2} = (-i)^{n^2}$, which equals to 1 when n is even and equals to $-i$ when n is odd. With direct computation, it will yield

$$
\begin{aligned}
|\alpha, T/4\rangle &= e^{-\frac{1}{2}|\alpha|^2} \left(\sum_{n\in\{0,2,\cdots\}} \frac{\alpha^n}{\sqrt{n!}} |n\rangle - i \sum_{n\in\{1,3,\cdots\}} \frac{\alpha^n}{\sqrt{n!}} |n\rangle \right) \\
&= \frac{1}{\sqrt{2}} \left(e^{-\frac{i\pi}{4}} |\alpha\rangle + e^{\frac{i\pi}{4}} |-\alpha\rangle \right),
\end{aligned}
\tag{3.11}
$$

which shows that the initial coherent state is now converted into a superposition of two coherent states with opposite amplitudes. When $|\alpha|$ is large, they are macroscopically distinguishable.

It turns out that the non-classical features of Schrödinger's cat states can be better understood by using Wigner's function which was introduced in last chapter. The Wigner function is a quasi-probability distribution which is non-negative everywhere if the quantum state has a classical analog but is negative somewhere if the quantum state possesses some non-classical features. The Wigner function can be written as the Fourier transform of a characteristic function

$$
W(z) \equiv \frac{1}{\pi^2} \int d^2\xi e^{-i(\xi z^* + \xi^* z)} C^{(W)}(\xi), \tag{3.12}
$$

where

$$
C^{(W)}(\xi) \equiv \mathrm{Tr}\{\rho e^{i(\xi a^\dagger + \xi^* a)}\} \tag{3.13}
$$

is the characteristic function of a density operator ρ. One may denote the superposition of two coherent states $|\alpha_1\rangle$ and $|\alpha_2\rangle$ as $|\psi\rangle = A^{-1/2}(c_1|\alpha_1\rangle + c_2|\alpha_2\rangle)$, with $A = |c_1|^2 + |c_2|^2 + 2\mathrm{Re}(c_1^* c_2 \langle\alpha_1|\alpha_2\rangle)$ as a normalization factor. Then, the density operator $\rho \equiv |\psi\rangle\langle\psi|$ can be written as

$$\rho = A^{-1}\left(|c_1|^2\rho_1 + |c_2|^2\rho_2 + c_1^*c_2|\alpha_2\rangle\langle\alpha_1| + c_2^*c_1|\alpha_1\rangle\langle\alpha_2|\right), \tag{3.14}$$

with $\rho_i \equiv |\alpha_i\rangle\langle\alpha_i|$ being the density operator for a coherent state $|\alpha_i\rangle$. Substitution of Eq. (3.14) into Eq. (3.13) immediately yields

$$\begin{aligned} C^{(W)}(\xi) = A^{-1}(&|c_1|^2\langle\alpha_1|D(i\xi)|\alpha_1\rangle + |c_2|^2\langle\alpha_2|D(i\xi)|\alpha_2\rangle \\ &+ c_1^*c_2\langle\alpha_1|D(i\xi)|\alpha_2\rangle + c_2^*c_1\langle\alpha_2|D(i\xi)|\alpha_1\rangle), \end{aligned} \tag{3.15}$$

where $D(\lambda) \equiv e^{\lambda a^\dagger - \lambda^* a}$ is the familiar displacement operator. Thus, the Wigner function for a superposition of two coherent states becomes

$$W(z) = A^{-1}(|c_1|^2 W_1(z) + |c_2|^2 W_2(z) + c_1^*c_2 W_{12}(z) + c_2^*c_1 W_{21}(z)), \tag{3.16}$$

where

$$W_i(z) = \frac{2}{\pi}e^{-2|z-\alpha_i|^2}, \tag{3.17a}$$

$$W_{12}(z) = W_{12}^*(z) = \frac{2}{\pi}e^{-\frac{1}{2}|\alpha_1|^2 - \frac{1}{2}|\alpha_2|^2} e^{\alpha_2\alpha_1^* - 2(z^* - \alpha_1^*)(z - \alpha_2)}. \tag{3.17b}$$

The first two terms in Eq. (3.16) are the Wigner functions for the constituent coherent states $|\alpha_i\rangle$, which are Gaussian bells centered on α_i with a width $1/\sqrt{2}$ and a maximum value $2/\pi$ at $z = \alpha_i$. The last two terms in Eq. (3.16) describe the quantum interference fringes between the two Gaussian bells. In particular, for a superposition of two coherent states with opposite amplitudes, substitution of $c_1^* = c_2 = e^{i\pi/4}$ and $\alpha_1 = -\alpha_2 = \alpha$ into Eqs. (3.17a)–(3.17b) yields

$$W(z) = \frac{1}{\pi}\left\{e^{-2|z-\alpha|^2} + e^{-2|z+\alpha|^2} + 2e^{-2|z|^2}\sin[4(\alpha_I x - \alpha_R p)]\right\}, \tag{3.18}$$

where $z \equiv x + ip$ and $\alpha \equiv \alpha_R + i\alpha_I$. The non-classical nature of Schrödinger's cat states is characterized by the negative peaks in the Wigner function, as shown in Fig. 3.1.

We now examine in more detail the Schrödinger's cat state as an equal-probability superposition of coherent states with opposite amplitudes, namely, a *phase-cat state* $|\psi_\theta\rangle \equiv N_\theta^{-1}(|\alpha\rangle + e^{i\theta}|-\alpha\rangle)$, where $N_\theta^2 \equiv 2(1 + \cos\theta e^{-2|\alpha|^2})$. The annihilation operator when applying on the phase-cat state yields

$$a|\psi_\theta\rangle = \frac{\alpha}{N_\theta}(|\alpha\rangle - e^{i\theta}|-\alpha\rangle), \quad a|\psi_{\pi+\theta}\rangle = \frac{\alpha}{N_{\pi+\theta}}(|\alpha\rangle + e^{i\theta}|-\alpha\rangle). \tag{3.19}$$

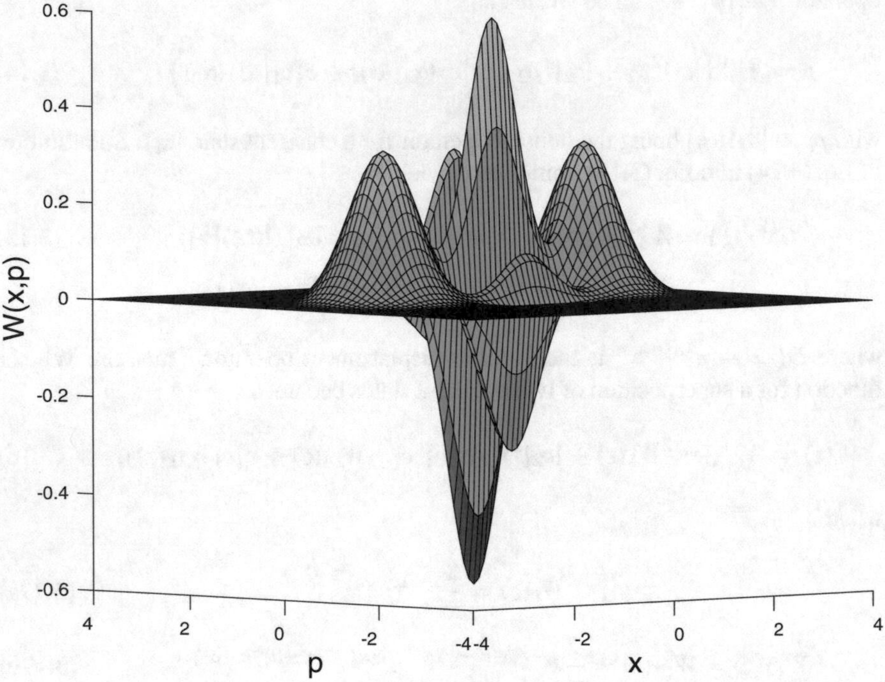

Fig. 3.1 Schematic of Schrödinger's cat state as a superposition of two coherent states with opposite amplitudes via the Wigner's function determined by Eq. (3.18) with $\alpha = 2$. The peak of the quantum interference fringe is about twice as large as the constituent coherent states, and the value pertains even when greatly increasing the distance between the constituent coherent states

Hence, the expectation values of the annihilation and creation operators for the electromagnetic field with respect to the phase-cat state are

$$\langle\psi_\theta|a|\psi_\theta\rangle = \frac{-2i\alpha}{N_\theta^2}\sin\theta e^{-2|\alpha|^2}, \ \langle\psi_\theta|a^\dagger|\psi_\theta\rangle = \frac{2i\alpha^*}{N_\theta^2}\sin\theta e^{-2|\alpha|^2}, \tag{3.20}$$

which vanishes for $\theta = 0$ or π, namely, *even*- or *odd-cat* states. The expectation values of the second-order moments for the annihilation and creation operators with respect to the phase-cat state are

$$\langle\psi_\theta|a^2|\psi_\theta\rangle = \alpha^2, \langle\psi_\theta|a^{\dagger 2}|\psi_\theta\rangle = \alpha^{*2}, \tag{3.21a}$$

$$\langle\psi_\theta|a^\dagger a|\psi_\theta\rangle = |\alpha|^2\frac{N_{\pi+\theta}^2}{N_\theta^2} = |\alpha|^2\frac{e^{2|\alpha|^2} - \cos\theta}{e^{2|\alpha|^2} + \cos\theta}. \tag{3.21b}$$

The quadrature operators of the electromagnetic field may be defined as $\hat{q} \equiv (a + a^\dagger)/\sqrt{2}$ and $\hat{p} \equiv -i(a - a^\dagger)/\sqrt{2}$, which obey the canonical commutation relation

$[\hat{q}, \hat{p}] = i$. Hence, the expectation values of the quadratures with respect to the phase-cat state are

$$\langle\psi_\theta|\hat{q}|\psi_\theta\rangle = \frac{\sqrt{2}|\alpha|\sin\vartheta\sin\theta}{e^{2|\alpha|^2}+\cos\theta}, \quad \langle\psi_\theta|\hat{p}|\psi_\theta\rangle = \frac{-\sqrt{2}|\alpha|\cos\vartheta\sin\theta}{e^{2|\alpha|^2}+\cos\theta}, \tag{3.22}$$

where $\alpha \equiv |\alpha|e^{i\vartheta}$ with ϑ being the phase of the coherent state amplitude. Then, the expectation values of second-order moments for the quadratures with respect to the phase-cat state become

$$\langle\psi_\theta|\hat{q}^2|\psi_\theta\rangle = \frac{1}{2} + |\alpha|^2\left(\frac{e^{2|\alpha|^2}-\cos\theta}{e^{2|\alpha|^2}+\cos\theta} + \cos 2\vartheta\right), \tag{3.23a}$$

$$\langle\psi_\theta|\hat{p}^2|\psi_\theta\rangle = \frac{1}{2} + |\alpha|^2\left(\frac{e^{2|\alpha|^2}-\cos\theta}{e^{2|\alpha|^2}+\cos\theta} - \cos 2\vartheta\right). \tag{3.23b}$$

Hence, the variances in the quadratures with respect to the even-cat states ($\theta = 0$) can be expressed as

$$(\Delta\hat{q})^2 = \frac{1}{2} + |\alpha|^2(\tanh|\alpha|^2 + \cos 2\vartheta), \tag{3.24a}$$

$$(\Delta\hat{p})^2 = \frac{1}{2} + |\alpha|^2(\tanh|\alpha|^2 - \cos 2\vartheta), \tag{3.24b}$$

whereas the variances in the quadratures with respect to the odd-cat states are obtained by replacing $\tanh|\alpha|^2$ by $\coth|\alpha|^2$. From Eqs. (3.24a) to (3.24b), we notice that for $\vartheta = 0$, we have $(\Delta\hat{q})^2 = \frac{1}{2} + |\alpha|^2(\tanh|\alpha|^2 + 1)$ and $(\Delta\hat{p})^2 = \frac{1}{2} - |\alpha|^2 e^{-|\alpha|^2}\mathrm{sech}|\alpha|^2 \leq 1/2$, and thus the variance in $\hat{p}$ is squeezed, whereas for $\theta = \pi/2$, we have $(\Delta\hat{q})^2 = \frac{1}{2} - |\alpha|^2 e^{-|\alpha|^2}\mathrm{sech}|\alpha|^2 \leq 1/2$ and $(\Delta\hat{p})^2 = \frac{1}{2} + |\alpha|^2(\tanh|\alpha|^2 + 1)$, and thus the variance in $\hat{q}$ is squeezed. For odd-cat states, we have $(\Delta\hat{q})^2 = \frac{1}{2} + |\alpha|^2(\coth|\alpha|^2 + 1)$ and $(\Delta\hat{p})^2 = \frac{1}{2} + |\alpha|^2 e^{-|\alpha|^2}\mathrm{csch}|\alpha|^2$ for $\vartheta = 0$ and $(\Delta\hat{q})^2 = \frac{1}{2} + |\alpha|^2 e^{-|\alpha|^2}\mathrm{csch}|\alpha|^2$ and $(\Delta\hat{p})^2 = \frac{1}{2} + |\alpha|^2(\coth|\alpha|^2 + 1)$ for $\vartheta = \pi/2$. Thus, both quadratures $\hat{q}$ and $\hat{p}$ do not exhibit squeezing for both $\vartheta = 0$ and $\pi/2$. Finally, for Yurke-Stoler cat states with $\theta = \pi/2$, the variances in the quadratures are given by

$$(\Delta\hat{q})^2 = \frac{1}{2} + 2|\alpha|^2\left(\cos^2\vartheta - \frac{\sin^2\vartheta}{e^{4|\alpha|^2}}\right), \tag{3.25a}$$

$$(\Delta\hat{p})^2 = \frac{1}{2} + 2|\alpha|^2\left(\sin^2\vartheta - \frac{\cos^2\vartheta}{e^{4|\alpha|^2}}\right). \tag{3.25b}$$

Hence, for $\vartheta = 0$, we have $(\Delta\hat{q})^2 = \frac{1}{2}+2|\alpha|^2$ and $(\Delta\hat{p})^2 = \frac{1}{2}-2|\alpha|^2 e^{-4|\alpha|^2} \leq 1/2$, and thus the variance in $\hat{p}$ is squeezed, whereas for $\vartheta = \pi/2$, we have $(\Delta\hat{q})^2 = \frac{1}{2} - 2|\alpha|^2 e^{-4|\alpha|^2}$ and $(\Delta\hat{p})^2 = \frac{1}{2} + 2|\alpha|^2$, and thus the variance in $\hat{q}$ is squeezed.

3.3 Application to Quantum Information

In the previous sections, we have discussed the Einstein-Podolsky-Rosen (EPR) paradox, and the Schrödinger cat states created from the superposition of well-separated coherent states. In this section, we will discuss how Schrödinger cat states in optical systems can be used for quantum information processing. We begin discussing quantum teleportation using the Schrödinger cat states. The quantum teleportation is a way of transferring a quantum state of a particle onto another particle over large distances without knowing any information about the state in the course of the transmission. This procedure usually requires the phenomenon of quantum entanglement to set up a quantum teleportation channel, so that the unknown quantum state can be disassembled into, and then later reconstructed from, purely classical information and non-classical EPR correlations [62].

Let us assume that the sender Alice and the receiver Bob share an entangled coherent state serve as the quantum channel for teleportation [63]

$$|C_\alpha\rangle \equiv \frac{1}{\sqrt{N_\alpha}} \left(|\alpha\rangle_a|\alpha\rangle_b - |-\alpha\rangle_a|-\alpha\rangle_b\right), \tag{3.26}$$

where $N_\alpha \equiv 2 - 2\exp(-4|\alpha|^2)$ is a normalization factor and $|\pm\alpha\rangle$ are coherent states with amplitudes $\pm\alpha$, which are non-orthogonal to one another but with an overlap $|\langle\alpha|-\alpha\rangle|^2 = \exp(-4|\alpha|^2)$ decrease exponentially with $|\alpha|^2$. For example, the overlap $|\langle\alpha|-\alpha\rangle|^2 \approx 10^{-7}$ when $|\alpha|$ is as small as 2. The quantum channel $|C_\alpha\rangle$ can be produced from a Schrödinger cat state $(|\sqrt{2}\alpha\rangle - |-\sqrt{2}\alpha\rangle)/\sqrt{N_\alpha}$ by splitting it at a lossless 50–50 beam splitter, and it can also be written as

$$|C_\alpha\rangle = \frac{1}{\sqrt{2}} \left(|\psi_+\rangle_a|\psi_-\rangle_b + |\psi_-\rangle_a|\psi_+\rangle_b\right), \tag{3.27}$$

where $|\psi_\pm\rangle \equiv (|\alpha\rangle \pm |-\alpha\rangle)/\sqrt{2 \pm 2e^{-2|\alpha|^2}}$ are the even- and odd-cat states, which are exactly orthogonal to one another, i.e., $\langle\psi_-|\psi_+\rangle = \langle\psi_+|\psi_-\rangle = 0$. When the amplitude $|\alpha|$ is large, one may identify the two coherent states $|\alpha\rangle$ and $|-\alpha\rangle$ as the basis states for a logical qubit, so that a qubit state with unknown amplitudes c_+ and c_- may be represented by

$$|\psi\rangle_a \equiv \frac{1}{\sqrt{N_a}}(c_+|\alpha\rangle_a + c_-|-\alpha\rangle_a), \tag{3.28}$$

where $N_a \equiv |c_+|^2 + |c_-|^2 + 2e^{-2|\alpha|^2}\mathrm{Re}(c_-^* c_+)$. Let us assume that Alice wants to quantum teleport the qubit state $|\psi\rangle_a$ to Bob by the quantum channel $|C_\alpha\rangle$. In order to perfectly teleport the qubit state $|\psi\rangle_a$ with a nonzero probability, Alice should mix her part of the entangled coherent state $|C_\alpha\rangle$ with $|\psi\rangle_a$ by using a lossless 50–50 beam splitter. From the action of a beam splitter on coherent states, $|\alpha\rangle|\beta\rangle \to |(\alpha+\beta)/\sqrt{2}\rangle|(\alpha-\beta)/\sqrt{2}\rangle$, Alice and Bob share a three-qubit entangled coherent state

$$|\Psi\rangle_{a,b} \equiv \frac{1}{\sqrt{N_a N_\alpha}}\Big[c_+\Big(|\sqrt{2}\alpha\rangle_a|0\rangle_a|\alpha\rangle_b - |0\rangle_a|\sqrt{2}\alpha\rangle_a|-\alpha\rangle_b\Big) \tag{3.29}$$
$$+c_-\Big(|0\rangle_a|-\sqrt{2}\alpha\rangle_a|\alpha\rangle_b - |-\sqrt{2}\alpha\rangle_a|0\rangle_a|-\alpha\rangle_b\Big)\Big],$$

where the first and second qubits both belong to the sender Alice and the third qubit belongs to the receiver Bob. To complete the quantum teleportation, Alice needs to perform a joint measurement on her two-qubit state. For example, she may measure the number of photons in the two modes on her side. Let us denote the probability of finding n and m photons in the two modes on Alice's side by $P(n,m) \equiv |{}_a\langle n|_a\langle m|\Psi\rangle_{a,b}|^2$. Then by construction, only one of the two measurement outcomes are nonzero.

Without loss of generality, one may assume that $n \neq 0$. According to the principles of quantum mechanics, the coherent state qubit on Bob's side after Alice's measurement should be

$$|\psi\rangle_b = \frac{1}{\sqrt{N_b}}(c_+|\alpha\rangle_b - (-1)^n c_-|-\alpha\rangle_b), \tag{3.30}$$

where $N_b \equiv |c_+|^2 + |c_-|^2 - 2(-1)^n e^{-2|\alpha|^2}\mathrm{Re}(c_-^* c_+)$. From Eq. (3.30), we see that Alice's original qubit state can be perfectly teleported to Bob when n is odd. When n is even, Bob is required to perform an additional transformation $|\alpha\rangle \to |\alpha\rangle$ and $|-\alpha\rangle \to -|-\alpha\rangle$ to perfectly teleport Alice's original unknown qubit state. For n odd, the probability of finding n and 0 photons in the first and second modes on Alice's side is

$$P(n,0) = \frac{|\langle n|\sqrt{2}\alpha\rangle|^2}{N_\alpha} = \frac{e^{-2|\alpha|^2}}{2(1-e^{-4|\alpha|^2})}\frac{(2|\alpha|^2)^n}{n!}, \tag{3.31}$$

which only depends on the amplitude $|\alpha|$ and is independent of the qubit state $|\psi\rangle_a$ being teleported. Finally, the success probability to perfectly teleport the unknown qubit state via the teleportation channel $|C_\alpha\rangle$ is

$$P = \sum_{n\in 2\mathbb{N}+1}(P(n,0) + P(0,n)) = \frac{1}{2}, \tag{3.32}$$

which is independent of the amplitude $|\alpha|$.

Besides applications in quantum teleportation, the Schrödinger cat states in optical systems can also be used to construct a discrete set of universal gates for quantum computation which involves only linear optics and photon detection. The universal set of quantum gates is a set of an entangling two-qubit gate assisted by single-qubit gates, which enables any n-qubit unitary transformation to be implemented to arbitrary accuracy for any n [64]. One such set can be constructed from the one-qubit Hadamard gate H, the one-qubit phase shift gate $R_z(\theta)$, and the two-qubit controlled-Z gate C_z. One may encode a logical qubit in the coherent states as $|0\rangle_L \equiv |0\rangle$ and $|1\rangle_L \equiv |\alpha\rangle$, where $|\alpha\rangle$ is a coherent state with a large amplitude α and $|0\rangle$ is the vacuum state. The action of the Hadamard gate, the phase shift gate, and the controlled-Z gate on the two-qubit computational state is specified by $H|x\rangle_L = ((-1)^x|x\rangle_L + |1-x\rangle_L)/\sqrt{2}$, $R_z(\theta)|x\rangle_L \equiv e^{ix\theta}|x\rangle_L$ and $C_z|x\rangle_L|y\rangle_L \equiv (-1)^{xy}|x\rangle_L|y\rangle_L$, respectively.

The advantage of encoding a logical qubit into coherent states is that one may implement an entangling two-qubit gate by using only a single beam splitter. To be specific, let us consider the following unitary beam splitter transformation $\hat{U}_{ab} = \exp[i\theta(\hat{a}\hat{b}^\dagger + \hat{a}^\dagger\hat{b})]$, where $\hat{a}$ and $\hat{b}$ are the annihilation operators for two coherent states $|\alpha\rangle$ and $|\beta\rangle$, respectively. Then one may show that the output state produced by the unitary transformation $\hat{U}_{ab}$ is

$$\hat{U}_{ab}|\alpha\rangle_a|\beta\rangle_b = |\cos\theta\alpha + i\sin\theta\beta\rangle_a\,|\cos\theta\beta + i\sin\theta\alpha\rangle_b\,, \tag{3.33}$$

where $\cos^2\theta$ and $\sin^2\theta$ are the reflectivity and transmissivity of the beam splitter, respectively. Now, using the relation $\langle\beta|\alpha\rangle = e^{\beta^*\alpha - \frac{1}{2}|\beta|^2 - \frac{1}{2}|\alpha|^2}$, one may easily obtain the overlap between the output and input states

$${}_b\langle\beta|_a\langle\alpha|\hat{U}_{ab}|\alpha\rangle_a|\beta\rangle_b = e^{(\cos\theta-1)(|\alpha|^2+|\beta|^2)+i\sin\theta(\alpha^*\beta+\beta^*\alpha)}. \tag{3.34}$$

In particular, for a beam splitter with a nearly perfect reflectivity satisfying $\theta^2 \ll \min\{|\alpha|^{-2}, |\beta|^{-2}\} \ll 1$, the phase of the output state after the beam splitter is shifted in an amount proportional to the coherent state amplitudes with respect to the input state [65]

$$\hat{U}_{ab}|\alpha\rangle_a|\beta\rangle_b \approx e^{i\theta(\alpha^*\beta+\beta^*\alpha)}|\alpha\rangle_a|\beta\rangle_b. \tag{3.35}$$

Hence, we see that when either or both of the input coherent state qubits $|\alpha\rangle_a$ and $|\beta\rangle_b$ are in the vacuum state $|0\rangle$, the beam splitter transformation produces no effects on the input state. In contrast, when both of the input coherent state qubits $|\alpha\rangle_a$ and $|\beta\rangle_b$ are in the same coherent state $|\alpha\rangle$ with $|\alpha| \gg 1$, the beam splitter transformation produces a sign change in the input state when the condition $\theta|\alpha|^2 = \pi/2$ is fulfilled. Under such conditions, the unitary beam splitter transformation $\hat{U}_{ab}$ realizes the two-qubit conditional sign flip (controlled-Z) gate [64], whose action on the two-qubit computational state is specified by $|a\rangle_L|b\rangle_L \rightarrow (-1)^{ab}|a\rangle_L|b\rangle_L$, where $|0\rangle_L \equiv |0\rangle$ and $|1\rangle_L \equiv |\alpha\rangle$ are the logical qubits encoded in the coherent states.

To implement the one-qubit Hadamard gate by use of optical Schrödinger cat states, one needs the above controlled-Z gate. When an arbitrary coherent state qubit $|\psi\rangle_1 \equiv (\mu|0\rangle + \nu|\alpha\rangle)/\sqrt{N_1}$ enters the input port 1 of a beam splitter with nearly perfect reflectivity $\cos^2\theta \approx 1$ and $\theta|\alpha|^2 = \pi/2$, while a resource state of an auxiliary qubit $|\Phi_+\rangle$ enters the input port 2 of the beam splitter, the output state of the beam splitter has the form

$$\begin{aligned} |\Psi\rangle_{1,2} &= \frac{\mu(|0\rangle_1|0\rangle_2 + |0\rangle_1|\alpha\rangle_2)}{\sqrt{N_1 N_+}} + \frac{\nu(|\alpha\rangle_1|0\rangle_2 - |\alpha\rangle_1|\alpha\rangle_2)}{\sqrt{N_1 N_+}} \\ &= \frac{\sqrt{N_+}|\Phi_+\rangle_1}{2}\left(\frac{\mu\sqrt{N_+}|\Phi_+\rangle_2 + \nu\sqrt{N_-}|\Phi_-\rangle_2}{\sqrt{N_1 N_+}}\right) \\ &\quad + \frac{\sqrt{N_-}|\Phi_-\rangle_1}{2}\left(\frac{\mu\sqrt{N_+}|\Phi_+\rangle_2 - \nu\sqrt{N_-}|\Phi_-\rangle_2}{\sqrt{N_1 N_+}}\right), \end{aligned} \tag{3.36}$$

where $N_1 \equiv |\mu|^2 + |\nu|^2 + 2\text{Re}(\nu^*\mu)e^{-|\alpha|^2/2}$, $|\Phi_\pm\rangle \equiv (|0\rangle \pm |\alpha\rangle)/\sqrt{N_\pm}$ and $N_\pm \equiv 2 \pm 2e^{-|\alpha|^2/2}$. Based on the above entangled coherent state $|\Psi\rangle_{1,2}$, one may reproduce the one-qubit Hadamard transformation by use of quantum teleportation. According to the laws of quantum mechanics, if one performs a cat basis experiment to test whether the output state of port 1 is the same as the auxiliary cat state $|\Phi_+\rangle$, the output state of port 2 should be

$$|\psi\rangle_2 = \frac{\mu\sqrt{N_+}|\Phi_+\rangle \pm \nu\sqrt{N_-}|\Phi_-\rangle}{\sqrt{N_1 N_+}} = \frac{\mu(|0\rangle + |\alpha\rangle) \pm \upsilon(|0\rangle - |\alpha\rangle)}{\sqrt{N_1 N_+}}, \tag{3.37}$$

where the plus and minus signs are for the experimental results $|\Phi_+\rangle$ and $|\Phi_-\rangle$, respectively. If the cat state $|\Phi_+\rangle$ is obtained, Eq. (3.37) reproduces the conventional one-qubit Hadamard transformation $|0\rangle \rightarrow (|0\rangle + |\alpha\rangle)/\sqrt{2}$ and $|\alpha\rangle \rightarrow (|0\rangle - |\alpha\rangle)/\sqrt{2}$ when $|\alpha| \gg 1$. In contrast, if the orthogonal cat state $|\Phi_-\rangle$ is obtained, one needs to perform an additional bit-flip operation $|0\rangle \leftrightarrow |\alpha\rangle$ on the output state $|\psi\rangle_2$ to reproduce the one-qubit Hadamard transformation. The real cat basis measurement can be experimentally implemented by applying the displacement operation $D(-\alpha/2)$ on the output state of port 1, which results in

$$D(-\frac{\alpha}{2})|\Psi\rangle_{1,2} = \left(\frac{\mu|\frac{-\alpha}{2}\rangle_1 + \nu|\frac{\alpha}{2}\rangle_1}{\sqrt{N_1 N_+}}\right)|0\rangle_2 + \left(\frac{\mu|\frac{-\alpha}{2}\rangle_1 - \nu|\frac{\alpha}{2}\rangle_1}{\sqrt{N_1 N_+}}\right)|\alpha\rangle_2, \tag{3.38}$$

where the displacement operation may be implemented by mixing the output state of port 1 with an auxiliary strong coherent state $|i\alpha/(2\theta)\rangle$ at a beam splitter with nearly perfect reflectivity $\cos^2\theta \approx 1$. One may measure the photon number in the displaced state. An even number result indicates the observation of an even cat state $D(-\alpha/2)|\Phi_+\rangle$, which implies that the output state of port 1 is the same cat state

$|\Phi_+\rangle$, whereas an odd number result indicates the observation of an odd cat state $D(-\alpha/2)|\Phi_-\rangle$, which implies that the output state of port 1 is the orthogonal cat state $|\Phi_-\rangle$.

To implement the bit-flip operation $|0\rangle \leftrightarrow |\alpha\rangle$, which is equivalent to the Pauli matrix σ_x in the computational basis, one may simply apply a displacement operation $D(-\alpha)$ followed by a π phase shift $U(\pi)$ of the coherent state amplitude, $\sigma_x = U(\pi)D(-\alpha)$, where the π phase shift $U(\pi) \equiv e^{i\pi\hat{a}^\dagger\hat{a}}$ may be experimentally implemented using a phase conjugate mirror, *e.g.*, a crystal formed from a photorefractive material such as $BaTiO_3$ [66].

Finally, to implement the one-qubit phase shift gate which equals the rotation matrix $R_z(\theta)$ in the computational basis, one only needs two displacement operations and single photon subtraction process. For an arbitrary coherent state qubit $|\psi\rangle \equiv \mu|0\rangle + \nu|\alpha\rangle$, one may first apply a displacement operation $D(\gamma)$ on it, then subtracts a photon from this state, and finally apply an inverse displacement $D(-\gamma)$ on it, which yields [67]

$$D(-\gamma)\hat{a}D(\gamma)|\psi\rangle = \mu\gamma|0\rangle + \nu(\alpha+\gamma)|\alpha\rangle. \tag{3.39}$$

Equation (3.39) realizes the one-qubit phase shift gate, provided that the complex amplitude γ satisfies $\alpha + \gamma = \gamma e^{i\theta}$. Then the output state after the one-qubit phase shift gate becomes

$$D(-\gamma)\hat{a}D(\gamma)|\psi\rangle = \frac{\alpha}{2i\sin\frac{\theta}{2}}(\mu e^{-\frac{i\theta}{2}}|0\rangle + \nu e^{\frac{i\theta}{2}}|\alpha\rangle). \tag{3.40}$$

The single photon subtraction process may be experimentally implemented by using a beam splitter with low reflectivity and an on-off photodetector [68], e.g., the avalanche photodiode.

Exercises

3.1. Using the relation $\langle\beta|D(\xi)|\alpha\rangle = e^{\beta^*\alpha-\frac{1}{2}|\alpha|^2-\frac{1}{2}|\beta|^2}e^{\xi\beta^*-\xi^*\alpha-\frac{1}{2}|\xi|^2}$, verify Eqs. (3.17a) and (3.17b).

3.2. Show that the average number of photons in the quantum teleportation channel $|C_\alpha\rangle$ is $2|\alpha|^2(1+e^{-4|\alpha|^2})/(1-e^{-4|\alpha|^2})$.

3.3. Prove the following identity: $R_y(\theta) = SR_x(\theta)S^\dagger$, where $R_x(\theta) \equiv e^{-i\theta\sigma_x/2}$ and $R_y(\theta) \equiv e^{-i\theta\sigma_y/2}$ are the rotation operators about the $\hat{x}$ and $\hat{y}$ axis and S is the phase gate defined by

$$S \equiv \begin{pmatrix} e^{-i\pi/4} & 0 \\ 0 & e^{i\pi/4} \end{pmatrix}.$$

3.4. Show that $R_x(\theta) = HR_z(\theta)H$, where H is the Hadamard gate and $R_z(\theta)$ is the Z-rotation gate

$$H \equiv \frac{1}{\sqrt{2}}\begin{pmatrix} 1 & 1 \\ 1 & -1 \end{pmatrix}.$$

3.5. Show that any single-qubit gate, i.e., an arbitrary unitary operation U on a single qubit, can be decomposed in terms of the Hadamard gate H and the Z-rotation gate $R_z(\theta)$.

3.6. Show that the output state produced by the following unitary transformation $\hat{U}_{ab} \equiv \exp[i\frac{\theta}{2}(\hat{a}\hat{b}^\dagger + \hat{a}^\dagger\hat{b})]$ is

$$\hat{U}_{ab}|\alpha\rangle_a|\beta\rangle_b = |\cos\theta\alpha + i\sin\theta\beta\rangle_a\,|\cos\theta\beta + i\sin\theta\alpha\rangle_b\,,$$

where $\hat{a}$ and $\hat{b}$ are the annihilation operators for the two coherent states $|\alpha\rangle_a$ and $|\beta\rangle_b$, respectively.

3.7. Show that the CNOT gate can be constructed from the controlled-Z gate and two Hadamard gates. The action of the CNOT gate on the two-qubit computational state is specified by $|x\rangle|y\rangle \to |x\rangle|x \oplus y\rangle$, where $x, y = 0, 1$ and labels have been omitted.

4 Coherent States for Fermions

4.1 Graßmann Algebra

Elementary particles are divided into two fundamental classes, in accordance to the integral or half-integral value of their spins. Bosons are particles with integral value of spins obeying the Bose-Einstein statistics, i.e., identical particles can be in the same states. Fermions are particles with half-integral value of spins obeying the Fermi-Dirac statistics, i.e., identical particles are forbidden to be in the same state. In Chap. 2, we showed that the coherent states of the bosonic fields are eigenstates of the annihilation operators for bosons. Hence, it is natural to ask is it possible to construct coherent states for fermionic fields?

At first sight, one may think that the fermionic coherent states may be constructed into the same manner as the bosonic ones. However, as a result of the Pauli exclusion principle, the creation and annihilation operators for a fermionic mode are required to obey the anti-commutation relations $\{a, a^\dagger\} \equiv aa^\dagger + a^\dagger a = 1$ and $\{a, a\} = \{a^\dagger, a^\dagger\} = 0$. Hence, if one were to define a fermionic coherent state $|\alpha\rangle$ as the eigenstate of the fermionic annihilation operator, labeled by a complex eigenvalue α, i.e., $a|\alpha\rangle = \alpha|\alpha\rangle$, we should have

$$a_1 a_2 |\alpha_1, \alpha_2\rangle = \alpha_2 \alpha_1 |\alpha_1, \alpha_2\rangle = \alpha_1 \alpha_2 |\alpha_1, \alpha_2\rangle = a_2 a_1 |\alpha_1, \alpha_2\rangle, \tag{4.1}$$

which is in contradiction to the anti-commutation relations $a_1 a_2 = -a_2 a_1$ which the fermionic annihilation operators must obey. This means that for fermions, one needs a new type of numbers, such that any numbers are anti-commuting to one another, i.e., $\alpha_1 \alpha_2 = -\alpha_2 \alpha_2$. Such a new type of numbers obeys a completely different algebra, which is opposition to the conventional algebra which complex numbers obey. Indeed, as was shown by Martin [69, 70] and independently by Ohnuki and Kashiwa [71], in analogy to the coherent states for the bosonic fields, coherent states for fermionic fields can be explicitly constructed with the aid of Graßmann numbers, which is a type of anti-commuting numbers first studied by the mathematician

C.-F. Kam et al., *Coherent States*, Lecture Notes in Physics 1011,
https://doi.org/10.1007/978-3-031-20766-2_4

Hermann Graßmann in his masterpiece *Die Lineale Ausdehnungslehre, ein neuer Zweig der Mathematik* (The Theory of Linear Extension, a New Branch of Mathematics) in 1844 [72], and were introduced in the context of a functional formulation of quantum field theory by Julian Schwinger in 1953 [73]. In the following, we shall discuss the properties of the Graßmann numbers and the Graßmann algebra in details.

Graßmann numbers are individual elements of the Graßmann algebra, the Graßmann algebra $\bigwedge(V)$ is a vector space over $\mathbb{R}$ or $\mathbb{C}$ spanned by products of Graßmann variables θ_i in the form $\theta_1 \wedge \theta_2 \wedge \cdots \wedge \theta_r$, and the Graßmann variables θ_i are basis elements of a vector space V of dimension n. Hence, a Graßmann number θ can be written as a linear combinations of products of Graßmann variables θ_i as

$$\theta = \sum_{k=0}^{n} \sum_{i_1, i_2, \ldots, i_k} \alpha_{i_1 i_2 \ldots i_k} \theta_{i_1} \wedge \theta_{i_2} \wedge \ldots \wedge \theta_{i_k}, \tag{4.2}$$

where $i_1 < i_2 < \cdots < i_k$, and $\alpha_{i_1 i_2 \ldots i_k} \in \mathbb{R}$ or $\mathbb{C}$ are totally antisymmetric in the indices. In other words, the set of products of degree k form a subspace $\bigwedge^k(V)$, where $k \equiv \deg(\theta_{i_1} \wedge \theta_{i_2} \wedge \ldots \wedge \theta_{i_k})$ is the number of Graßmann variables in the product, and the Graßmann algebra is a vector space of dimension 2^n, which can be written as a direct sum $\bigwedge(V) = \bigwedge^0(V) \oplus \bigwedge^1(V) \oplus \cdots \oplus \bigwedge^n(V)$, where $\bigwedge^0(V) = \mathbb{R}$ or $\mathbb{C}$. In the following, we may omit the wedge symbol $\wedge$ when writing products of Graßmann variables.

By definition, the products of Graßmann variables are subjected to the axioms: (i) the product is *associative*: $((\theta_1 \cdots \theta_i)(\theta_{i+1} \cdots \theta_j))(\theta_{j+1} \cdots \theta_k) = (\theta_1 \cdots \theta_i)((\theta_{i+1} \cdots \theta_j)(\theta_{j+1} \cdots \theta_k))$ for any i, j, k with $1 < i < j < k$; (2) the product is *bilinear*: $\theta_1 \cdots \theta_i (\alpha \eta_1 + \beta \eta_2) \theta_{i+1} \cdots \theta_j = \alpha \theta_1 \cdots \theta_i \eta_1 \theta_{i+1} \cdots \theta_j + \beta \theta_1 \cdots \theta_i \eta_2 \theta_{i+1} \cdots \theta_j$ for any i, j with $1 < i < j$.

From the two axioms, it follows that the Graßmann algebra is an *associative* algebra. On the other hand, an additional structure of the Graßmann algebra is that, as opposed to the usual associative algebra, such as the multivariate polynomial algebra, it makes a different assumption on the multiplication law: the Graßmann variables are required to satisfy the anti-commutation relations

$$\theta_i \theta_j + \theta_j \theta_i = 0, \ (i, j = 1, 2, \ldots, n). \tag{4.3}$$

Hence, an arbitrary product of an assembly of n Graßmann variables satisfies a simple relation $\theta_{\sigma(1)} \theta_{\sigma(2)} \ldots \theta_{\sigma(n)} = \mathrm{sgn}(\sigma) \theta_1 \theta_2 \ldots \theta_n$, where $\mathrm{sgn}(\sigma)$ is the signature of the permutation σ. From the anti-commutation relations, there are several corollaries:

(1) All elements of the Graßmann algebra are *affine functions*, i.e., linear polynomials, of a Graßmann variable θ_i, after bring θ_i to the leftmost or rightmost position by use of the anti-commutation relations.

(2) The Graßmann variables are *nilpotent*, i.e., $\theta_i^2 = 0$ for $i = 1, 2, \ldots, n$. In other words, the Graßmann variables are nonzero square-root of zero. As a result, if $f(\theta)$ is a function of one Graßmann variable, it has a simple expansion $f(\theta) = f_0 + f_1\theta$. For example, the exponential function of one Graßmann variable is $e^\theta = 1 + \theta$, where all higher-order terms vanish.

In general, the expansion of a function $f(\theta)$ of n Graßmann variables takes on the form $f(\theta) = f_0 + \sum_i f_i\theta_i + \sum_{i,j} f_{ij}\theta_i\theta_j + \cdots + f_{12\cdots n}\theta_1\theta_2\cdots\theta_n$. The Graßmann function $f(\theta)$ is even (or odd) if it contains terms with even (or odd) numbers of Graßmann variables, i.e., $f_i = f_{ijk} = \cdots = 0$ (or $f_0 = f_{ij} = \cdots = 0$). By grouping the terms in the expansion, one may express any Graßmann function as a sum of an even and an odd Graßmann functions, $f = f_e + f_o$. For a pair of functions f and g, one can show that the commutator $[f, g] = [f_o, g_o] = 2 f_o g_o$, as even Graßmann functions commute with both even and odd Graßmann functions, and odd Graßmann functions anti-commute with each other. As the commutator $[f, g] = 2 f_o g_o$ is even, it commutes with any other Graßmann function h, i.e., $[[f, g], h] = 0$.

As Graßmann variables θ_i are the basis elements of a vector space over real or complex numbers, they commute with real or complex numbers, i.e, $\theta_i z = z\theta_i$ for $z \in \mathbb{R}$ or $\mathbb{C}$. A complex Graßmann variable can be written as $\theta \equiv \theta_R + i\theta_I$, where θ_R and θ_I are two real Graßmann variables. Then, one may define the complex conjugation of θ by $\theta^* \equiv \theta_R - i\theta_I$, which keeps the two real Graßmann variables θ_R and θ_I invariant. By requiring the product $\theta^*\theta$ to be real, i.e., $(\theta^*\theta)^* = \theta^*\theta$, one obtains

$$\theta^*\theta = i\theta_R\theta_I - i\theta_I\theta_R = i(\theta_I\theta_R)^* - i(\theta_R\theta_I)^* = (\theta^*\theta)^*, \tag{4.4}$$

which is satisfied if $(\eta\xi)^* = \xi\theta$ for two real Graßmann variables η and ξ. A direct calculation would yield $(\theta_1\theta_2\cdots\theta_n)^* = \theta_n^*\cdots\theta_2^*\theta_1^*$ for n complex Graßmann variables θ_i, which is similar to the Hermitian adjoint of a product of n matrices.

In the following, we discuss some fundamental notions of calculus, i.e., differentiation and integration, for the Graßmann variables. Unlike conventional calculus for real or complex numbers which involve the concept of a limit, such as the Riemann sum, the Berezin calculus for anti-commuting Graßmann variables, invented and developed by the Soviet Russian mathematician Felix Berezin [74], has to be formulated in an abstract algebraic manner, as the anti-commuting nature of the Graßmann variables makes the standard analysis constructions unavailable.

The derivative of a single Graßmann variable is defined as $\partial\theta_i/\partial\theta_j \equiv \delta_{ij}$. Since the Graßmann variables are anti-commuting, therefore when one takes derivatives of their products, one has to carefully decide the order of the variables and specify the direction in which the derivatives operate. For example, a right derivative of a product of two Graßmann variables would yield

$$\frac{\partial}{\partial\theta_i}(\theta_j\theta_k) = \theta_j\frac{\partial\theta_k}{\partial\theta_i} - \theta_k\frac{\partial\theta_j}{\partial\theta_i} = \delta_{ik}\theta_j - \delta_{ij}\theta_k, \tag{4.5}$$

and a left derivative of a product of two Graßmann variables would yield

$$\frac{\partial}{\partial \theta_i}(\theta_j\theta_k) = \frac{\partial \theta_j}{\partial \theta_i}\theta_k - \frac{\partial \theta_k}{\partial \theta_i}\theta_j = \delta_{ij}\theta_k - \delta_{ik}\theta_j, \tag{4.6}$$

Hence, the results of the left and right derivatives differ by an overall sign. In general, for an element θ of the Graßmann algebra generated by n Graßmann variables θ_i, one can always write $\theta = c_0 + \theta_j c_1$, where c_0 and c_1 are independent of θ_j. Thus, the left derivative of θ with respect to θ_j is defined by c_1. Similarly, one can write $\theta = c_0 + c_2\theta_j$ with c_0 and c_2 being independent of θ_j. Then, the right derivative of θ with respect to θ_j is defined by c_2. In the following, we use exclusively the left derivatives, where the left derivative of a product of n Graßmann variables is given by

$$\frac{\partial}{\partial \theta_i}(\theta_1 \cdots \theta_{i-1}\theta_i\theta_{i+1}\cdots\theta_n) = (-1)^{i-1}(\theta_1\cdots\theta_{i-1}\theta_{i+1}\cdots\theta_n). \tag{4.7}$$

In order to introduce the Berezin integral for Graßmann variables, one may first analyze the properties of the differential of a Graßmann variable. For a Graßmann algebra generated by n Graßmann variables, the differential operator is assumed to have the same form as that in ordinary calculus

$$d \equiv \sum_{i=1}^{n} d\theta_i \frac{\partial}{\partial \theta_i}. \tag{4.8}$$

Consequently, one has $d(\theta_1\theta_2) = d\theta_1\theta_2 - d\theta_2\theta_1$ for a product of two Graßmann variables. By requiring the derivative of a product of two Graßmann variables satisfies the Leibniz rule, $d(\theta_1\theta_2) = d\theta_1\theta_2 + \theta_1 d\theta_2$, one immediately obtains $\theta_1 d\theta_2 = -d\theta_2\theta_1$, i.e., θ_1 and $d\theta_2$ are anti-commute. Hence, $d\theta_2$ should be regarded as a Graßmann variable. In other words, the differentials of the Graßmann variables satisfy the anti-commutation relations

$$\{\theta_i, d\theta_j\} = 0, \{d\theta_i, d\theta_j\} = 0, (i, j = 1, 2, \cdots, n). \tag{4.9}$$

In particular, $(d\theta_i)^2 = 0$ for the differentials of the Graßmann variables.

Although the Berezin integrals of functions of anti-commutating Graßmann variables are defined via formal operations, it may be instructive to compare them with the Riemann integrals, which have the properties (1) *linearity*, $\int (af(x) + bg(x))dx = a\int f(x)dx + b\int g(x)dx$, which yields a linear functional on the vector space of Riemann-integrable functions, and (2) *translation invariance*, $\int f(x + x_0)dx = \int f(x)dx$. Hence, it is natural to require that the Berezin integrals keep these properties, i.e., a Berezin integral over a Graßmann variable should be (1) *a linear functional*, $\int (af(\theta) + bg(\theta))d\theta = a\int f(\theta)d\theta + b\int g(\theta)d\theta$; (2) *translation invariant*, $\int f(\theta + \theta_0)d\theta = \int f(\theta)d\theta$; and (3) *independent* of the integration variable θ.

For a Graßmann variable, there are two basis integrals $\int d\theta$ and $\int d\theta\theta$. From the assumptions of *linearity* and *translation invariance*, one obtains $\int d\theta' = \int d\theta$ and $\int d\theta'\theta' = \int d\theta\theta + \theta_0 \int d\theta = \int d\theta\theta$, where $\theta' \equiv \theta + \theta_0$ and θ_0 is a constant. Consequently, one obtains $\int d\theta = 0$. From the assumption that $\int d\theta f(\theta) \equiv \int d\theta(c_0 + c_1\theta)$ is independent of the integration variable θ, one is forced to assume that $\int d\theta\theta$ is a constant. Following Berezin's convention, one fixes the constant by requiring $\int d\theta\theta = 1$. As a result, the Berezin integral $\int d\theta$ is equivalent to the derivative ∂_θ for a single Graßmann variable θ, $\int d\theta f(\theta) = \partial_\theta f(\theta) = c_1$, where $f(\theta) \equiv c_0 + c_1\theta$. For a change of variable $\eta \equiv a\theta$, one would expect from standard calculus that $\int d\theta f(a\theta) = \frac{1}{a} \int d\theta f(\theta)$. But one may show that the Berezin integral satisfies instead $\int d\theta f(a\theta) = a \int d\theta f(\theta)$. A direct computation yields $\int d\eta\eta = \int d\eta a\theta = \int d\theta\theta = 1$. Hence, $d\theta = a d\eta$, or equivalently $d\eta = \frac{1}{a} d\theta$. In other words, the differentials of Graßmann variables scale opposite to what appear in standard calculus. In general, the Berezin integral for a Graßmann algebra generated by n Graßmann variables is defined as a linear functional which has the properties

$$\int d\theta_i = 0, \int d\theta_i\theta_j = \delta_{jk}, (i, j = 1, \cdots, n). \tag{4.10}$$

Hence, to integral a monomial of the Graßmann variables with respect to θ_i, one may first use the anti-commutation relations to bring θ_i to the leftmost position and then drops it. For example, the Berezin integral of a product of two Graßmann variables would yield

$$\int d\theta_i\theta_j\theta_k = \delta_{ij}\theta_k - \delta_{ik}\theta_j, \tag{4.11}$$

which is the same as the left derivative of the same product. Similar results held for higher monomials.

We now consider multiple Berezin integrals for n Graßmann variables, which is defined to be a linear functional which has the property

$$\int d\theta_n \cdots d\theta_1 (f_0 + \sum_i f_i\theta_i + \sum_{i,j} \theta_i\theta_j + \cdots + f_{12\cdots n}\theta_1\theta_2 \cdots \theta_n) \equiv f_{12\cdots n}, \tag{4.12}$$

where we adapted to the convention that one must perform the innermost integral first, i.e., $\int d\theta_2 d\theta_1\theta_1\theta_2 \equiv \int d\theta_2(\int d\theta_1\theta_1)\theta_2$. We now perform a change of variables $\eta_i \equiv a_{ij}\theta_j$, where $a_{ij} \equiv (A)_{ij}$ are elements of an invertible $n \times n$ c-number matrix. Then, the Jacobian J accompanied with this transformation is determined by requiring an invariant integration result

$$\int d\eta_n \cdots d\eta_1 f(\eta_1, \cdots, \eta_n) \equiv J \int d\theta_n \cdots d\theta_1 f(\eta_1, \cdots, \eta_n), \tag{4.13}$$

where η_i on the right-hand side are regarded as functions of θ_i. From the definition of the multiple Berezin integrals, only the product of all n Graßmann variables would contribute to Eq. (4.13). Under the change of variables $\eta_i \equiv a_{ij}\theta_j$, the product of all Graßmann variables transforms as

$$\begin{aligned}\eta_1 \cdots \eta_n &= a_{1\sigma(1)} \cdots a_{n\sigma(n)} \theta_{\sigma(1)} \cdots \theta_{\sigma(n)} \\ &= \operatorname{sgn}(\sigma) a_{1\sigma(1)} \cdots a_{n\sigma(n)} \theta_1 \cdots \theta_n \\ &= \det(A) \theta_1 \cdots \theta_n,\end{aligned} \tag{4.14}$$

which scales opposite to what appears in standard calculus. Hence, we obtain $J = (\det A)^{-1}$, and the formula for the change of variables

$$\int d\eta_n \cdots d\theta_1 f(\eta_1, \cdots, \eta_n) = \det A \int d\theta_n \cdots d\theta_1 f(\theta_1, \cdots, \theta_n), \tag{4.15}$$

where the Graßmann differentials transform as

$$d\eta_n \cdots d\eta_1 = (\det A)^{-1} d\theta_n \cdots d\theta_1. \tag{4.16}$$

Finally, the multiple Berezin integrals over a pair of mutually conjugate Graßmann variables θ and θ^* are treated as integrating over independent ones

$$\int d\theta d\theta^* e^{-a\theta\theta^*} = \int d\theta d\theta^* (1 - a\theta\theta^*) = a \int d\theta d\theta^* \theta^* \theta = a. \tag{4.17}$$

As an example, we may consider a Gaussian Berezin integral over two sets of Graßmann variables $\{\theta_i, \theta_i^*\}$ $(i = 1, \cdots, n)$, in analogue to the complex Gaussian integrals. In particular, we consider the following integral

$$\mathcal{Z}(M) \equiv \int d\theta_n d\theta_n^* \cdots d\theta_1 d\theta_1^* \exp\left(\sum_{i,j=1}^{n} \theta_i^* M_{ij} \theta_j\right). \tag{4.18}$$

According to the properties of the multiple Berezin integrals, only terms that are proportional to $\theta_1^*\theta_1 \cdots \theta_n^*\theta_n$ in the expansion of the integrand would contribute to Eq. (7.10). As products of an even number of Graßmann variables commute with each other, the integrand can thus be expanded as

$$\exp\left(\sum_{i,j=1}^{n} \theta_i^* M_{ij} \theta_j\right) = \prod_{i=1}^{n} \exp\left(\sum_{j=1}^{n} \theta_i^* M_{ij} \theta_j\right) = \prod_{i=1}^{n}\left(1 + \theta_i^* \sum_{j=1}^{n} M_{ij} \theta_j\right).$$

Since only the terms that are proportional to θ_i^* in each factor would contribute to the integral, it remains to integrate

$$\prod_{i=1}^{n}\theta_i^*\left(\sum_{j=1}^{n}M_{ij}\theta_j\right) = \sum_{\sigma\in S_n} M_{1\sigma(1)}\cdots M_{n\sigma(n)}\theta_1^*\theta_{\sigma(1)}\cdots\theta_n^*\theta_{\sigma(n)} \tag{4.19}$$

$$= \sum_{\sigma\in S_n}\operatorname{sgn}(\sigma)M_{1\sigma(1)}\cdots M_{n\sigma(n)}\theta_1^*\theta_1\cdots\theta_n^*\theta_n,$$

where we have used the relation $\theta_{\sigma(1)}\theta_{\sigma(2)}\ldots\theta_{\sigma(n)} = \operatorname{sgn}(\sigma)\theta_1\theta_2\ldots\theta_n$, with $\operatorname{sgn}(\sigma)$ being the signature of the permutation σ. From Eq. (4.19), one recognizes that the coefficient of the product $\theta_1^*\theta_1\cdots\theta_n^*\theta_n$ is exactly the determinant of the matrix M. Hence, one immediately obtains

$$\mathcal{Z}(M) = \int d\theta_n d\theta_n^*\cdots d\theta_1 d\theta_1^* \exp\left(\sum_{i,j=1}^{n}\theta_i^* M_{ij}\theta_j\right) = \det M. \tag{4.20}$$

In the above derivation, as the knowledge that θ_i and θ_i^* are mutually conjugate has not been used, we actually proved a slightly more general result

$$\int d\theta_n d\eta_n\cdots d\theta_1 d\eta_1 \exp\left(\sum_{i,j=1}^{n}\eta_i M_{ij}\theta_j\right) = \det M, \tag{4.21}$$

where θ_i and η_i are two set of independent Graßmann variables. As another example, we may evaluate the general Gaussian Berezin integral

$$\mathcal{Z}(\eta,\eta^*) \equiv \int d\theta_n d\theta_n^*\cdots d\theta_1 d\theta_1^* \exp\left[\sum_{i,j=1}^{n}(\theta_i^* M_{ij}\theta_j + \eta_i^*\theta_i + \theta_i^*\eta_i)\right], \tag{4.22}$$

where η_i and η_i^* are another two sets of Graßmann variables. After the change of variables $\theta_i' \equiv \theta_i + \sum_{j=1}^{n}\Delta_{ij}\eta_j$ and $\theta_i'^* = \theta_i^* + \sum_{j=1}^{n}\eta_j^*\Delta_{ji}$, we obtain

$$\mathcal{Z}(\eta,\eta^*) = \int d\theta_n' d\theta_n'^*\cdots d\theta_1' d\theta_1'^* \exp\left(\sum_{i,j=1}^{n}\theta_i'^* M_{ij}\theta_j' - \eta_i^*\Delta_{ij}\eta_j\right) \tag{4.23}$$

$$= \det M \exp\left(-\sum_{i,j=1}^{n}\eta_i^*\Delta_{ij}\eta_j\right),$$

where $\Delta \equiv M^{-1}$ is the inverse of the matrix M. As a final example, we may evaluate the **Pfaffian** of a skew-symmetric $2n \times 2n$ matrix A, i.e., $A_{ij} = -A_{ji}$, by using the Berezin integral

$$\mathrm{Pf}(A) \equiv \int d\theta_{2n} \cdots d\theta_1 \exp\left(\frac{1}{2}\sum_{i,j=1}^{2n} \theta_i A_{ij} \theta_j\right). \tag{4.24}$$

Interestingly, the Pfaffian of a skew-symmetric matrix is closely related to its determinant. To see this, let us consider a Gaussian Berezin integral over two sets of Graßmann variables $\{\eta_i, \xi_i\}$ and perform the change of variables $\eta_i = \frac{1}{\sqrt{2}}(\theta_i' + i\theta_i'')$ and $\xi_i = \frac{1}{\sqrt{2}}(\theta_i' - i\theta_i'')$, then we obtain

$$\int d\eta d\xi \exp\left(\sum_{i,j=1}^{2n} \eta_i A_{ij} \xi_j\right) = J \int d\theta' d\theta'' \exp\left[\frac{1}{2}\sum_{i,j=1}^{2n} (\theta_i' A_{ij} \theta_j' + \theta_i'' A_{ij} \theta_j'')\right],$$

where $d\eta d\xi \equiv d\eta_{2n} d\xi_{2n} \cdots d\eta_1 d\xi_1$, $d\theta' d\theta'' \equiv d\theta_{2n}' d\theta_{2n}'' \cdots d\theta_1' d\theta_1''$ and $J = i^{2n}$. According to the properties of Graßmann variables, we have

$$d\theta' d\theta'' = (-1)^{n(2n-1)} d\theta_{2n}' \cdots d\theta_1' d\theta_{2n}'' \cdots d\theta_1''. \tag{4.25}$$

As a consequence, we obtain

$$\int d\eta d\xi \exp\left(\sum_{i,j=1}^{2n} \eta_i A_{ij} \xi_j\right) = \left[\int d\theta_{2n} \cdots d\theta_1 \exp\left(\frac{1}{2}\sum_{i,j=1}^{2n} \theta_i A_{ij} \theta_j\right)\right]^2,$$

which implies that the Pfaffian of a skew-symmetric matrix is a square root of its determinant: $\det A = \mathrm{Pf}(A)^2$.

4.2 Coherent States for Fermions

In the last section, we discussed some basic properties of the Graßmann numbers and the Graßmann algebra. In this section, we will introduce the fermionic coherent states based on the Graßmann algebra. Unlike the familiar bosonic coherent states which are defined in the bosonic Fock space, the fermionic coherent states, which appear in the functional formulation of quantum field theory, are *not* defined in the fermionic Fock space but in a **super Hilbert space** over the Graßmann algebra, i.e., an enlarged Hilbert space spanned by linear combinations of vectors with Graßmann number coefficients.

To begin with, let us introduce the fermionic coherent states for a single fermion, which is defined as the *displaced* vacuum state in analogous to the bosonic coherent states [75]

$$|\theta\rangle \equiv \exp(a^\dagger\theta - \theta^* a)|0\rangle \equiv D(\theta)|0\rangle, \tag{4.26}$$

where a and $a^\dagger$ are the fermionic annihilation and creation operators which satisfy the anti-commutation relations $\{a, a^\dagger\} = 1$ and $\{a, a\} = \{a^\dagger, a^\dagger\} = 0$, θ and θ^* are a pair of mutually conjugated Graßmann variables which satisfy the anti-commutation relations $\{\theta, \theta^*\} = \{\theta, \theta\} = \{\theta^*, \theta^*\} = 0$, and $|0\rangle$ is the vacuum state which satisfies $a|0\rangle = 0$.

For definiteness, any Graßmann variables are assumed to be anti-commute with the fermionic operators $\{\theta, a\} = \{\theta, a^\dagger\} = 0$ and are assumed to commute with the vacuum states $\theta|0\rangle = |0\rangle\theta$ and $\theta\langle 0| = \langle 0|\theta$. The fermionic coherent states are normalized to unity, as the displacement operator $D(\theta)$ is a unitary operator in the super Hilbert space, which satisfies the translation properties $D^\dagger(\theta)aD(\theta) = a + \theta$ and $D^\dagger(\theta)a^\dagger D(\theta) = a^\dagger + \theta^*$ in analogous to those for bosonic coherent states.

From now on, the fermionic coherent state $|\theta\rangle$ is defined as an eigenstate of the fermionic annihilation operator

$$\begin{aligned} a|\theta\rangle &\equiv aD(\theta)|0\rangle = D(\theta)D^\dagger(\theta)aD(\theta)|0\rangle \\ &= D(\theta)(a+\theta)|0\rangle = D(\theta)\theta|0\rangle = \theta D(\theta)|0\rangle = \theta|\theta\rangle, \end{aligned} \tag{4.27}$$

where we have used the fact that $[D(\theta), \theta] = 0$. One may apply the Baker-Campbell-Hausdorff formula $e^A e^B = e^{A+B} e^{\frac{1}{2}[A,B]}$ to the case $A = a^\dagger\theta - \theta^* a$ and $B = a^\dagger\eta - \eta^* a$ and obtains $D(\theta)D(\eta) = D(\theta+\eta)e^{\frac{1}{2}(\eta^*\theta - \theta^*\eta)}$. From the definition of the displacement operator, Eq. (4.26), we can write the fermionic coherent states as $|\theta\rangle \equiv D(\theta)|0\rangle = (1 + a^\dagger\theta - \frac{1}{2}\theta^*\theta)|0\rangle$ and, similarly, the adjoint of the fermionic coherent states as $\langle\theta| \equiv \langle 0|D^\dagger(\theta) = \langle 0|1 + \theta^* a - \frac{1}{2}\theta^*\theta)$. Hence, the inner product of two fermionic coherent states becomes

$$\langle\eta|\theta\rangle = e^{\eta^*\theta - \frac{1}{2}(\eta^*\eta + \theta^*\theta)}, \tag{4.28}$$

which yields $\langle\theta|\eta\rangle\langle\eta|\theta\rangle = e^{-(\theta^*-\eta^*)(\theta-\eta)} = 1 - (\theta^* - \eta^*)(\theta - \eta)$. Using the properties of the Berezin integrals, one may readily show that the fermionic coherent states $|\theta\rangle$ are *over-complete* in the super Hilbert space, i.e., one may expand any fermionic state $|\psi\rangle \equiv (c_1 + c_2 a^\dagger)|0\rangle$ with $c_1, c_2 \in \mathbb{C}$ in terms of the fermionic coherent states $|\theta\rangle$ as

$$\begin{aligned} \int d\theta^* d\theta \psi(\theta)|\theta\rangle &\equiv \int d\theta^* d\theta \langle\theta|\psi\rangle|\theta\rangle \\ &= \int d\theta^* d\theta \left(c_1 + c_2\theta^* - \frac{c_1}{2}\theta^*\theta\right)|\theta\rangle = (c_1 + c_2 a^\dagger)|0\rangle = |\psi\rangle. \end{aligned} \tag{4.29}$$

It shows that the fermionic coherent states $|\theta\rangle$, similar to their bosonic counterparts, resolve the identity, $\int d\theta^* d\theta |\theta\rangle\langle\theta| = I$, and thus the function $\psi(\theta) \equiv \langle\theta|\psi\rangle$ provides a continuous representation of the super Hilbert space. However, it should be underscored that the fermionic coherent states are over-complete in the sense that they are not linearly independent to each other. For example, we have the identity $\int d^*\theta d\theta\theta|\theta\rangle = 0$.

Similar to the single mode case, the fermionic coherent states for multi fermionic modes can be also defined as the displaced vacuum state [75]

$$|\boldsymbol{\theta}\rangle = \exp\left\{\sum_{i=1}^{n}\left(a_i^\dagger\theta_i - \theta_i^* a_i\right)\right\}|\mathbf{0}\rangle \equiv D(\boldsymbol{\theta})|\mathbf{0}\rangle, \tag{4.30}$$

where $\boldsymbol{\theta} \equiv \{\theta_1, \cdots, \theta_n\}$ is a set of Graßmann variables, $|\mathbf{0}\rangle \equiv |0\cdots 0\rangle$ is the multimodes vacuum state, $a_i^\dagger$ and a_i are the creation and annihilation operators which satisfy the anti-commutation relations: $\{a_i, a_j^\dagger\} = \delta_{ij}$, $\{a_i, a_j\} = \{a_i^\dagger, a_j^\dagger\} = 0$, and $a_i|\mathbf{0}\rangle = 0$. The Graßmann variables and their complex conjugation satisfy the anti-commutation relations: $\{\theta_i, \theta_j\} = \{\theta_i^*, \theta_j\} = \{\theta_i^*, \theta_j^*\} = 0$. The Graßmann variables are assumed to be anti-commute with the fermionic operators, $\{\theta_i, a_j\} = \{\theta_i, a_j^\dagger\} = 0$, and commute with the vacuum state, $\theta_i|\mathbf{0}\rangle = |\mathbf{0}\rangle\theta_i$.

As the operators $a_i^\dagger\theta_i$ and $\theta_j^* a_j$ commute for $i \neq j$, one may write the displacement operator as a product

$$\begin{aligned} D(\boldsymbol{\theta}) \equiv \exp\left\{\sum_{i=1}^{n}\left(a_i^\dagger\theta_i - \theta_i^* a_i\right)\right\} &= \prod_{i=1}^{n}\exp\left(a_i^\dagger\theta_i - \theta_i^* a_i\right) \\ &= \prod_{i=1}^{n}\left[1 + a_i^\dagger\theta_i - \theta_i^* a_i + (a_i^\dagger a_i - \frac{1}{2})\theta_i^*\theta_i\right]. \end{aligned} \tag{4.31}$$

Hence, one may calculate the displaced annihilation operators as

$$\begin{aligned} D^\dagger(\boldsymbol{\theta}) a_j D(\boldsymbol{\theta}) &\equiv \prod_{i=1}^{n}\exp\left(\theta_i^* a_i - a_i^\dagger\theta_i\right) a_j \prod_{k=1}^{n}\exp\left(a_k^\dagger\theta_k - \theta_k^* a_k\right) \\ &= \exp\left(\theta_j^* a_j - a_j^\dagger\theta_j\right) a_j \exp\left(a_j^\dagger\theta_j - \theta_j^* a_j\right). \\ &= a_j + \theta_j, \end{aligned} \tag{4.32}$$

and similarly, the displaced creation operators, $D^\dagger(\boldsymbol{\theta}) a_j^\dagger D(\boldsymbol{\theta}) = a_j^\dagger + \theta_j^*$. Here, the coherent state representation for a general fermionic state $|\psi\rangle \equiv$

$\sum_n c(n_1 \cdots n_k) a_{n_1}^\dagger \cdots a_{n_k}^\dagger |\mathbf{0}\rangle$ has the form

$$\begin{aligned}
\langle \boldsymbol{\theta} | \psi \rangle &= \sum_n c(n_1 \cdots n_k) \langle \mathbf{0} | D^\dagger(\boldsymbol{\theta}) a_{n_1}^\dagger \cdots a_{n_k}^\dagger |\mathbf{0}\rangle \qquad (4.33) \\
&= \sum_n c(n_1 \cdots n_k) \langle \mathbf{0} | D^\dagger(\boldsymbol{\theta}) a_{n_1}^\dagger D(\boldsymbol{\theta}) \cdots D^\dagger(\boldsymbol{\theta}) a_{n_k}^\dagger D(\boldsymbol{\theta}) D^\dagger(\boldsymbol{\theta}) |\mathbf{0}\rangle \\
&= \sum_n c(n_1 \cdots n_k) \langle \mathbf{0} | (a_{n_1}^\dagger + \theta_{n_1}^*) \cdots (a_{n_k}^\dagger + \theta_{n_k}^*) D^\dagger(\boldsymbol{\theta}) |\mathbf{0}\rangle \\
&= \sum_n c(n_1 \cdots n_k) \theta_{n_1}^* \cdots \theta_{n_k}^* \langle \mathbf{0} | D^\dagger(\boldsymbol{\theta}) |\mathbf{0}\rangle \\
&= \sum_n c(n_1 \cdots n_k) \theta_{n_1}^* \cdots \theta_{n_k}^* \exp\left\{ - \sum_{i=1}^n \frac{1}{2} \theta_i^* \theta_i \right\},
\end{aligned}$$

where $n = \{n_1, n_2, \cdots, n_k\}$ lists a set of occupied modes. Although the displacement operator is defined in a symmetric ordered form, it is sometimes useful to consider the *normally* and *anti-normally* ordered forms. Applying the Baker-Campbell-Hausdorff formula $e^A e^B = e^{A+B} e^{\frac{1}{2}[A,B]}$ to the case $A = \sum_i a_i^\dagger \theta_i$ and $B = -\sum_i \theta_i^* a_i$, one obtains the normally ordered displacement operator

$$\begin{aligned}
D_N(\boldsymbol{\theta}) &\equiv \exp\left\{ \sum_i a_i^\dagger \theta_i \right\} \exp\left\{ - \sum_i \theta_i^* a_i \right\} \qquad (4.34) \\
&= D(\boldsymbol{\theta}) \exp\left\{ \frac{1}{2} \sum_i \theta_i^* \theta_i \right\},
\end{aligned}$$

and the anti-normally ordered displacement operator

$$\begin{aligned}
D_A(\boldsymbol{\theta}) &\equiv \exp\left\{ - \sum_i \theta_i^* a_i \right\} \exp\left\{ \sum_i a_i^\dagger \theta_i \right\} \qquad (4.35) \\
&= D(\boldsymbol{\theta}) \exp\left\{ -\frac{1}{2} \sum_i \theta_i^* \theta_i \right\}.
\end{aligned}$$

Applying the Baker-Campbell-Hausdorff formula again, one finds the multiplication formula for the fermion displacement operators

$$D(\boldsymbol{\theta}) D(\boldsymbol{\eta}) = D(\boldsymbol{\theta} + \boldsymbol{\eta}) \exp\left\{ \frac{1}{2} \sum_i (\eta_i^* \theta_i - \theta_i^* \eta_i) \right\}, \qquad (4.36)$$

where $\boldsymbol{\theta}$ and $\boldsymbol{\eta}$ are two sets of Graßmann variables. Equation (4.36) shows that a product of two displacement operators always gives another displacement operator multiplied by a phase factor. By use of the displacement relation (4.32), one may show that the fermionic coherent state $|\boldsymbol{\theta}\rangle$ is an eigenstate of all annihilation operators a_i

$$\begin{aligned} a_i|\boldsymbol{\theta}\rangle &= D(\boldsymbol{\theta})D^\dagger(\boldsymbol{\theta})a_i D(\boldsymbol{\theta})|\mathbf{0}\rangle = D(\boldsymbol{\theta})(a_i+\theta_i)|\mathbf{0}\rangle \\ &= D(\boldsymbol{\theta})\theta_i|\mathbf{0}\rangle = \theta_i D(\boldsymbol{\theta})|\mathbf{0}\rangle = \theta_i|\boldsymbol{\theta}\rangle, \end{aligned} \tag{4.37}$$

where we have used the relation $D(\boldsymbol{\theta})\theta_i = \theta_i D(\boldsymbol{\theta})$. By use of the product formula (4.31), one may write the fermionic coherent state as

$$\begin{aligned} |\boldsymbol{\theta}\rangle \equiv D(\boldsymbol{\theta})|\mathbf{0}\rangle &= \prod_{i=1}^{n}\left[1 + a_i^\dagger\theta_i - \theta_i^* a_i + (a_i^\dagger a_i - \frac{1}{2})\theta_i^*\theta_i\right]|\mathbf{0}\rangle \\ &= \prod_{i=1}^{n}\left(1 + a_i^\dagger\theta_i - \frac{1}{2}\theta_i^*\theta_i\right)|\mathbf{0}\rangle = \exp\left\{\sum_{i=1}^{n}\left(a_i^\dagger\theta_i - \frac{1}{2}\theta_i^*\theta_i\right)\right\}|\mathbf{0}\rangle. \end{aligned} \tag{4.38}$$

Similarly, the adjoint of the coherent state is

$$\begin{aligned} \langle\boldsymbol{\theta}| \equiv \langle\mathbf{0}|D^\dagger(\boldsymbol{\theta}) &= \langle\mathbf{0}|\prod_{i=1}^{n}\left(1 + \theta_i^* a_i - \frac{1}{2}\theta_i^*\theta_i\right) \\ &= \langle\mathbf{0}|\exp\left\{\sum_{i=1}^{n}\left(\theta_i^* a_i - \frac{1}{2}\theta_i^*\theta_i\right)\right\}, \end{aligned} \tag{4.39}$$

which satisfies the relation $\langle\boldsymbol{\theta}|a_i^\dagger = \langle\boldsymbol{\theta}|\theta_i^*$. A direct computation yields the inner product between two arbitrary fermionic coherent states

$$\begin{aligned} \langle\boldsymbol{\theta}|\boldsymbol{\eta}\rangle &\equiv \langle\mathbf{0}|D^\dagger(\boldsymbol{\theta})D(\boldsymbol{\eta})|\mathbf{0}\rangle = \langle\mathbf{0}|D(-\boldsymbol{\theta})D(\boldsymbol{\eta})|\mathbf{0}\rangle \\ &= \langle\mathbf{0}|D(-\boldsymbol{\theta}+\boldsymbol{\eta})|\mathbf{0}\rangle \exp\left\{\frac{1}{2}\sum_{i=1}^{n}(\theta_i^*\eta_i - \eta_i^*\theta_i)\right\} \\ &= \exp\left\{-\frac{1}{2}\sum_{i=1}^{n}(\eta_i^* - \theta_i^*)(\eta_i - \theta_i)\right\}\exp\left\{\frac{1}{2}\sum_{i=1}^{n}(\theta_i^*\eta_i - \eta_i^*\theta_i)\right\} \\ &= \exp\left\{\sum_{i=1}^{n}\left[\theta_i^*\eta_i - \frac{1}{2}\left(\theta_i^*\theta_i + \eta_i^*\eta_i\right)\right]\right\}, \end{aligned} \tag{4.40}$$

and hence

$$\langle \boldsymbol{\theta}|\boldsymbol{\eta}\rangle\langle \boldsymbol{\eta}|\boldsymbol{\theta}\rangle = \exp\left\{-\sum_{i=1}^{n}(\eta_i^* - \theta_i^*)(\eta_i - \theta_i)\right\} \tag{4.41}$$

$$= \prod_{i=1}^{n}\left[1 - (\eta_i^* - \theta_i^*)(\eta_i - \theta_i)\right].$$

In the following, we derive some useful integral formulas based on properties of the Berezin integrals. From the formula for the Gaussian Berezin integral, Eq. (4.23), one obtains

$$\int d^2\boldsymbol{\theta}|\boldsymbol{\theta}\rangle\langle \boldsymbol{\theta}|\boldsymbol{\eta}\rangle = e^{\frac{1}{2}\eta_i\eta_i^*}\int d^2\boldsymbol{\theta}\exp\left\{\sum_{i=1}^{n}\left(\theta_i\theta_i^* + a_i^\dagger\theta_i + \theta_i^*\eta_i\right)\right\}|0\rangle \tag{4.42}$$

$$= \exp\left\{\sum_{i=1}^{n}\left(a_i^\dagger\eta_i + \frac{1}{2}\eta_i\eta_i^*\right)\right\}|0\rangle = |\boldsymbol{\eta}\rangle,$$

where $d^2\boldsymbol{\theta} \equiv \prod_{i=1}^{n} d^2\theta_i$ and $d^2\theta_i \equiv d\theta_i^* d\theta_i$. It follows that the identity operator may be written as a Berezin integral in the coherent state representation

$$I = \int d^2\boldsymbol{\theta}|\boldsymbol{\theta}\rangle\langle \boldsymbol{\theta}|. \tag{4.43}$$

Interestingly, the trace of an arbitrary operator G, i.e., $\mathrm{Tr}G \equiv \sum_{\boldsymbol{n}}\langle \boldsymbol{n}|G|\boldsymbol{n}\rangle$ with $|\boldsymbol{n}\rangle \equiv |n_1\rangle \otimes |n_2\rangle \otimes \cdots \otimes |n_n\rangle$, can also be written as a Berezin integral in the coherent state representation. A direct computation yields

$$\int d^2\boldsymbol{\theta}\langle -\boldsymbol{\theta}|G|\boldsymbol{\theta}\rangle = \int d^2\boldsymbol{\theta}e^{-\boldsymbol{\theta}^*\boldsymbol{\theta}}\langle 0|e^{-\sum_i\theta_i^* a_i}Ge^{\sum_j a_j^\dagger\theta_j}|0\rangle \tag{4.44}$$

$$= \int d^2\boldsymbol{\theta}(1 - \boldsymbol{\theta}^*\boldsymbol{\theta})\langle 0|\prod_i(1 + a_i\theta_i^*)G\prod_j(1 - \theta_j a_j^\dagger)|0\rangle$$

$$= \int d^2\boldsymbol{\theta}\boldsymbol{\theta}\boldsymbol{\theta}^*\sum_{\boldsymbol{n}}\langle \boldsymbol{n}|G|\boldsymbol{n}\rangle = \mathrm{Tr}G.$$

As an example, the trace of the coherent state dyadic $|\boldsymbol{\theta}\rangle\langle \boldsymbol{\eta}|$ has the form

$$\mathrm{Tr}(|\boldsymbol{\theta}\rangle\langle \boldsymbol{\eta}|) = \int d^2\boldsymbol{\beta}\langle -\boldsymbol{\beta}|\boldsymbol{\theta}\rangle\langle \boldsymbol{\eta}|\boldsymbol{\beta}\rangle = \int d^2\boldsymbol{\beta}\langle \boldsymbol{\eta}|\boldsymbol{\beta}\rangle\langle -\boldsymbol{\beta}|\boldsymbol{\theta}\rangle \tag{4.45}$$

$$= \int d^2\boldsymbol{\beta}\langle \boldsymbol{\eta}|\boldsymbol{\beta}\rangle\langle \boldsymbol{\beta}| - \boldsymbol{\theta}\rangle = \langle \boldsymbol{\eta}| - \boldsymbol{\theta}\rangle = \langle -\boldsymbol{\eta}|\boldsymbol{\theta}\rangle.$$

and similarly

$$\mathrm{Tr}(G|\boldsymbol{\theta}\rangle\langle\boldsymbol{\eta}|) = \langle-\boldsymbol{\eta}|G|\boldsymbol{\theta}\rangle. \tag{4.46}$$

As another application of the Berezin integrals, one may verify that the Dirac delta function may be expressed as

$$\begin{aligned}\delta(\boldsymbol{\theta}-\boldsymbol{\eta}) &\equiv \int d^2\boldsymbol{\alpha}\exp\left[\sum_{i=1}^{n}\left(\alpha_i(\theta_i^*-\eta_i^*)-(\theta_i-\eta_i)\alpha_i^*\right)\right] \\ &= \prod_{i=1}^{n}(\theta_i-\eta_i)(\theta_i^*-\eta_i^*),\end{aligned} \tag{4.47}$$

which satisfies the relation $\delta(\boldsymbol{\theta}-\boldsymbol{\eta}) = \delta(\boldsymbol{\eta}-\boldsymbol{\theta})$. From the above result, one may introduce the Fourier transform of an arbitrary function $f(\boldsymbol{\theta})$ as

$$\tilde{f}(\boldsymbol{\eta}) \equiv \int d^2\boldsymbol{\theta}e^{\boldsymbol{\eta}\boldsymbol{\theta}^*-\boldsymbol{\theta}\boldsymbol{\eta}^*}f(\boldsymbol{\theta}), \tag{4.48}$$

and express the inverse Fourier transform of $\tilde{f}(\boldsymbol{\eta})$ as

$$f(\boldsymbol{\theta}) = \int d^2\boldsymbol{\eta}e^{\boldsymbol{\theta}\boldsymbol{\eta}^*-\boldsymbol{\eta}\boldsymbol{\theta}^*}\tilde{f}(\boldsymbol{\eta}). \tag{4.49}$$

As a result of Eq. (4.47), one may also derive the fermionic analogue of the convolution theorem

$$\begin{aligned}&\int d^2\boldsymbol{\theta}e^{\boldsymbol{\eta}\boldsymbol{\theta}^*-\boldsymbol{\theta}\boldsymbol{\eta}^*}f(\boldsymbol{\theta})g(\boldsymbol{\theta}) \\ &\quad= \int d^2\boldsymbol{\theta}e^{\boldsymbol{\eta}\boldsymbol{\theta}^*-\boldsymbol{\theta}\boldsymbol{\eta}^*}f(\boldsymbol{\theta})\int d^2\boldsymbol{\beta}\int d^2\boldsymbol{\alpha}e^{\boldsymbol{\alpha}(\boldsymbol{\beta}^*-\boldsymbol{\theta}^*)-(\boldsymbol{\beta}-\boldsymbol{\theta})\boldsymbol{\alpha}^*}g(\boldsymbol{\beta}) \\ &\quad= \int d^2\boldsymbol{\alpha}\int d^2\boldsymbol{\theta}e^{(\boldsymbol{\eta}-\boldsymbol{\alpha})\boldsymbol{\theta}^*-\boldsymbol{\theta}(\boldsymbol{\eta}^*-\boldsymbol{\alpha}^*)}f(\boldsymbol{\theta})\int d^2\boldsymbol{\beta}e^{\boldsymbol{\alpha}\boldsymbol{\beta}^*-\boldsymbol{\beta}\boldsymbol{\alpha}^*}g(\boldsymbol{\beta}) \\ &\quad= \int d^2\boldsymbol{\alpha}f(\boldsymbol{\eta}-\boldsymbol{\alpha})\tilde{g}(\boldsymbol{\beta}).\end{aligned} \tag{4.50}$$

From the normally ordered form of the displacement operator, Eq. (4.34), one may express the Dirac delta function via the displacement operator as

$$\begin{aligned}\int d^2\boldsymbol{\eta}\langle\boldsymbol{\eta}|D(\boldsymbol{\theta})|\boldsymbol{\eta}\rangle &= \int d^2\boldsymbol{\eta}\langle\boldsymbol{\eta}|e^{\sum_i a_i^\dagger\theta_i}e^{-\sum_i\theta_i^*a_i}|\boldsymbol{\eta}\rangle e^{-\frac{1}{2}\boldsymbol{\theta}^*\boldsymbol{\theta}} \\ &= \int d^2\boldsymbol{\eta}e^{\sum_i(\eta_i^*\theta_i-\theta_i^*\eta_i)}e^{-\frac{1}{2}\boldsymbol{\theta}^*\boldsymbol{\theta}} = \delta(\boldsymbol{\theta})e^{-\frac{1}{2}\boldsymbol{\theta}^*\boldsymbol{\theta}},\end{aligned} \tag{4.51}$$

and similarly

$$\int d^2\boldsymbol{\eta}\langle\boldsymbol{\eta}|D(\boldsymbol{\alpha})D(-\boldsymbol{\beta})|\boldsymbol{\eta}\rangle = \delta(\boldsymbol{\alpha}-\boldsymbol{\beta})e^{\boldsymbol{\alpha}^*\boldsymbol{\beta}-\frac{1}{2}\boldsymbol{\alpha}^*\boldsymbol{\alpha}-\frac{1}{2}\boldsymbol{\beta}^*\boldsymbol{\beta}}. \tag{4.52}$$

It follows that one may express an arbitrary operator G as a Berezin integral over the displacement operator as

$$G \equiv \int d^2\boldsymbol{\theta}\, g(\boldsymbol{\theta})D(-\boldsymbol{\theta}), \tag{4.53}$$

where $g(\boldsymbol{\theta})$ is a weight function that can be expressed as a Berezin integral over the operator G and the displacement operator

$$\int d^2\boldsymbol{\eta}\langle\boldsymbol{\eta}|GD(\boldsymbol{\alpha})|\boldsymbol{\eta}\rangle = \int d^2\boldsymbol{\eta}\int d^2\boldsymbol{\theta}\, g(\boldsymbol{\theta})\langle\boldsymbol{\eta}|D(-\boldsymbol{\theta})D(\boldsymbol{\alpha})|\boldsymbol{\eta}\rangle \tag{4.54}$$

$$= \int d^2\boldsymbol{\theta}\, g(\boldsymbol{\theta})\delta(\boldsymbol{\alpha}-\boldsymbol{\theta})e^{\boldsymbol{\theta}^*\boldsymbol{\alpha}-\frac{1}{2}\boldsymbol{\theta}^*\boldsymbol{\theta}-\frac{1}{2}\boldsymbol{\alpha}^*\boldsymbol{\alpha}} = g(\boldsymbol{\alpha}).$$

Hence, the final expression of an arbitrary operator G in the fermionic coherent state representation is

$$G = \int d^2\boldsymbol{\theta}\int d^2\boldsymbol{\eta}\langle\boldsymbol{\eta}|GD(\boldsymbol{\theta})|\boldsymbol{\eta}\rangle D(-\boldsymbol{\theta}). \tag{4.55}$$

Using the trace formula, Eq. (4.46), one may interpret the Dirac delta function $\delta(\boldsymbol{\theta}-\boldsymbol{\eta})$ as a trace identity as

$$\delta(\boldsymbol{\theta}-\boldsymbol{\eta}) = \int d^2\boldsymbol{\alpha}\, e^{\boldsymbol{\alpha}\boldsymbol{\theta}^*-\boldsymbol{\theta}\boldsymbol{\alpha}^*}e^{\boldsymbol{\eta}\boldsymbol{\alpha}^*-\boldsymbol{\alpha}\boldsymbol{\eta}^*} \tag{4.56}$$

$$= \int d^2\boldsymbol{\alpha}\, e^{\boldsymbol{\eta}\boldsymbol{\alpha}^*-\boldsymbol{\alpha}\boldsymbol{\eta}^*}\langle\boldsymbol{\alpha}|D_N(\boldsymbol{\theta})|\boldsymbol{\alpha}\rangle$$

$$= \int d^2\boldsymbol{\alpha}\, e^{\boldsymbol{\eta}\boldsymbol{\alpha}^*-\boldsymbol{\alpha}\boldsymbol{\eta}^*}\mathrm{Tr}[D_N(\boldsymbol{\theta})|\boldsymbol{\alpha}\rangle\langle-\boldsymbol{\alpha}|]$$

$$= \mathrm{Tr}[D_N(\boldsymbol{\theta})E_A(\boldsymbol{\eta})],$$

where $E_A(\boldsymbol{\eta})$ is the Fourier transform of the coherent state dyadic $|\boldsymbol{\alpha}\rangle\langle-\boldsymbol{\alpha}|$. As one may verify in the exercises, the displacement operators are complete, and hence, one may expand an arbitrary operator G in terms of the normally order displacement operator $D_N(\boldsymbol{\theta})$ as

$$G = \int d^2\boldsymbol{\theta}\, g(\boldsymbol{\theta})D_N(\boldsymbol{\theta}), \tag{4.57}$$

where the function $g(\boldsymbol{\theta})$ may be solved by

$$\begin{aligned}\mathrm{Tr}[GE_A(\boldsymbol{\eta})] &= \int d^2\boldsymbol{\theta}\, g(\boldsymbol{\theta})\mathrm{Tr}[D_N(\boldsymbol{\theta})E_A(\boldsymbol{\eta})] \\ &= \int d^2\boldsymbol{\theta}\, g(\boldsymbol{\theta})\delta(\boldsymbol{\theta}-\boldsymbol{\eta}) = g(\boldsymbol{\eta}).\end{aligned} \tag{4.58}$$

Hence, an arbitrary operator G can be expanded as

$$G = \int d^2\boldsymbol{\theta}\mathrm{Tr}[GE_A(\boldsymbol{\theta})]D_N(\boldsymbol{\theta}). \tag{4.59}$$

Using the properties of the Berezin integrals, one may easily express $E_A(\boldsymbol{\eta})$, i.e., the Fourier transform of the coherent state dyadic $|\boldsymbol{\alpha}\rangle\langle-\boldsymbol{\alpha}|$ as

$$\begin{aligned}E_A(\boldsymbol{\eta}) &\equiv \int d^2\boldsymbol{\alpha}\, e^{\boldsymbol{\eta}\boldsymbol{\alpha}^* - \boldsymbol{\alpha}\boldsymbol{\eta}^*}|\boldsymbol{\alpha}\rangle\langle-\boldsymbol{\alpha}| \\ &= |\mathbf{0}\rangle\langle\mathbf{0}| - \prod_{i=1}^{n}(a_i^\dagger + \eta_i^*)|\mathbf{0}\rangle\langle\mathbf{0}|(a_i + \eta_i).\end{aligned} \tag{4.60}$$

One may show that the operators $E_A(\boldsymbol{\eta})$ are complete. For simplicity, here we only verify the case of a single mode. The key is that the operators $|0\rangle\langle 0|$, $|0\rangle\langle 0|a$, $a^\dagger|0\rangle\langle 0|$, and $a^\dagger|0\rangle\langle 0|a$ form a complete set of operators and can be expressed as Berezin integrals over the operators $E_A(\eta)$

$$\begin{aligned}&\int d^2\eta E_A(\eta) = |0\rangle\langle 0|, \int d^2\eta(-\eta)E_A(\eta) = |0\rangle\langle 0|a, \\ &\int d^2\eta(-\eta^*)E_A(\eta) = a^\dagger|0\rangle\langle 0|, \int d^2\eta(1-\eta\eta^*)E_A(\eta) = a^\dagger|0\rangle\langle 0|a.\end{aligned} \tag{4.61}$$

One may verify the multimode case similarly. Hence, an arbitrary operator G may be expanded in terms of $E_A(\boldsymbol{\eta})$ as

$$G = \int d^2\boldsymbol{\eta}\, g(\boldsymbol{\eta})E_A(\boldsymbol{\eta}), \tag{4.62}$$

where the function $g(\boldsymbol{\eta})$ may be solved by

$$\begin{aligned}\mathrm{Tr}[D_N(\boldsymbol{\theta})G] &= \int d^2\boldsymbol{\eta}\, g(\boldsymbol{\eta})\mathrm{Tr}[D_N(\boldsymbol{\theta})E_A(\boldsymbol{\eta})] \\ &= \int d^2\boldsymbol{\eta}\, g(\boldsymbol{\eta})\delta(\boldsymbol{\theta}-\boldsymbol{\eta}) = g(\boldsymbol{\theta}).\end{aligned} \tag{4.63}$$

Hence, any arbitrary operator G may be expanded as

$$G = \int d^2\boldsymbol{\eta} \mathrm{Tr}[D_N(\boldsymbol{\theta})G]E_A(\boldsymbol{\eta}). \tag{4.64}$$

In the following, it is convenient to introduce a characteristic function $\chi(\boldsymbol{\eta})$ for a system described by a density ρ

$$\begin{aligned}\chi(\boldsymbol{\eta}) &\equiv \mathrm{Tr}\left[\exp\left(\sum_n(\eta_n a_n^\dagger - a_n\eta_n^*)\right)\rho\right] \qquad (4.65)\\ &= \mathrm{Tr}\left[\prod_n(1+\eta_n a_n^\dagger - a_n\eta_n^* + \eta_n^*\eta_n(a_n^\dagger a_n - \frac{1}{2}))\rho\right].\end{aligned}$$

Similarly, one may define the normally ordered characteristic function as

$$\begin{aligned}\chi_N(\boldsymbol{\eta}) &\equiv \mathrm{Tr}\left[\exp\left(\sum_m \eta_m a_m^\dagger\right)\exp\left(-\sum_n a_n\eta_n^*\right)\rho\right] \qquad (4.66)\\ &= \mathrm{Tr}\left[\prod_n(1+\eta_n a_n^\dagger - a_n\eta_n^* + \eta_n^*\eta_n a_n^\dagger a_n)\rho\right],\end{aligned}$$

and the anti-normally characteristic function as

$$\begin{aligned}\chi_A(\boldsymbol{\eta}) &\equiv \mathrm{Tr}\left[\exp\left(-\sum_n a_n\eta_n^*\right)\exp\left(\sum_m \eta_m a_m^\dagger\right)\rho\right] \qquad (4.67)\\ &= \mathrm{Tr}\left[\prod_n(1+\eta_n a_n^\dagger - a_n\eta_n^* + \eta_n^*\eta_n(a_n^\dagger a_n - 1))\rho\right].\end{aligned}$$

One may readily show that the anti-normally characteristic function $\chi_A(\boldsymbol{\eta})$ is the Fourier transform of the matrix element $\langle\boldsymbol{\theta}|\rho|-\boldsymbol{\theta}\rangle$

$$\begin{aligned}\chi_A(\boldsymbol{\eta}) &\equiv \mathrm{Tr}\left[\exp\left(-\sum_n \theta_n\eta_n^*\right)\int d^2\boldsymbol{\theta}|\boldsymbol{\theta}\rangle\langle\boldsymbol{\theta}|\exp\left(\sum_m \eta_m\theta_m^*\right)\rho\right] \qquad (4.68)\\ &= \int d^2\boldsymbol{\theta}\exp\left(\sum_n(\eta_n\theta_n^* - \theta_n\eta_n^*)\right)\langle\boldsymbol{\theta}|\rho|-\boldsymbol{\theta}\rangle.\end{aligned}$$

A direct computation yields

$$\rho = \int d^2\boldsymbol{\eta}\mathrm{Tr}[D_N(\boldsymbol{\theta})\rho]E_A(\boldsymbol{\eta}) \tag{4.69}$$
$$= \int d^2\boldsymbol{\eta}\chi(-\boldsymbol{\eta})E_A(\boldsymbol{\eta}) = \int d^2\boldsymbol{\eta}\chi(\boldsymbol{\eta})E_A(-\boldsymbol{\eta}).$$

which shows that the characteristic function $\chi(\boldsymbol{\eta})$ is the weight function for the density operator ρ in the above expansion. It turns out that one may define a quasi-probability distribution $P(\boldsymbol{\theta})$ as the Fourier transform of the characteristic function $\chi_N(\boldsymbol{\eta})$

$$P(\boldsymbol{\theta}) \equiv \int d^2\boldsymbol{\eta} \exp\left(\sum_n (\theta_n\eta_n^* - \eta_n\theta_n^*)\right)\chi_N(\boldsymbol{\eta}), \tag{4.70}$$

and show that the quasi-probability distribution $P(\boldsymbol{\theta})$ is the weight function for the density operator ρ in the expansion

$$\rho = \int d^2\boldsymbol{\theta}\, P(\boldsymbol{\theta})|\boldsymbol{\theta}\rangle\langle -\boldsymbol{\theta}|. \tag{4.71}$$

Indeed, one may directly verify that

$$\mathrm{Tr}\left[e^{\sum_m \eta_m a_m^\dagger} e^{-\sum_n a_n\eta_n^*}\rho\right] \tag{4.72}$$
$$= \int d^2\boldsymbol{\alpha} d^2\boldsymbol{\theta} e^{\boldsymbol{\theta}(\boldsymbol{\alpha}^* - \boldsymbol{\eta}^*) - (\boldsymbol{\alpha}-\boldsymbol{\eta})\boldsymbol{\theta}^*}\chi(\boldsymbol{\alpha})$$
$$= \int d^2\boldsymbol{\alpha}\delta(\boldsymbol{\alpha}-\boldsymbol{\eta})\chi_N(\boldsymbol{\alpha}) = \chi_N(\boldsymbol{\eta}).$$

Equation (4.71) shows that the quasi-probability distribution $P(\boldsymbol{\theta})$ is the fermionic analogue of the Glauber-Sudarshan P-representation in quantum optics. As an immediately consequence of Eq. (4.71), one may show that the fermionic quasi-probability distribution $P(\boldsymbol{\theta})$ is normalized to unity

$$\int d^2\boldsymbol{\theta}\, P(\boldsymbol{\theta}) = \mathrm{Tr}\rho = 1. \tag{4.73}$$

Similarly, one may define another quasi-probability distribution $Q(\boldsymbol{\theta})$ using the anti-normally characteristic function $\chi_A(\boldsymbol{\eta})$

$$Q(\boldsymbol{\theta}) \equiv \int d^2\boldsymbol{\eta} \exp\left(\sum_n (\theta_n\eta_n^* - \eta_n\theta_n^*)\right)\chi_A(\boldsymbol{\eta}). \tag{4.74}$$

Substitution of Eq. (4.68) into Eq. (4.74) immediately yields

$$Q(\boldsymbol{\theta}) = \int d^2\boldsymbol{\alpha} d^2\boldsymbol{\eta} e^{\boldsymbol{\eta}(\boldsymbol{\alpha}^* - \boldsymbol{\theta}^*) - (\boldsymbol{\alpha} - \boldsymbol{\theta})\boldsymbol{\eta}^*} \langle \boldsymbol{\alpha} | \rho | - \boldsymbol{\alpha} \rangle \tag{4.75}$$

$$= \int d^2\boldsymbol{\alpha} \delta(\boldsymbol{\alpha} - \boldsymbol{\theta}) \langle \boldsymbol{\alpha} | \rho | - \boldsymbol{\alpha} \rangle = \langle \boldsymbol{\theta} | \rho | - \boldsymbol{\theta} \rangle.$$

Hence, the quasi-probability distribution $Q(\boldsymbol{\theta})$ is the fermionic analogue of the Husimi Q-representation in quantum optics. Similarly, one may show that the quasi-probability distribution $Q(\boldsymbol{\theta})$ is normalized to unity

$$\int d^2\boldsymbol{\theta}\, Q(\boldsymbol{\theta}) = \int d^2\boldsymbol{\theta} \langle \boldsymbol{\theta} | \rho | - \boldsymbol{\theta} \rangle = \mathrm{Tr}\rho = 1. \tag{4.76}$$

As an application, one may use the fermionic P-representation to evaluate the mean values of normally ordered products of monomials such as $a_i^{\dagger m_i} a_i^{n_i}$. For example, for the case of a single mode, one may obtain

$$\mathrm{Tr}\left(a^{\dagger m} a^n \rho\right) = \int d^2\theta \mathrm{Tr}\left(a^{\dagger m} a^n P(\theta) |\theta\rangle\langle -\theta|\right) \tag{4.77}$$

$$= \int d^2\theta \langle \theta | a^{\dagger m} a^n | \theta \rangle P(\theta) = \int d^2\theta \theta^{*m} \theta^n P(\theta),$$

where the cases of multimodes are left for the readers.

As an important application, one may also use the P-representation for fermionic fields to evaluate normally ordered correlation functions. Similar to the case of a bosonic field, one may denote the positive-frequency part of a fermionic field, regarded as a function of a space-time variable x as $\psi(x)$. Then, the nth-order correlation function can be defined as

$$G^{(n)}(x_1, \cdots, x_n, y_n, \cdots, y_1) \equiv \mathrm{Tr}[\psi^\dagger(x_1) \cdots \psi^\dagger(x_n) \psi(y_n) \cdots \psi(y_1) \rho]. \tag{4.78}$$

When the positive-frequency part of the fermionic field is expanded in terms of its mode functions as $\psi(x) = \sum_k a_k \phi_k(x)$, one could simply introduce the fermionic coherent states $|\boldsymbol{\alpha}\rangle$ via $\psi(x)|\boldsymbol{\alpha}\rangle = \varphi(x)|\boldsymbol{\alpha}\rangle$, where $\varphi(x) = \sum_k \alpha_k \phi_k(x)$ is the corresponding Graßmann field and $\boldsymbol{\alpha} = \{\alpha_k\}$ is an assembly of Graßmann variables. Then one may use the P-representation to express the nth correlation function as a

Berezin integral

$$G^{(n)}(x_1, \cdots, x_n, y_n, \cdots, y_1) \equiv \mathrm{Tr}[\psi^\dagger(x_1)\cdots\psi^\dagger(x_n)\psi(y_n)\cdots\psi(y_1)\rho] \tag{4.79}$$

$$= \int d^2\boldsymbol{\alpha}\langle\boldsymbol{\alpha}|\psi^\dagger(x_1)\cdots\psi^\dagger(x_n)\psi(y_1)\cdots\psi(y_n)P(\boldsymbol{\alpha})|\boldsymbol{\alpha}\rangle.$$

$$= \int d^2\boldsymbol{\alpha}\varphi^*(x_1)\cdots\varphi^*(x_n)\varphi(y_n)\cdots\varphi(y_1)P(\boldsymbol{\alpha}).$$

Exercises

4.1. Prove the Baker-Campbell-Hausdorff formula for Graßmann variables, $e^{\theta+\eta} = e^\theta e^\eta e^{-\frac{1}{2}[\theta,\eta]}$, where θ and η are two arbitrary Graßmann variables.

4.2. Verify the result $(\theta_1\theta_2\cdots\theta_n)^* = \theta_n^*\cdots\theta_2^*\theta_1^*$ for a product of n complex Graßmann variables.

4.3. Show that the Dirac delta function for a single Graßmann variable has the form $\delta(\theta-\theta') = \theta - \theta'$, which satisfies

$$\int d\eta\delta(\eta-\eta')f(\eta) = f(\eta').$$

4.4. Verify that the Dirac delta function for a single Graßmann variable has the following integral representation

$$\delta(\eta-\eta') = \int d\xi\, e^{\xi(\eta-\eta')}.$$

4.5. Verify that the Jacobian associated with the change of variables $\eta_i = \frac{1}{\sqrt{2}}(\theta_i' + i\theta_i'')$ and $\xi_i = \frac{1}{\sqrt{2}}(\theta_i' - i\theta_i'')$ is $J = i^{2n}$.

4.6. Verify that $d\theta' d\theta'' = (-1)^{n(2n-1)} d\theta'_{2n}\cdots d\theta'_1 d\theta''_{2n}\cdots d\theta''_1$.

4.7. By using the Berezin integral, Eq. (4.24), show that the Pfaffian of the skew-symmetric $2n \times 2n$ matrix A can be evaluated by the formula $\mathrm{Pf}(A) = \sum_{\alpha\in\Pi} A_\alpha$, where Π is the set of all partitions of $\{1, 2, \cdots, 2n\}$ into pairs without order and an element $\alpha \in \Pi$ is written as $\alpha = \{(i_1, j_1), (i_2, j_2), \cdots, (i_n, j_n)\}$ subjected to the constraints $i_k < j_k$ and $i_1 < i_2 < \cdots < i_n$. Let

$$\pi_\alpha = \begin{bmatrix} 1 & 2 & 3 & 4 & \cdots & 2n-1 & 2n \\ i_1 & j_1 & i_2 & j_2 & \cdots & i_n & j_n \end{bmatrix}$$

be the corresponding permutation. Then, the formula for the Pfaffian can be explicitly written as

$$\mathrm{Pf}(A) = \sum_{\alpha \in \Pi} \mathrm{sgn}(\pi_\alpha) a_{i_1 j_1} a_{i_2 j_2} \cdots a_{i_n j_n}.$$

4.8. Prove that $D(\theta) \equiv \exp(a^\dagger\theta - \theta^* a) = 1 + a^\dagger\theta - \theta^* a + (a^\dagger a - \frac{1}{2})\theta^*\theta$ and show that $[D(\theta), \theta] = 0$.

4.9. Verify the relations $D^\dagger(\theta) a D(\theta) = a + \theta$ and $D^\dagger(\theta) a^\dagger D(\theta) = a^\dagger + \theta^*$.

4.10. Show that $[a^\dagger\theta - \theta^* a, a^\dagger\eta - \eta^* a] = \eta^*\theta - \eta\theta^*$.

4.11. Verify Eq. (4.29).

4.12. Show that the operators I, a, $a^\dagger$, and $\frac{1}{2} - a^\dagger a$ can be written as an Berezin integral over the displacement operators as

$$\begin{aligned} I &= \int d^2\theta\theta\theta^* D(\theta), \\ a &= \int d^2\theta(-\theta) D(\theta), \\ a^\dagger &= \int d^2\theta\theta^* D(\theta), \\ \frac{1}{2} - a^\dagger a &= \int d^2\theta\, D(\theta), \end{aligned}$$

which implies the completeness of the displacement operators.

4.13. Verify Eq. (4.36).

4.14. Verify that

$$\langle 0|D(\boldsymbol{\theta})|0\rangle = \prod_{i=1}^{n}\left(1 - \frac{1}{2}\theta_i^*\theta_i\right) = \exp\left\{-\frac{1}{2}\sum_{i=1}^{n}\theta_i^*\theta_i\right\}.$$

4.15. Verify the relation

$$\int d^2\boldsymbol{\theta}\delta(\boldsymbol{\theta}-\boldsymbol{\eta})f(\boldsymbol{\theta}) = f(\boldsymbol{\eta}),$$

where $f(\boldsymbol{\theta})$ is an arbitrary function of an assembly of Graßmann variables $\boldsymbol{\theta} \equiv \{\theta_1, \cdots, \theta_n\}$.

4.16. Verify the following two forms of Parseval's relations

$$\int d^2\boldsymbol{\eta}\tilde{f}(\boldsymbol{\eta})\tilde{g}^*(\boldsymbol{\eta}) = \int d^2\boldsymbol{\theta} f(\boldsymbol{\theta})g^*(\boldsymbol{\theta}),$$

and

$$\int d^2\boldsymbol{\eta}\tilde{f}(\boldsymbol{\eta})\tilde{g}(-\boldsymbol{\eta}) = \int d^2\boldsymbol{\theta} f(\boldsymbol{\theta})g(\boldsymbol{\theta}).$$

4.17. Verify Eq. (4.44).

4.18. Verity that $E_A(\boldsymbol{\eta}) = |\mathbf{0}\rangle\langle\mathbf{0}| - \prod_{i=1}^n (a_i^\dagger + \eta_i^*)|\mathbf{0}\rangle\langle\mathbf{0}|(a_i + \eta_i)$.

4.19. Prove that the fermionic P- and Q-representations are related by

$$P(\boldsymbol{\theta}) = \int \prod_n (-d^2\eta_n) e^{-(\boldsymbol{\theta}-\boldsymbol{\eta})(\boldsymbol{\theta}^*-\boldsymbol{\eta}^*)} Q(\boldsymbol{\eta}).$$

4.20. Show that for the density operator $\rho \equiv |0\cdots 0\rangle\langle 0\cdots 0|$ which represents the multimodes vacuum state, the normally ordered characteristic function is $\chi_N(\boldsymbol{\theta}) = 1$, the weight function of the P-representation is $P(\boldsymbol{\theta}) = \delta(\boldsymbol{\theta})$, and the quasi-probability distribution $Q(\boldsymbol{\theta}) = \exp(-\boldsymbol{\theta}^*\boldsymbol{\theta})$.

4.21. Show that for the most general physical two-mode fermionic density operator

$$\begin{aligned} \rho = r|00\rangle\langle 00| + u|10\rangle\langle 10| + v|01\rangle\langle 01| + w|10\rangle\langle 01|| \\ + w^*|01\rangle\langle 10 + x|00\rangle\langle 11| + x^*|11\rangle\langle 00| + t|11\rangle\langle 11| \end{aligned}$$

in which $|10\rangle \equiv a_1^\dagger|00\rangle$, $|11\rangle \equiv a_2^\dagger a_1^\dagger|00\rangle$, etc., the normally ordered characteristic function $\chi_N(\boldsymbol{\theta})$ is

$$\begin{aligned} \chi_N(\boldsymbol{\theta}) = 1 + w\theta_1^*\theta_2 + w^*\theta_2^*\theta_1 + (u+t)\theta_1^*\theta_1 + (v+t)\theta_2^*\theta_2 \\ + x\theta_1\theta_2 + x^*\theta_2^*\theta_1^* + t\theta_1^*\theta_1\theta_2^*\theta_2. \end{aligned}$$

4.22. Show that the P-representation may be used to evaluate the mean values of normally ordered products of monomials such as $\prod_i\{a_i^{\dagger n_i} a_i^{m_i}\}$, i.e., one has the following relation

$$\mathrm{Tr}\left[\prod_i \left\{a_i^{\dagger n_i} a_i^{m_i}\right\} \rho\right] = \int d^2\boldsymbol{\alpha} \prod_i \left\{\alpha_i^{*n_i} \alpha_i^{m_i}\right\} P(\boldsymbol{\alpha}).$$

Coherent States Path Integrals 5

5.1 Path Integrals Formalism

Quantum mechanics was originally proposed to understand the stability of atoms and their discrete spectra. The key is that material particles such as electrons behave like waves in the atomic length scale. Such matter waves cannot be squeezed into an arbitrary small volume, unless their frequencies and energies increase unlimitedly, which is a physical impossibility and ensures atomic stability. In the conventional canonical formalism of quantum mechanics, the wave function of a nonrelativistic particle evolves over time according to the Schrödinger equation $(i\hbar\partial_t - \hat{H})\Psi(\mathbf{x}, t) = 0$, where $\hat{H}(\hat{\mathbf{p}}, \mathbf{x}, t)$ is a differential operator obtained from the classical Hamiltonian $H(\mathbf{p}, \mathbf{x}, t)$ by replacing $\mathbf{p}$ as $\hat{\mathbf{p}} \equiv -i\hbar\nabla$, so that the Schrödinger operators of momentum satisfy with positions the canonical commutation relations $[\hat{p}_i, \hat{x}_j] = -i\hbar\delta_{ij}$.

In terms of the Dirac's bra-ket notation, the Schrödinger equation may be expressed in a basis-independent way as an operator equation $i\hbar\partial_t|\Psi(t)\rangle = H(\hat{\mathbf{p}}, \hat{\mathbf{x}}, t)|\Psi(t)\rangle$, supplemented by the specifications of the canonical operators $\langle\mathbf{x}|\hat{\mathbf{p}} \equiv -i\hbar\nabla\langle\mathbf{x}|$ and $\langle\mathbf{x}|\hat{\mathbf{x}} \equiv \mathbf{x}\langle\mathbf{x}|$. The above canonical formalism is widely taught at the undergraduate level in a first course of quantum mechanics. However, this traditional approach to quantum mechanics may not always lead to either the simplest solution or the most natural understanding of quantum phenomena.

An equivalent formalism of quantum mechanics using infinite products of integrals, called "path integrals," in which operators are not taken in the first place, was developed by Richard Feynman in his 1942 thesis [19] and was published in 1948 in a seminal paper titled *Space-Time Approach to Non-relativistic Quantum Mechanics* [18]. The key points of Feynman's path integral approach are the following:

C.-F. Kam et al., *Coherent States*, Lecture Notes in Physics 1011,
https://doi.org/10.1007/978-3-031-20766-2_5

(i) The probability for a particle to move from a space-time point (x_a, t_a) to another space-time point (x_b, t_b) is given by the squared modulus of a complex **transition amplitude** $K(x_b, t_b; x_a, t_a)$.
(ii) The transition amplitude is the sum of contributions from all paths in the configuration space.
(iii) All the paths contribute equally in amplitude, but the phase for each path is proportional to the classical action along the path. In contrast to the conventional formalism of quantum mechanics that builds on the Schrödinger equation involving only the knowledge of the system at an earlier time, Feynman's path integrals formalism determines the transition amplitude in a global manner that all the histories of the system are involved. In the following, we will discuss in more detail the basic properties of the path integrals formalism.

To start with, we shall consider the simplest case that a point particle moving in a one-dimensional Cartesian space, which yields a transition amplitude of the time evolution operator between the localized states of the particle as

$$K(x_b, t_b; x_a, t_a) \equiv \langle x_b|\hat{U}(t_b, t_a)|x_a\rangle, \tag{5.1}$$

where $\hat{U}(t_b, t_a) \equiv e^{-i(t_b-t_a)\hat{H}/\hbar}$ is the time evolution operator for a time-independent Hamiltonian $\hat{H} \equiv \hat{T} + \hat{V}$. As a consequence of the **Trotter product formula** which reads

$$e^{-i(t_b-t_a)\hat{H}/\hbar} = e^{-i\epsilon(N+1)(\hat{T}+\hat{V})/\hbar} = \lim_{N\to\infty} \left(e^{-i\epsilon\hat{T}/\hbar} e^{-i\epsilon\hat{V}/\hbar}\right)^{N+1}, \tag{5.2}$$

the transition amplitude can be sliced into an infinity number of time evolution operators, each acting on an infinitesimal slice of time of thickness $\epsilon \equiv (t_b - t_a)/(N+1) > 0$

$$K(x_b, t_b; x_a, t_a) = \langle x_b| \lim_{N\to\infty} \prod_{n=1}^{N+1} \hat{U}(t_n, t_{n-1})|x_a\rangle, \tag{5.3}$$

where $t_0 \equiv t_a$ and $t_{N+1} \equiv t_b$. The Trotter product formula is valid for potentials which are bounded from below, i.e., there exists a real number C such that $\hat{V} + C$ is a positive operator, and that is applicable for most physically relevant potentials such as the harmonic potential. However, it should be stressed that it is not directly applicable for singular potentials, such as the attractive Coulomb potential. For a proof, readers are referred to the end of chapter problems.

We may now insert a series of resolution of identity in terms of the complete set of position states between each pair of $\hat{U}(t_{n+1}, t_n)$ and $\hat{U}(t_n, t_{n-1})$

$$\int_{-\infty}^{\infty} dx_n|x_n\rangle\langle x_n| = I, \text{ for } n = 1, \cdots, N, \tag{5.4}$$

so that the transition amplitude becomes an infinite product of integrals

$$K(x_b, t_b; x_a, t_a) = \lim_{N\to\infty} \prod_{n=1}^{N} \int_{-\infty}^{\infty} dx_n \prod_{n=1}^{N+1} K(x_n, t_n; x_{n-1}, t_{n-1}), \tag{5.5}$$

where $x_0 \equiv x_a$, $x_{N+1} \equiv x_b$, and $K(x_n, t_n; x_{n-1}, t_{n-1}) \equiv \langle x_n|\hat{U}(t_n, t_{n-1})|x_{n-1}\rangle$ is the infinitesimal transition amplitude, which may be evaluated as

$$\begin{aligned} K(x_n, t_n; x_{n-1}, t_{n-1}) &= \langle x_n|\hat{U}(t_n, t_{n-1})|x_{n-1}\rangle = \langle x_n|e^{-i\epsilon\hat{H}/\hbar}|x_{n-1}\rangle \\ &= \int_{-\infty}^{\infty} dx \langle x_n|e^{-i\epsilon\hat{V}/\hbar}|x\rangle\langle x|e^{-i\epsilon\hat{T}/\hbar}|x_{n-1}\rangle. \end{aligned} \tag{5.6}$$

Evaluating the local matrix elements $\langle x_n|e^{-i\epsilon\hat{V}/\hbar}|x\rangle$ and $\langle x|e^{-i\epsilon\hat{T}/\hbar}|x_{n-1}\rangle$

$$\langle x_n|e^{-i\epsilon\hat{V}/\hbar}|x\rangle = \delta(x_n - x)e^{\ i\epsilon V(x_n)/\hbar}, \tag{5.7a}$$

$$\langle x|e^{-i\epsilon\hat{T}/\hbar}|x_{n-1}\rangle = \frac{1}{2\pi\hbar}\int_{-\infty}^{\infty} dp_n e^{ip_n(x-x_{n-1})/\hbar} e^{-i\epsilon T(p_n)/\hbar}, \tag{5.7b}$$

one immediately obtains the infinitesimal transition amplitude as

$$K(x_n, t_n; x_{n-1}, t_{n-1}) = \int_{-\infty}^{\infty} \frac{dp_n}{2\pi\hbar} e^{ip_n(x_n-x_{n-1})/\hbar} e^{-i\epsilon[V(x_n)+T(p_n)]/\hbar}, \tag{5.8}$$

Substitution of Eq. (5.8) into Eq. (5.5) will yield **Feynman's path integral formula** of the transition amplitude, which consists of an infinite product of integrals

$$\begin{aligned} K(x_b,\ t_b; x_a, t_a) &= \lim_{N\to\infty} \prod_{n=1}^{N} \int_{-\infty}^{\infty} dx_n \prod_{n=1}^{N+1} \int_{-\infty}^{\infty} \frac{dp_n}{2\pi\hbar} e^{\frac{i}{\hbar}[p_n(x_n-x_{n-1})-\epsilon H(p_n,x_n)]} \\ &= \lim_{N\to\infty} \prod_{n=1}^{N} \left[\int_{-\infty}^{\infty} dx_n\right] \prod_{n=1}^{N+1} \left[\int_{-\infty}^{\infty} \frac{dp_n}{2\pi\hbar}\right] e^{\frac{i\epsilon}{\hbar}\sum_{n=1}^{N+1}[p_n \frac{x_n-x_{n-1}}{\epsilon} - H(p_n,x_n)]}, \end{aligned}$$

where $H(p_n, x_n) \equiv T(p_n) + V(x_n)$.

As can be seen from the above formula, since the position variables x_{N+1} and x_0 are fixed at the initial values x_b and x_a, all the paths satisfy the boundary conditions $x(t_a) = x_a$ and $x(t_b) = x_b$. The phase of the transition amplitude in the continuous limit $N \to \infty$ and $\epsilon \to 0$ is precisely the classical canonical action for a path $(x(t), p(t))$ in phase space.

In fact, we may go one step further and obtain a configuration space path integral. As $H(p, x) = \frac{1}{2m}p^2 + V(x)$, the phase of the transition amplitude

$$\frac{\epsilon}{\hbar}\sum_{n=1}^{N+1}\left[p_n\left(\frac{x_n - x_{n-1}}{\epsilon}\right) - \frac{p_n^2}{2m} - V(x_n)\right], \tag{5.9}$$

may be quadratically completed to

$$\frac{\epsilon}{\hbar}\sum_{n=1}^{N+1}\left[\frac{m}{2}\left(\frac{x_n - x_{n-1}}{\epsilon}\right)^2 - \frac{1}{2m}\left(p_n - \frac{x_n - x_{n-1}}{\epsilon}m\right)^2 - V(x_n)\right]. \tag{5.10}$$

The momentum integrals in Feynman's path integrals formula may be evaluated by using the **complex Fresnel integrals**:

$$\int_{-\infty}^{\infty} dp e^{\pm\frac{i}{2}ap^2} = \sqrt{\pm\frac{2\pi i}{a}} \quad \text{for } a > 0, \tag{5.11}$$

so that after integration, the transition amplitude becomes

$$K(x_b, t_b; x_a, t_a) = \sqrt{\frac{m}{2\pi\hbar i\epsilon}} \lim_{N\to\infty}\prod_{n=1}^{N}\left[\int_{-\infty}^{\infty}\frac{dx_n}{\sqrt{2\pi\hbar i\epsilon/m}}\right] e^{\sum_{i=1}^{N+1}\frac{i\epsilon}{\hbar}L(x_n, x_{n-1})},$$

where $x_{N+1} \equiv x_b$ and $x_0 \equiv x_a$ and $L(x_n, x_{n-1}) \equiv \frac{m}{2}(\frac{x_n - x_{n-1}}{\epsilon})^2 - V(x_n)$ is the Lagrangian of the one-dimensional point particle evaluated on the infinitesimal line connecting x_n and x_{n-1} in the configuration space.

In the continuum limit, one may write the transition amplitude as

$$K(x_b, t_b; x_a, t_a) = \int_{x(t_a)=x_a}^{x(t_b)=x_b} \mathcal{D}x e^{\frac{i}{\hbar}\int_{t_a}^{t_b} dt L(\dot{x}, x)}, \tag{5.12}$$

where $L(\dot{x}, x) \equiv \frac{m}{2}\dot{x}^2 - V(x)$ is the Lagrangian of a point particle and

$$\mathcal{D}x \equiv \sqrt{\frac{m}{2\pi\hbar i\epsilon}} \lim_{N\to\infty}\prod_{n=1}^{N}\left[\int_{-\infty}^{\infty}\frac{dx_n}{\sqrt{2\pi\hbar i\epsilon/m}}\right] \tag{5.13}$$

is the limiting measure of integration.

As a first example, one may calculate the transition amplitude for a free particle with $V(x) = 0$ in terms of Feynman's path integral formula. One may first evaluate the path integral evaluated on the broken line which connects the points x_{n+1}, x_n, and x_{n-1}

$$\sqrt{\frac{m}{2\pi\hbar i\epsilon}}\int_{-\infty}^{\infty}\frac{dx_n}{\sqrt{2\pi\hbar i\epsilon/m}}e^{\frac{i\epsilon}{\hbar}\frac{m}{2}\left(\frac{x_{n+1}-x_n}{\epsilon}\right)^2}e^{\frac{i\epsilon}{\hbar}\frac{m}{2}\left(\frac{x_n-x_{n-1}}{\epsilon}\right)^2} \tag{5.14}$$
$$=\sqrt{\frac{m}{2\pi\hbar i\cdot 2\epsilon}}e^{\frac{i\cdot 2\epsilon}{\hbar}\frac{m}{2}\left(\frac{x_{n+1}-x_{n-1}}{2\epsilon}\right)^2},$$

which equals the transition amplitude for the straight line which connects the two end points x_{n+1} and x_{n-1}. As a result, one obtains the full expression of the free particle transition amplitude

$$K(x_b,t_b;x_a,t_a)=\sqrt{\frac{m}{2\pi\hbar i(N+1)\epsilon}}e^{\frac{i(N+1)\epsilon}{\hbar}\frac{m}{2}\left(\frac{x_{N+1}-x_0}{(N+1)\epsilon}\right)^2} \tag{5.15}$$
$$=\sqrt{\frac{m}{2\pi\hbar i(t_b-t_a)}}e^{\frac{i}{\hbar}\frac{m}{2}\frac{(x_b-x_a)^2}{t_b-t_a}}\equiv F_0(t_b-t_a)e^{\frac{i}{\hbar}S[x_c]}.$$

Equation (5.15) shows that the transition amplitude for a free particle may be factorized into a product of a classical amplitude $e^{iS[x_c]/\hbar}$ and a **quantum fluctuation** factor $F_0(t_b-t_a)$, where $S[x_c]$ is the action integral evaluated on the classical path $x_c(t)$ which satisfies the equation of motion $\ddot{x}_c(t)=0$ and the boundary conditions $x(t_b)=x_b$ and $x(t_a)=x_a$. Due to the vanishing of deviations with respect to the classical paths at the end points, the quantum fluctuation factor does not depend on the end points x_b and x_a: $\delta x(t_b)=\delta x(t_a)=0$, with $\delta x(t)\equiv x(t)-x_c(t)$.

As another example, one may evaluate Feynman's path integral for a one-dimensional harmonic oscillator with $V(x)\equiv\frac{m}{2}\omega^2x^2$. As in the case of a free particle, the transition for a harmonic oscillator can be split into a product of a classical amplitude $e^{iS[x_c]/\hbar}$ and a quantum fluctuation factor $F_\omega(t_b-t_a)$. Obviously, the classical path which satisfies the equation of motion $\ddot{x}_c(t)=-\omega^2x_c(t)$ and the boundary conditions $x(t_b)=x_b$ and $x(t_a)=x_a$ has the form

$$x_c(t)=\frac{x_b\sin\omega(t-t_a)+x_a\sin\omega(t_b-t)}{\sin\omega(t_b-t_a)}. \tag{5.16}$$

Notice that Eq. (5.16) is available only when $t_b-t_a\neq k\pi/\omega$, with k being an arbitrary integer. The cases for $t_b-t_a=k\pi/\omega$, known as **caustic phenomena**, have to be treated separately, since the classical path $x_c(t)$ no longer exists.

With the above as preamble, we may rewrite the classical action integral by using integration by parts as

$$\mathcal{S}[x_c] \equiv \frac{m}{2}\int_{t_a}^{t_b} dt\left(\dot{x}_c^2(t) - \omega^2 x_c(t)\right) \tag{5.17}$$

$$= -\frac{m}{2}\int_{t_a}^{t_b} x_c(t)\left[\ddot{x}_c(t) + \omega^2 x_c(t)\right] + \frac{m}{2}x_c(t)\dot{x}_c(t)\Big|_{t_a}^{t_b}.$$

Since the classical path for a harmonic oscillator satisfies the equation of motion $\ddot{x}_c(t) + \omega^2 x_c(t) = 0$, the first term in Eq. (5.17) vanishes identically. By using Eq. (5.16), one can directly calculate the second term in Eq. (5.17) as

$$\mathcal{S}[x_c] = \frac{m}{2}\left[x_c(t_b)\dot{x}_c(t_b) - x_c(t_a)\dot{x}_c(t_a)\right] \tag{5.18}$$

$$= \frac{m\omega}{2\sin\omega(t_b - t_a)}\left((x_b^2 + x_a^2)\cos\omega(t_b - t_a) - 2x_b x_a\right).$$

We now evaluate the quantum fluctuation factor $F_\omega(t_b - t_a)$ for the quantum fluctuations $\delta x(t)$ with respect to the classical path $x_c(t)$ which is given by Eq. (5.16) based on the boundary conditions $\delta x(t_a) = \delta x(t_b) = 0$

$$F_\omega(t_b - t_a) = \int_{\delta x(t_a)=0}^{\delta x(t_b)=0} \mathcal{D}\delta x \exp\left\{\frac{i}{\hbar}\frac{m}{2}\int_{t_a}^{t_b} dt\left[\left(\frac{d\delta x}{dt}\right)^2 - \omega^2 \delta x^2\right]\right\} \tag{5.19}$$

$$\equiv \sqrt{\frac{m}{2\pi\hbar i\epsilon}}\lim_{N\to\infty}\prod_{n=1}^{N}\left[\int_{-\infty}^{\infty}\frac{d\delta x_n}{\sqrt{2\pi\hbar i\epsilon/m}}\right]\exp\left(\frac{i}{\hbar}\mathcal{S}^N[\delta x]\right),$$

where $\mathcal{S}^N[\delta x]$ is the time-sliced quadratic fluctuation expansion around the classical action, which is given by

$$\mathcal{S}^N[\delta x] = \frac{m}{2}\epsilon\sum_{n=1}^{N+1}\left[\left(\frac{\delta x_n - \delta x_{n-1}}{\epsilon}\right)^2 - \omega^2\delta x_n^2\right] \tag{5.20}$$

$$\equiv \frac{m}{2}\epsilon\sum_{n=1}^{N+1}\left[(\overline{\nabla}\delta x_n)^2 - \omega^2\delta x_n^2\right] = \frac{m}{2}\epsilon\sum_{n=0}^{N}\left[(\nabla\delta x_n)^2 - \omega^2\delta x_n^2\right],$$

where $\overline{\nabla}\delta x_n$ and $\nabla\delta x_n$ are the **lattice derivatives** acting on the quantum fluctuations δx_n at times t_n, defined by

$$\overline{\nabla}\delta x_n \equiv \frac{\delta x_n - \delta x_{n-1}}{\epsilon}, \quad \text{for } 1 \le n \le N+1, \tag{5.21a}$$

$$\nabla\delta x_n \equiv \frac{\delta x_{n+1} - \delta x_n}{\epsilon}, \quad \text{for } 0 \le n \le N. \tag{5.21b}$$

In the continuum limit $\epsilon \to 0$, both the two lattice derivatives $\overline{\nabla}$ and ∇ are reduced to the ordinary time derivative d/dt:

$$\lim_{\epsilon\to 0} \overline{\nabla}\delta x(t) \equiv \lim_{\epsilon\to 0} \frac{\delta x(t) - \delta x(t-\epsilon)}{\epsilon} = \frac{dx(t)}{dt}, \tag{5.22a}$$

$$\lim_{\epsilon\to 0} \nabla\delta x(t) \equiv \lim_{\epsilon\to 0} \frac{\delta x(t+\epsilon) - \delta x(t)}{\epsilon} = \frac{dx(t)}{dt}. \tag{5.22b}$$

After carrying out a summation by parts, the time-sliced action can now be expressed as

$$\begin{aligned} \mathcal{S}^N[\delta x] &= \frac{m}{2}\epsilon \sum_{n=1}^{N+1} \left[(\overline{\nabla}\delta x_n)^2 - \omega^2 \delta x_n^2 \right] \\ &= -\frac{m}{2}\epsilon \sum_{n=1}^{N} \delta x_n \left(\nabla\overline{\nabla} + \omega^2 \right) \delta x_n, \end{aligned} \tag{5.23}$$

where $\nabla\overline{\nabla}$ is the **lattice Laplacian** acting on the quantum fluctuations δx_n at times t_n, defined by $\nabla\overline{\nabla}\delta x_n \equiv (\delta x_{n+1} - 2\delta x_n + \delta x_{n-1})/\epsilon^2$. Notice that the boundary terms have been dropped in the last step of Eq. (5.23), as the quantum fluctuations will vanish at the end points: $\delta x(t_b) = \delta x(t_a) = 0$. As all the paths $\delta x(t)$ are from $\delta x = 0$ at t_a and arrive at $\delta x = 0$ at t_b, one may expand $\delta x(t)$ at discrete points $\delta x(t_n)$ into **discrete Fourier series** as

$$\delta x(t_n) = \sum_{m=1}^{N} \sqrt{\frac{2}{N+1}} \sin \nu_m(t_n - t_a)\delta x(\nu_m), \tag{5.24}$$

where $\nu_m \equiv m\pi/T$ and $T \equiv t_b - t_a$ is the period of the Fourier series. A key property of the discrete Fourier transform is that the expansion functions are orthogonal to each other

$$\frac{2}{N+1} \sum_{n=1}^{N} \sin \nu_m(t_n - t_a) \sin \nu_{m'}(t_n - t_a) = \delta_{mm'}, \tag{5.25}$$

After using the orthogonal relation Eq. (5.25), the time-sliced fluctuation action $\mathcal{S}^N[\delta x]$ may be expressed as a sum of independent quadratic terms

$$\mathcal{S}^N[\delta x] = \frac{m}{2}\epsilon \sum_{m=1}^{N} \left(\Omega_m^2 - \omega^2 \right) [\delta x(\nu_m)]^2 \equiv \frac{m}{2}\epsilon \sum_{m=1}^{N} \tilde{\Omega}_m^2 [\delta x(\nu_m)]^2, \tag{5.26}$$

where $\Omega_m^2 \equiv (2 - 2\cos(\nu_m \epsilon))/\epsilon^2$ are the real and non-negative eigenvalues of the negative lattice laplacian, i.e., $-\nabla\overline{\nabla}\delta x_n = \Omega_m^2 \delta x_n$. Hence, the quantum fluctuation factor $F_\omega(t_b - t_a)$ can be expressed as a product of N independent Fresnel integrals

$$F_\omega(t_b - t_a) = \sqrt{\frac{m}{2\pi\hbar i\epsilon}} \lim_{N\to\infty} \prod_{m=1}^{N} \left[\int_{-\infty}^{\infty} \frac{d\delta x(\nu_m)}{\sqrt{2\pi\hbar i\epsilon/m}} e^{\frac{i}{\hbar}\frac{m}{2}\epsilon\tilde{\Omega}_m^2 [\delta x(\nu_m)]^2} \right], \tag{5.27}$$

where we have used the fact that the transformation from the variables δx_n to the Fourier components $x(\nu_m)$ has a unit determinant due to the orthogonality relation, so that $\prod_{n=1}^{N} d\delta x(t_n) = \prod_{m=1}^{N} d\delta x(\nu_m)$. For the case when $\Omega_m^2 > \omega^2$ for all m, the coefficients $\tilde{\Omega}_m^2 \equiv \Omega_m^2 - \omega^2$ are all positive, and hence a direct computation would yield

$$F_\omega(t_b - t_a) = \sqrt{\frac{m}{2\pi\hbar i\epsilon}} \lim_{N\to\infty} \prod_{m=1}^{N} \frac{1}{\sqrt{\epsilon^2\tilde{\Omega}_m^2}}, \tag{5.28}$$

where the infinite product involved in the fluctuation factor $F_\omega(t_b - t_a)$ can be evaluated as

$$\begin{aligned} \prod_{m=1}^{N} \epsilon^2\tilde{\Omega}_m^2 &\equiv \prod_{m=1}^{N} \left[\epsilon^2(\Omega_m^2 - \omega^2)\right] \\ &= \prod_{m=1}^{N} \left(2 - 2\cos\frac{m\pi}{N+1}\right) \prod_{m=1}^{N} \left[1 - \frac{\sin^2\frac{\epsilon\tilde{\omega}}{2}}{\sin^2\frac{m\pi}{2(N+1)}}\right] \\ &= (N+1)\cdot\frac{\sin(N+1)\tilde{\omega}\epsilon}{(N+1)\sin\tilde{\omega}\epsilon} = \frac{\sin\tilde{\omega}(t_b - t_a)}{\sin\tilde{\omega}\epsilon}, \end{aligned} \tag{5.29}$$

where $\tilde{\omega}$ is an auxiliary frequency defined by $\sin(\tilde{\omega}\epsilon/2) \equiv \omega\epsilon/2$. Substitution of Eq. (5.29) into (5.28) immediately yields

$$\begin{aligned} F_\omega(t_b - t_a) &= \lim_{\epsilon\to 0} \sqrt{\frac{m}{2\pi\hbar i\epsilon}} \sqrt{\frac{\sin\tilde{\omega}\epsilon}{\sin\tilde{\omega}(t_b - t_a)}} \\ &= \sqrt{\frac{m}{2\pi\hbar i}} \sqrt{\frac{\omega}{\sin\omega(t_b - t_a)}}, \end{aligned} \tag{5.30}$$

where we have used the fact that $\tilde{\omega} \to \omega$ in the continuum limit $\epsilon \to 0$. A direct computation shows that the condition for $\Omega_m^2 > \omega^2$ is $t_b - t_a < \pi/\tilde{\omega}$. When $t_b - t_a$ grows larger than $\pi/\tilde{\omega}$, the coefficient $\Omega_1^2 - \omega^2$ becomes negative, and there will be a $e^{-i\pi/2}$ phase factor in the Fresnel integral of the associated Fourier component $\delta x(\nu_1)$.

Similarly, when $t_b - t_a$ continues to grow until $t_b - t_a > 2\pi/\tilde{\omega}$, the coefficient $\Omega_2^2 - \omega^2$ becomes negative too, and there will be an extra phase factor $e^{-i\pi/2}$ in the associate Fresnel integral. Hence, for $k\pi/\tilde{\omega} < t_b - t_a < (k+1)\pi/\tilde{\omega}$, an extra phase factor $e^{-ik\pi/2}$, known as the **Maslov-Morse index**, will appear in the quantum fluctuation factor $F_\omega(t_b - t_a)$, so that in the continuum limit $\epsilon \to 0$, the full expression of the transition amplitude becomes

$$K(x_b, t_b; x_a, t_a) = \sqrt{\frac{m}{2\pi\hbar i}}\sqrt{\frac{\omega}{|\sin\omega(t_b - t_a)|}}e^{-ik\pi/2} \times \exp\left\{\frac{i}{\hbar}\frac{m\omega}{2\sin\omega(t_b - t_a)}\left((x_b^2 + x_a^2)\cos\omega(t_b - t_a) - 2x_b x_a\right)\right\}, \tag{5.31}$$

for $k\pi/\omega < t_b - t_a < (k+1)\pi/\omega$. Clearly, for the special case when $\omega \to 0$, the transition amplitude for a harmonic oscillator, Eq. (5.31), reduces to that for a free particle, Eq. (5.15). To evaluate the path integral for the harmonic oscillator at the caustics $t_b - t_a = k\pi/\omega$, one needs to notice that the transition amplitude connects the wave functions at different times as follows

$$\psi_{t_b}(x_b) = \int_{-\infty}^{\infty} K(x_b, t_b; x_a, t_a)\psi_{t_a}(x_a)dx_a. \tag{5.32}$$

Hence, for $t_b - t_a = \pi/(2\omega)$, one may evaluate the wave function $\psi_{t_b}(x_b)$ at t_b by using the path integral formula Eq. (5.31) as

$$\psi_{t_b}(x_b) = \sqrt{\frac{m\omega}{2\pi\hbar i}}\int_{-\infty}^{\infty} e^{-\frac{i}{\hbar}m\omega x_b x_a}\psi_{t_a}(x_a)dx_a. \tag{5.33}$$

One may introduce a dimensionless variable $u \equiv x/L$ with $L \equiv \sqrt{\hbar/(m\omega)}$ being the natural length for the harmonic oscillator, and then from Eq. (5.33), one sees that the wave function $\psi_{t_b}(Lu_b)$ is the Fourier transform of the wave function $\psi_{t_a}(Lu_a)$ multiplied by a phase factor $e^{-i\pi/4}$:

$$\psi_{t_b}(Lu_b) = \frac{e^{-i\pi/4}}{\sqrt{2\pi}}\int_{-\infty}^{\infty} e^{-iu_b u_a}\psi_{t_a}(Lu_a)du_a. \tag{5.34}$$

Similarly, for $t_c - t_b = \pi/(2\omega)$, one has the following relation between the wave functions at times t_c and t_b

$$\begin{aligned}\psi_{t_c}(Lu_c) &= \frac{e^{-i\pi/4}}{\sqrt{2\pi}}\int_{-\infty}^{\infty} e^{-iu_c u_b}\psi_{t_b}(Lu_b)du_b \\ &= \frac{e^{-i\pi/2}}{2\pi}\int_{-\infty}^{\infty}\int_{-\infty}^{\infty} e^{-i(u_c+u_a)u_b}\psi_{t_a}(Lu_a)du_b du_a. \\ &= e^{-i\pi/2}\int_{-\infty}^{\infty}\delta(u_c + u_a)\psi_{t_a}(Lu_a)du_a = e^{-i\pi/2}\psi_{t_a}(-Lu_c).\end{aligned} \tag{5.35}$$

Hence for $t_b - t_a = k\pi/(2\omega)$, one immediately obtains

$$\psi_{t_b}(x_b) = e^{-ik\pi/2}\psi_{t_a}((-1)^k x_b), \tag{5.36}$$

which implies that the transition amplitude $K(x_b, t_b; x_a, t_a)$ for $t_b - t_a = k\pi/\omega$ has the form

$$K(x_b, t_b; x_a, t_a) = e^{-ik\pi/2}\delta(x_a - (-1)^k x_b). \tag{5.37}$$

5.2 Coherent States Path Integral

In the previous section, we showed that the transition matrix element $\langle x_b|\hat{U}(t_b, t_a)|x_a\rangle$ can be expressed in the form of path integrals. This provides an alternative approach to quantum mechanics in addition to the traditional canonical approach. Historically, Feynman's original formulation of the path integral was restricted to the position and momentum representations in quantum mechanics, which cannot be applied to some kinematical systems including the spin systems. In order to overcome the problem, in 1960, John Klauder developed a general path integral formulation of both bosonic fields and spinor fields in terms of the coherent states, in a seminal paper titled "The action option and a Feynman quantization of spinor fields in terms of ordinary c-numbers" [21]. In fact, Klauder's coherent states, which was introduced as an over-complete basis in the Hilbert space, appeared 3 years earlier than Glauber's coherent states. In the following, we will present in detail Klauder's construction of coherent state path integrals.

As an analogue of Eq. (5.38), one can begin with the transition amplitude $\langle z_b|\hat{U}(t_b, t_a)|z_a\rangle$ of the time evolution operator $\hat{U}(t_b, t_a) \equiv e^{-i(t_b-t_a)\hat{H}(a^\dagger,a)/\hbar}$ between two coherent states $|z_a\rangle$ and $|z_b\rangle$, where $\hat{H}(a^\dagger, a)$ is a Hamiltonian of a bosonic system, which is assumed to be a normal ordered operator of the annihilation and creation operators, i.e., all creation operators are to the left of all annihilation operators in the product. As a consequence of the Trotter product formula, one may still express the transition amplitude into an infinite product of time evolution operators

$$K(z_b, t_b; z_a, t_a) \equiv \langle z_b|\hat{U}(t_b, t_a)|z_a\rangle = \langle z_b| \lim_{N\to\infty} \prod_{n=1}^{N+1} \hat{U}(t_n, t_{n-1})|z_a\rangle, \tag{5.38}$$

where $t_n \equiv t_a + n\epsilon$ and $\epsilon \equiv (t_b - t_a)/(N+1)$. To generate a path integral formulation for the transition amplitude, one may use the resolution of identity in terms of the coherent states, Eq. (2.51), and repeatedly insert the identity operator into Eq. (5.38), which yields

$$K(z_b, t_b; z_a, t_a) = \lim_{N\to\infty} \int \cdots \int \prod_{n=1}^{N} \frac{d^2 z_n}{\pi} \prod_{n=1}^{N+1} \langle z_n|\hat{U}(t_n, t_{n-1})|z_{n-1}\rangle, \tag{5.39}$$

where $d^2 z_n \equiv d\Re z_n d\Im z_n$ and the integration extends over the entire complex z-plane. In the limit $N \to \infty$ and $\epsilon \to 0$, one may evaluate each term in the integrand in Eq. (5.39) as follows (up to the first order in ϵ)

$$\begin{aligned}
\langle z_n|\hat{U}(t_n, t_{n-1})|z_{n-1}\rangle &\equiv \langle z_n|e^{-i\frac{\epsilon}{\hbar}\hat{H}(a^\dagger,a)}|z_{n-1}\rangle \\
&\approx \langle z_n|I - i\frac{\epsilon}{\hbar}\hat{H}(a^\dagger, a)|z_{n-1}\rangle \\
&= \langle z_n|z_{n-1}\rangle \left(1 - i\frac{\epsilon}{\hbar}\frac{\langle z_n|\hat{H}(a^\dagger, a)|z_{n-1}\rangle}{\langle z_n|z_{n-1}\rangle}\right), \\
&\approx \langle z_n|z_{n-1}\rangle \exp\left(-i\frac{\epsilon}{\hbar}\frac{\langle z_n|\hat{H}(a^\dagger, a)|z_{n-1}\rangle}{\langle z_n|z_{n-1}\rangle}\right), \\
&\equiv \langle z_n|z_{n-1}\rangle \exp\left(-i\frac{\epsilon}{\hbar}H(z_{n-1}, z_n^*)\right),
\end{aligned} \tag{5.40}$$

where $H(z_{n-1}, z_n^*)$ is a function obtained from the normal ordered Hamiltonian by using the substitutions $a^\dagger \to z_n^*$ and $a \to z_{n-1}$, e.g., $H(z_{n-1}, z_n^*) = g z_n^{*p} z_{n-1}^q$ for $\hat{H}(a^\dagger, a) = g a^{\dagger p} a^q$, and $\langle z_n|z_{n-1}\rangle = e^{z_n^* z_{n-1} - \frac{1}{2}|z_n|^2 - \frac{1}{2}|z_{n-1}|^2}$ is the overlap between two adjacent coherent states $|z_n\rangle$ and $|z_{n-1}\rangle$. Substituting Eq. (5.40) into Eq. (5.39) will yield

$$\begin{aligned}
K(z_b, t_b; z_a, t_a) = \lim_{N\to\infty} \int \cdots \int \prod_{n=1}^{N} \frac{d^2 z_n}{\pi} \exp\Bigg\{ i \sum_{n=1}^{N+1} \\
\left[\frac{i}{2}(z_n^*(z_n - z_{n-1}) - z_{n-1}(z_n^* - z_{n-1}^*)) - \frac{\epsilon}{\hbar}H(z_{n-1}, z_n^*)\right]\Bigg\}.
\end{aligned} \tag{5.41}$$

Assuming the existence of the derivatives $\dot{z}_n \equiv (z_n - z_{n-1})/\epsilon$ and $\dot{z}_n^* \equiv (z_n^* - z_{n-1}^*)/\epsilon$ in the continuous limit $N \to \infty$ and $\epsilon \to 0$, and using the approximations $\epsilon H(z_{n-1}, z_n^*) \approx \epsilon H(z_n, z_n^*) - \epsilon^2 \partial_z H(z_n, z_n^*)\dot{z}_n = \epsilon H(z_n, z_n^*) + \mathcal{O}(\epsilon^2)$ and $z_{n-1}(z_n^* - z_{n-1}^*) \approx \epsilon z_n \dot{z}_n^* - \epsilon^2 |\dot{z}_n|^2 = \epsilon z_n \dot{z}_n^* + \mathcal{O}(\epsilon^2)$, one obtains the final expression of the transition amplitude as a path integral

$$K(z_b, t_b; z_a, t_a) = \int \mathcal{D}z^* \mathcal{D}z \exp\left\{\frac{i}{\hbar}\mathcal{S}(z(t), z^*(t))\right\}, \tag{5.42}$$

where $z(t)$ and $z^*(t)$ are subjected to the boundary conditions $z(t_a) = z_a$ and $z^*(t_b) = z_b^*$ and $\mathcal{S}(z(t), z^*(t))$ is the classical action defined by

$$\mathcal{S}(z(t), z^*(t)) \equiv \int_{t_a}^{t_b} \left[\frac{i\hbar}{2}(z^*\dot{z} - z\dot{z}^*) - H_Q(z, z^*) \right] dt. \tag{5.43}$$

Here, $H_Q(z, z^*) \equiv \langle z|\hat{H}(a^\dagger, a)|z\rangle$ is the expectation value of the Hamiltonian operator $\hat{H}(a^\dagger, a)$ evaluated in the coherent states—the Husimi Q-representation of the Hamiltonian operator, e.g., $H_Q(z, z^*) = gz^{*p}z^q$ for $\hat{H}(a^\dagger, a) = ga^{\dagger p}a^q$.

It must be underscored that the appearance of the Q-representation in the coherent state path integrals is not a coincidence. Both Klauder and Skagerstam [76] had shown that one can calculate the infinitesimal time evolution operator $e^{-\frac{i\epsilon}{\hbar}\hat{H}(a^\dagger, a)}$ via the Glauber-Sudarshan P-representation of the Hamiltonian operator, $\hat{H}(a^\dagger, a) \equiv \int \frac{d^2z}{\pi} H_P(z, z^*)|z\rangle\langle z|$. More precisely, one may express the infinitesimal time evolution operator as

$$\begin{aligned} e^{-\frac{i\epsilon}{\hbar}\hat{H}(a^\dagger,a)} &\approx I - \frac{i\epsilon}{\hbar}\hat{H}(a^\dagger, a) \\ &= I - \frac{i\epsilon}{\hbar}\int \frac{d^2z_n}{\pi} H_P(z_n, z_n^*)|z_n\rangle\langle z_n| \\ &= \int \frac{d^2z_n}{\pi}\left(1 - \frac{i\epsilon}{\hbar}H_P(z_n, z_n^*)\right)|z_n\rangle\langle z_n| \\ &\approx \int \frac{d^2z_n}{\pi} e^{-\frac{i\epsilon}{\hbar}H_P(z_n,z_n^*)}|z_n\rangle\langle z_n|. \end{aligned} \tag{5.44}$$

Hence, the transition amplitude can be written into an infinite product as

$$\begin{aligned} K(z_b, t_b; z_a, t_a) &\equiv \langle z_b|\hat{U}(t_b, t_a)|z_a\rangle = \langle z_b| \lim_{N\to\infty} \prod_{n=1}^{N} \hat{U}(t_n, t_{n-1})|z_a\rangle \\ &= \lim_{N\to\infty} \prod_{n=1}^{N} \int \cdots \int \frac{d^2z_n}{\pi} \prod_{n=1}^{N+1} \langle z_n|z_{n-1}\rangle e^{-\frac{i\epsilon}{\hbar}H_P(z_n,z_n^*)}, \end{aligned} \tag{5.45}$$

where $z_0 \equiv z_a$ and $z_{N+1} \equiv z_b$. Using the formula of the overlap between two adjacent coherent states, $\langle z_n|z_{n-1}\rangle = e^{z_n^* z_{n-1} - \frac{1}{2}|z_n|^2 - \frac{1}{2}|z_{n-1}|^2}$, one immediately obtains the discrete version of the coherent state path integral

$$\begin{aligned} K(z_b, t_b; z_a, t_a) = \lim_{N\to\infty} \int \cdots \int \prod_{n=1}^{N} \frac{d^2z_n}{\pi} \exp\left\{ i \sum_{n=1}^{N+1} \right. \\ \left. \left[\frac{i}{2}(z_n^*(z_n - z_{n-1}) - z_{n-1}(z_n^* - z_{n-1}^*)) - \frac{\epsilon}{\hbar}H_P(z_n, z_n^*) \right] \right\}. \end{aligned} \tag{5.46}$$

Assuming the existence of the derivatives $\dot{z}_n \equiv (z_n - z_{n-1})/\epsilon$ and $\dot{z}_n^* \equiv (z_n^* - z_{n-1}^*)/\epsilon$ in the continuous limit $N \to \infty$ and $\epsilon \to 0$, and using the approximation $z_{n-1}(z_n^* - z_{n-1}^*) \approx \epsilon z_n \dot{z}_n^* - \epsilon^2 |\dot{z}_n|^2 = \epsilon z_n \dot{z}_n^* + \mathcal{O}(\epsilon^2)$, one obtains the final expression of the transition amplitude as a coherent state path integral

$$K(z_b, t_b; z_a, t_a) = \int \mathcal{D}z^* \mathcal{D}z \exp\left\{\frac{i}{\hbar}\mathcal{S}(z(t), z^*(t))\right\}, \tag{5.47}$$

where $z(t)$ and $z^*(t)$ are subjected to the boundary conditions $z(t_a) = z_a$ and $z^*(t_b) = z_b^*$ and $\mathcal{S}(z(t), z^*(t))$ is the classical action defined by

$$\mathcal{S}(z(t), z^*(t)) \equiv \int_{t_a}^{t_b} \left[\frac{i\hbar}{2}(z^*\dot{z} - z\dot{z}^*) - H_P(z, z^*)\right] dt. \tag{5.48}$$

In contrast to the previous approach, the advantage of using the Glauber-Sudarshan P-representation to evaluate the transition amplitude is twofold: not only the two arguments of $H_P(z_n, z_n^*)$ in the discrete coherent state path integral belong to the same time in the mesh, but also the approximation $\epsilon H(z_{n-1}, z_n^*) \approx \epsilon H(z_n, z_n^*) + \mathcal{O}(\epsilon^2)$ used to derive the continuous limit Eq. (5.42) is avoided.

As an example, one may consider the coherent state path integral for a harmonic oscillator with a Hamiltonian $\hat{H}(a^\dagger, a) = \hbar\omega a^\dagger a$. A direct computation yields $H(z_{n-1}, z_n^*) \equiv \hbar\omega z_n^* z_{n-1}$. According to Eq. (5.41), the discrete coherent state path integral for a harmonic oscillator has the form

$$K(z_b, t_b; z_a, t_a) = \lim_{N\to\infty} \int \cdots \int \prod_{n=1}^{N} \frac{d^2 z_n}{\pi} \exp\left\{ i \sum_{n=1}^{N+1} \left[\frac{i}{2}(z_n^*(z_n - z_{n-1}) - z_{n-1}(z_n^* - z_{n-1}^*)) - \epsilon\omega z_n^* z_{n-1}\right]\right\}. \tag{5.49}$$

Using the identity $\sum_{n=1}^{N+1} |z_{n-1}|^2 = \sum_{n=1}^{N+1} |z_n|^2 + |z_a|^2 - |z_b|^2$, one may write the transition amplitude for a harmonic oscillator as

$$K(z_b, t_b; z_a, t_a) = e^{-\frac{1}{2}|z_a|^2 - \frac{1}{2}|z_b|^2} \lim_{N\to\infty} \int \cdots \int \prod_{n=1}^{N} \frac{d^2 z_n}{\pi} \exp\left\{ \sum_{n=1}^{N+1} (1 - i\epsilon\omega) z_n^* z_{n-1} - \sum_{n=1}^{N} |z_n|^2 \right\}. \tag{5.50}$$

Since the exponential of the transition amplitude $K(z_b, t_b; z_a, t_a)$ for a harmonic oscillator involves only quadratic forms, one may obtain $K(z_b, t_b; z_a, t_a)$ by using a series of complex Gaussian integrals. To begin with, one needs to evaluate the complex Gaussian integral for the non-diagonal quadratic form $-|z_1|^2 + f(z_a z_1^* +$

$z_2^* z_1$) with $f \equiv 1 - i\epsilon\omega$. Using the formula

$$\int \frac{d^2z}{\pi} e^{a_1 z^2 + a_2 z^{*2} + a_3 z z^* + b_1 z + b_2 z^*} = \frac{\exp\left(\frac{b_1^2 a_2 + b_2^2 a_1 - b_1 b_2 a_3}{a_3^2 - 4a_1 a_2}\right)}{\sqrt{a_3^2 - 4a_1 a_2}}, \tag{5.51}$$

one immediately obtains

$$\int \frac{d^2 z_1}{\pi} e^{-z_1 z_1^* + f z_2^* z_1 + f z_a z_1^*} = e^{f^2 z_a z_2^*}. \tag{5.52}$$

Hence, as the next step, one needs to evaluate the complex Gaussian integral for the non-diagonal quadratic form $-|z_2|^2 + f z_3^* z_2 + f^2 z_a z_2^*$, which results in an exponential factor $e^{f^3 z_a z_3^*}$. After integrating all the variables z_n for $n = 1, 2, \cdots N$, one obtains an exponential factor $e^{f^{N+1} z_a z_b^*}$. Hence, the final expression for the transition amplitude for a harmonic oscillator is

$$\begin{aligned} K(z_b, t_b; z_a, t_a) &= e^{-\frac{1}{2}|z_a|^2 - \frac{1}{2}|z_b|^2} \lim_{N\to\infty} e^{f^{N+1} z_b^* z_a} \\ &= e^{-\frac{1}{2}|z_a|^2 - \frac{1}{2}|z_b|^2} \lim_{N\to\infty} e^{\left(1 - \frac{i\omega(t_b - t_a)}{N+1}\right)^{N+1} z_b^* z_a} \\ &= \exp\left(-\frac{1}{2}|z_a|^2 - \frac{1}{2}|z_b|^2 + e^{-i\omega(t_b - t_a)} z_b^* z_a\right). \end{aligned} \tag{5.53}$$

5.3 Functional Quantum Field Theory

In this section, we discuss the functional integral formalism of quantum field theory which describes a system with an infinite number of degrees of freedom, based on the method of coherent state path integration introduced in the last section. To begin with, we shall consider coherent state path integrals over a $d+1$ dimensional scalar field $\phi(\mathbf{x}, t)$. Here, the scalar field $\phi(\mathbf{x}, t)$ is a function of $\mathbf{x}$ and t defined in a $(d+1)$-dimensional space-time. A scalar field in quantum field theory is the dynamical variable and the coordinates $\mathbf{x}$ are regarded as labels, which simply specify the scalar field at a given point in space. The action for a free scalar field $\phi(\mathbf{x}, t)$ is given by

$$S[\phi(\mathbf{x}, t)] = \int dt \int d^d x \mathcal{L}(\phi(\mathbf{x}, t), \partial_\mu \phi(\mathbf{x}, t)), \tag{5.54}$$

where $\mathcal{L}(\phi, \partial_\mu \phi) \equiv -\frac{1}{2}(\partial_\mu \phi)^2 - \frac{1}{2}m^2\phi^2 \equiv \frac{1}{2}(\partial_t \phi)^2 - \frac{1}{2}(\nabla\phi)^2 - \frac{1}{2}m^2\phi^2$ is the Lagrangian density, which gives the Lagrangian after integration over space. Here, we adopted a metric with signature $(-1, 1, \cdots, 1)$. In general, the

Lagrangian density for a scalar field with a self-interaction is given by $\mathcal{L}(\phi, \partial_\mu\phi) = -\frac{1}{2}(\partial_\mu\phi)^2 - V(\phi)$, with $V(\phi) \equiv \frac{1}{2}m^2\phi^2 + \sum_{n=3}^{\infty}\frac{1}{n!}\lambda_n\phi^n$.

The canonical quantization procedure starts with a scalar field $\phi(\mathbf{x}, t)$ and its conjugate momentum field $\pi(\mathbf{x}, t)$, which satisfy the equal time commutation relations. In particular, the conjugate momentum field of $\phi(\mathbf{x}, t)$ is defined by

$$\pi(\mathbf{x}, t) \equiv \frac{\delta\mathcal{L}}{\delta\dot{\phi}(\mathbf{x}, t)} = \dot{\phi}(\mathbf{x}, t). \tag{5.55}$$

Then one may introduce the Hamiltonian as

$$H[\phi, \pi] = \int d^d x \left(\pi\dot{\phi} - \mathcal{L}\right) = \frac{1}{2}\int d^d x \left(\pi^2 + (\nabla\phi)^2 + m^2\phi^2\right), \tag{5.56}$$

which is a functional of $\phi(\mathbf{x}, t)$ and $\pi(\mathbf{x}, t)$. In quantum field theory, the field is an infinite set of harmonic oscillator operators acting on the quantum mechanical Hilbert space, which may be decomposed in terms of the creation and annihilation operators as

$$\hat{\phi}(\mathbf{x}, t) = \sqrt{\frac{\hbar}{(2\pi)^d}}\int \frac{d^d k}{\sqrt{2\omega_\mathbf{k}}}[\hat{a}_\mathbf{k} e^{i\mathbf{k}\cdot\mathbf{x}} + \hat{a}_\mathbf{k}^\dagger e^{-i\mathbf{k}\cdot\mathbf{x}}], \tag{5.57a}$$

$$\hat{\pi}(\mathbf{x}, t) = -i\sqrt{\frac{\hbar}{(2\pi)^d}}\int d^d k\sqrt{\frac{\omega_\mathbf{k}}{2}}[\hat{a}_\mathbf{k} e^{i\mathbf{k}\cdot\mathbf{x}} - \hat{a}_\mathbf{k}^\dagger e^{-i\mathbf{k}\cdot\mathbf{x}}]. \tag{5.57b}$$

where the frequencies are given by $\omega_\mathbf{k} = \sqrt{\mathbf{k}^2 + m^2}$. Using the commutation relations between the creation and annihilation operators, i.e., $[\hat{a}_\mathbf{k}, \hat{a}_\mathbf{q}^\dagger] = \delta^d(\mathbf{k} - \mathbf{q})$, one obtains the equal time commutation relations between the field operator and its conjugate momentum field operator

$$[\hat{\phi}(\mathbf{x}, t), \hat{\pi}(\mathbf{x}', t')]_{t=t'} = i\hbar\delta^d(\mathbf{x} - \mathbf{x}'). \tag{5.58}$$

Substitution of Eqs. (5.57a)–(5.57b) into Eq. (5.56) immediately yields the Hamiltonian for the free scalar field $\hat{\phi}(\mathbf{x}, t)$

$$\hat{H} = \frac{\hbar}{2}\int d^d k\omega_\mathbf{k}(\hat{a}_\mathbf{k}^\dagger\hat{a}_\mathbf{k} + \hat{a}_\mathbf{k}\hat{a}_\mathbf{k}^\dagger). \tag{5.59}$$

One may now define scalar field coherent states as the common eigenstates of the field operators and the conjugate momentum field operators, i.e., $\hat{\phi}(\mathbf{x}, t)|\phi, \pi\rangle =$

$\phi(\mathbf{x},t)|\phi,\pi\rangle$ and $\hat{\pi}(\mathbf{x},t)|\phi,\pi\rangle = \pi(\mathbf{x},t)|\phi,\pi\rangle$, where

$$\phi(\mathbf{x},t) = \sqrt{\frac{\hbar}{(2\pi)^d}} \int \frac{d^d k}{\sqrt{2\omega_{\mathbf{k}}}} [z_{\mathbf{k}} e^{i\mathbf{k}\cdot\mathbf{x}} + z_{\mathbf{k}}^* e^{-i\mathbf{k}\cdot\mathbf{x}}], \tag{5.60a}$$

$$\pi(\mathbf{x},t) = -i\sqrt{\frac{\hbar}{(2\pi)^d}} \int d^d k \sqrt{\frac{\omega_{\mathbf{k}}}{2}} [z_{\mathbf{k}} e^{i\mathbf{k}\cdot\mathbf{x}} - z_{\mathbf{k}}^* e^{-i\mathbf{k}\cdot\mathbf{x}}]. \tag{5.60b}$$

In fact, the scalar field coherent states may be obtained from the vacuum state through the application of a unitary operator, i.e., $|\phi,\pi\rangle \equiv \exp[\frac{i}{\hbar}\int d^d x(\pi\hat{\phi} - \phi\hat{\pi})]|0\rangle$. A direct computation will yield

$$\begin{aligned}\langle\phi,\pi|\hat{\phi}|\phi,\pi\rangle &= \langle 0|e^{-\frac{i}{\hbar}\int d^d x(\pi\hat{\phi}-\phi\hat{\pi})}\hat{\phi}e^{\frac{i}{\hbar}\int d^d x(\pi\hat{\phi}-\phi\hat{\pi})}|0\rangle \\ &= \langle 0|\hat{\phi} + \frac{i}{\hbar}\int d^d x\phi[\hat{\pi},\hat{\phi}]|0\rangle = \phi,\end{aligned} \tag{5.61}$$

and similarly $\langle\phi,\pi|\hat{\pi}|\phi,\pi\rangle = \pi$, where the vacuum state satisfies the relations $\langle 0|\hat{\phi}|0\rangle = \langle 0|\hat{\pi}|0\rangle = 0$. Substituting Eqs. (5.57a)–(5.57b) and (5.60a)–(5.60b) into the definition of the scalar field coherent states $|\phi,\pi\rangle$ will yield

$$|\phi,\pi\rangle \equiv e^{\frac{i}{\hbar}\int d^d x(\pi\hat{\phi}-\phi\hat{\pi})}|0\rangle = \exp\left[\int d^d k(z_k a_k^\dagger - z_k^* a_k)\right]|0\rangle, \tag{5.62}$$

where the overlap between two scalar field coherent states $|\phi,\pi\rangle$ and $|\phi',\pi'\rangle$ has the form $\langle\phi,\pi|\phi',\pi'\rangle = \exp\left[\int d^d k\left(z_k^* z_k' - \frac{1}{2}|z_k|^2 - \frac{1}{2}|z_k'|^2\right)\right]$. Clearly, the scalar field coherent state $|\phi,\pi\rangle$ is normalized to one, i.e., $\langle\phi,\pi|\phi,\pi\rangle = 1$.

Based on the scalar field coherent states defined above, a functional integral formalism of quantum field theory may be obtained with the same procedure as in the single-particle case. As the coherent states for a degree of freedom form an over-complete basis of the single-particle Hilbert space, the scalar field coherent states form an over-complete basis of the many-particle Fock space, which is a direct sum of the tensor products of copies of single-particle Hilbert spaces. In fact, one may introduce a resolution of identity in terms of the scalar field coherent states by discretizing the space onto a lattice

$$\mathcal{N}\int \frac{[d\phi(\mathbf{x})][d\pi(\mathbf{x})]}{(2\pi)^{(2L+1)d}}|\phi,\pi\rangle\langle\phi,\pi| = I, \tag{5.63}$$

with the measures $[d\phi(\mathbf{x})]$ and $[d\pi(\mathbf{x})]$ defined by

$$[d\phi(\mathbf{x})] \equiv \prod_{k=1}^{d} \prod_{n_k=-\frac{L}{a}}^{\frac{L}{a}} d\phi(n_1 a, \cdots, n_d a), \tag{5.64a}$$

$$[d\pi(\mathbf{x})] \equiv \prod_{k=1}^{d} \prod_{n_k=-\frac{L}{a}}^{\frac{L}{a}} d\pi(n_1 a, \cdots, n_d a), \tag{5.64b}$$

where a is the size of the lattice spacing, L is a cutoff scale, and $\mathcal{N}$ is a normalization constant that depends on both a and L. Here, we have discretized the space so that the continuous variables $(x_1, \cdots, x_d)$ become the discrete variables $(n_1 a, \cdots, n_d a)$. The harmonic chain is recovered in the limit $a \to 0$. The procedure of discretizing the space onto a lattice is essential for a proper definition of the over-complete relation Eq. (5.63), as well as the functional integration measures appeared in the coherent state path integration for a scalar field.

In order to derive a functional integral formalism for a quantum scalar field, one may evaluate the Green's function $G(t_b, t_a)$, i.e., the matrix element of the evolution operator $\hat{U}(t_b, t_a)$ in the coherent state basis

$$\begin{aligned} G(t_b, t_a) &\equiv \langle \phi_b, \pi_b | \hat{U}(t_b, t_a) | \phi_a, \pi_a \rangle \\ &= \langle \phi_b, \pi_b | \hat{\mathcal{T}} \exp\left\{ -\frac{i}{\hbar} \int_{t_a}^{t_b} dt\, \hat{H}(t) \right\} | \phi_a, \pi_a \rangle, \end{aligned} \tag{5.65}$$

where $\hat{\mathcal{T}}$ is the time-ordering operator defined through

$$\hat{\mathcal{T}}\{\mathcal{O}_1(t_1) \cdots \mathcal{O}_n(t_n)\} = \mathcal{O}_{\pi(1)}(t_{\pi(1)}) \cdots \mathcal{O}_{\pi(n)}(t_{\pi(n)}) \tag{5.66}$$

when acting on a product of n field operators $\mathcal{O}_1, \cdots, \mathcal{O}_n$. Here, π denotes a permutation of $1, \cdots, n$ such that $t_{\pi(1)} > \cdots > t_{\pi(n)}$. One may now define a coherent state path integral for a quantum scalar field similar to that for a harmonic oscillator in the last section. In particular, one may divide the underlying time interval $t_b - t_a$ into $N+1$ subintervals $[t_n, t_{n+1}]$ with equal lengths, i.e., $t_n = t_a + n\epsilon$ and $\epsilon \equiv (t_b - t_a)/(N+1)$. As a result of the Lie-Trotter product formula, the evolution operator $\hat{U}(t_a, t_b)$ can be written as a product of $N+1$ evolution operators acting on the subintervals $[t_n, t_{n+1}]$ in the limit $N \to \infty$

$$\hat{U}(t_b, t_a) = \lim_{N\to\infty} \prod_{n=1}^{N+1} \hat{U}(t_n, t_{n-1}). \tag{5.67}$$

Now, one may insert the resolution of identity for scalar field coherent states, namely, Eq. (5.63) at each interval points, which will yield

$$G(t_b, t_a) = \lim_{N\to\infty} \lim_{L\to\infty} \lim_{a\to 0} \prod_{n=1}^{N} \frac{\mathcal{N}}{(2\pi)^{(2L+1)d}} \int [d\phi_n(\mathbf{x})][d\pi_n(\mathbf{x})] \cdot \prod_{n=1}^{N+1} \langle \phi_n, \pi_n | \hat{U}_{t_n, t_{n-1}} | \phi_{n-1}, \pi_{n-1} \rangle, \tag{5.68}$$

where $\hat{U}(t_n, t_{n-1}) \approx \exp[-\frac{i\epsilon}{\hbar}\hat{H}(t_n)]$. Hence, up to the first order in ϵ, the matrix elements of the evolution operator in the coherent state basis may be approximated by

$$\langle \phi_n, \pi_n | \hat{U}(t_n, t_{n-1}) | \phi_{n-1}, \pi_{n-1} \rangle \approx \langle \phi_n, \pi_n | \phi_{n-1}, \pi_{n-1} \rangle \cdot \exp\left(\frac{-i\epsilon}{\hbar} \frac{\langle \phi_n, \pi_n | \hat{H}(t_n) | \phi_{n-1}, \pi_{n-1} \rangle}{\langle \phi_n, \pi_n | \phi_{n-1}, \pi_{n-1} \rangle} \right). \tag{5.69}$$

As the scalar field coherent state is normalized, in the limit $\epsilon \to 0$ and $N \to \infty$, one obtains

$$\langle \phi_n, \pi_n | \phi_{n-1}, \pi_{n-1} \rangle = 1 - \epsilon \langle \phi_n, \pi_n | \bar{\nabla} | \phi_n, \pi_n \rangle \approx \exp(-\epsilon \langle \phi_n, \pi_n | \bar{\nabla} | \phi_n, \pi_n \rangle), \tag{5.70}$$

where $\bar{\nabla} |\phi_n, \pi_n\rangle \equiv \frac{1}{\epsilon}(|\phi_n, \pi_n\rangle - |\phi_{n-1}, \pi_{n-1}\rangle)$ is the lattice derivative acting on the scalar field coherent states at times t_n with $1 \le n \le N+1$. Substitution of Eqs. (5.69) and (5.70) into (5.68) immediately yields

$$\begin{aligned} G(t_b, t_a) &= \lim_{N\to\infty} \lim_{L\to\infty} \lim_{a\to 0} \prod_{n=1}^{N} \frac{\mathcal{N}}{(2\pi)^{(2L+1)d}} \int [d\phi_n(\mathbf{x})][d\pi_n(\mathbf{x})] \\ &\quad \cdot \exp\left\{ \sum_{n=1}^{N+1} \frac{i\epsilon}{\hbar} \left(i\hbar \langle \phi_n, \pi_n | \bar{\nabla} | \phi_n, \pi_n \rangle - \langle \phi_n, \pi_n | \hat{H}(t_n) | \phi_n, \pi_n \rangle \right) \right\} \\ &= \int \mathcal{D}[\phi, \pi] \exp\left\{ \frac{i}{\hbar} \int_{t_a}^{t_b} dt \left(i\hbar \langle \phi, \pi | \frac{d}{dt} | \phi, \pi \rangle - \langle \phi, \pi | \hat{H} | \phi, \pi \rangle \right) \right\} \\ &= \int \mathcal{D}[\phi, \pi] \exp\left\{ \frac{i}{\hbar} \int_{t_a}^{t_b} dt \int d^d x \left[\frac{1}{2}(\pi\dot{\phi} - \phi\dot{\pi}) - \mathcal{H} \right] \right\}, \end{aligned} \tag{5.71}$$

where $\mathcal{D}[\phi, \pi]$ is a function integration measure for the scalar field ϕ and the conjugate momentum field π

$$\mathcal{D}[\phi, \pi] \equiv \lim_{N\to\infty} \lim_{L\to\infty} \lim_{a\to 0} \prod_{n=1}^{N} \frac{\mathcal{N}}{(2\pi)^{(2L+1)d}} \int [d\phi_n(\mathbf{x})][d\pi_n(\mathbf{x})], \tag{5.72}$$

and $\mathcal{H} \equiv \frac{1}{2}(\pi^2 + (\nabla\phi)^2 + m^2\phi^2)$ is the Hamiltonian density for a free scalar field. For a scalar field with a self-interaction, the scalar potential $\frac{1}{2}m^2\phi^2$ is replaced by $V(\phi) \equiv \frac{1}{2}m^2\phi^2 + \sum_{n=3}^{\infty} \frac{1}{n!}\lambda_n \phi^n$. Notice that as the phase of the Green's function $G(t_b, t_a)$ is a quadratic form in the conjugate momentum field π, it may be integrated out. From Eq. (5.71), one may write the discrete coherent state path integral as

$$G(t_b, t_a) = \int \mathcal{D}[\phi, \pi] \exp\left\{ \frac{i\epsilon}{\hbar} \int d^d x \sum_{n=1}^{N+1} \left[\frac{\phi_n \pi_{n-1} - \phi_{n-1}\pi_n}{2\epsilon} \right.\right. \tag{5.73}$$
$$\left.\left. - \frac{1}{2}\pi_n^2 - \frac{1}{2}(\nabla\phi_n)^2 - \frac{1}{2}m^2\phi_n^2 \right]\right\},$$

where the phase of the Green's function may be quadratically completed to

$$-\frac{1}{2}\sum_{n=1}^{N} \left(\pi_n - \frac{\phi_{n+1} - \phi_n}{2\epsilon} \right)^2 + \frac{\phi_1 \pi_a - \phi_N \pi_b}{2\epsilon} - \mathcal{H}(\phi_b, \pi_b) \tag{5.74}$$
$$+ \frac{1}{2}\sum_{n=1}^{N} \left[\left(\frac{\phi_{n+1} - \phi_{n-1}}{2\epsilon} \right)^2 - (\nabla\phi_n)^2 - V(\phi_n) \right].$$

After integrating out the conjugate momentum field π, the Green's function $G(t_b, t_a)$ in the continuous limit $N \to \infty$ and $\epsilon \to 0$ becomes

$$G(t_b, t_a) = \exp\left\{ \frac{i}{2\hbar} \int d^d x \, (\phi_a \pi_a - \phi_b \pi_b) \right\} \tag{5.75}$$
$$\cdot \int \mathcal{D}[\phi] \exp\left\{ \frac{i}{2\hbar} \int_{t_a}^{t_b} dt \int d^d x [\dot{\phi}^2 - (\nabla\phi)^2 - V(\phi)] \right\},$$

where $\mathcal{D}[\phi]$ is a functional integration measure for the scalar field ϕ

$$\mathcal{D}[\phi] \equiv \lim_{N\to\infty} \lim_{L\to\infty} \lim_{a\to 0} \prod_{n=1}^{N} \frac{\mathcal{N}}{(2\pi i\epsilon/\hbar)^{(2L+1)d/2}} \int [d\phi_n(\mathbf{x})]. \tag{5.76}$$

Exercises

5.1. Prove that there exists a constant c such that for all $n \times n$ matrices B with complex entries with $\|B\| < 1/2$, we have

$$\|\log(I + B) - B\| \leq c\|B\|^2,$$

where $\|B\| \equiv (\mathrm{Tr}(B^\dagger B))^{1/2}$ is the Hilbert-Schmidt norm of B.

5.2. Based on the about result, prove the *Lie Product Formula* for matrices: let X and Y be two arbitrary $n \times n$ matrices with complex entries and then

$$e^{X+Y} = \lim_{N\to\infty} \left(e^{\frac{X}{N}} e^{\frac{Y}{N}}\right)^N .$$

5.3. Verify the relation

$$\langle x|e^{-i\epsilon\hat{T}/\hbar}|x_{n-1}\rangle = \frac{1}{2\pi\hbar}\int_{-\infty}^{\infty} dp_n e^{ip_n(x-x_{n-1})/\hbar} e^{-i\epsilon T(p_n)/\hbar},$$

where $\hat{T} \equiv \hat{p}^2/2m$ is the kinetic energy operator and $T(p) \equiv p^2/2m$ is the kinetic energy.

5.4. Verify the free particle integral, Eq. (5.14).

5.5. Verify that the classical action integral $S[x_c]$ for a free particle is

$$S[x_c] \equiv \frac{m}{2}\int_{t_a}^{t_b} dt\dot{x}_c^2(t) = \frac{m}{2}\frac{(x_b - x_a)^2}{t_b - t_a},$$

where $x_c(t)$ is the classical path which satisfies the equation of motion $\ddot{x}_c(t) = 0$ and the boundary conditions $x(t_b) = x_b$ and $x(t_a) = x_a$.

5.6. Verify the expression of the classical path, Eq. (5.18).

5.7. Verity the orthogonal relation, Eq. (5.25).

5.8. Prove that the transformation from the variables δx_n to the Fourier components $x(\nu_m)$ has a unit determinant due to the orthogonality relation, Eq. (5.25), so that $\prod_{n=1}^{N} d\delta x(t_n) = \prod_{m=1}^{N} d\delta x(\nu_m)$.

5.9. Verify the following relations

$$\prod_{n=1}^{N}\left(2-2\cos\frac{m\pi}{N+1}\right)=N+1,$$

$$\prod_{m=1}^{N}\left[1-\frac{\sin^2 x}{\sin^2\frac{m\pi}{2(N+1)}}\right]=\frac{\sin 2(N+1)x}{(N+1)\sin 2x}.$$

5.10. Verify Eq. (5.58) by using Eqs. (5.57a) and (5.57b).

5.11. Calculate the commutator $[\phi(\mathbf{x},t),\phi(\mathbf{x}',t')]$ for the field operators at two different space-time points by using Eqs. (5.57a) and (5.57b). Show that the commutator vanishes when the two space-time points $(\mathbf{x},t)$ and $(\mathbf{x}',t')$ are separated by a space-like distance, i.e., $(\mathbf{x}-\mathbf{x}')^2>(t-t')^2$. (Hint: Perform a Lorentz transformation such that $t=t'$ by exploiting the Lorentz invariance and then show that the commutator vanishes because of an odd integrand.)

5.12. Verify Eq. (5.59). (Hint: use the integral $\frac{1}{(2\pi)^d}\int d^d x e^{i(\mathbf{k}-\mathbf{q})\cdot\mathbf{x}}=\delta^d(\mathbf{k}-\mathbf{q})$.)

5.13. Verify Eq. (5.62) by using Eqs. (5.57a)–(5.57b) and (5.60a)–(5.60b).

5.14. Prove that $\langle\phi,\pi|\phi',\pi'\rangle=\exp[\int d^d k(z_k^* z_k'-\frac{1}{2}|z_k|^2-\frac{1}{2}|z_k'|^2)]$.

5.15. Show that for $(\phi',\pi')\to(\phi,\pi)$, the overlap between the two scalar field coherent states $|\phi,\pi\rangle$ and $|\phi',\pi'\rangle$ may be written as $\langle\phi,\pi|\phi',\pi'\rangle=\exp\left\{-\frac{1}{2}\int d^d k(z_k^*\dot{z}_k-z_k\dot{z}_k^*)\right\}$.

5.16. Verify the relation $i\hbar\langle\phi,\pi|\frac{d}{dt}|\phi,\pi\rangle=\frac{1}{2}\int d^d x(\pi\dot{\phi}-\phi\dot{\pi})$.

End of Chapter Problems

1. *Stationary-phase approximation* [77]: The coherent state path integral provides a natural framework to approximate the transition amplitude in the semiclassical limit via the stationary-phase approximation. In the stationary-phase approximation or steepest-descent approximation, one may estimate the coherent state path integral by looking for the stationary points of the exponent in Eq. (5.41) and expanding the exponent in the vicinity of the stationary trajectory up to quadratic terms, so that one obtains an analytical approximation of the quantum fluctuation factor after performing a series of Gaussian integrals. To be precise, let us denote

the transition amplitude as

$$K(z_b, t_b; z_a, t_a) = \lim_{N\to\infty} \int \cdots \int \prod_{n=1}^{N} \frac{d^2 z_n}{\pi} \exp\{f(z, z^*)\},$$

$$f(z, z^*) \equiv \sum_{n=1}^{N+1} [\frac{\epsilon}{2}(z_{n-1}\overline{\nabla} z_n^* - z_n^* \overline{\nabla} z_n) - \frac{i\epsilon}{\hbar} H(z_{n-1}, z_n^*)],$$

where $\overline{\nabla} z_n \equiv (z_n - z_{n-1})/\epsilon$ and $\overline{\nabla} z_n^* \equiv (z_n^* - z_{n-1}^*)/\epsilon$ are the lattice derivatives of $z(t)$ and $z^*(t)$ at time t_n.

(a) Show that the stationary points $(\bar{z}, \bar{z}^*)$ of the exponent $f(z, z^*)$ are determined by

$$z_n^* - z_{n-1}^* + \frac{i\epsilon}{\hbar} \frac{\partial H(z_{n-1}, z_n^*)}{\partial z_{n-1}} = 0,$$

$$-(z_n - z_{n-1}) - \frac{i\epsilon}{\hbar} \frac{\partial H(z_{n-1}, z_n^*)}{\partial z_n^*} = 0.$$

(b) Let us denote the quantum fluctuations around the stationary trajectory as $\eta \equiv z - \bar{z}$ and $\eta^* \equiv z^* - \bar{z}^*$, which satisfy the boundary conditions $\eta_0 = \eta_{N+1}^* = 0$. Show that the exponent $f(z, z^*)$ can be expanded into a Taylor series in η and η^* around the stationary trajectory up to second order in η and η^* as

$$f(z, z^*) = f(\bar{z}, \bar{z}^*) + \sum_{n=1}^{N+1} \Bigg[\eta_n^* \eta_{n-1} - \frac{1}{2}|\eta_n|^2 - \frac{1}{2}|\eta_{n-1}|^2$$
$$- \frac{i\epsilon}{2\hbar} \left(\frac{\partial^2 H_{n-1,n}}{\partial z_{n-1}^2} \eta_{n-1}^2 + 2 \frac{\partial^2 H_{n-1,n}}{\partial z_{n-1} \partial z_n^*} \eta_n^* \eta_{n-1} + \frac{\partial^2 H_{n-1,n}}{\partial z_n^{*2}} \eta_n^{*2} \right) \Bigg],$$

where $H_{i,j} \equiv H(z_i, z_j^*)$, and show that the transition amplitude may be approximated by

$$K(z_b, t_b; z_a, t_a) \approx e^{f(\bar{z}, \bar{z}^*)} \lim_{N\to\infty} \int \cdots \int \prod_{n=1}^{N} \frac{d^2 \eta_n}{\pi}$$
$$\exp\Bigg\{ \sum_{n=1}^{N} \Bigg[-\frac{i\epsilon}{2\hbar} \frac{\partial^2 H_{n,n+1}}{\partial z_n^2} \eta_n^2 - \frac{i\epsilon}{2\hbar} \frac{\partial^2 H_{n-1,n}}{\partial z_n^{*2}} \eta_n^{*2}$$
$$- \eta_n^* \eta_n + \left(1 - \frac{i\epsilon}{\hbar} \frac{\partial^2 H_{n,n+1}}{\partial z_n \partial z_n^*}\right) \eta_{n+1}^* \eta_n \Bigg] \Bigg\}.$$

(c) Verify the following formula

$$\int \frac{d\eta^* d\eta}{2\pi i} e^{a_1\eta^2 + a_2\eta^{*2} + a_3\eta\eta^* + b_1\eta + b_2\eta^*} = \frac{\exp\left(\frac{b_1^2 a_2 + b_2^2 a_1 - b_1 b_2 a_3}{a_3^2 - 4a_1 a_2}\right)}{\sqrt{a_3^2 - 4a_1 a_2}}$$

by using the general formula for n-dimensional Gaussian integral

$$\int e^{-\frac{1}{2}x^T \mathbf{A} x + B^T x} d^n x = \sqrt{\frac{(2\pi)^n}{\det \mathbf{A}}} e^{\frac{1}{2} B^T \mathbf{A}^{-1} B}.$$

(d) Show that applying the above formula to evaluate the complex Gaussian integral for η_1 and η_1^* in the transition amplitude yields

$$a_1 = -\frac{i\epsilon}{2\hbar}\frac{\partial^2 H_{1,2}}{\partial z_1^2},\ a_2 = -\frac{i\epsilon}{2\hbar}\frac{\partial^2 H_{0,1}}{\partial z_1^{*2}} \equiv X_1,\ a_3 = -1,$$

$$b_1 \equiv \left(1 - \frac{i\epsilon}{\hbar}\frac{\partial^2 H_{1,2}}{\partial z_1 \partial z_2^*}\right)\eta_2^*,\ b_2 = 0,$$

and

$$K(z_b, t_b; z_a, t_a) \approx \frac{e^{f(\bar{z},\bar{z}^*)}}{\sqrt{1 + \frac{2i\epsilon}{\hbar}\frac{\partial^2 H_{1,2}}{\partial z_1^2} X_1}} \lim_{N\to\infty} \int \cdots \int \prod_{n=1}^{N} \frac{d^2\eta_n}{\pi}$$

$$\exp\left\{\sum_{n=2}^{N}\left[-\frac{i\epsilon}{2\hbar}\frac{\partial^2 H_{n-1,n}}{\partial z_{n-1}^2}\eta_{n-1}^2 - \frac{i\epsilon}{2\hbar}\frac{\partial^2 H_{n-1,n}}{\partial z_n^{*2}}\eta_n^{*2} - \eta_n^*\eta_n\right.\right.$$

$$\left.\left. + \left(1 - \frac{i\epsilon}{\hbar}\frac{\partial^2 H_{n-1,n}}{\partial z_{n-1}\partial z_n^*}\right)\eta_n^*\eta_{n-1}\right]\right\} \exp\left(\frac{(1 - \frac{i\epsilon}{\hbar}\frac{\partial H_{1,2}}{\partial z_1 \partial z_2^*})^2 X_1 \eta_2^{*2}}{1 + \frac{2i\epsilon}{\hbar}\frac{\partial^2 H_{1,2}}{\partial z_1^2} X_1}\right).$$

(e) Show that the transition amplitude, after repeating the above process, may be evaluated by

$$K(z_b, t_b; z_a, t_a) \approx \frac{e^{f(\bar{z},\bar{z}^*)}}{\prod_{n=1}^{N}\sqrt{1 + \frac{2i\epsilon}{\hbar}\frac{\partial^2 H_{n,n+1}}{\partial z_n^2} X_n}},$$

where X_n is determined by the following recursive relation

$$X_n = -\frac{i\epsilon}{2\hbar}\frac{\partial^2 H_{n-1,n}}{\partial z_n^{*2}} + \frac{(1 - \frac{i\epsilon}{\hbar}\frac{\partial H_{n-1,n}}{\partial z_{n-1}\partial z_n^*})^2 X_{n-1}}{1 + \frac{2i\epsilon}{\hbar}\frac{\partial^2 H_{n-1,n}}{\partial z_{n-1}^2} X_{n-1}},\ X_0 = 0.$$

Spin Coherent States

6

6.1 Spin Coherent States in Quantum Optics

As we have discussed at the end of chapter problem 2.31 of Chap. 2, a classical prescribed atomic source of radiation, when applied to the vacuum state of a quantized electromagnetic field, will always produce electromagnetic field coherent states. The electromagnetic field coherent states may be obtained from the vacuum state by a unitary displacement operator and are classical-like minimum uncertainty states. With this as background, there would bring up a natural question: What will happen when one applies a classical electromagnetic field to an assembly of N atoms? The concept of spin coherent states, along this line of thought, was first proposed by Radcliffe in a seminal paper titled *Some properties of coherent spin states* [78] published in 1971, in analogous to coherent states for harmonic oscillators, and was subsequently reexamined thoroughly by Arrechi, Courtens, Gilmore, and Thomas in a paper titled *Atomic Coherent States in Quantum Optics* published in 1972 [79].

In quantum optics, if the electromagnetic field operators were to be replaced by their classical average values, the Hamiltonian which describes the interaction between an assembly of N two-level atoms and a classical electromagnetic field can then be written as

$$\begin{aligned}\mathcal{H}_A &= \hbar \sum_i \Omega_i \sigma_z^i + \hbar \sum_k \omega_k \langle a_k^\dagger a_k \rangle + \hbar \sum_{i,k} (g_{ik} \sigma_+^i \langle a_k \rangle + g_{ik}^* \sigma_-^i \langle a_k^\dagger \rangle) \qquad (6.1)\\ &= \hbar \sum_i \Omega_i \sigma_z^i + \hbar \sum_i (\lambda_i(t) \sigma_+^i + \lambda_i^*(t) \sigma_-^i),\end{aligned}$$

where the constant energy term $\hbar \sum_k \omega_k \langle a_k^\dagger a_k \rangle$ is omitted in $\mathcal{H}_A$ and σ_+^i, σ_-^i, and σ_z^i are the single-atomic operators which obey the conventional angular momentum commutation relations $[\sigma_z^i, \sigma_\pm^j] = \pm \sigma_\pm^i \delta_{ij}$ and $[\sigma_+^i, \sigma_-^j] = 2\sigma_z^i \delta_{ij}$. In particular,

C.-F. Kam et al., *Coherent States*, Lecture Notes in Physics 1011,
https://doi.org/10.1007/978-3-031-20766-2_6

for the case that $\lambda_i(t) = \lambda(t)$ and $\Omega_i = \Omega$, we may introduce the many-atomic operators $J_z \equiv \sum_i \sigma_z^i$ and $J_\pm \equiv \sum_i \sigma_\pm^i$, so that

$$\mathcal{H}_A = \hbar\Omega J_z + \hbar\lambda(t)J_+ + \hbar\lambda^*(t)J_-, \tag{6.2}$$

where J_z, J_+, and J_- satisfy the conventional angular momentum commutation relations $[J_z, J_\pm] = \pm J_\pm$ and $[J_+, J_-] = 2J_z$ and the Hamiltonian 6.2 describes a collective pseudo-spin with quantum number $j = N/2$. In the absence of driving field, the free atoms are described by the Hamiltonian $\mathcal{H}_0 = \hbar\Omega J_z$. The eigenstates of $\mathcal{H}_0$ are the angular momentum eigenstates $|jm\rangle$ which satisfy $\mathcal{H}_0|jm\rangle = m\hbar\Omega|jm\rangle$. Among the angular momentum eigenstates, the state $|jm\rangle$ with $m = -j$ has the lowest energy, $\mathcal{H}_0|j-j\rangle = -j\hbar\Omega|j-j\rangle$. When we apply a classical electromagnetic field to an assembly of atoms, the equation of motion for the atomic system becomes $i\hbar\partial_t|\psi(t)\rangle = (\hbar\Omega J_z + \hbar\lambda(t)J_+ + \hbar\lambda^*(t)J_-)|\psi(t)\rangle$. In the interaction picture, the collective pseudo-spin Hamiltonian takes on the form

$$\begin{aligned}\mathcal{H}_I &\equiv e^{i\mathcal{H}_0 t/\hbar}(\hbar\lambda(t)J_+ + \hbar\lambda^*(t)J_-)e^{i\mathcal{H}_0 t/\hbar} \\ &= \hbar\lambda(t)e^{i\Omega t}J_+ + \hbar\lambda^*(t)e^{-i\Omega t}J_-,\end{aligned} \tag{6.3}$$

and the equation of motion for the atomic system in the interaction picture becomes $i\hbar\partial_t|\psi_I(t)\rangle = (\hbar\lambda(t)e^{i\Omega t}J_+ + \hbar\lambda^*(t)e^{-i\Omega t}J_-)|\psi_I(t)\rangle$, where $|\psi_I(t)\rangle \equiv e^{i\mathcal{H}_0 t/\hbar}|\psi(t)\rangle = e^{i\Omega J_z t}|\psi(t)\rangle$. If one prepares the atomic system initially in the lowest energy state $|j-j\rangle$, one could obtain $|\psi_I(t)\rangle = e^{\zeta(t)J_+ - \zeta^*(t)J_-}|j-j\rangle$, where $\zeta(t) \equiv -i\lambda(t)e^{i\Omega t}$.

Comparing the electromagnetic field coherent states $|\alpha\rangle$ obtained from the vacuum state $|0\rangle$ by a unitary displacement operator $e^{\alpha a^\dagger - \alpha^* a}$, one can now define the spin coherent states $|j,\zeta\rangle$ of an assembly of atoms as the state obtained from the lowest energy state $|j-j\rangle$ by a unitary operator $e^{\zeta J_+ - \zeta^* J_-}$, where ζ is in general a time-dependent complex number.

One can expand the spin coherent states $|j,\zeta\rangle$ in terms of the angular momentum eigenstates $|jm\rangle$. Using the disentangling theorem for the angular momentum operators, the unitary operator $e^{\zeta J_+ - \zeta^* J_-}$ can be written as $e^{\tau J_+}e^{\ln(1+|\tau|^2)J_z}e^{-\tau^* J_-}$, where $\tau \equiv \tan|\zeta|e^{i\arg\zeta}$. Hence, one obtains

$$\begin{aligned}|j,\zeta\rangle &\equiv e^{\zeta J_+ - \zeta^* J_-}|j-j\rangle \\ &= e^{\tau J_+}e^{\ln(1+|\tau|^2)J_z}e^{-\tau^* J_-}|j-j\rangle \\ &= (1+|\tau|^2)^{-j}e^{\tau J_+}|j-j\rangle \\ &= (1+|\tau|^2)^{-j}\sum_{m=-j}^{j}\frac{(\tau J_+)^{j+m}}{(j+m)!}|j-j\rangle \\ &= (1+|\tau|^2)^{-j}\sum_{m=-j}^{j}\sqrt{\binom{2j}{j+m}}\tau^{j+m}|jm\rangle,\end{aligned} \tag{6.4}$$

by using the formula

$$|jm\rangle = \sqrt{\frac{(j-m)!}{(2j)!(j+m)!}} J_+^{j+m} |j-j\rangle \tag{6.5}$$

in the last step, where $\binom{2j}{j+m}$ is the binomial coefficient.

In analogy to the relationship between electromagnetic field coherent states and points on the complex plane, there also is a one-to-one correspondence between spin coherent states and points on the two-dimensional unit sphere, which is commonly known as the **Bloch representation**, or **Bloch sphere** (see Fig. 6.1). To establish the Bloch representation of spin coherent states, one can write the unitary operator $e^{\zeta J_+ - \zeta^* J_-}$ as $R_{\theta,\phi} = e^{-i\theta \mathbf{J}\cdot\hat{\mathbf{n}}} = e^{-i\theta(J_x \sin\phi - J_y \cos\phi)}$, where $J_\pm \equiv J_x \pm i J_y$, $\hat{\mathbf{n}} = (\sin\phi, -\cos\phi, 0)$, and $\zeta \equiv \frac{1}{2}\theta e^{-i\phi}$. Visually, the spin coherent states can be obtained from a rotation of the lowest energy state $|j-j\rangle$ through an angle θ about the $\hat{\mathbf{n}}$-axis. In other words, the spin coherent states $|j,\zeta\rangle$ can also be represented as $|\theta,\phi\rangle \equiv R_{\theta,\phi}|j-j\rangle$. As $|j-j\rangle$ is the common eigenstate of J^2 and J_z, the spin coherent state $|\theta,\phi\rangle$ is the common eigenstate of the spin operators J^2 and $R_{\theta,\phi} J_z R_{\theta,\phi}^{-1} = -J_k \sin\theta + J_z \cos\theta$:

$$(-J_k \sin\theta + J_z \cos\theta)|\theta,\phi\rangle = -j|\theta,\phi\rangle \text{ and } J^2|\theta,\phi\rangle = j(j+1)|\theta,\phi\rangle, \tag{6.6}$$

where $J_k \equiv J_x \cos\phi + J_y \sin\phi$. Hence, the spin coherent state $|\theta,\phi\rangle$ is simply the eigenstate of the spin operator $\mathbf{J}\cdot\hat{\mathbf{R}}$, where $\hat{\mathbf{R}} \equiv -\sin\theta\cos\phi\hat{\mathbf{x}} - \sin\theta\sin\phi\hat{\mathbf{y}} + \sin\theta\hat{\mathbf{z}}$ is a unit vector pointing in the direction of $(\pi+\theta,\phi)$ on the Bloch sphere. Now,

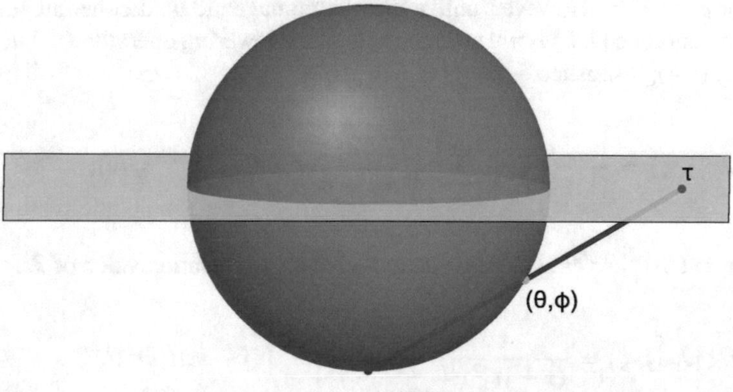

Fig. 6.1 Bloch representation of spin coherent states. The spin coherent state $|\theta,\phi\rangle$ is obtained from a rotation of the lowest energy state $|j-j\rangle$ through an angle θ about the $\hat{\mathbf{n}}$-axis, where $\hat{\mathbf{n}} \equiv (\sin\phi, -\cos\phi, 0)$ and $|j-j\rangle$ is represented by the north pole. Hence, the spin coherent state $|\theta,\phi\rangle$ corresponds one to one to a point (θ,ϕ) on the Bloch sphere, which is the inverse stereographic projection from $\tau \equiv \tan\frac{\theta}{2}e^{-i\phi}$ on the complex-τ plane onto the Bloch sphere

substitution of $\zeta = \frac{1}{2}\theta e^{-i\phi}$ into Eq. (6.4) yields $\tau = \tan\frac{\theta}{2}e^{-i\phi}$ and

$$|\theta,\phi\rangle = (1+|\tau|^2)^{-j}\sum_{m=-j}^{j}\tau^{j+m}\sqrt{\binom{2j}{j+m}}|jm\rangle \tag{6.7}$$

$$= \sum_{m=-j}^{j}\sqrt{\binom{2j}{j+m}}\left(\cos\frac{\theta}{2}\right)^{j-m}\left(\sin\frac{\theta}{2}e^{-i\phi}\right)^{j+m}|jm\rangle.$$

Finally, as we have previously discussed, if an assembly of atoms is initially prepared in the lowest energy state, the state of the collective pseudo-spin of the atoms in the interaction picture $|\psi_I(t)\rangle$ is a spin coherent state $|j,\zeta(t)\rangle$, where $\zeta(t) \equiv -i\lambda(t)e^{i\Omega t}$. Hence, a direct computation yields

$$\theta(t) = 2|\lambda(t)| \text{ and } \phi(t) = \pi/2 - \arg\lambda(t) - \Omega t. \tag{6.8}$$

Using the formula $|\psi_I(t)\rangle = e^{i\Omega J_z t}|\psi(t)\rangle$ and Eq. (6.7), we obtain

$$|\psi(t)\rangle = e^{-i\Omega J_z t}|\theta(t),\phi(t)\rangle = |\theta(t),\phi(t)+\Omega t\rangle e^{ij\Omega t}, \tag{6.9}$$

Clearly, as long as the atomic system is initially prepared in the lowest energy state $|j-j\rangle$, the state of the collective pseudo-spin of the atoms to within a time-dependent phase factor will always be a spin coherent state.

In the above, we demonstrated that a spin coherent state $|j,\zeta\rangle$ for an assembly of atoms can be generated in a similar way as the coherent states for an electromagnetic field, from the lowest energy state $|j,-j\rangle$ by the application of the displacement operator $e^{\zeta J_+ - \zeta^* J_-}$. However, unlike the electromagnetic field coherent states, the spin coherent state $|j,\zeta\rangle$ *is not* an eigenstate of the lowering operator J_-, but instead from Eq. (6.4); it satisfies

$$J_-|j,\zeta\rangle = \frac{\tau}{(1+|\tau|^2)^j}\sum_{m=-j}^{j-1}\sqrt{\binom{2j}{j+m}}(j-m)\tau^{j+m}|jm\rangle, \tag{6.10}$$

where $\tau \equiv \tan|\zeta|e^{i\arg\zeta}$. One can easily derive the expectation value of J_-

$$\langle j,\zeta|J_-|j,\zeta\rangle = \frac{\tau}{(1+|\tau|^2)^{2j}}\sum_{m=-j}^{j-1}\binom{2j}{j+m}(j-m)(|\tau|^2)^{j+m} \tag{6.11}$$

$$= \frac{2j\tau}{(1+|\tau|^2)^{2j}}\sum_{m=-j}^{j-1}\binom{2j-1}{j+m}(|\tau|^2)^{j+m}$$

$$= \frac{2j\tau}{1+|\tau|^2},$$

and similarly the expectation values of J_+ and J_z

$$\langle j, \zeta | J_+ | j, \zeta \rangle = \frac{2j\tau^*}{1+|\tau|^2}, \tag{6.12a}$$

$$\langle j, \zeta | J_z | j, \zeta \rangle = \frac{j(|\tau|^2 - 1)}{1+|\tau|^2}. \tag{6.12b}$$

The expectation values of J_x, J_y, and J_z in the Bloch sphere representation of spin coherent states are then given by

$$\langle \theta, \phi | J_x | \theta, \phi \rangle = j \sin\theta \cos\phi, \tag{6.13a}$$

$$\langle \theta, \phi | J_y | \theta, \phi \rangle = j \sin\theta \sin\phi, \tag{6.13b}$$

$$\langle \theta, \phi | J_z | \theta, \phi \rangle = -j \cos\theta. \tag{6.13c}$$

One of the intriguing properties of the field coherent states is that they are necessarily all minimum uncertainty states which satisfy Heisenberg's uncertainty relation $\Delta x \Delta p = 1/2$. For general spin states, a similar uncertainty relation exists, $\Delta J_x \Delta J_y \geq \frac{1}{2}|\langle J_z \rangle|$. Hence, one can ask whether the spin coherent states satisfy this uncertainty relation for the spin operators. To this end, one notices that in the Bloch sphere representation of spin coherent states, the expectations values of J_x^2 and J_y^2 are

$$\langle \theta, \phi | J_x^2 | \theta, \phi \rangle = \frac{j}{2} + \frac{j(2j-1)}{2} (\sin\theta \cos\phi)^2, \tag{6.14a}$$

$$\langle \theta, \phi | J_y^2 | \theta, \phi \rangle = \frac{j}{2} + \frac{j(2j-1)}{2} (\sin\theta \sin\phi)^2. \tag{6.14b}$$

Comparison between Eqs. (6.13a)–(6.13b) and Eqs. (6.14a)–(6.14b) yields

$$\Delta J_x^2 \equiv \langle \theta, \phi | J_x^2 | \theta, \phi \rangle - \langle \theta, \phi | J_x | \theta, \phi \rangle^2 = \frac{j}{2} \left[1 - (\sin\theta \cos\phi)^2 \right], \tag{6.15a}$$

$$\Delta J_y^2 \equiv \langle \theta, \phi | J_y^2 | \theta, \phi \rangle - \langle \theta, \phi | J_y | \theta, \phi \rangle^2 = \frac{j}{2} \left[1 - (\sin\theta \sin\phi)^2 \right]. \tag{6.15b}$$

Hence, we obtain $\Delta J_x^2 \Delta J_y^2 - \frac{1}{4}|\langle \theta, \phi | J_z | \theta, \phi \rangle|^2 = \frac{1}{4} j^2 \sin^4\theta \cos^2\phi \sin^2\phi$, which vanishes only when $\theta = 0$, or $\phi = n\pi/2$, where n is an arbitrary integer. In other words, unlike the field coherent states, which satisfy Heisenberg's uncertainty relation without any condition, the spin coherent state $|j, \zeta\rangle$ satisfies the uncertainty relation $\Delta J_x \Delta J_y \geq \frac{1}{2}|\langle J_z \rangle|$ only when $\zeta = \frac{1}{2}\theta e^{-i\phi}$ is real or purely imaginary.

As we have shown in the last chapter, any two field coherent states are strictly non-orthogonal in the sense that their scalar product never vanishes. But for two spin coherent states $|j, \zeta\rangle$ and $|j, \zeta'\rangle$, we have

$$\langle j, \zeta | j, \zeta'\rangle = \frac{1}{(1+|\tau|^2)^j(1+|\tau'|^2)^j} \sum_{m=-j}^{j} \binom{2j}{j+m} (\tau^* \tau')^{j+m} \tag{6.16}$$

$$= \frac{(1+\tau^*\tau')^{2j}}{(1+|\tau|^2)^j(1+|\tau'|^2)^j},$$

which implies that the two states $|j, \zeta\rangle$ and $|j, \zeta'\rangle$ are orthogonal when $\tau' = -1/\tau^*$. In other words, there is only one single point on the Bloch sphere representing the orthogonal state of a given spin coherent state, which is the antipodal point $(\theta', \phi') = (\pi - \theta, \pi + \phi)$. Substitution of $\tau' = \tan\frac{\theta'}{2} e^{-i\phi'}$ and $\tau = \tan\frac{\theta}{2} e^{-i\phi}$ into Eq. (6.16) yields

$$|\langle j, \zeta | j, \zeta'\rangle| = \left| \cos\frac{\theta'}{2}\cos\frac{\theta}{2} + \sin\frac{\theta'}{2}\sin\frac{\theta}{2} e^{-i(\phi'-\phi)} \right|^{2j} \tag{6.17}$$

$$= \left\{ \frac{1}{2}\left[1 + \cos\theta'\cos\theta + \sin\theta'\sin\theta\cos(\phi'-\phi)\right] \right\}^j = \cos^{2j}\left(\frac{\Theta}{2}\right),$$

where Θ is the angle between the two vectors (θ', ϕ') and (θ, ϕ) on the Bloch sphere. With the above discussions, we see clearly that two spin coherent states $|j, \zeta\rangle$ and $|j, \zeta'\rangle$ are orthogonal only when $\Theta = \pi$, that is, (θ', ϕ') is the antipodal direction of (θ, ϕ). We also note that, for large quantum number j, the overlap between two spin coherent states can be very small, even when $\Theta \ll 1$. Hence, in the large-j limit, different spin coherent states are well-separated. They have small overlap and tend to approximately orthogonal to one another.

We now further examine the completeness properties of the spin coherent states. One may show that the identity operator can be resolved with respect to the spin coherent states. Using Eq. (6.7), one obtains

$$\frac{1}{4\pi}\int d\Omega |\theta, \phi\rangle\langle\theta, \phi| \tag{6.18}$$

$$= \frac{1}{4\pi}\int d\Omega \sum_{m'=-j}^{j}\sum_{m=-j}^{j} e^{i(m'-m)\phi} \sqrt{\binom{2j}{j+m}}\sqrt{\binom{2j}{j+m'}}$$

$$\left(\cos\frac{\theta}{2}\right)^{2j-m'-m}\left(\sin\frac{\theta}{2}\right)^{2j+m'+m} |jm'\rangle\langle jm|.$$

where $d\Omega = \sin\theta d\theta d\phi$ is the differential solid angle in spherical coordinates. Using the integral $\int_0^{2\pi} e^{i(m-m')\phi} d\phi = 2\pi\delta_{mm'}$, and performing a change of variable $s =$

$\sin^2 \frac{\theta}{2}$, Eq. (6.18) becomes

$$\begin{aligned}
&\frac{1}{4\pi}\int d\Omega|\theta,\phi\rangle\langle\theta,\phi| \qquad (6.19)\\
&=\int_0^{\pi}\frac{\sin\theta}{2}d\theta\sum_{m=-j}^{j}\binom{2j}{j+m}\left(\cos^2\frac{\theta}{2}\right)^{j-m}\left(\sin^2\frac{\theta}{2}\right)^{j+m}|jm\rangle\langle jm|\\
&=\int_0^1 ds\sum_{m=-j}^{j}\binom{2j}{j+m}s^{j+m}(1-s)^{j-m}|jm\rangle\langle jm|\\
&=\sum_{m=-j}^{j}\binom{2j}{j+m}\frac{(j+m)!(j-m)!}{(2j+1)!}|jm\rangle\langle jm|\\
&=\frac{1}{2j+1}\sum_{m=-j}^{j}|jm\rangle\langle jm|.
\end{aligned}$$

Using the completeness properties of the angular momentum eigenstates, the identity operator is resolved in terms of the spin coherent states

$$\frac{2j+1}{4\pi}\int d\Omega|\theta,\phi\rangle\langle\theta,\phi| = \sum_{m=-j}^{j}|jm\rangle\langle jm| = I_{2j+1}. \qquad (6.20)$$

Hence, a general spin state $|\psi\rangle$ can always be expanded in terms of the spin coherent states as

$$\begin{aligned}
|\psi\rangle &= \sum_{m=-j}^{j} c_m|jm\rangle = \frac{2j+1}{4\pi}\int d\Omega\sum_{m=-j}^{j}|j,\zeta\rangle\langle j,\zeta|jm\rangle \qquad (6.21)\\
&= \frac{2j+1}{4\pi}\int d\Omega\frac{f(\tau^*)}{(1+|\tau|^2)^j}|j,\zeta\rangle,
\end{aligned}$$

where $f(\tau^*) \equiv (1+|\tau|^2)^j\langle j,\zeta|\psi\rangle = \sum_{m=-j}^{j} c_m\binom{2j}{j+m}^{1/2}(\tau^*)^{j+m}$ is the spin coherent state representation of $|\psi\rangle$ and is a polynomial of τ^* of degree $2j$.

6.2 Geometric Phases for Spin Coherent States

In the last section, we discussed some basic properties of spin coherent states. In the following, we will use these properties to introduce the important concept of geometric phases for spin coherent states.

The concept of geometric phases was first introduced by Michael Berry in his paper of 1984 titled “Quantal phase factors accompanying adiabatic changes” [80], in which he showed that a quantum state undergoing cyclic adiabatic evolution will accumulate a geometric phase factor in addition to the conventional dynamical phase factor, which depends only on the contour accompanying adiabatic changes in the parameter space. Berry’s phase for cyclic adiabatic changes of parameters was soon used to provide an elegant explanation of the integer quantization of Hall conductance. The assumption of adiabatic changes was later removed by Yakir Aharonov and Jeeva Anandan in their paper titled “Phase change during a cyclic quantum evolution” [81] published in 1987. In particular, for spin coherent states, as we will show below, the geometric phase equals to the solid angle subtended by the tip of the spin on Bloch sphere multiplied by the spin quantum number.

We now introduce Aharonov and Anandan’s geometric phase in more detail. Let $|\psi(t)\rangle$ be a quantum state undergoing cyclic evolution, and then the initial and final states are related by $|\psi(T)\rangle = e^{i\Phi}|\psi(0)\rangle$, where Φ is the total phase associated with the cyclic evolution of $|\psi(t)\rangle$. To evaluate Φ, we set $|\psi(t)\rangle \equiv e^{if(t)}|\phi(t)\rangle$ with $f(T) - f(0) \equiv \Phi$. It follows that $|\phi(t)\rangle$ undergoes a periodic motion with $|\phi(T)\rangle = |\phi(0)\rangle$. Substituting $|\psi(t)\rangle \equiv e^{if(t)}|\phi(t)\rangle$ into the Schrödinger equation of $|\psi(t)\rangle$, we obtain

$$-\dot{f} = \hbar^{-1}\langle\psi(t)|\hat{H}|\psi(t)\rangle - i\langle\phi(t)|\dot{\phi}(t)\rangle. \tag{6.22}$$

After integration, the initial and final states are found to be related by

$$|\psi(T)\rangle = \exp\left\{-\frac{i}{\hbar}\int_0^T \langle\psi|\hat{H}|\psi\rangle dt\right\}\exp\left\{-\int_0^T \langle\phi|\dot{\phi}\rangle dt\right\}|\psi(0)\rangle, \tag{6.23}$$

where the first exponential is the conventional dynamical phase factor and the second exponential is a geometric phase factor which only depends on the closed trajectory swept out by $|\phi\rangle$ in the Hilbert space.

As we shall see, the meaning of the geometric phase factor will be clearer by considering spin coherent states. Using the definition of spin coherent states, $|j,\zeta\rangle = (1+|\tau|^2)^{-j}e^{\tau J_+}|j-j\rangle$, a direct computation yields

$$\frac{d}{dt}|j,\zeta\rangle = \left(\frac{-j(\tau^*\dot{\tau}+\tau\dot{\tau}^*)}{1+|\tau|^2} + \dot{\tau}J_+\right)|j,\zeta\rangle. \tag{6.24}$$

Using Eq. (6.12a), we obtain

$$\langle j,\zeta|\frac{d}{dt}|j,\zeta\rangle = \frac{-j(\tau^*\dot{\tau}+\tau\dot{\tau}^*)}{1+|\tau|^2} + \frac{2j\tau^*\dot{\tau}}{1+|\tau|^2} = \frac{j(\tau^*\dot{\tau}-\tau\dot{\tau}^*)}{1+|\tau|^2}. \tag{6.25}$$

Using the relation $\tau = \tan\frac{\theta}{2}e^{-i\phi}$, the geometric phase factor becomes

$$e^{i\gamma} \equiv \exp\left\{-j\int_0^T \frac{\tau^*\dot{\tau} - \tau\dot{\tau}^*}{1+|\tau|^2}dt\right\} = \exp\left\{ij\int_C (1-\cos\theta)d\phi\right\}, \tag{6.26}$$

where C is the contour traced by the spin on the Bloch sphere. Equation (6.26) shows that the geometric phase γ for spin coherent states equals to the spin quantum number j times the solid angle $\Omega(C) \equiv \int_C(1-\cos\theta)d\phi$ enclosed by the spin trajectory on the Bloch sphere.

We now proceed to the geometric phases for a superposition of two spin coherent states, $|\phi\rangle = c_1|j,\zeta_1\rangle + c_2|j,\zeta_2\rangle$. A direct computation yields

$$\begin{aligned}\langle\phi|\dot{\phi}\rangle = &\sum_i |c_i|^2\langle j,\zeta_i|\frac{d}{dt}|j,\zeta_i\rangle + c_1^*c_2\frac{d}{dt}(\langle j,\zeta_1|j,\zeta_2\rangle) \\ &+ 2i\Im[c_2^*c_1\langle j,\zeta_2|\frac{d}{dt}|j,\zeta_1\rangle],\end{aligned} \tag{6.27}$$

where the first term in Eq. (6.27) is contributed from the trajectories of the individual spin coherent states on Bloch sphere, the second term is a total derivative which vanishes after integration over a whole period, and the third term is contributed from the interference between the spin coherent states $|j,\zeta_1\rangle$ and $|j,\zeta_2\rangle$. Using Eqs. (6.10) and (6.16), we obtain

$$\begin{aligned}\langle j,\zeta_2|\frac{d}{dt}|j,\zeta_1\rangle &= \frac{-j(\tau_1^*\dot{\tau}_1 + \tau_1\dot{\tau}_1^*)}{1+|\tau_1|^2}\langle j,\zeta_2|j,\zeta_1\rangle + \dot{\tau}_1\langle j,\zeta_2|J_+|j,\zeta_1\rangle \\ &= \left(\frac{-j(\tau_1^*\dot{\tau}_1 + \tau_1\dot{\tau}_1^*)}{1+|\tau_1|^2} + \frac{2j\tau_2^*\dot{\tau}_1}{1+\tau_2^*\tau_1}\right)\langle j,\zeta_2|j,\zeta_1\rangle.\end{aligned} \tag{6.28}$$

From Eqs. (6.27) and (6.28), we see that when two spin coherent states $|j,\zeta_1\rangle$ and $|j,\zeta_2\rangle$ are orthogonal, the geometric phase for their superposition simply equals to the summation of the solid angles subtended by the individual spin coherent state trajectories on the Bloch sphere multiplied by the spin quantum number, $\gamma \equiv i\int_0^T\langle\phi|\dot{\phi}\rangle dt = j\sum_i |c_i|^2\Omega(C_i)$. For the general case, using the relations $\tau_i \equiv \tan\frac{\theta_i}{2}e^{-i\phi_i}$ and $\langle j,\zeta_2|j,\zeta_1\rangle = \cos\frac{\Theta}{2}e^{iA}$, we obtain

$$\begin{aligned}\langle j,\zeta_2|\frac{d}{dt}|j,\zeta_1\rangle = j\left(\cos\frac{\Theta}{2}e^{\frac{iA}{2}}\right)^{2j-1}&\left[\sin\frac{\Theta}{2}e^{\frac{iA'}{2}}d\theta_1\right. \\ &\left.-2i\sin\frac{\theta_1}{2}\sin\frac{\theta_2}{2}e^{i(\phi_2-\phi_1)}d\phi_1\right],\end{aligned} \tag{6.29}$$

where Θ is the angle between the two vectors (θ_1,ϕ_1) and (θ_2,ϕ_2) on the Bloch sphere; A is the area of the spherical triangle with points (θ_1,ϕ_1), (θ_2,ϕ_2), and the north pole; and A' is the area of the spherical triangle with points $(\pi - \theta_1, \pi + \phi_1)$, (θ_2, ϕ_2), and the north pole. For a superposition of two spin coherent states with opposite amplitudes, $|\phi\rangle = c_1|j, \zeta_1\rangle + c_2|j, -\zeta_1\rangle$, we have $A = A' = 0$ and $\Theta = 2\theta_1$. Hence, the geometric phase for $|\phi\rangle$ becomes

$$\gamma = j \sum_i |c_i|^2 \Omega(C_i) - 2j\Re(c_2^* c_1) \int_{C_1} \left(\cos\frac{\Theta}{2}\right)^{2j-1} d\Omega_1, \tag{6.30}$$

where $d\Omega_1 = (1 - \cos\theta_1)d\phi_1$ is the differential solid angle subtended by the vector (θ_1, ϕ_1) on the Bloch sphere. From Eq. (6.30), we notice that the geometric phase for a superposition of spin coherent states with opposite amplitudes equals to the summation of solid angles subtended by the individual spin coherent state trajectories plus an additional solid angle weighted by the spherical distance between their Bloch representations.

6.3 Spin Cat States

Over the last three decades, there has been a steady growth of interest in the generation of different kinds of quantum states that consist of a superposition of macroscopic distinct states, which are known as the **Schrödinger cat states**. These states reveal the unique peculiar aspect of quantum mechanics—the preparation of a state smeared over two and more distinct values is feasible but is very unlikely to happen at the classical level. Thus, the preparation and manipulation of such superposition states are of fundamental importance and will enhance our understanding of the foundations of quantum mechanics. In Chap. 3, we have already discussed some properties of the Schrödinger cat states in quantum optics which consist of superpositions of two distinct field coherent states with opposite amplitudes. In the following, we will discuss another kind of cat states—the superposition of spin coherent states which involves an assembly of atoms.

An experimental proposal for the generation of spin cat states was first presented by J. I. Cirac and P. Zoller in a seminal paper titled "Preparation of macroscopic superpositions in many-atom systems" [82] published in 1994. It is based on the interaction between the atoms and a cavity mode. The Cirac-Zoller proposal on generating spin cat states by using cavity quantum electrodynamics was later refined in the published papers [83–85].

To begin, let us consider a collection of N two-level atoms with a transition frequency ω_a confined to a cavity which supports a single mode of quantized electromagnetic field of frequency ω_c. We introduce the many-atomic operators $J_\pm \equiv \sum_{i=1}^N \sigma_\pm^{(i)}$ and $J_z \equiv \frac{1}{2}\sum_{i=1}^N \sigma_z^{(i)}$ as in Chap. 6, where $\sigma_\pm^{(i)}$ and $\sigma_z^{(i)}$ are the Pauli operators for a single atom. The many-atomic operators obey the usual angular momentum commutation relations: $[J_+, J_-] = 2J_z$ and $[J_z, J_\pm] = \pm J_\pm$.

The atom-field interaction is described by

$$H = \hbar\omega_c a^\dagger a + \frac{\hbar\omega_a}{2} J_z + \hbar g(a^\dagger J_- + a J_+). \tag{6.31}$$

The Hamiltonian (6.31) describes the exchange of energy between the internal states of the atoms in a cavity and the photon confined to the cavity and is commonly known as the **Jaynes-Cummings Hamiltonian** [86–88], where g is the atom-field coupling constant proportional to the dipole moment of the atoms. In the dispersive regime where the atomic transition frequency is far out of resonance with the cavity field, we may assume that the detuning $\Delta \equiv \omega_a - \omega_c$ is much larger than the atom-field coupling constant g. Under this assumption, the Jaynes-Cummings Hamiltonian can be approximately diagonalized using a unitary transformation $H' = UHU^\dagger$, where $U = \exp\left[\frac{g}{\Delta}(aJ_+ - a^\dagger J_-)\right]$. After expanding to second order in g, we obtain the effective Hamiltonian in the dispersive regime [89,90]

$$H' = \hbar\,(\omega_c + \chi J_z)\, a^\dagger a + \frac{\hbar}{2}\,(\omega_a + \chi)\, J_z, \tag{6.32}$$

where $\chi \equiv g^2/\Delta$ is the dispersive coupling constant. From Eq. (6.32), we see that the atomic transition frequency is shifted by χ, known as the **Lamb shift**, and the cavity frequency is shifted by χm, depending on the internal states of the atoms, and is known as the **Stark shift**. The linear dispersive approximation is valid only when $4g^2\bar{n} \ll \Delta^2$, i.e., the mean photon number does not exceed the critical value $n_c \equiv \frac{\Delta^2}{4g^2}$. Let us denote $H' = H_0' + H_1'$ with $H_0' \equiv \hbar\omega_c a^\dagger a + \frac{\hbar}{2}(\omega_a + \chi)J_z$ and $H_1' \equiv \hbar\chi J_z a^\dagger a$. Then we see that the dispersive coupling H_1' commutes with the atomic and cavity Hamiltonian H_0'. Hence, in the interaction picture, the atom-field interaction $H_I' \equiv e^{iH_0't/\hbar} H_1' e^{-iH_0't/\hbar}$ still equals to H_1'. If the quantized cavity field is initially prepared in a coherent state, and the atoms are prepared in a spin coherent state, the initial state $|\psi_I(t=0)\rangle = |j,\zeta\rangle|\alpha\rangle = \sum_{m=-j}^{j} c_m|jm\rangle|\alpha\rangle$ in the interaction picture will evolve into the state after time t

$$\begin{aligned} |\psi_I(t)\rangle &\equiv e^{-iH_1't/\hbar}|\psi_I(t=0)\rangle = \sum_{m=-j}^{j} c_m e^{-i\chi m t a^\dagger a}|jm\rangle|\alpha\rangle \\ &= \sum_{m=-j}^{j} c_m|jm\rangle|\alpha e^{-i\chi m t}\rangle, \end{aligned} \tag{6.33}$$

where $j \equiv N/2$ and $c_m \equiv (1+|\tau|^2)^{-j}\sqrt{\binom{2j}{j+m}}\tau^{j+m}$ for spin coherent states. Equation (6.33) shows that the atoms and the cavity mode are entangled through the dispersive interaction.

Let us first discuss the case where both j and m are integers. If we stop the dispersive interaction after time $T \equiv \pi/\chi$, the state $|\psi_I(T)\rangle$ then has the form

$$|\psi_I(T)\rangle = \sum_{m=-j}^{j} c_m|jm\rangle|(-1)^m\alpha\rangle \tag{6.34}$$
$$= \frac{1}{2N_+}|j,\zeta\rangle_+|\alpha\rangle + \frac{1}{2N_-}|j,\zeta\rangle_-|-\alpha\rangle,$$

where $|j,\zeta\rangle_\pm \equiv N_\pm(|j,\zeta\rangle \pm (-1)^j|j,-\zeta\rangle)$ are the atomic counterparts of the even and odd cat states of the harmonic oscillator and hence can be referred to the even and odd spin cat states. Here, $N_\pm$ are the normalization factor given by

$$N_\pm = \frac{1}{\sqrt{2}}\left[1 \pm (-1)^j \left(\frac{1-|\tau|^2}{1+|\tau|^2}\right)^{2j}\right]^{-1/2}. \tag{6.35}$$

Now, the remaining task is to detect the cavity field and projects the atoms into the corresponding spin coherent states. However, here is a difficulty arising: as the quantized electromagnetic field is trapped in a low-loss cavity during the entire operating time to form the cat states, direct access to the field is forbidden, and one can only infer the properties of the field from the detection of auxiliary atoms coupled to the field when sent through the cavity. One way to solve the issue is described by Brune et al. [91]: after the generation of collective spin cat states, the cavity will be coupled to the same classical radiation source as the one which has used initially to generate the coherent state $|\alpha\rangle$, so that a reference coherent state $|\alpha_r\rangle$ is added to the field coherent state in the cavity mode, i.e., $|\alpha\rangle \to |\alpha+\alpha_r\rangle$ and $|-\alpha\rangle \to |-\alpha+\alpha_r\rangle$, where the amplitude of the reference coherent states can be adjusted by varying the interaction time interval. The coupling between the classical source and the cavity has to be weak enough to ensure low cavity losses. This field amplitude superposition mechanism has already emphasized by Glauber in his remarkable paper titled "*Coherent and Incoherent States of the Radiation Field*" [4]. In his original words, he wrote that: "The amplitudes of successive coherent excitations of the mode add as complex numbers in quantum theory, just as they do in classical theory." To this end, let us choose $\alpha_r = \alpha$, and then after the superposition, the entangled atomic-field state becomes

$$|\psi_I'(T)\rangle = \frac{1}{2N_+}|j,\zeta\rangle_+|2\alpha\rangle + \frac{1}{2N_-}|j,\zeta\rangle_-|0\rangle. \tag{6.36}$$

Now, the quantized cavity field is in a superposition of the vacuum state with a classical state. In other words, either empty or filled with coherent field, depending on whether the atoms are in even or odd spin cat states. Finally, we send through the cavity a reference two-level atom which has an atomic transition frequency resonant with the cavity field. If we assume the reference atom is in the excited state when it enters the cavity, then according to the Jaynes-Cummings model [86], we obtain

$$|\psi_I''(T)\rangle = \frac{1}{2N_+}|j,\zeta\rangle_+(|\psi_g(t)\rangle|g_r\rangle + |\psi_e(t)\rangle|e_r\rangle) + \frac{1}{2N_-}|j,\zeta\rangle_-|0\rangle|g_r\rangle, \tag{6.37}$$

where $|g_r\rangle$ and $|e_r\rangle$ are the ground and excited states of the reference atom respectively, t is the interaction time, and

$$|\psi_g(t)\rangle = -i\sum_{n=0}^{\infty} C_n(2\alpha)\sin(\lambda t\sqrt{n+1})|n+1\rangle, \tag{6.38a}$$

$$|\psi_e(t)\rangle = \sum_{n=0}^{\infty} C_n(2\alpha)\cos(\lambda t\sqrt{n+1})|n\rangle, \tag{6.38b}$$

where $C_n(z) = e^{-\frac{|z|^2}{2}} z^n/\sqrt{n!}$. Hence, the detection of the reference atom in the excited state reduces the collective atomic state into the even spin cat state $|j,\zeta\rangle_+$. Similarly, if one were to choose the amplitude of the reference coherent state as $\alpha_r = -\alpha$, one will obtain the odd spin cat state $|j,\zeta\rangle_-$, provided that the reference atom is detected in the excited state.

In the above, we have discussed a simple experimental scheme for the generation of Schrödinger cat states for an assembly of two-level atoms based on dispersive interaction between the atoms and the field in a low-loss cavity along with state reduction. In the following, we will discuss in detail the intriguing properties for these states by use of quasi-probability distributions such as the Husimi Q- and the Wigner functions [92, 93].

We first discuss the Husimi Q-distribution $Q(\theta,\phi) \equiv \frac{2j+1}{4\pi}|\langle\theta,\phi|j,\zeta\rangle_\pm|^2$ for even and odd spin cat states $|j,\zeta\rangle_\pm \equiv N_\pm(|j,\zeta\rangle \pm (-1)^j|j,-\zeta\rangle)$, where $N_\pm$ is a normalization constant given by Eq. (6.35) and the Q-function is normalized as $\int Q(\theta,\phi)\sin\theta d\theta d\phi = 1$. A direct computation yields

$$\begin{aligned} Q(\theta,\phi) = \frac{2j+1}{4\pi}N_\pm^2(&|\langle\theta,\phi|j,\zeta\rangle|^2 \pm (-1)^j\langle\theta,\phi|j,-\zeta\rangle\langle j,\zeta|\theta,\phi\rangle \\ &\pm (-1)^j\langle\theta,\phi|j,\zeta\rangle\langle j,-\zeta|\theta,\phi\rangle + (-1)^{2j}|\langle\theta,\phi|j,-\zeta\rangle|^2). \end{aligned} \tag{6.39}$$

Without loss of generality, we could assume $\zeta = \beta/2$, so that $N_\pm^2 = \frac{1}{2}[1 \pm (-\cos^2\beta)^j]^{-1}$. Using Eqs. (6.16) and (6.17), we obtain

$$\langle\theta,\phi|j,\pm\zeta\rangle\langle j,\mp\zeta|\theta,\phi\rangle = \left\{\frac{1}{2}(\cos\beta + \cos\theta \pm i\sin\beta\sin\theta\sin\phi)\right\}^{2j}, \tag{6.40a}$$

$$|\langle\theta,\phi|j,\pm\zeta\rangle|^2 = \left\{\frac{1}{2}(1+\cos\beta\cos\theta \pm \sin\beta\sin\theta\cos\phi)\right\}^{2j}. \tag{6.40b}$$

In particular, for $\beta = \pi/2$, we have $N_{\pm}^2 = 1/2$, and hence the Q-function for even and odd spin cat states can be explicitly written as [85]

$$\begin{aligned} Q(\theta,\phi) = \frac{2j+1}{4\pi}\frac{1}{2^{N+1}}\{&(\sin\theta\cos\phi+1)^N + (\sin\theta\cos\phi-1)^N \\ &\pm(-1)^{N/2}[(\cos\theta+i\sin\theta\sin\phi)^N + (\cos\theta - i\sin\theta\sin\phi)^N]\}, \end{aligned} \tag{6.41}$$

where $N = 2j$ is the number of atoms. In Eq. (6.41), the first two terms are the contributions from the two spin coherent states, respectively, and the last two terms represent the quantum interference between them. The existence of the quantum interference terms clearly reveals the quantum nature of the spin cat states. Experimentally, one may obtain the atomic quasi-probability Q-function by performing a spin-coherent states positive operator-valued measure (POVM) via a sequence of weak measurements [94]. In Fig. 6.2, we show the spherical polar plot of the quasi-probability Q-function for even spin cat states with $N = 10$ and

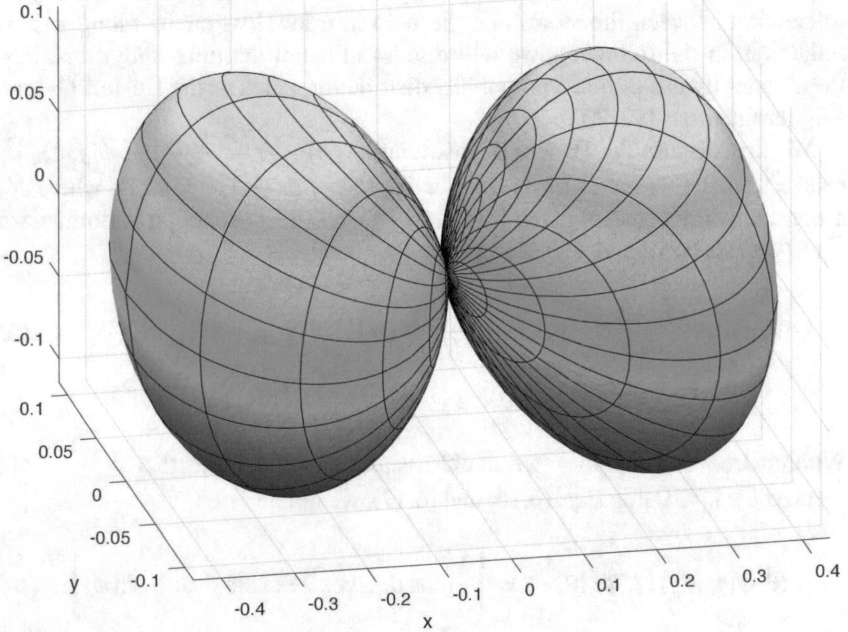

Fig. 6.2 Spherical polar plot of the quasi-probability Q-function $Q(\theta,\phi) \equiv \frac{N+1}{4\pi}|\langle\theta,\phi|j,\zeta\rangle_+|^2$ for even spin cat states with $N = 10$ and $\zeta = \pi/4$, where $x \equiv Q(\theta,\phi)\sin\theta\cos\phi$, $y \equiv Q(\theta,\phi)\sin\theta\sin\phi$, and $z \equiv Q(\theta,\phi)\cos\theta$

$\beta = \pi/2$, i.e., a unit sphere deforms into the isosurface of the quasi-probability Q-function. As the Q-function is non-negative and real, the nonclassical properties of the spin cat states is manifested by those values of $Q(\theta, \phi)$ that approaches zero, or equivalently by those regions of sphere which dramatically shrink to the origin. In Fig. 6.2, the quantum nature of the spin cat states is clearly seen from the dumbbell-like shape in the quasi-probability distribution $Q(\theta, \phi)$. Let us denote $Q_0(\theta, \phi) \equiv \frac{2j+1}{4\pi} N_{\pm}^2 (|\langle \theta, \phi | j, \zeta \rangle|^2 + (-1)^N |\langle \theta, \phi | j, -\zeta \rangle|^2)$ as the summation of the Q-functions for the spin coherent states $|j, \zeta\rangle$ and $|j, -\zeta\rangle$, which corresponds to the contribution to $Q(\theta, \phi)$ without quantum interferences. In Fig. 6.3, we show the spherical polar plot of the ratio $\eta(\theta, \phi) \equiv Q(\theta, \phi)/Q_0(\theta, \phi)$ for even spin cat states with $N = 10$ and $\beta = \pi/2$. The delicate quantum interference structure in the quasi-probability distribution $Q(\theta, \phi)$ is clearly displayed in the decagram-like shape ratio between $Q(\theta, \phi)$ and $Q_0(\theta, \phi)$, where there are well-separated humps and hollows on the surface of the unit sphere.

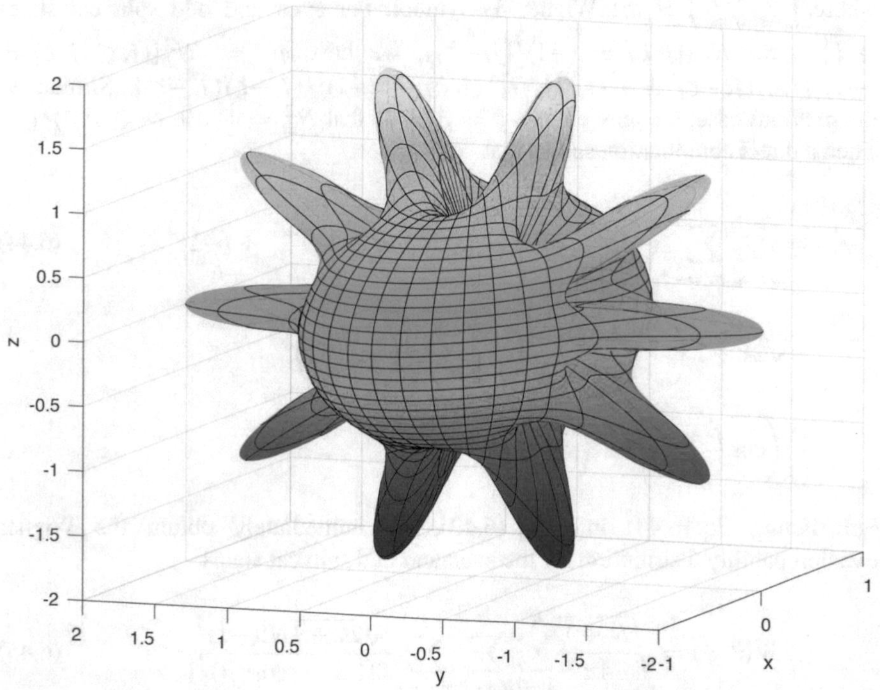

Fig. 6.3 Spherical polar plot of the ratio $\eta(\theta, \phi)$ between $Q(\theta, \phi)$ and $Q_0(\theta, \phi)$ for even spin cat states with $N = 10$ and $\zeta = \pi/4$, where $x \equiv \eta(\theta, \phi) \sin\theta \cos\phi$, $y \equiv \eta(\theta, \phi) \sin\theta \sin\phi$, and $z \equiv \eta(\theta, \phi) \cos\theta$

We now discuss an equally important quasi-probability distribution—the Wigner distribution—for spin cat states. The Wigner function $W(\theta, \phi)$ for an arbitrary density operator ρ may be expanded in terms of the spherical harmonics $Y_{kq}(\theta, \phi)$

$$W(\theta, \phi) \equiv \sqrt{\frac{N+1}{4\pi}} \sum_{k=0}^{N} \sum_{q=-k}^{k} \rho_{kq} Y_{kq}(\theta, \phi), \tag{6.42}$$

where $N = 2j$ is the number of atoms, $\rho_{kq} \equiv \mathrm{Tr}(T_{kq}^{\dagger}\rho)$ are the characteristic matrices of ρ, and T_{kq} are the spherical tensor operators defined by [95]

$$T_{kq} \equiv \sum_{m,m'=-j}^{j} (-1)^{j-m}\sqrt{2k+1} \begin{pmatrix} j & k & j \\ -m & q & m' \end{pmatrix} |jm\rangle\langle jm'|, \tag{6.43}$$

where $\begin{pmatrix} j & k & j \\ -m & q & m' \end{pmatrix}$ is the Wigner $3j$-symbol. For even and odd spin cat states $|j, \zeta\rangle_{\pm} \equiv N_{\pm}(|j, \zeta\rangle \pm (-1)^j |j, -\zeta\rangle)$, we have $\rho = N_{\pm}^2[|j, \zeta\rangle\langle j, \zeta| \pm (-1)^j |j, \zeta\rangle\langle j, -\zeta| \pm (-1)^j |j, -\zeta\rangle\langle j, \zeta| + (-1)^{2j}|j, -\zeta\rangle\langle j, -\zeta|]$. Similar to the previous case, we may choose $\zeta = \beta/2$, so that $N_{\pm}^2 = \frac{1}{2}[1 \pm (-\cos^2\beta)^j]^{-1}$. Then a direct computation shows that

$$\begin{aligned}\rho_{kq} =& N_{\pm}^2 \sum_{m,m'=-j}^{j} (-1)^{j-m}[1 \pm (-1)^{N+m'} \pm (-1)^{N+m} + (-1)^{m+m'}] \\ & \sqrt{2k+1} \begin{pmatrix} j & k & j \\ -m & q & m' \end{pmatrix} \sqrt{\begin{pmatrix} N \\ j+m \end{pmatrix}} \sqrt{\begin{pmatrix} N \\ j+m' \end{pmatrix}} \\ & \left(\cos\frac{\beta}{2}\right)^{N-m-m'} \left(\sin\frac{\beta}{2}\right)^{N+m+m'}.\end{aligned} \tag{6.44}$$

Substituting Eq. (6.44) into Eq. (6.42), we immediately obtain the Wigner quasipropability distribution for the even and odd spin cat states

$$\begin{aligned} W(\theta, \phi) =& \sqrt{\frac{N+1}{4\pi}} \sum_{k=0}^{N} \sum_{q=-k}^{k} \sum_{m=-j}^{j} \frac{\sqrt{2k+1}N!(-1)^j}{2[1 \pm (-\cos^2\beta)^j]} \\ & \frac{[(-1)^m \pm (-1)^q][1 \pm (-1)^m]}{\sqrt{(j+m)!(j-m)!(j+m-q)!(j-m+q)!}} \begin{pmatrix} j & k & j \\ -m & q & m-q \end{pmatrix} \\ & \left(\cos\frac{\beta}{2}\right)^{2(j-m)+q} \left(\sin\frac{\beta}{2}\right)^{2(j+m)-q} Y_{kq}(\theta, \phi), \end{aligned} \tag{6.45}$$

where we have used the fact that the Wigner $3j$-symbol $\begin{pmatrix} j_1 & j_2 & j_3 \\ m_1 & m_2 & m_3 \end{pmatrix}$ vanishes when $m_1 + m_2 + m_3 \neq 0$. In particular, for $\beta = \pi$, we have $\rho = \frac{1}{2}(|jj\rangle\langle jj| \pm (-1)^j |jj\rangle\langle j-j| \pm (-1)^j |j-j\rangle\langle jj| + (-1)^N |j-j\rangle\langle j-j|)$, which yields

$$\rho_{kq} = \frac{1}{2}\left\{\frac{[1+(-1)^{N+k}]\sqrt{2k+1}N!\delta_{q,0}}{\sqrt{(N+k+1)!(N-k)!}} \pm(-1)^j[\delta_{q,N} + (-1)^N\delta_{q,-N}]\delta_{k,N}\right\}, \tag{6.46a}$$

$$W(\theta,\phi) = \frac{1}{2}\sqrt{\frac{N+1}{4\pi}}\left\{\sum_{k=0}^{N}\frac{[1+(-1)^{N+k}]\sqrt{2k+1}N!}{\sqrt{(N+k+1)!(N-k)!}}Y_{k0}(\theta,\phi) \pm(-1)^j[Y_{NN}(\theta,\phi) + (-1)^N Y_{N-N}(\theta,\phi)]\right\}. \tag{6.46b}$$

In Fig. 6.4, we show the spherical polar plot of the quasi-probability Wigner function for even spin cat states with $N = 4$ and $\beta = \pi$. The top and bottom bumps on the sphere correspond to the statistical mixture of the two spin coherent states, for which quantum interference effects are absent, while the humps and hollows on the equator correspond to the interference effects between the two spin coherent states and the number of humps or hollows equals to the number of atoms.

Exercises

6.1. Verify Eq. (6.3) by using the formula

$$e^A B e^{-A} = B + [A, B] + \frac{1}{2!}[A, [A, B]] + \frac{1}{3!}[A, [A, [A, B]]] + \cdots.$$

6.2. Prove the disentangling theorem for the quantum angular momentum operators

$$e^{\zeta J_+ - \zeta^* J_-} = e^{\tau J_+} e^{\ln(1+|\tau|^2) J_z} e^{-\tau^* J_-},$$

where $\tau \equiv \tan|\zeta| e^{i \arg\zeta}$. (Hint: The 2×2 matrix representations of the angular momentum operators J_+, J_- and J_z may help.)

6.3. Verify Eq. (6.5) by using the formula

$$J_+|jm\rangle = \sqrt{(j-m)(j+m+1)}|jm+1\rangle.$$

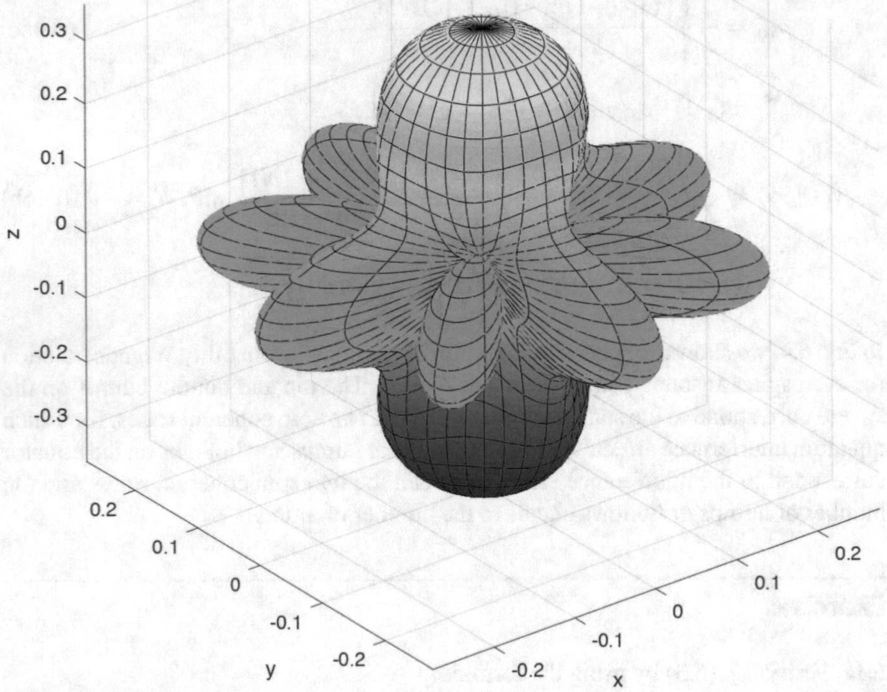

Fig. 6.4 Wigner distribution $W(\theta, \phi)$ for even spin cat state with $N = 4$ and $\zeta = \pi/2$, where $x \equiv W(\theta, \phi) \sin\theta \cos\phi$, $y \equiv W(\theta, \phi) \sin\theta \sin\phi$, and $z \equiv W(\theta, \phi) \cos\theta$

6.4. Prove the following relations

$$R_{\theta,\phi} J_n R_{\theta,\phi}^{-1} = J_n,$$

$$R_{\theta,\phi} J_k R_{\theta,\phi}^{-1} = J_k \cos\theta + J_z \sin\theta,$$

$$R_{\theta,\phi} J_z R_{\theta,\phi}^{-1} = -J_k \sin\theta + J_z \cos\theta,$$

where $J_n \equiv J_x \sin\phi - J_y \cos\phi$ and $J_k = J_x \cos\phi + J_y \sin\phi$.

6.5. By using the results from Exercise 6.4, verify Eq. (6.6).

6.6. From Eq. (6.4), show that the expectation values of J_x, J_y, and J_z are

$$\langle j, \zeta | J_x | j, \zeta \rangle = \frac{2j\Re\tau}{1+|\tau|^2},$$

$$\langle j, \zeta | J_y | j, \zeta \rangle = \frac{-2j\Im\tau}{1+|\tau|^2},$$

$$\langle j, \zeta | J_z | j, \zeta \rangle = \frac{j(|\tau|^2-1)}{1+|\tau|^2}.$$

6.7. By using the results from Exercise 6.6, show that

$$\langle \theta, \phi | J_x | \theta, \phi \rangle = j \sin\theta \cos\phi,$$

$$\langle \theta, \phi | J_y | \theta, \phi \rangle = j \sin\theta \sin\phi,$$

$$\langle \theta, \phi | J_z | \theta, \phi \rangle = -j \cos\theta.$$

6.8. Using the resolution of identity formula, Eq. (6.20), show that

$$\langle \theta, \phi | J_x^2 | \theta, \phi \rangle = \frac{j}{2} + \frac{j(2j-1)}{2} (\sin\theta \cos\phi)^2,$$

$$\langle \theta, \phi | J_y^2 | \theta, \phi \rangle = \frac{j}{2} + \frac{j(2j-1)}{2} (\sin\theta \sin\phi)^2.$$

6.9. Let A and B be two non-commuting Hermitian operators and $|\psi\rangle$ be a general quantum state; prove Schrödinger's uncertainty relation

$$\Delta A^2 \Delta B^2 \geq |\mathrm{C}(A, B)|^2 + \frac{1}{4} |\langle\psi|[A, B]|\psi\rangle|^2,$$

where $\Delta A^2 \equiv \langle\psi|A^2|\psi\rangle - \langle\psi|A|\psi\rangle^2$ is the variance of the operator A and $\mathrm{C}(A, B) \equiv \frac{1}{2}\langle\psi|\{A, B\}|\psi\rangle - \langle\psi|A|\psi\rangle\langle\psi|B|\psi\rangle$ is the covariance between the operators A and B. When the covariance $C(A, B)$ vanishes, Schrödinger's uncertainty relation becomes Robertson's uncertainty relation, $\Delta A^2 \Delta B^2 \geq \frac{1}{4} |\langle\psi|[A, B]|\psi\rangle|^2$. (Hint: Use Schwartz's inequality.)

6.10. Verify the relation $\langle j, \zeta_2 | j, \zeta_1 \rangle = \cos\frac{\Theta}{2} e^{iA}$, where Θ is the angle between the vectors (θ_1, ϕ_1) and (θ_2, ϕ_2) on the Bloch sphere and A is the area of a spherical triangle with points (θ_1, ϕ_1), (θ_2, ϕ_2) and the north pole.

6.11. Verify Eq. (6.32) by using the operator expansion theorem

$$e^{sA}Be^{-sA} = B + s[A, B] + \frac{s^2}{2!}[A, [A, B]] + \ldots.$$

6.12. Verify Eqs. (6.38a) and (6.38b) by using the Jaynes-Cummings Hamiltonian (6.31).

6.13. The Wigner $3j$-symbol vanishes if either of the conditions $|j_1 - j_2| \leq j_3 \leq j_1 + j_2$, $|m_i| \leq j_i$, or $m_1 + m_2 + m_3 = 0$ is not satisfied. When all the conditions are satisfied, the Wigner $3j$-symbol can be expressed as a finite sum

$$\begin{pmatrix} j_1 & j_2 & j_3 \\ m_1 & m_2 & m_3 \end{pmatrix} = (-1)^{j_1-j_2-m_3}\Delta(j_1 j_2 j_3)\Big(\prod_{i=1}^{3}(j_i + m_i)!(j_i - m_i)!\Big)^{\frac{1}{2}}$$

$$\times \sum_t \frac{(-1)^t}{t!(t-t_1)!(t-t_2)!(t+t_3)!(t+t_4)!(t+t_5)!},$$

where

$$\Delta(j_1 j_2 j_3) \equiv \left(\frac{(j_1 + j_2 - j_3)!(j_1 - j_2 + j_3)!(-j_1 + j_2 + j_3)!}{(j_1 + j_2 + j_3 + 1)!}\right)^{\frac{1}{2}},$$

and $t_m \leq t \leq t_M$. Here $t_m \equiv \max\{0, t_1, t_2\}$, $t_M \equiv \min\{t_3, t_4, t_5\}$, and

$$t_1 \equiv j_2 - m_1 - j_3,\ t_2 \equiv j_1 + m_2 - j_3,$$

$$t_3 \equiv j_1 + j_2 - j_3,\ t_4 \equiv j_1 - m_1,\ t_5 \equiv j_2 + m_2.$$

By using the above finite sum formula, verify the following relation

$$\begin{pmatrix} j_1 & j_2 & j_3 \\ j_1 & m_2 & -j_1 - m_2 \end{pmatrix} = \frac{(-1)^{2j_1-j_2+m_2}}{\sqrt{(J+1)!}}$$

$$\times \sqrt{\frac{(2j_1)!(j_2 - m_2)!(j_3 + j_1 + m_2)!(-j_1 + j_2 + j_3)!}{(j_1 + j_2 - j_3)!(j_2 + m_2)!(j_3 - j_2 + j_1)!(j_3 - j_1 - m_2)!}},$$

where $J \equiv j_1 + j_2 + j_3$. In particular, for $j_1 = j_2$, show that

$$\begin{pmatrix} j_1 & j_1 & j_3 \\ j_1 & m_2 & -j_1 - m_2 \end{pmatrix} = \frac{(-1)^{j_1+m_2}}{\sqrt{(2j_1 + j_3 + 1)!}}$$

$$\times \sqrt{\frac{(2j_1)!(j_1 - m_2)!(j_3 + j_1 + m_2)!}{(2j_1 - j_3)!(j_1 + m_2)!(j_3 - j_1 - m_2)!}}.$$

6.14. The Wigner $3j$-symbol has many symmetry properties, e.g., it is invariant under an even permutation of its columns

$$\begin{pmatrix} j_1 & j_2 & j_3 \\ m_1 & m_2 & m_3 \end{pmatrix} = \begin{pmatrix} j_3 & j_1 & j_2 \\ m_3 & m_1 & m_2 \end{pmatrix} = \begin{pmatrix} j_2 & j_3 & j_1 \\ m_2 & m_3 & m_1 \end{pmatrix},$$

and will yield a phase by changing the sign of the m quantum numbers

$$\begin{pmatrix} j_1 & j_2 & j_3 \\ m_1 & m_2 & m_3 \end{pmatrix} = (-1)^{j_1+j_2+j_3} \begin{pmatrix} j_1 & j_2 & j_3 \\ -m_1 & -m_2 & -m_3 \end{pmatrix}$$

By using the above symmetry properties, verify the following relations

$$\begin{pmatrix} j & k & j \\ -j & 0 & j \end{pmatrix} = \frac{(2j)!}{\sqrt{(2j+k+1)!(2j-k)!}},$$

$$\begin{pmatrix} j & k & j \\ j & 0 & -j \end{pmatrix} = (-1)^{2j+k} \begin{pmatrix} j & k & j \\ -j & 0 & j \end{pmatrix},$$

$$\begin{pmatrix} j & 2j & j \\ -j & 2j & -j \end{pmatrix} = \begin{pmatrix} j & 2j & j \\ j & -2j & j \end{pmatrix} = \frac{1}{\sqrt{4j+1}},$$

6.15. Using the above results, verify the following relations

$$\langle jj|T_{kq}^{\dagger}|jj\rangle = \frac{\sqrt{2k+1}(2j)!\delta_{q,0}}{\sqrt{(2j+k+1)!(2j-k)!}},$$

$$\langle j-j|T_{kq}^{\dagger}|jj\rangle = \delta_{k,2j}\delta_{q,2j},$$

$$\langle jj|T_{kq}^{\dagger}|j-j\rangle = (-1)^{2j}\delta_{k,2j}\delta_{q,-2j},$$

$$\langle j-j|T_{kq}^{\dagger}|j-j\rangle = \frac{(-1)^k\sqrt{2k+1}(2j)!\delta_{q,0}}{\sqrt{(2j+k+1)!(2j-k)!}}.$$

6.16. The spherical harmonics can be written in terms of the associated Legendre polynomials as

$$Y_{lm} = \begin{cases} (-1)^m \sqrt{2}\sqrt{\dfrac{2l+1}{4\pi}\dfrac{(l-|m|)!}{(l+|m|)!}} P_l^{|m|}(\cos\theta)\sin(|m|\phi), & (m<0), \\ \sqrt{\dfrac{2l+1}{4\pi}} P_l(\cos\theta), & (m=0), \\ (-1)^m \sqrt{2}\sqrt{\dfrac{2l+1}{4\pi}\dfrac{(l-m)!}{(l+m)!}} P_l^{m}(\cos\theta)\cos(m\phi), & (m>0). \end{cases}$$

By using the above formula and identity

$$P_l^l(\cos\theta) = (-1)^l \frac{(2l)!}{2^l l!} \sin^l\theta,$$

show that Eq. (6.46b) can be explicitly written as

$$W(\theta,\phi) = \frac{1}{2}\sqrt{\frac{N+1}{4\pi}}\sqrt{\frac{2N+1}{4\pi}} \left\{ \sum_{k=0}^{N} \frac{[1+(-1)^{N+k}]\sqrt{2k+1}N!}{\sqrt{(N+k+1)!(N-k)!}} \right.$$

$$\left. \times P_k(\cos\theta) \pm (-1)^j \frac{\sqrt{2(2N)!}}{2^N N!} \sin^N\theta[\cos(N\phi) + (-1)^N \sin(N\phi)] \right\}.$$

7 Squeezed Coherent States

7.1 Squeezed Coherent States in Quantum Optics

A key characteristic of the coherent state of light is that it minimizes the product of uncertainty in a pair canonically conjugate variables such as position and momentum. However, there is a larger class of states which are unitarily equivalent to the coherent states, minimizing the uncertainty product. In 1970, this fact was first pointed out by Stoler [96, 97]. Subsequently in 1976, Yuen [98] had shown in his seminal paper titled "Two-photon coherent states of the radiation field" that these new minimum uncertainty states are physically the radiation states of ideal two-photon lasers operating far above threshold. Unlike Glauber's coherent states which have equal fluctuations in both position and momentum, the two-photon coherent states have a unique property that the **quantum noise**, or zero-point fluctuations, in one field component can be significantly reduced below the quantum limit, at the expense of enhancing fluctuations in the conjugate one. In other words, these minimum uncertainty states are special states of light where the noise caused by quantum effects has been largely squeezed out. Thus, they are then referred to as **squeezed coherent states** of light. In the late 1980s, squeezed states of light were first demonstrated by three independent experimental groups using different methods, such as four-wave mixing in optical cavities [99] or single-mode optical fibers [100], or parametric down conversion in nonlinear optical crystals [101].

The quantum noise suppression technique via squeezed coherent states makes it indispensable when ultraprecision measurements are called for, like the ground-based laser interferometric gravitational-wave observatories, such as **LIGO** interferometers [102] in the United States and **Virgo** interferometer [103] in Italy. In both these measurements, in order to detect the faintest gravitational radiation from the merger of compact astrophysical systems such as binary neutron stars or black holes, the detector must be able to measure strains at the level of 10^{-21} or less [104]. Such a tiny strain only distorts the shape of the Earth by 10^{-14} meters, or about thousandth of the size of an atom. Hence, any vacuum fluctuations which

C.-F. Kam et al., *Coherent States*, Lecture Notes in Physics 1011,
https://doi.org/10.1007/978-3-031-20766-2_7

limit the phase sensitivity of the gravitational-wave interferometric detector have to be kept at an absolute minimum. The idea of phase sensitive gravitational-wave detection was first revealed by Hollenhorst [105] in 1979 in the context of resonant-mass antenna. Subsequently in the early 1980s, by replacing coherent vacuum states by squeezed vacuum states, Caves [106, 107] discussed in-depth phase sensitive gravitational wave detection in the context of laser interferometric detectors. Unlike producing squeezed coherent states in optical frequency region, the production of squeezed coherent states in audio frequency region that is related to gravitational-wave detection was not successfully demonstrated until the mid-2000s [108–110]. Since then, squeezed light for quantum noise reduction becomes a standard experimental tool in interferometric gravitational wave observations [111]. In the following, we discuss the basic properties of squeezed coherent states.

In order to generate a squeezed coherent state of light, one can consider a unitary **squeeze operator** in the form

$$\hat{S}(\xi) \equiv \exp\left\{\frac{1}{2}\left(\xi^*\hat{a}^2 - \xi\hat{a}^{\dagger 2}\right)\right\}, \tag{7.1}$$

where $\xi \equiv re^{i\theta}$ is an arbitrary complex number. By a straightforward application of the operator expansion formula

$$e^{\hat{A}}\hat{B}e^{-\hat{A}} = \hat{B} + [\hat{A}, \hat{B}] + \frac{1}{2!}[\hat{A}, [\hat{A}, \hat{B}]] + \cdots, \tag{7.2}$$

one immediately obtains

$$\hat{A} \equiv \hat{S}(\xi)\hat{a}\hat{S}^\dagger(\xi) = \mu\hat{a} + \nu\hat{a}^\dagger, \tag{7.3a}$$

$$\hat{A}^\dagger \equiv \hat{S}(\xi)\hat{a}^\dagger\hat{S}^\dagger(\xi) = \mu\hat{a}^\dagger + \nu^*\hat{a}, \tag{7.3b}$$

where $\mu \equiv \cosh r$ and $\nu \equiv \sinh r e^{i\theta}$. The inverse relations of Eqs. (7.3a)–(7.3b) are

$$\hat{a} = \mu\hat{A} - \nu\hat{A}^\dagger, \hat{a}^\dagger = \mu\hat{A}^\dagger - \nu^*\hat{A}. \tag{7.4}$$

As $|\mu|^2 - |\nu|^2 = 1$, one can verify that such transformation preserves the commutation relation, $[\hat{A}, \hat{A}^\dagger] = [\hat{a}, \hat{a}^\dagger]$. Hence, the change of variables from $\hat{a}$ and $\hat{a}^\dagger$ to $\hat{A}$ and $\hat{A}^\dagger$ is a canonical transformation. A squeezed coherent state is then obtained by applying the squeeze operator on a coherent state $|\alpha\rangle$

$$|\alpha, \xi\rangle \equiv \hat{S}(\xi)|\alpha\rangle = \hat{S}(\xi)\hat{D}(\alpha)|0\rangle, \tag{7.5}$$

where $\hat{D}(\alpha) \equiv e^{\alpha\hat{a}^\dagger - \alpha^*\hat{a}}$ is the displacement operator. From Eq. (7.5), we readily find that $\hat{A}|\alpha, \xi\rangle = \hat{S}(\xi)\hat{a}\hat{S}^\dagger(\xi)\hat{S}(\xi)|\alpha\rangle = \hat{S}(\xi)\hat{a}|\alpha\rangle = \alpha|\alpha, \xi\rangle$. Hence, the squeezed coherent state can also be regarded as the eigenstate of $\hat{A} = \mu\hat{a} + \nu\hat{a}^\dagger$ with eigenvalue α. In order to examine the properties of squeezed coherent states, we will introduce the two **quadratures** of the electromagnetic field, $\hat{Q} \equiv (\hat{a}^\dagger + \hat{a})/\sqrt{2}$ and $\hat{P} \equiv i(\hat{a}^\dagger - \hat{a})/\sqrt{2}$, so that the expectation values and variances for the squeezed coherent states $|\alpha, \xi\rangle$ are

$$\langle \alpha, \xi | \hat{Q} | \alpha, \xi \rangle = \frac{1}{\sqrt{2}} \left[(\mu - \nu)\alpha^* + (\mu - \nu^*)\alpha \right], \tag{7.6a}$$

$$\langle \alpha, \xi | \hat{P} | \alpha, \xi \rangle = \frac{i}{\sqrt{2}} \left[(\mu + \nu)\alpha^* - (\mu + \nu^*)\alpha \right], \tag{7.6b}$$

$$\langle \alpha, \xi | (\Delta \hat{Q})^2 | \alpha, \xi \rangle = \frac{1}{2} |\mu - \nu|^2 = \frac{1}{2}(\cosh 2r - \sinh 2r \cos\theta), \tag{7.6c}$$

$$\langle \alpha, \xi | (\Delta \hat{P})^2 | \alpha, \xi \rangle = \frac{1}{2} |\mu + \nu|^2 = \frac{1}{2}(\cosh 2r + \sinh 2r \cos\theta), \tag{7.6d}$$

$$\langle \alpha, \xi | (\Delta \hat{P})(\Delta \hat{Q}) | \alpha, \xi \rangle = \frac{i}{2}\mu(\nu - \nu^*) - \frac{i}{2} = -\frac{1}{2} \sinh 2r \sin\theta - \frac{i}{2}, \tag{7.6e}$$

$$\langle \alpha, \xi | (\Delta \hat{Q})(\Delta \hat{P}) | \alpha, \xi \rangle = \frac{i}{2}\mu(\nu - \nu^*) + \frac{i}{2} = -\frac{1}{2} \sinh 2r \sin\theta + \frac{i}{2}, \tag{7.6f}$$

which yields

$$\sqrt{\langle \alpha, \xi | (\Delta \hat{Q})^2 | \alpha, \xi \rangle} \sqrt{\langle \alpha, \xi | (\Delta \hat{P})^2 | \alpha, \xi \rangle} = \frac{1}{2}(1 + \sinh^2 2r \sin^2\theta) \geq \frac{1}{2}. \tag{7.7}$$

Hence, the squeezed coherent state $|\alpha, \xi\rangle$ is a minimum uncertainty state for $r = 0$, and $\theta = 0$ or π. Clearly, the former case $r = 0$ corresponds to the conventional coherent states without squeezing. For the latter case, we have

$$\Delta \hat{Q} = \frac{e^{-r}}{\sqrt{2}}, \; \Delta \hat{P} = \frac{e^{r}}{\sqrt{2}}, \; \text{for } \xi = r, \tag{7.8}$$

$$\Delta \hat{Q} = \frac{e^{r}}{\sqrt{2}}, \; \Delta \hat{P} = \frac{e^{-r}}{\sqrt{2}}, \; \text{for } \xi = -r.$$

From the above, we see that in the squeezed coherent states with $\xi = \pm r$, the fluctuations of one quadrature are reduced below the quantum limit at the expense of an increase in the fluctuations of another quadrature, so that the uncertainty product remains at the minimum value. As the reduction of fluctuations of the quadratures depends only on the parameter r, it is therefore referred to as a **squeeze parameter**.

Through the above discussions, we have shown that there is a kind of special coherent states—the squeezed coherent state, which minimizes the uncertainty product in such a way that one part of the field fluctuates less and another part of the field fluctuates more. In the following, we shall discuss in detail the basic properties of the squeezed coherent state, including its photon statistics and quasi-probability distributions.

To begin with, let us derive the coherent state representation of the squeezed coherent states $|\beta, \xi\rangle$, i.e., the projection of $\langle\alpha|\beta, \xi\rangle$ on the Glauber coherent state $|\alpha\rangle$. Using the relation $\hat{A}|\beta, \xi\rangle \equiv (\mu\hat{a} + \nu\hat{a}^\dagger)|\beta, \xi\rangle = \beta|\beta, \xi\rangle$, we obtain a differential equation for $|\beta, \xi\rangle$

$$\left[\mu\left(\frac{\partial}{\partial\alpha^*} + \frac{\alpha}{2}\right) + \nu\alpha^*\right]\langle\alpha|\beta, \xi\rangle = \beta\langle\alpha|\beta, \xi\rangle, \tag{7.9}$$

which is solved by $\langle\alpha|\beta, \xi\rangle = \mathcal{N}(\beta, \beta^*)\exp\{\frac{1}{\mu}(\beta\alpha^* - \frac{\nu}{2}\alpha^{*2}) - \frac{1}{2}|\alpha|^2\}$, where $\mathcal{N}(\beta, \beta^*)$ is arbitrary function of β and β^* chosen to ensure the normalization of the squeezed coherent states, $\langle\beta, \xi|\beta, \xi\rangle = \frac{1}{\pi}\int|\langle\alpha|\beta, \xi\rangle|^2 d^2\alpha = 1$. A direct computation yields

$$\frac{1}{\pi}\int|\langle\alpha|\beta, \xi\rangle|^2 d^2\alpha = |\mathcal{N}(\beta, \beta^*)|^2\mu\exp\left\{|\beta|^2 - \frac{1}{2\mu}(\nu\beta^{*2} + \nu^*\beta^2)\right\}. \tag{7.10}$$

Equation (7.10) then immediately yields, apart from an arbitrary phase factor, the coherent state representation of the squeezed coherent state $|\beta, \xi\rangle$

$$\langle\alpha|\beta, \xi\rangle = \frac{1}{\sqrt{\mu}}\exp\left\{\frac{1}{\mu}(\beta\alpha^* - \frac{\nu}{2}\alpha^{*2} + \frac{\nu^*}{2}\beta^2) - \frac{1}{2}|\alpha|^2 - \frac{1}{2}|\beta|^2\right\}. \tag{7.11}$$

As $e^{-z^2+2sz} = \sum_n \frac{1}{n!}H_n(s)z^n$ is the generating function of the Hermite polynomials $H_n(s)$, one immediately obtains

$$\begin{aligned}\langle\alpha|\beta, \xi\rangle &= \mathcal{N}(\beta, \beta^*)e^{-\frac{1}{2}|\alpha|^2}e^{-z^2+2sz} \\ &= \mathcal{N}(\beta, \beta^*)e^{-\frac{1}{2}|\alpha|^2}\sum_{n=0}^{\infty}\frac{1}{n!}H_n(s)z^n,\end{aligned} \tag{7.12}$$

where $s \equiv \frac{\beta}{\sqrt{2\mu\nu}}$ and $z \equiv \sqrt{\frac{\nu}{2\mu}}\alpha^*$. Comparing Eq. (7.12) to the coherent state representation of $|\beta, \xi\rangle$ in terms of the projections $\langle n|\beta, \xi\rangle$

$$\langle\alpha|\beta, \xi\rangle = \sum_{n=0}^{\infty}\langle\alpha|n\rangle\langle n|\beta, \xi\rangle = e^{-\frac{1}{2}|\alpha|^2}\sum_{n=0}^{\infty}\frac{\alpha^{*n}}{\sqrt{n!}}\langle n|\beta, \xi\rangle, \tag{7.13}$$

one finally obtains the projections $\langle n|\beta, \xi\rangle$

$$\langle n|\beta, \xi\rangle = \mathcal{N}(\beta, \beta^*)\frac{1}{\sqrt{n!}}\left(\frac{\nu}{2\mu}\right)^{n/2} H_n\left(\frac{\beta}{\sqrt{2\mu\nu}}\right). \tag{7.14}$$

The probability that there are n photons in the squeezed coherent state $|\beta, \xi\rangle$ is then

$$p(n) \equiv |\langle n|\beta, \xi\rangle|^2 = e^{-|\beta|^2 + \frac{1}{2\mu}(\nu\beta^{*2} + \nu^*\beta^2)} \frac{1}{\mu n!} \left|\frac{\nu}{2\mu}\right|^n \left| H_n\left(\frac{\beta}{\sqrt{2\mu\nu}}\right)\right|^2, \tag{7.15}$$

where $\mu \equiv \cosh r$, $\nu \equiv \sinh r e^{i\theta}$, and r is the squeeze parameter.

Leveraging the relations $\hat{a} = \mu\hat{A} - \nu\hat{A}^\dagger$ and $\hat{A}|\beta, \xi\rangle = \beta|\beta, \xi\rangle$, we readily find the mean photon number in the squeezed coherent states

$$\begin{aligned} \langle\hat{n}\rangle &\equiv \langle\beta, \xi|\hat{a}^\dagger\hat{a}|\beta, \xi\rangle \\ &= \langle\beta, \xi|(\mu\hat{A}^\dagger - \nu^*\hat{A})(\mu\hat{A} - \nu\hat{A}^\dagger)|\beta, \xi\rangle \\ &= \left|\mu\beta^* - \nu^*\beta\right|^2 + |\nu|^2, \end{aligned} \tag{7.16}$$

which equals to $|\beta|^2$ for the Glauber coherent state with $\nu = 0$ and equals to $|\nu|^2 = \sinh^2 r$ for the squeezed vacuum state with $\beta = 0$. After examining the second-order correlation function

$$\begin{aligned} \langle(\hat{a}^\dagger)^2\hat{a}^2\rangle &= \langle\beta, \xi|(\mu\hat{A}^\dagger - \nu^*\hat{A})^2(\mu\hat{A} - \nu\hat{A}^\dagger)^2|\beta, \xi\rangle \\ &= \langle\hat{n}\rangle^2 + |\nu|^2[(6\mu^2 + 2|\nu|^2)|\beta|^2 + \mu^2 + |\nu|^2] \\ &\quad - \mu\nu(4|\nu|^2 + 1)\beta^{*2} - \mu\nu^*(4|\nu|^2 + 1)\beta^2, \end{aligned} \tag{7.17}$$

we immediately obtain the photon number variance $\langle(\Delta\hat{n})^2\rangle$ in the squeezed coherent states as

$$\begin{aligned} \langle\hat{n}^2\rangle - \langle\hat{n}\rangle^2 &= \langle\hat{n}\rangle + |\nu|^2[(6 + 8|\nu|^2)|\beta|^2 + 2|\nu|^2 + 1] \\ &\quad - (4|\nu|^2 + 1)(\mu\nu\beta^{*2} + \mu\nu^*\beta^2). \end{aligned} \tag{7.18}$$

In particular, for the Glauber coherent state with $\nu = 0$, we recover the result $\langle(\Delta\hat{n})^2\rangle = \langle\hat{n}\rangle = |\beta|^2$.

With the about results, one may use the **Mandel Q parameter** [112], defined by $Q \equiv (\langle\Delta\hat{n}^2\rangle - \langle\hat{n}\rangle)/\langle\hat{n}\rangle$, to measure the departure of the occupation number distribution from Poissonian statistics. When Q is negative, i.e., $-1 \leq Q < 0$, the photon number variance is less than the mean photon number and is characterized by sub-Poissonian statistics. The extremal case $Q = -1$ is attained for Fock states with definite number of photons. In Fig. 7.1, we plot the Mandel Q parameter for squeezed coherent states as a function of the squeeze parameter, from which we see that $Q \leq 0$ for $r \leq 0.413$. In general, the photon statistics of the squeezed coherent state $|\beta, \xi\rangle$ can be both super- or sub-Poissonian, depending on both β and ξ.

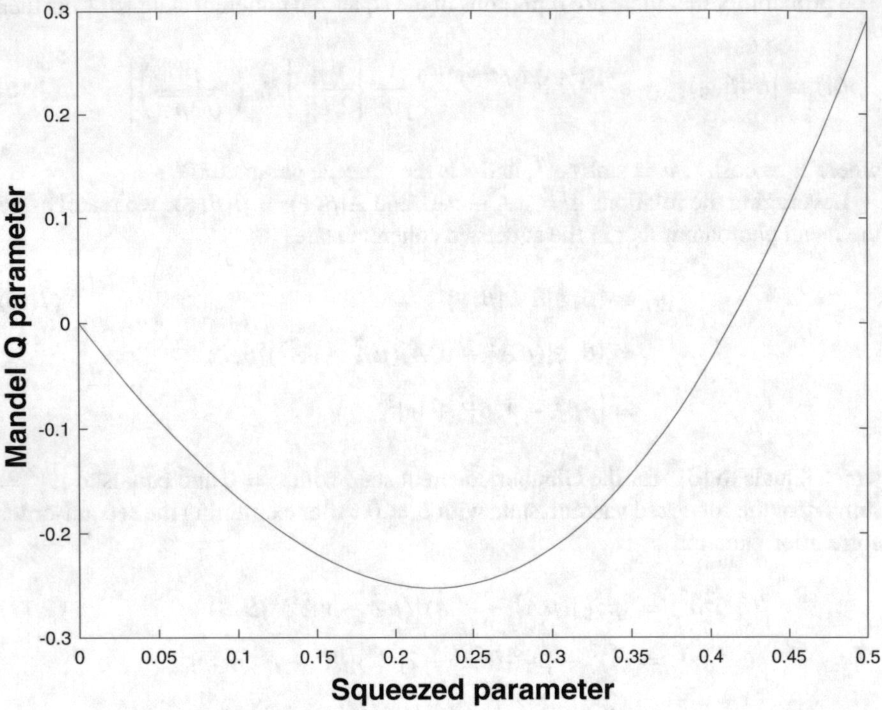

Fig. 7.1 Mandel Q parameter for the squeezed coherent state $|\beta, \xi\rangle \equiv \hat{S}(\xi)|\beta\rangle$ as a function of the squeeze parameter r. Here $\beta = 1$ and $\xi = r$

We now discuss some quasi-probability functions for squeezed coherent states. We begin by studying the Wigner function $W(\alpha)$ defined as the Fourier transform of the characteristic function $C_W(\eta) \equiv \mathrm{Tr}[\hat{\rho}\hat{D}(\eta)]$

$$W(\alpha) \equiv \frac{1}{\pi^2} \int d^2\eta e^{\alpha\eta^* - \alpha^*\eta} C_W(\eta), \tag{7.19}$$

where $\hat{D}(\eta) \equiv e^{\eta\hat{a}^\dagger - \eta^*\hat{a}}$ is the displacement operator and $\hat{\rho} \equiv |\beta, \xi\rangle\langle\beta, \xi|$ is the squeezed coherent state density matrix. Using the relations $S^\dagger(\xi)\hat{a}\hat{S}(\xi) = \mu\hat{a} - \nu\hat{a}^\dagger$ and $S^\dagger(\xi)\hat{a}^\dagger\hat{S}(\xi) = \mu\hat{a} - \nu^*\hat{a}^\dagger$, we obtain $S^\dagger(\xi)\hat{D}(\eta)\hat{S}(\xi) = \hat{D}(\zeta)$ from the Taylor expansion of the displacement operator, where $\zeta \equiv \eta\mu + \eta^*\nu$. Hence, the characteristic function for the squeezed coherent state becomes

$$\begin{aligned} C_W(\eta) &= \langle\beta|S^\dagger(\xi)\hat{D}(\eta)\hat{S}(\xi)|\beta\rangle \\ &= \langle\beta|\hat{D}(\zeta)|\beta\rangle = e^{\zeta\beta^* - \zeta^*\beta - \frac{1}{2}|\zeta|^2}. \end{aligned} \tag{7.20}$$

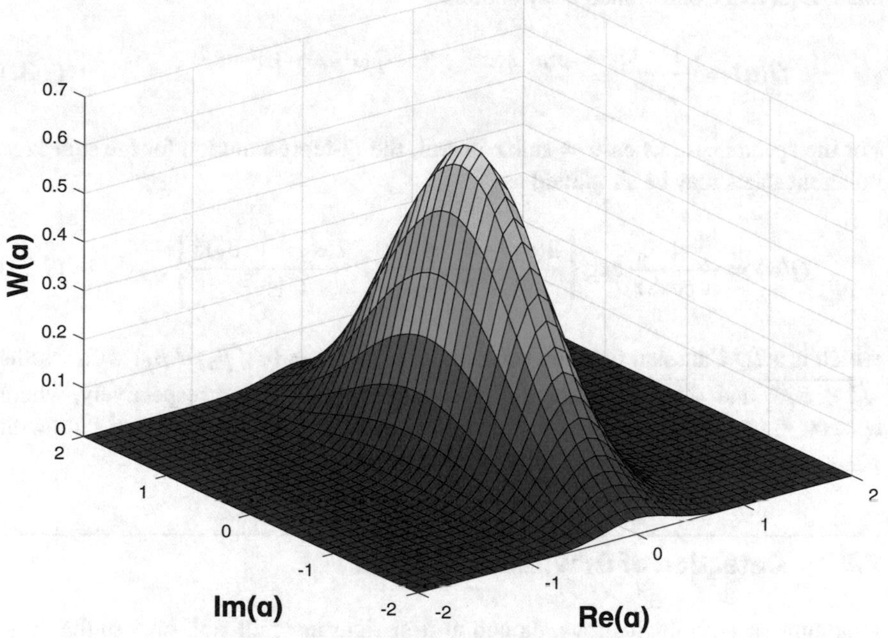

Fig. 7.2 Schematic of the Wigner quasi-probability distribution for the squeezed coherent states $|\beta, \xi\rangle$. Here we plot the squeezed vacuum state with $\beta = 0$ and $\xi = 0.5$

With a straightforward calculation, we obtain

$$
\begin{aligned}
W(\alpha) &= \frac{1}{\pi^2}\int d^2\eta e^{\tilde{\alpha}\eta^* - \tilde{\alpha}^*\eta - \frac{\mu^2+|\nu|^2}{2}|\eta|^2 - \frac{\mu\nu^*}{2}\eta^2 - \frac{\mu\nu}{2}\eta^{*2}} \\
&= \frac{2}{\pi}\exp\{-2e^{2r}\Re(e^{-\frac{i\theta}{2}}\tilde{\alpha})^2 - 2e^{-2r}\Im(e^{-\frac{i\theta}{2}}\tilde{\alpha})^2\}.
\end{aligned}
\tag{7.21}
$$

where $\tilde{\alpha} \equiv \alpha + \nu\beta^* - \mu\beta$. In particular, for the Glauber coherent states $|\beta\rangle$, we recover the result $W(\alpha) = \frac{2}{\pi}e^{-2|\alpha-\beta|^2}$. From Eq. (7.21), we noticed that the Wigner distribution for a squeezed coherent state is a 2D Gaussian function, which is a product of two Gaussian functions with widths e^{-r} and e^{r}, respectively. In Fig. 7.2, we plot the Wigner distribution for the squeezed vacuum state with $\beta = 0$. Unlike the Glauber coherent states which possess a rotational symmetric Wigner distribution, the Wigner distribution for squeezed coherent states is stretched along the imaginary axis and is squeezed along the real axis.

Similarly, one could calculate the Q-representation for the squeezed coherent state $Q(\alpha) \equiv \frac{1}{\pi}\langle\alpha|\hat{\rho}|\alpha\rangle$, where $\hat{\rho} \equiv |\beta, \xi\rangle\langle\beta, \xi|$ is the squeezed coherent state density matrix. Using the coherent state representation of the squeezed coherent state, Eq. (7.11), one immediately obtains

$$Q(\alpha) = \frac{1}{\pi\mu} e^{\frac{1}{\mu}\left(\beta\alpha^* + \beta^*\alpha - \frac{\nu}{2}(\alpha^{*2} - \beta^{*2}) - \frac{\nu^*}{2}(\alpha^2 - \beta^2) - |\alpha|^2 - |\beta|^2\right)}. \tag{7.22}$$

For the special case when $\nu = \sinh r$ is real, the Q-representation for the squeezed coherent states may be simplified as

$$Q(\alpha) = \frac{1}{\pi \cosh r} \exp\left\{\frac{2(\alpha_x - e^{-r}\beta_x)^2}{1 + e^{-2r}} + \frac{2(\alpha_y - e^{r}\beta_y)^2}{1 + e^{2r}}\right\}, \tag{7.23}$$

which is a 2D Gaussian function centered at $(\alpha_x, \alpha_y) = (e^{-r}\beta_x, e^r\beta_y)$ with widths $\sqrt{1 + e^{-2r}}$ and $\sqrt{1 + e^{2r}}$ along the real and imaginary axis, respectively, where $\alpha \equiv \alpha_x + i\alpha_y$ and $\beta \equiv \beta_x + i\beta_y$. In particular, for the Glauber coherent state with $r = 0$, we recover the result $Q(\alpha) = \frac{1}{\pi} e^{|\alpha - \beta|^2}$.

7.2 Detection of Gravitational Wave

Quantum vacuum fluctuations, though at first sight insignificant, impose the most severe limit on the sensitivity of present-day high-precision measurements such as interferometric gravitational wave observation. Fortunately, with the use of squeezed coherent states, it may help us to mitigate this seemingly fundamental quantum interferometry bound. It was generally proven that the combination of a bright coherent state and a squeezed vacuum state is optimal for interferometric gravitational wave observations [113]. After decades of proof-of-principle experiments [108, 110, 114–119], squeezed coherent states of light have been successfully implemented to enhance the sensitivity of the gravitational-wave detector GEO600 [111, 120] and the initial LIGO detector at Hanford, WA [104].

The prototypical gravitational wave detectors such as LIGO's interferometers are essentially kilometer-scale Michelson interferometers. An ideal version of such an interferometric detector is operated as follows [121]: a continuous-wave laser light is split at a lossless 50-50 beam splitter into two beams propagating in the nearly equal length perpendicular arms of the interferometer, which are reflected by perfectly reflecting mirrors and are finally recombined at the beam splitter. The two beams generate interference patterns which are detected by a photodiode. As shown in Fig. 7.3, we have used the Latin letters ($a, b, c, \ldots$) to designate the field amplitude in each light path and use letters with hat ($\hat{a}, \hat{b}, \hat{c}, \ldots$) to designate the annihilation operators for the associated field modes propagating in the given direction.

The central target for interferometric gravitational wave observation is the phase difference between the output fields c' and d' before recombination at the beam splitter. As these fields are related to the fields c and d by a phase change, one has to explore how the phase between the fields c and d affected by noise. Classically, if there is no light incident from the direction of b, the two beams c and d are split from the same source field a at the beam splitter, which implies that all their fluctuations would be perfectly correlated, referring to the fluctuations in the source field a. In other words, even if the phase of the source field is random, the phase difference between fields c and d is definite. But for a quantum field, some kind of uncorrelated fluctuations always exists in the phases of the beams c and d, which will yield an indefinite phase difference between them.

The origin of these uncorrelated fluctuations may be understood by taking into account the inevitably vacuum fluctuations incident from the other port of the interferometer [106, 107], i.e., a mode of the electromagnetic field in the vacuum state marked as b in Fig. 7.3. As the beam splitter will combine the two incident fields a and b, we may write the annihilation operators for modes c and d as follows:

$$\hat{c} = (\hat{a} + \hat{b})/\sqrt{2}, \tag{7.24a}$$

$$\hat{d} = (\hat{a} - \hat{b})/\sqrt{2}, \tag{7.24b}$$

where the π phase shift upon reflection is expected from the time-reversal arguments, hence the minus sign in front of $\hat{b}$ in Eq. (7.24b). Now, we may see that although the fluctuations in the field a are perfectly correlated in c and d, the fluctuations in the field b are in fact *anticorrelated* in c and d, causing them to *add* instead of cancel when the two fields subtracted. The annihilation operations for modes c' and d' are then given by $\hat{c}' = e^{i\varphi_c}\hat{c}$ and $\hat{d}' = e^{i\varphi_d}\hat{d}$, with a phase difference $\varphi \equiv \varphi_c - \varphi_d$. The two beams c' and d' are then recombined at the beam splitter along with a phase shift of π in the field mode d'. Hence, the annihilation operator for the field mode arriving the photodiode at the bottom port is

$$\begin{aligned} \hat{b}' &= [(a+b)e^{i\varphi_c} - (a-b)e^{i\varphi_d}]/2 \\ &= e^{i(\varphi_c+\varphi_d)/2}[\cos(\varphi/2)\hat{b} + i\sin(\varphi/2)\hat{a}]. \end{aligned} \tag{7.25}$$

The intensity of the light arriving the photodiode is then given by

$$\begin{aligned} \langle \hat{b}'^{\dagger}\hat{b}' \rangle &= \sin^2(\varphi/2)\langle \hat{a}^{\dagger}\hat{a} \rangle + \cos^2(\varphi/2)\langle \hat{b}^{\dagger}\hat{b} \rangle \\ &\quad + i\cos(\varphi/2)\sin(\varphi/2)(\langle \hat{b}^{\dagger}\hat{a} \rangle - \langle \hat{a}^{\dagger}\hat{b} \rangle). \end{aligned} \tag{7.26}$$

Clearly, when the mode b is in the vacuum state, the last two terms have zero expectation value. Besides this, the second term does not contribute to the fluctuations of the photon number in the output mode. Hence, the photon-number fluctuations in the output mode are due to those of the input laser mode and the interference of light coming from the two input ports. In other words, although the vacuum fluctuations are not directly observable, they give a nonzero contribution to photon fluctuations in the output light through interference. For the case when the input laser in a coherent state $|\alpha\rangle$ with $\alpha = |\alpha|e^{i\delta}$, we obtain the second-order correlation function

$$\langle(\hat{b}^\dagger\hat{a} - \hat{b}\hat{a}^\dagger)^2\rangle = |\alpha|^2\langle(\hat{b}^\dagger e^{i\delta} - \hat{b}e^{-i\delta})^2\rangle - \langle\hat{b}^\dagger\hat{b}\rangle. \quad (7.27)$$

Let us introduce two field quadratures as follows: $\hat{P}_\delta \equiv i(\hat{b}^\dagger e^{i\delta} - be^{-i\delta})/\sqrt{2} = -\hat{Q}\sin\delta + \hat{P}\cos\delta$ and $\hat{Q}_\delta \equiv (\hat{b}^\dagger e^{i\delta} + be^{-i\delta})/\sqrt{2} = \hat{Q}\cos\delta + \hat{P}\sin\delta$, where $\hat{P} \equiv i(\hat{b}^\dagger - b)/\sqrt{2}$ and $\hat{Q} \equiv (\hat{b}^\dagger + b)/\sqrt{2}$ are the conventional field quadratures which satisfy the commutation relation $[\hat{Q}, \hat{P}] = i$. Clearly, the field quadratures $\hat{P}_\delta$ and $\hat{Q}_\delta$ are related to $\hat{P}$ and $\hat{Q}$ by a rotation angle δ. Now, using Eq. (7.27), the fluctuations in the last term in Eq. (7.26) become $\cos^2(\varphi/2)\sin^2(\varphi/2)[2|\alpha|^2\langle(\Delta\hat{P}_\delta)^2\rangle + \langle\hat{b}^\dagger\hat{b}\rangle]$, of which the dominant term is proportional to the variance in the field quadrature $\hat{P}_\delta$ for a strong input laser field ($|\alpha| \gg 1$). It implies that one may apply the squeezed vacuum state in another input port to significantly reduce the quantum noise in the output beam, i.e., $\langle(\Delta\hat{P}_\delta)^2\rangle \ll 1/2$. For a squeezed vacuum state $|0, re^{i\theta}\rangle$, one obtains $\langle\hat{b}\rangle = \langle\hat{b}^\dagger\rangle = \langle\hat{b}^{\dagger 2}b\rangle = \langle\hat{b}^\dagger\hat{b}^2\rangle = 0$, which implies that the cross terms do not account for the photon-number fluctuations in the output mode when calculating the square of Eq. (7.26). A direct computation yields

$$\begin{aligned}\langle(\Delta\hat{b}'^\dagger\hat{b}')^2\rangle &= \sin^4(\varphi/2)\langle(\Delta\hat{a}^\dagger\hat{a})^2\rangle + \cos^4(\varphi/2)\langle(\Delta\hat{b}^\dagger\hat{b})^2\rangle \\ &\quad - \cos^2(\varphi/2)\sin^2(\varphi/2)\langle(\hat{b}^\dagger\hat{a} - \hat{b}\hat{a}^\dagger)^2\rangle.\end{aligned} \quad (7.28)$$

In order to maximally reduce the fluctuations in the output photon number, one requires $\theta = 2\delta + \pi$, so that $\langle(\hat{b}^\dagger e^{i\delta} - \hat{b}e^{-i\delta})^2\rangle = -e^{-2r}$. Then using the relations $\langle(\Delta\hat{a}^\dagger\hat{a})^2\rangle = |\alpha|^2$, $\langle(\Delta\hat{b}^\dagger\hat{b})^2\rangle = 2\cosh^2 r\sinh^2 r$, and $\langle\hat{b}^\dagger\hat{b}\rangle = \sinh^2 r$, one obtains the final expression for the photon-number fluctuations in the output mode [107, 121]

$$\begin{aligned}\langle(\Delta\hat{b}'^\dagger\hat{b}')^2\rangle &= \sin^4(\varphi/2)|\alpha|^2 + 2\cos^4(\varphi/2)\cosh^2 r\sinh^2 r \\ &\quad + \cos^2(\varphi/2)\sin^2(\varphi/2)(|\alpha|^2 e^{-2r} + \sinh^2 r).\end{aligned} \quad (7.29)$$

At this point, one may assume that the net phase difference φ consists of a reference phase φ_0 and an extremely small phase shift $\delta\varphi$ caused by the gravitational wave, where $\delta\varphi \ll \varphi_0 \ll 1$. Then, the contribution of the signal $\delta\varphi$ to the intensity of light arriving the photodiode is

$$\langle\hat{b}'^\dagger\hat{b}'\rangle = \sin^2(\varphi/2)|\alpha|^2 + \cos^2(\varphi/2)\sinh^2 r \approx \frac{\varphi_0}{2}|\alpha|^2\delta\varphi, \quad (7.30)$$

where we have neglected the contributions from the squeezed vacuum state, as long as $|\alpha| \gg \cot(\varphi/2)\sinh r \approx 2\sinh r/\varphi_0$ for a strong input laser field and a moderate value of the squeeze parameter. Besides, under the conditions $\varphi_0 \ll 2e^{-r}$ and $|\alpha| \gg \sqrt{2}e^r\cot(\varphi/2)\cosh r\sinh r \approx \sqrt{2}e^r\sinh 2r/\varphi_0$, the photon-number fluctuations in

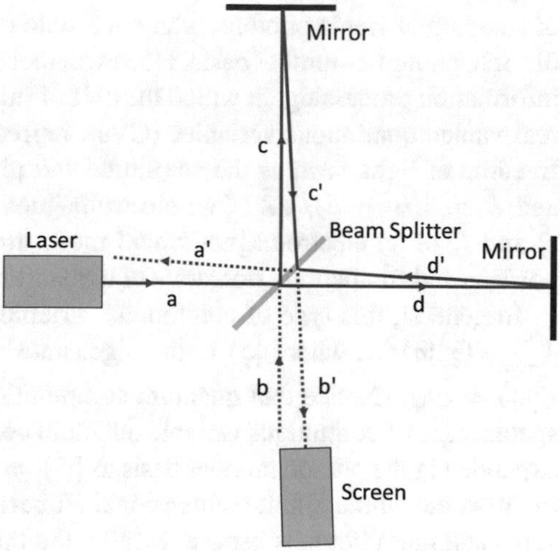

Fig. 7.3 Schematic of a gravitational wave Michelson interferometer. Typically, a is a strong laser field in a coherent state, and b is either a vacuum or a squeezed vacuum state. The amplitudes of the fields in each arm of the interferometer just behind the beam splitter are c and d, and the amplitudes of the fields just before recombination are c' and d', which produce the output beams a' and b'

the output mode are dominated by

$$\sqrt{\langle(\Delta\hat{b}'^{\dagger}\hat{b}')^2\rangle} \approx \cos(\varphi/2)\sin(\varphi/2)|\alpha|e^{-r} \approx \frac{\varphi_0}{2}|\alpha|e^{-r}. \tag{7.31}$$

Comparing Eqs. (7.30) and (7.31), the signal from the gravitational wave is large than the noise only when

$$\delta\varphi > \frac{e^{-r}}{|\alpha|} = \frac{e^{-r}}{\sqrt{\langle\hat{n}_a\rangle}}, \tag{7.32}$$

which reduces to the standard shot noise without squeezing. On the other hand, in the present of squeezing, the inevitable vacuum fluctuations incident from the other port of the interferometer will be attenuated by a factor e^{-r}.

7.3 Continuous Variable Quantum Information

Besides the detection of gravitational wave, another major application of squeezed coherent states is in continuous variable quantum information. The concept of continuous variable quantum information was initially proposed by Lloyd and Braunstein in 1999 [122], and the associated error correction codes were proposed even earlier by Braunstein in 1998 [123]. Traditionally, quantum information processing concerns the application and manipulation of **qubits**, which is the quantum analogue of the classical bits, described by a linear superposition of two orthonormal basis states $|0\rangle$ and $|1\rangle$, i.e., $|\psi\rangle = \alpha|0\rangle + \beta|1\rangle$, where the quantum information is encoded in the complex coefficients α and β. In particular, for photonic quantum information processing, the information is carried by the degrees of freedom of single photons, where the state of photonic qubits is described by the discrete photonic-number basis. However, there exists another approach to quantum information processing, in which the unit of information is a linear superposition of real-valued continuous variables (CVs), represented by the continuous degrees of freedom of light, such as the amplitude and phase quadratures $\hat{Q} \equiv (\hat{a} + \hat{a}^\dagger)/\sqrt{2}$ and $\hat{P} \equiv i(\hat{a}^\dagger - \hat{a})/\sqrt{2}$ of an electromagnetic field mode. Here, the quadratures $\hat{P}$ and $\hat{Q}$ of an electromagnetic field mode are the mathematical analogues of the position and momentum operators of a quantum harmonic oscillator.

In general, this type of continuous variable information is described by $|\psi\rangle = \int_{-\infty}^{\infty} \psi(q)|q\rangle dq$, where $|q\rangle$ is the eigenstate of the amplitude quadrature $\hat{Q}$, i.e., $\hat{Q}|q\rangle = q|q\rangle$. Notice that quantum computation with discrete photonic qubits is a special case of continuous variable quantum computation, since the state $|\psi\rangle$ can be expanded in the photon number basis as $|\psi\rangle = \sum_{n=0}^{\infty} c_n|n\rangle$ with $c_n \equiv \langle n|\psi\rangle$, which involves the whole infinite dimensional Hilbert space of a light mode rather than its zero- and one-photon subspace. Ideally, the basic state $|q\rangle$ for continuous variable quantum information processing can be realized by an infinitely squeezed coherent state

$$|q\rangle \equiv \lim_{r\to\infty} \frac{\sqrt{\mu}}{\pi^{1/4}} |\sqrt{2}\mu q, r\rangle, \tag{7.33}$$

where $\mu \equiv \cosh r$ and r is the squeeze parameter. Here, one can show that two arbitrary basic states satisfy the orthogonality relation $\langle q'|q\rangle = \delta(q' - q)$.

Similarly to a universal quantum computer over discrete variables which applies local operations to execute any desired unitary transformation over those variables to an arbitrary accuracy, a continuous variable quantum computer is *universal* if it can simulate any desired unitary quantum gates which transform the initial superposition of CVs, $|\psi\rangle = \int_{-\infty}^{\infty} \psi(q)|q\rangle dq$, into another superposition of CVs, $\hat{U}|\psi\rangle = \int_{-\infty}^{\infty} \psi(q)\hat{U}|q\rangle dq$.

In order to construct an arbitrary unitary quantum gate $\hat{U} \equiv e^{i\hat{H}t}$, it is required that the Hamiltonian $\hat{H}$ consists arbitrary polynomials of the quadratures $\hat{Q}$ and $\hat{P}$ of an electromagnetic field mode. Practically, the numbers of possible quantum gates over discrete or continuous variables are both uncountable, and thus it is impossible to exactly simulate arbitrary quantum logic gates. However, one may still find a finite set of **universal quantum gates**, so that any desired unitary operations can be approximated up to an arbitrary accuracy by a sequence of gates from this finite set.

For traditional quantum information processing with discrete variables such as qubits, the single-qubit rotation gates $R_x(\theta)$, $R_y(\theta)$, $R_z(\theta)$ which rotate the qubit on the Bloch sphere by an angle θ about the x-, y-, or z-axis, respectively, the single-qubit phase shift gate $P(\varphi)$ which maps the basis states $|0\rangle \to |0\rangle$ and $|1\rangle \to e^{i\varphi}|1\rangle$, and the two-qubit controlled NOT gate which maps the basics states $|a, b\rangle$ to $|a, a \oplus b\rangle$ form a universal set of quantum gates, where $\oplus$ denotes the exclusive disjunction operation. Similarly, for quantum information processing with continuous variables, one may show that a universal quantum computer over continuous variables such as the quadratures of the electromagnetic field can be constructed via simple devices including beam splitters, phase shifters, or squeezers and single nonlinear devices such as optical fibers with Kerr-like nonlinearity. Unitary transformations which involve Hamiltonians that are linear or quadratic in the quadratures $\hat{Q}$ and $\hat{P}$ are called **Gaussian gates**. This implies that an arbitrary Gaussian gate and at least one non-Gaussian gate which involve higher-order Hamiltonians are sufficient to simulate arbitrary unitary transformations. The simplest logic gates among this finite set are the position translation and momentum boost operators defined by $\hat{X}(s) \equiv e^{-is\hat{P}}$ and $\hat{Z}(s) \equiv e^{is\hat{Q}}$, respectively, where the associated Hamiltonians are linear in the quadratures of an electromagnetic field mode. On the computational basis of the amplitude quadrature eigenstates $|q\rangle$ with $q \in \mathbb{R}$, the actions of the position translation and momentum boost operators are

$$\hat{X}(s)|q\rangle = |q+s\rangle, \; \hat{Z}|q\rangle = e^{isq}|q\rangle. \tag{7.34}$$

In optical systems, one can implement the displacement operations by use of the linear elector-optic effects (*Pockels effects*), i.e., by modulating the amplitude and phase of optical beams with an electro-optic modulator (EOM), which is described by a Hamiltonian linear in the annihilation and creation operators, or equivalently, in the amplitude and phase quadratures of an electromagnetic field mode: $\hat{H} \propto a\hat{Q} - b\hat{P}$. Another two important Gaussian logic gates over continuous variables are the rotation gate $\hat{R}(s) \equiv e^{is(\hat{P}^2+\hat{Q}^2)/2}$ with $\hat{H} \propto \hat{Q}^2 + \hat{P}^2$ and the squeezing gate $\hat{S}(s) \equiv e^{-is(\hat{Q}\hat{P}+\hat{P}\hat{Q})/2}$ with $\hat{H} \propto \hat{Q}\hat{P} + \hat{P}\hat{Q}$, which can be implemented in optical systems by introducing a phase shift in the beams and by using the second-order nonlinear effects, respectively. In particular, a rotation operation by $\theta \equiv \pi/2$ is called the **Fourier transform gate** $\hat{F} \equiv \hat{R}(\pi/2)$. It is a continuous variable quantum information analogue of the **Hadamard gate** acting on a single qubit, which maps between the quadrature operators of an electromagnetic field mode as

$$\hat{F}^\dagger\hat{Q}\hat{F} = -\hat{P}, \; \hat{F}^\dagger\hat{P}\hat{F} = \hat{Q}. \tag{7.35}$$

To simulate arbitrary Gaussian gates, one also needs the phase gate $P(s) \equiv e^{s\hat{Q}^2/2}$ with $\hat{H} \propto \hat{Q}^2$, which is the continuous variable quantum information analogue of the phase gate acting on a single qubit, and can be also implemented by use of second-order nonlinear effects in optical systems. Finally, in order to perform arbitrary unitary operations, one needs an additional non-Gaussian gate such as the Kerr gate $V(s) \equiv e^{is\hat{Q}^3/3}$ with $\hat{H} \propto \hat{Q}^3$, which can be implemented in optical systems by imposing a cubic phase shift on the beams based on the Kerr-like third-order nonlinearity effects. Hence, any arbitrary unitary operation can be expressed as a finite sequence of gates from the universal set of gates specified by

$$\hat{R}(s) \equiv e^{is(\hat{Q}^2+\hat{P}^2)/2}, \tag{7.36a}$$

$$\hat{Z}(s) \equiv e^{is\hat{Q}}, \tag{7.36b}$$

$$\hat{P}(s) \equiv e^{is\hat{Q}^2/2}, \quad (7.36c)$$

$$\hat{V}(s) \equiv e^{is\hat{Q}^3/3}. \quad (7.36d)$$

In general, for multimode optical systems of which the quadrature operators satisfy the commutation relations $[\hat{Q}_j, \hat{Q}_k] = [\hat{P}_j, \hat{P}_k] = 0$ and $[\hat{Q}_j, \hat{P}_k] = i\delta_{jk}$, the universal set of gates is specified by the multimode analogues of the above gates plus the multi-mode controlled phase gate $\hat{C}_z(s) \equiv e^{is\hat{Q}_i\hat{Q}_j}$

$$\hat{R}_j(s) \equiv e^{is(\hat{Q}_j^2+\hat{P}_j^2)/2}, \quad (7.37a)$$

$$\hat{Z}_j(s) \equiv e^{is\hat{Q}_j}, \quad (7.37b)$$

$$\hat{P}_j(s) \equiv e^{is\hat{Q}_j^2/2}, \quad (7.37c)$$

$$\hat{V}_j(s) \equiv e^{is\hat{Q}_j^3/3}, \quad (7.37d)$$

$$\hat{C}_z(s) \equiv e^{is\hat{Q}_i\hat{Q}_j}. \quad (7.37e)$$

7.4 Spin Squeezed States

In the last few sections, we have shown that squeezing in bosonic systems is a promising method to reduce quantum fluctuations while preserving the minimum uncertainty product. An electromagnetic field is squeezed if the uncertainty measured by the variance of one field quadrature is smaller than the standard quantum limit of 1/4.

Similar to bosonic systems, atomic systems which consist of $N = 2j$ spin-1/2 particles also exhibit quantum squeezing. Based on the observation that a spin-j coherent state $|\theta, \phi\rangle$ is equivalent to a set of N elementary spins all pointing along the same mean direction (θ, ϕ), in a paper entitled "Squeezed spin states" published in 1993 [124], Kitagawa and Ueda argued that the reduction of quantum fluctuations in one direction at the expense of those enhanced in the other direction is realizable, as long as the collective spin system is constructed from correlated or entangled elementary spin-1/2 particles. As we will show below, in contrast to the spin coherent states which have an isotropic quasi-probability distribution in a spherical space, the spin squeezed states have an elliptical quasi-probability distribution. As linear Hamiltonians only rotate the elementary spin-1/2 particles collectively and do not produce correlations and entanglements among them, the spin squeezing must be established by nonlinear interactions. Utilizing different nonlinear Hamiltonians, Kitagawa and Ueda showed that there are two types of spin squeezed states in correlated spin ensembles, namely, one- and two-axis spin squeezed states.

The one-axis spin squeezed states $|j, \zeta, \mu\rangle \equiv \hat{U}|j, \zeta\rangle$, with $\hat{U} \equiv e^{-\frac{i}{2}\mu\hat{J}_z^2}$ and $\mu \equiv 2\chi t$, are generated by the lowest-order nonlinear interaction $\hat{H} \equiv \chi\hat{J}_z^2$, where $|j, \zeta\rangle$ is an initial spin coherent state. In the Heisenberg picture, the spin operators obey the precession equations

$$\dot{\hat{J}}_x = i\chi(\hat{J}_z\hat{J}_y + \hat{J}_y\hat{J}_z),\ \dot{\hat{J}}_y = -i\chi(\hat{J}_z\hat{J}_x + \hat{J}_x\hat{J}_z),\ \dot{\hat{J}}_z = 0. \tag{7.38}$$

One sees that the precession of the collective spin is modulated by $\hat{J}_z$, which is a twisting effect analogous to the self-phase modulation in photonic systems. As the nonlinear twisting Hamiltonian $\hat{H}(\hat{J}_z^2)$ is along the z-axis, one may choose the initial state as a spin coherent state along the x-axis, so that the one-axis spin squeezed state has the form

$$|j, \mu\rangle \equiv e^{-\frac{i}{2}\mu\hat{J}_z^2}\left|\frac{\pi}{2}, 0\right\rangle = 2^{-j}\sum_{m=-j}^{j}\sqrt{\binom{2j}{j+m}}e^{-\frac{i}{2}\mu m^2}|j, m\rangle. \tag{7.39}$$

A direct calculation shows that the mean spin direction of the one-axis spin squeezed state is still long the x-axis, i.e., $\langle\hat{J}_x\rangle = j\cos^{2j-1}\frac{\mu}{2}$, $\langle\hat{J}_y\rangle = \langle\hat{J}_z\rangle = 0$, but the magnitude of the mean spin vector is reduced by a factor $\cos^{2j-1}\frac{\mu}{2}$, compared with that of the spin coherent state $|\pi/2, 0\rangle$. A direct computation yields the variances and covariances for the one-axis spin squeezed states

$$(\Delta\hat{J}_x)^2 = \frac{j}{2}\left[j + \frac{1}{2} + \left(j - \frac{1}{2}\right)\cos^{2j-2}\mu - 2j\cos^{2(2j-1)}\frac{\mu}{2}\right], \tag{7.40}$$

$$(\Delta\hat{J}_y)^2 = \frac{j}{2}\left[j + \frac{1}{2} - \left(j - \frac{1}{2}\right)\cos^{2j-2}\mu\right],\ (\Delta\hat{J}_z)^2 = \frac{j}{2},$$

$$\mathrm{Cov}(\hat{J}_y, \hat{J}_z) = j\left(j - \frac{1}{2}\right)\sin\frac{\mu}{2}\cos^{2j-2}\frac{\mu}{2},\ \mathrm{Cov}(\hat{J}_z, \hat{J}_x) = \mathrm{Cov}(\hat{J}_x, \hat{J}_y) = 0,$$

where $\mathrm{Cov}(\hat{A}, \hat{B}) \equiv \frac{1}{2}(\hat{A}\hat{B} + \hat{B}\hat{A})$ is the covariance between the operators $\hat{A}$ and $\hat{B}$. Notice that in the absence of the nonlinear twisting effects, i.e., $\mu \equiv 2\chi t = 0$, one immediately recovers the variances for the spin coherent states, $(\Delta\hat{J}_x)^2 = 0$, $(\Delta\hat{J}_y)^2 = (\Delta\hat{J}_z)^2 = j/2$.

In order to visualize the one-axis spin squeezed state $|j, \mu\rangle$, one may plot the quasi-probability distribution $Q(\theta, \phi)$ defined by

$$Q(\theta, \phi) \equiv \frac{2j+1}{4\pi}|\langle\theta, \phi|j, \mu\rangle|^2 \tag{7.41}$$

$$= \frac{2j+1}{4\pi \cdot 2^{2j}}\left|\sum_{m=-j}^{j}\binom{2j}{j+m}\left(\cos\frac{\theta}{2}\right)^{j-m}\left(e^{i\phi}\sin\frac{\theta}{2}\right)^{j+m}e^{-\frac{i}{2}\mu m^2}\right|^2.$$

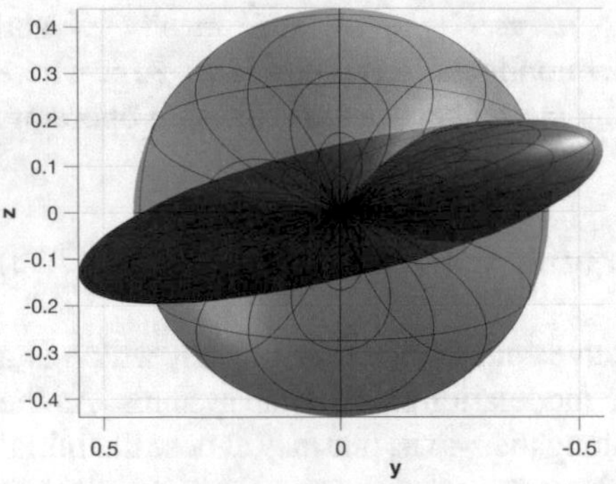

(A) $\mu = 0.2$

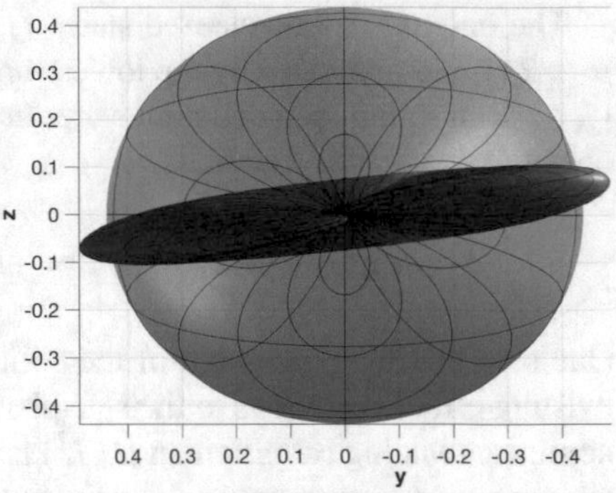

(B) $\mu = 0.4$

Fig. 7.4 Schematic of the Husimi-Q quasi-probability distribution $Q(\theta, \phi)$ for the one-axis spin squeezed state $|j, \mu\rangle$ with $j = 20$ and $\mu = 0.2$ or 0.4, respectively, where $x \equiv Q(\theta, \phi) \sin\theta \cos\phi$, $y \equiv Q(\theta, \phi) \sin\theta \sin\phi$, and $z \equiv Q(\theta, \phi) \cos\theta$. The spin squeezed states $|j, \mu\rangle$ are shown in viridian, and the initial spin coherent state $|\pi/2, 0\rangle$ is shown in half-transparent moss green and lavender

In Fig. 7.4, the Husimi-Q quasi-probability distribution $Q(\theta, \phi)$ for the one-axis spin squeezed state $|j, \mu\rangle$ manifests the spin squeezing effects when μ is nonzero. One sees that the spin squeezing effect is larger when μ increases from 0.2 to 0.4, while the optimal squeezing angle depends on the evolution time t via $\mu = 2\chi t$.

Unlike the one-axis spin squeezed states $|j, \mu\rangle$, there is another type of spin squeezed states, namely, the two-axis spin squeezed states, of which the optimal squeezing angles are invariant under time evolution. The two-axis spin squeezed states can be generated by applying nonlinear twisting simultaneously clockwise and counterclockwise along the two orthonormal axes in the $\theta = \pi/2$ and $\phi = \pm\pi/4$ directions, which are both in the plane normal to the mean spin direction. Hence, the nonlinear twisting Hamiltonians for the two-axis spin squeezed states $|j, \nu\rangle \equiv e^{-it\hat{H}(\hat{J}_+, \hat{J}_-)}|j, -j\rangle = e^{-\frac{\nu}{2}(\hat{J}_+^2 - \hat{J}_-^2)}|j, -j\rangle$ are given by $\hat{H} \equiv \chi(\hat{J}_x\hat{J}_y + \hat{J}_y\hat{J}_x) = \frac{\chi}{2i}(\hat{J}_+^2 - \hat{J}_-^2)$, where $\nu \equiv \chi t$. One characteristic of the two-axis spin squeezed states is that the covariance $\langle \hat{J}_x\hat{J}_y + \hat{J}_x\hat{J}_y\rangle$ vanishes, as $\hat{J}_x\hat{J}_y + \hat{J}_x\hat{J}_y$ is conserved in time, i.e., $[\hat{J}_x\hat{J}_y + \hat{J}_x\hat{J}_y, \hat{H}] = 0$, and the covariances $\langle \hat{J}_\pm^2\rangle$ vanish for the initial lowest weight state $|j, -j\rangle$.

Moreover, as $\hat{J}_\pm^2|j, m\rangle \propto |j, m \pm 2\rangle$, the two-axis twisting Hamiltonian change m in the Dicke states $|j, m\rangle$ by 2. Hence, the two-axis spin squeezed state $|j, \nu\rangle$ must be an even parity state, i.e., $|j, \nu\rangle$ is spanned only by Dicke states $|j, -j + n\rangle$ with n even, and the states $\hat{J}_\pm|j, \nu\rangle$ must be odd parity states, i.e., $\hat{J}_\pm|j, \nu\rangle$ are spanned only be Dicke states $|j, -j + n\rangle$ with n odd. In other words, the two-axis spin squeezed states $|j, \nu\rangle$ as well as $\hat{J}_\pm^2|j, \nu\rangle$ are eigenstates of the parity operator $(-1)^{\hat{J}_z+j}$ with an eigenvalue 1, while the states $\hat{J}_\pm|j, \nu\rangle$ are the eigenstate of the parity operator $(-1)^{\hat{J}_z+j}$ with eigenvalues -1. In this regard, both the expectations $\langle \hat{J}_x\rangle \equiv \frac{1}{2}(\langle \hat{J}_+\rangle + \langle \hat{J}_-\rangle)$ and $\langle \hat{J}_y\rangle \equiv \frac{1}{2i}(\langle \hat{J}_+\rangle - \langle \hat{J}_-\rangle)$ vanish, as they involve inner products between Dicke states with different parities. Unlike the one-axis spin squeezed state, as both the expectations $\langle \hat{J}_x \rangle$ and $\langle \hat{J}_y \rangle = 0$ vanish, the mean spin direction always points along the z-direction. For the same reason, both the expectations $\langle \hat{J}_\pm \hat{J}_z \rangle$ and $\langle \hat{J}_z \hat{J}_\pm \rangle$, as well as the covariances $\langle \hat{J}_x \hat{J}_z + \hat{J}_z \hat{J}_x \rangle$ and $\langle \hat{J}_y \hat{J}_z + \hat{J}_z \hat{J}_y \rangle$, vanish identically. But unfortunately, the two-axis twisting model cannot be solved analytically for arbitrary j. As examples, one may compute the two-axis spin squeezed states $|j, \nu\rangle$ explicitly for $j \leq 2$:

$$|1, \nu\rangle = -\sin \nu |1, 1\rangle + \cos \nu |1, -1\rangle, \tag{7.42a}$$

$$\left|\frac{3}{2}, \nu\right\rangle = -\sin(\sqrt{3}\nu)|\frac{3}{2}, \frac{1}{2}\rangle + \cos(\sqrt{3}\nu)\left|\frac{3}{2}, -\frac{3}{2}\right\rangle, \tag{7.42b}$$

$$|2, \nu\rangle = \sin^2(\sqrt{3}\nu)|2, 2\rangle - \frac{1}{\sqrt{2}}\sin(2\sqrt{3}\nu)|2, 0\rangle + \cos^2(\sqrt{3}\nu)|2, -2\rangle, \tag{7.42c}$$

which yields the following variances with respect to $|j, \nu\rangle$ for $j \leq 2$:

$$\langle \hat{J}_x^2 \rangle = \sin^2 \frac{\nu}{2}, \langle \hat{J}_y^2 \rangle = \cos^2 \frac{\nu}{2}, \text{ for } j = 1; \tag{7.43a}$$

$$\begin{aligned}\langle \hat{J}_x^2 \rangle &= \frac{3}{4} + \sin^2(\sqrt{3}\nu) + \frac{\sqrt{3}}{2}\sin(2\sqrt{3}\nu), \\ \langle \hat{J}_y^2 \rangle &= \frac{3}{4} + \sin^2(\sqrt{3}\nu) - \frac{\sqrt{3}}{2}\sin(2\sqrt{3}\nu), \text{ for } j = \frac{3}{2};\end{aligned} \tag{7.43b}$$

$$\begin{aligned}\langle \hat{J}_x^2 \rangle &= 1 + \sin^2(2\sqrt{3}\nu) - \sqrt{3}\sin(2\sqrt{3}\nu), \\ \langle \hat{J}_y^2 \rangle &= 1 + \sin^2(2\sqrt{3}\nu) + \sqrt{3}\sin(2\sqrt{3}\nu), \text{ for } j = 2.\end{aligned} \tag{7.43c}$$

In general, one may expand the two-axis spin squeezed states $|j, \nu\rangle$ as $\sum_{m=-j}^{j} c_m |j, m\rangle$, where the coefficients c_{-j+n} vanish with n odd. Then one may visualize the two-axis spin squeezed state $|j, \mu\rangle$ by plotting the quasi-probability distribution $Q(\theta, \phi)$

$$Q(\theta, \phi) = \frac{2j+1}{4\pi}\left|\sum_{m=-j}^{j}\sqrt{\binom{2j}{j+m}}\left(\cos\frac{\theta}{2}\right)^{j-m}\left(\sin\frac{\theta}{2}e^{i\phi}\right)^{j+m} c_m\right|^2. \tag{7.44}$$

In Fig. 7.5, the Husimi-Q quasi-probability distribution for the two-axis spin squeezed state $|j, \nu\rangle$ manifests the spin squeezing effects when ν is nonzero. From Figs. 7.5a and b, the xy plane projections of $Q(\theta, \phi)$, one sees that unlike the one-axis spin squeezed states, the optimal squeezing direction for two-axis spin squeezed states is always along the x axis for small ν. From Figs. 7.5c and d, the full three-

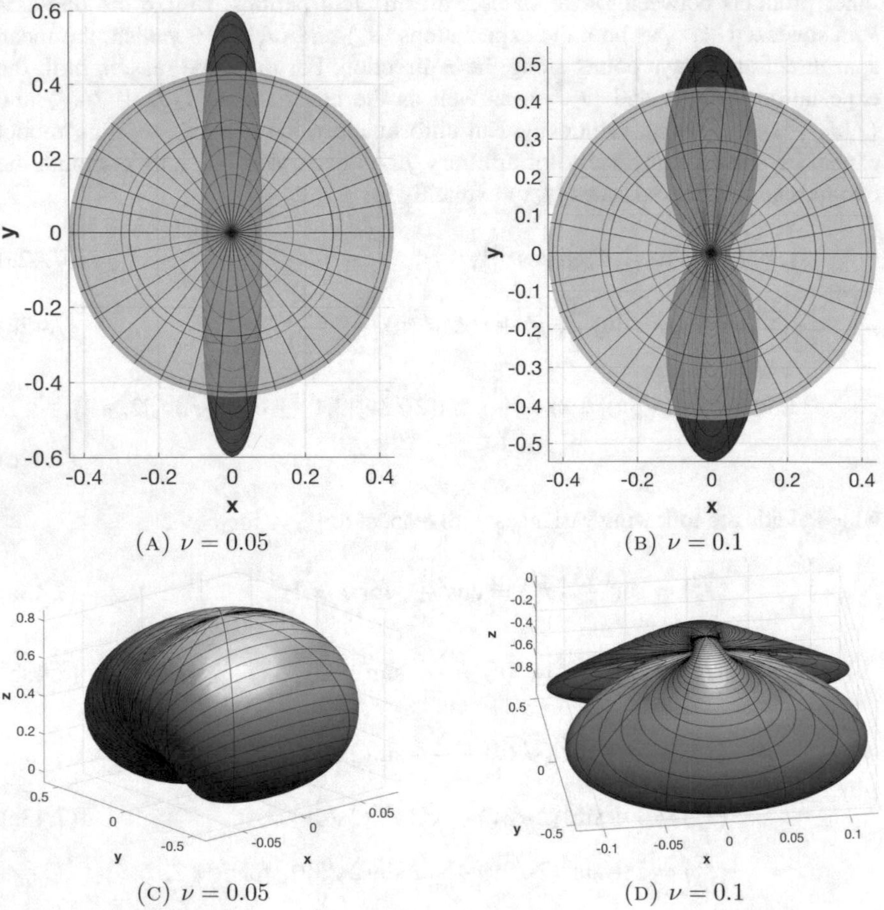

(A) $\nu = 0.05$ (B) $\nu = 0.1$

(C) $\nu = 0.05$ (D) $\nu = 0.1$

Fig. 7.5 Schematic of the Husimi-Q quasi-probability distribution $Q(\theta, \phi)$ for the two-axis spin squeezed state $|j, \nu\rangle$ with $j = 20$, and $\nu = 0.05$ or 0.1, respectively. In Figs. 7.5a and b, the xy plane projections of the Q distribution for the two-axis spin squeezed states are shown in blue, where the Q distribution for the initial state $|j, -j\rangle$ is shown in half-transparent grass green. In Figs. 7.5c and d, the Q distributions for the two-axis squeezed states are shown. Here, $x \equiv Q(\theta, \phi) \sin\theta \cos\phi$, $y \equiv Q(\theta, \phi) \sin\theta \sin\phi$, and $z \equiv Q(\theta, \phi) \cos\theta$

dimensional Q distribution, one observes that the Q distribution changes from the shape of a pea to the shape of a cell when ν increases from 0.05 to 0.1, indicating larger spin squeezing effects along the x direction.

For a spin coherent state, the variance $(\Delta \hat{J}_{\mathbf{n}})^2$ of the spin operator $\hat{J}_{\mathbf{n}} \equiv \hat{J} \cdot \mathbf{n}$ will in general vary for different directions $\mathbf{n}$. However, it remains the same along the directions $\mathbf{n}_\perp$ perpendicular to the mean spin direction, i.e., $(\Delta \hat{J}_{\mathbf{n}_\perp})^2 = j/2$. Kitagawa and Ueda argued that, the criteria for spin squeezing is that the variance $(\Delta \hat{J}_{\mathbf{n}_\perp})^2 < j/2$ for some specific directions $\mathbf{n}_\perp$ perpendicular to the mean spin direction [124]. In other words, one may define the spin squeezing parameter ξ_s via

$$\xi_s^2 \equiv \frac{\min((\Delta \hat{J}_{\mathbf{n}_\perp})^2)}{j/2}, \tag{7.45}$$

where the minimization is over all directions $\mathbf{n}_\perp$ perpendicular to the mean spin direction. Let us denote the mean spin direction for a general spin-j state as $\mathbf{n}_0 \equiv (\theta, \phi)$. For example, the mean spin direction for the one-axis spin squeezed state $|j, \mu\rangle \equiv e^{-\frac{i}{2}\mu \hat{J}_z^2}|\pi/2, 0\rangle$ is the x direction. With respect to the mean spin direction $\mathbf{n}_0$, the other two orthonormal bases are

$$\mathbf{n}_1 \equiv (-\sin\phi, \cos\phi, 0) \text{ and } \mathbf{n}_2 \equiv (\cos\theta\cos\phi, \cos\theta\sin\phi, -\sin\theta). \tag{7.46}$$

where the above expressions are only valid for $\theta \neq 0$ or π. For $\theta = 0$ or π, the mean spin direction is along the $\pm z$ directions, and ϕ can be fixed as 0 or π respectively. An arbitrary direction perpendicular to the mean spin direction can be represented as $\mathbf{n}_\perp \equiv \mathbf{n}_1 \cos\varphi + \mathbf{n}_2 \sin\varphi$. As $\langle \hat{J}_{\mathbf{n}_1}\rangle = \langle \hat{J}_{\mathbf{n}_2}\rangle = 0$, the variance $\langle \hat{J}^2_{\mathbf{n}_\perp}\rangle$ can be written as

$$\begin{aligned}\langle \hat{J}^2_{\mathbf{n}_\perp}\rangle &= \langle \hat{J}^2_{\mathbf{n}_1}\rangle \cos^2\varphi + \langle\{\hat{J}_{\mathbf{n}_1}, \hat{J}_{\mathbf{n}_2}\}\rangle \cos\varphi\sin\varphi + \langle \hat{J}^2_{\mathbf{n}_2}\rangle \sin^2\varphi \\ &= \frac{1}{2}\left(\langle \hat{J}^2_{\mathbf{n}_1} + \hat{J}^2_{\mathbf{n}_2}\rangle\right) + \frac{A}{2}\cos 2\varphi + \frac{B}{2}\sin 2\varphi,\end{aligned} \tag{7.47}$$

where $A \equiv \langle \hat{J}^2_{\mathbf{n}_1} - \hat{J}^2_{\mathbf{n}_2}\rangle$, $B \equiv 2\,\mathrm{Cov}(\hat{J}_{\mathbf{n}_1}, \hat{J}_{\mathbf{n}_2}) \equiv \langle\{\hat{J}_{\mathbf{n}_1}, \hat{J}_{\mathbf{n}_2}\}\rangle$, and $\mathrm{Cov}(\hat{J}_{\mathbf{n}_1}, \hat{J}_{\mathbf{n}_2})$ are the covariances between $\hat{J}_{\mathbf{n}_1}$ and $\hat{J}_{\mathbf{n}_2}$. Hence, one immediately obtains

$$\frac{(\Delta \hat{J}_{\mathbf{n}_\perp})^2}{j/2} \geq \frac{1}{j}\left(\langle \hat{J}^2_{\mathbf{n}_1} + \hat{J}^2_{\mathbf{n}_2}\rangle - \sqrt{A^2 + B^2}\right), \tag{7.48}$$

where the minimum value of the variance $(\Delta \hat{J}_{\mathbf{n}_\perp})^2$ occurs when the relations $\cos 2\varphi = -A/\sqrt{A^2 + B^2}$ and $\sin 2\varphi \equiv -B/\sqrt{A^2 + B^2}$ are satisfied. In other words, the spin squeezing parameter is determined by

$$\xi_s^2 = \frac{1}{j}\left\{\langle \hat{J}^2_{\mathbf{n}_1} + \hat{J}^2_{\mathbf{n}_2}\rangle - \sqrt{\left(\langle \hat{J}^2_{\mathbf{n}_1} - \hat{J}^2_{\mathbf{n}_2}\rangle\right)^2 + 4\,\mathrm{Cov}(\hat{J}_{\mathbf{n}_1}, \hat{J}_{\mathbf{n}_2})^2}\right\}, \tag{7.49}$$

where the optimal squeezing angle is given by

$$\varphi = \begin{cases} \dfrac{1}{2}\arccos\left(\dfrac{-A}{\sqrt{A^2 + B^2}}\right) & \text{for } B \leq 0, \\ \pi - \dfrac{1}{2}\arccos\left(\dfrac{-A}{\sqrt{A^2 + B^2}}\right) & \text{for } B > 0. \end{cases} \tag{7.50}$$

In particular, from Eq. (7.40), one immediately obtains the spin squeezing parameter for the one-axis spin squeezed states

$$\xi_s^2 = 1 + \frac{N-1}{2}\left(\frac{1-\cos^{N-2}\mu}{2} - \sqrt{\frac{(1-\cos^{N-2}\mu)^2}{4} + 4\sin^2\frac{\mu}{2}\cos^{2(N-2)}\frac{\mu}{2}}\right), \tag{7.51}$$

where $N \equiv 2j$. Clearly, for $j = 1/2$ or $N = 1$, the spin squeezing parameter always equals to 1, and hence spin squeezing effects only exist in higher spin systems with $j > 1/2$. Here, the spin squeezing parameters ξ_s for one-axis spin squeezed states $|j, \mu\rangle$ with $j \le 2$ are shown as explicit examples:

$$\xi_s^2 = 1 - \left|\sin\frac{\mu}{2}\right| \ge 0, \text{ for } j = 1, \tag{7.52a}$$

$$\xi_s^2 = 1 + \sin^2\frac{\mu}{2} - \sqrt{\sin^4\frac{\mu}{2} + 4\sin^2\frac{\mu}{2}\cos^2\frac{\mu}{2}} \ge \frac{1}{3}, \text{ for } j = \frac{3}{2}, \tag{7.52b}$$

$$\xi_s^2 = 1 + 3\cos^2\frac{\mu}{2}\left(\sin^2\frac{\mu}{2} - \sqrt{\sin^4\frac{\mu}{2} + \sin^2\frac{\mu}{2}}\right) \tag{7.52c}$$

$$\ge 1 + 3(1-r)(r - \sqrt{r^2 + r}) \approx 0.3025, \text{ for } j = 2,$$

where $r \approx 0.2228$ is the smallest positive root of the cubic equation $8r^3 + 5r^2 - 6r + 1 = 0$. As a comparison, one can also compute the spin squeezing parameters ξ_s for two-axis spin squeezed state $|j, \nu\rangle$ with $j \le 2$:

$$\xi_s^2 = 1 - \sin(2\nu), \text{ for } j = 1, \tag{7.53a}$$

$$\xi_s^2 = 1 + \frac{4}{3}\sin^2(\sqrt{3}\nu) \mp \frac{2\sqrt{3}}{3}\sin(2\sqrt{3}\nu) \ge \frac{1}{3}, \text{ for } j = \frac{3}{2}, \tag{7.53b}$$

$$\xi_s^2 = 1 + \sin^2(2\sqrt{3}\nu) \mp \sqrt{3}\sin(2\sqrt{3}\nu) \ge \frac{1}{4}, \text{ for } j = 2. \tag{7.53c}$$

Exercises

7.1. Verify Eq. (7.10) by using the Gaussian integral

$$\int_{-\infty}^{\infty} \exp\left(-\frac{1}{2}\sum_{i,j=1}^{n} A_{ij}x_i x_j\right) d^n x = \sqrt{\frac{(2\pi)^n}{\det A}}.$$

7.2. Show that the eigenstates $|q\rangle$ and $|p\rangle$ of the amplitude and phase quadratures which satisfy $\hat{Q}|q\rangle = q|q\rangle$ and $\hat{P}|p\rangle = p|p\rangle$ can be constructed from the following infinitely squeezed coherent states

$$|q\rangle \equiv \lim_{r\to\infty} \frac{\sqrt{\mu}}{\pi^{1/4}} |\sqrt{2}\mu q, r\rangle, \; |p\rangle \equiv \lim_{r\to\infty} \frac{\sqrt{\mu}}{\pi^{1/4}} |i\sqrt{2}\mu p, r\rangle,$$

where $\mu \equiv \cosh r$ and r is the squeeze parameter.

7.3. Verity the relation $\langle q|p\rangle = \frac{1}{\sqrt{2\pi}} e^{iqp}$, where $|q\rangle$ and $|p\rangle$ are eigenstates of the quadrature operators of an electromagnetic field mode.

7.4. Show that the operators $\hat{X}(q)$ and $\hat{Z}(p)$ are non-commutative and satisfy the identity

$$\hat{X}(q)\hat{Z}(p) = e^{-iqp}\hat{Z}(p)\hat{X}(q).$$

7.5. Show that in the computational basis of the quadrature eigenstates, the actions of $\hat{X}(q)$ and $\hat{Z}(p)$ are

$$\hat{X}(s)|q\rangle = |q+s\rangle, \; \hat{Z}|q\rangle = e^{isq}|q\rangle,$$

$$\hat{Z}(s)|p\rangle = |p+s\rangle, \; \hat{X}|p\rangle = e^{-isp}|p\rangle.$$

7.6. Verify that the Fourier transform gate F maps between the quadrature operators of an electromagnetic field mode as

$$\hat{F}^\dagger \hat{Q} \hat{F} = -\hat{P}, \; \hat{F}^\dagger \hat{P} \hat{F} = \hat{Q}.$$

7.7. For the one-axis spin squeezed state $|j,\mu\rangle \equiv e^{-\frac{i}{2}\mu \hat{J}_z^2}|\pi/2, 0\rangle$, verify the relations $\langle \hat{J}_x\rangle = j\cos^{2j-1}\frac{\mu}{2}$, $\langle \hat{J}_y\rangle = \langle \hat{J}_z\rangle = 0$.

7.8. Verify the following relations for the one-axis spin squeezed state

$$(\Delta \hat{J}_x)^2 = \frac{j}{2}\left[j + \frac{1}{2} + \left(j - \frac{1}{2}\right)\cos^{2j-2}\mu - 2j\cos^{2(2j-1)}\frac{\mu}{2}\right],$$

$$(\Delta \hat{J}_y)^2 = \frac{j}{2}\left[j + \frac{1}{2} - \left(j - \frac{1}{2}\right)\cos^{2j-2}\mu\right], (\Delta \hat{J}_z)^2 = \frac{j}{2},$$

$$\mathrm{Cov}(\hat{J}_y, \hat{J}_z) = j\left(j - \frac{1}{2}\right)\sin\frac{\mu}{2}\cos^{2j-2}\frac{\mu}{2}, \mathrm{Cov}(\hat{J}_z, \hat{J}_x) = \mathrm{Cov}(\hat{J}_x, \hat{J}_y) = 0.$$

7.9. Show the expectations $\langle \hat{J}_{\mathbf{n}_1}\rangle$ and $\langle \hat{J}_{\mathbf{n}_2}\rangle$ for spin squeezed states vanish in the two orthonormal directions $\mathbf{n}_1$ and $\mathbf{n}_2$ perpendicular to the mean spin direction.

Examples of Coherent States Beyond SU(2) 8

8.1 SU(1,1) Coherent States

In the last few chapters, we have discussed Glauber's coherent states for optical systems and spin coherent states for atomic systems which possess collective SU(2) symmetries. Of course, the applications of coherent states are definitely not limited to these two symmetries. In this chapter, we will discuss examples of coherent states which are beyond Glauber's coherent states and spin coherent states. One of the first nontrivial example of generated coherent states is that for the Lie group SU(1, 1), which is the simplest non-compact non-Abelian simple Lie group. In fact, the earliest application of this coherent state was in nonlinear quantum optics, where the Hamiltonian of a degenerate parametric oscillator is described by [125]

$$\hat{H} = \omega \hat{a}^\dagger \hat{a} + \kappa (e^{-2i\omega t} \hat{a}^{\dagger 2} + e^{2i\omega t} \hat{a}^2), \tag{8.1}$$

where κ is a coupling constant. One can show that the generators of the Lie group SU(1, 1) are realized by

$$K_+ \equiv \frac{1}{2}\hat{a}^{\dagger 2},\ K_- \equiv \frac{1}{2}\hat{a}^2,\ K_0 \equiv \frac{1}{4}(a^\dagger a + aa^\dagger). \tag{8.2}$$

As such, the Hamiltonian of a degenerate parametric oscillator becomes a linear combination of the SU(1, 1) generators K_0, K_+, and K_-

$$\hat{H} = 2\omega K_0 + 2\kappa (e^{-2i\omega t} K_+ + e^{2i\omega t} K_-), \tag{8.3}$$

which implies that the Hamiltonian (8.1) will preserve the SU(1, 1) coherent states under time evolution. The SU(1, 1) coherent states also appear in the Jaynes-Cummings model in quantum optics modified by intensity-dependent couplings,

C.-F. Kam et al., *Coherent States*, Lecture Notes in Physics 1011,
https://doi.org/10.1007/978-3-031-20766-2_8

so that the SU(1, 1) generators are realized by [126]

$$K_+ \equiv \sqrt{a^\dagger a} a^\dagger, K_- \equiv a\sqrt{a^\dagger a}, K_0 \equiv a^\dagger a + \frac{1}{2}. \tag{8.4}$$

In the following, we discuss in details the properties of the SU(1, 1) coherent states.

As we mentioned above, the Lie algebra $\mathfrak{su}(1, 1)$ corresponding to the Lie group SU(1, 1) has three generators K_0, K_+, and K_-, which obey the commutation relations

$$[K_-, K_+] = 2K_0, [K_0, K_\pm] = \pm K_\pm. \tag{8.5}$$

As for the Lie algebra $\mathfrak{su}(2)$, one may introduce another basis $\{K_0, K_1, K_2\}$ via $K_\pm = \pm i(K_1 \pm iK_2)$, so that the associated commutation relations are

$$[K_1, K_2] = -iK_0, [K_0, K_1] = iK_2, [K_2, K_0] = iK_1. \tag{8.6}$$

Let us consider an irreducible representation of SU(1, 1), marked by a number k which acquires discrete values. In order to specific its rows, one needs another number μ which is the eigenvalue of the operator K_0, i.e.,

$$K_0|k, \mu\rangle = \mu|k, \mu\rangle. \tag{8.7}$$

In fact, there are two discrete series of representations D_k^+ and D_k^-, where

$$D_k^+ : k > 0, \mu = k + m, m \in \{0, 1, 2, \cdots\}, \tag{8.8a}$$

$$D_k^- : k > 0, \mu = -k - m, m \in \{0, 1, 2, \cdots\}. \tag{8.8b}$$

Hence, it suffices to discuss the positive discrete series of representations D_k^+ instead of D_k^-, since one can transfer all the results immediately from one to another. As the Lie group SU(1, 1) is non-compact, the representations of the discrete series are all infinite-dimensional. However, in many respects, they are similar to the finite-dimensional representations of SU(2), as the basic vector $|k, k + n\rangle$ may be determined by an integer n ranging from zero to infinity.

From Eqs. (8.5) and (8.7), one may check that the raising and lowering operators K_+ and K_- obey the identities

$$K_0K_+|k, \mu\rangle = (\mu + 1)K_+|k, \mu\rangle, \tag{8.9a}$$

$$K_0K_-|k, \mu\rangle = (\mu - 1)K_-|k, \mu\rangle. \tag{8.9b}$$

One may also verify that the quadratic Casimir operator

$$\hat{C}_2 \equiv K_0^2 - \frac{1}{2}(K_+K_- + K_-K_+) = K_0^2 - K_1^2 - K_2^2 \tag{8.10}$$

is an invariant which commutes with all the generators K_0, K_+, and K_-. By virtue of Schur's lemma, the Casimir operator $\hat{C}_2$ is a multiple of the identity, $\hat{C}_2 = \lambda \hat{I}$. Then Eqs. (8.5) and (8.10) imply that the raising and lowering operators have to obey the identities

$$K_+ K_- = -\lambda \hat{I} + K_0(K_0 - 1), \tag{8.11a}$$

$$K_- K_+ = -\lambda \hat{I} + K_0(K_0 + 1). \tag{8.11b}$$

In the positive discrete series of representations D_k^+, the norm of the states $K_\pm |k, \mu\rangle$ must be non-negative since

$$\| K_\pm |k, \mu\rangle \|^2 = \langle k, \mu | K_\mp K_\pm | k, \mu\rangle \geq 0. \tag{8.12}$$

This implies that the operators $K_+ K_-$ and $K_- K_+$ must be positive Hermitian, and hence the eigenvalues of the operators

$$-\lambda \hat{I} + K_0(K_0 - 1), \tag{8.13a}$$

$$-\lambda \hat{I} + K_0(K_0 + 1), \tag{8.13b}$$

must be non-negative. By contrast, in the compact SU(2) case, these eigenvalues must be non-positive. Applying the lowering operators K_- repeatedly on $|k, \mu\rangle$, one of the following alternatives occurs: (i) the chain never terminates; (ii) the chain terminates. In case (i), the positivity conditions hold for all μ implies that $\lambda < -\frac{1}{4}$, where μ is a half-integer. In case (ii), by letting $|k, k\rangle$ be the last nonvanishing state in the descending chain, it follows that

$$K_- |k, k\rangle = 0, \tag{8.14}$$

and thus $K_+ K_- |k, k\rangle = [-\lambda \hat{I} + K_0(K_0 - 1)]|k, k\rangle = 0$, which implies that

$$\lambda = k(k - 1). \tag{8.15}$$

The case $k = 0$ corresponds to the identity representation with $K_\pm |k, k\rangle = 0$ and $\hat{C}_2 = 0$. The case $k > 0$ corresponds to an infinite chain of states $|k, k\rangle$, $|k, k + 1\rangle$, $\cdots$, obtained by applying the raising operator K_+ repeatedly on the state $|k, k\rangle$, which all exist and are different from zero since

$$\begin{aligned} \| K_+ |k, \mu\rangle \|^2 &= \langle k, \mu | K_- K_+ | k, \mu\rangle \\ &= \langle k, \mu | (2K_0 + K_+ K_-) | k, \mu\rangle \\ &= 2(k + m) \| |k, \mu\rangle \|^2 + \| K_+ |k, \mu\rangle \|^2 > 0, \end{aligned} \tag{8.16}$$

where $\mu \equiv k + m > 0$ in the positive discrete series of representations. From Eqs. (8.9a)–(8.9b) and Eqs. (8.11a)–(8.11b), one could write

$$K_+|k,\mu\rangle = \sqrt{(\mu+k)(\mu-k+1)}|k,\mu+1\rangle, \tag{8.17a}$$

$$K_-|k,\mu\rangle = \sqrt{(\mu-k)(\mu-k-1)}|k,\mu-1\rangle. \tag{8.17b}$$

As such, all the states can be obtained from $|k,k\rangle$ by the application of the raising operator K_+

$$|k,k+m\rangle = \sqrt{\frac{\Gamma(2k)}{m!\Gamma(2k+m)}}(K_+)^m|k,k\rangle. \tag{8.18}$$

In order to construct the SU(1, 1) coherent states, one needs to choose a fixed reference state. One may choose the lowest-weight state $|k,k\rangle$. By the extremal state $|k,k\rangle$, one can find the U(1) isotropy subgroup of SU(1, 1) which leaves $|k,k\rangle$ invariant, i.e.,

$$h|k,k\rangle = e^{i\varphi}|k,k\rangle, h \in \mathrm{U}(1), \tag{8.19}$$

where the general form of the subgroup element h can be expressed as

$$h = e^{i\alpha K_0}, \tag{8.20}$$

so that $\varphi = k\alpha$. From this one obtains a unique coset decomposition with respect to the isotropy subgroup $U(1)$, $g = \Omega h$, where $g \in \mathrm{SU}(1,1)$, $h \in \mathrm{U}(1)$, and Ω are coset representatives of $\mathrm{SU}(1,1)/\mathrm{U}(1)$ given by

$$\Omega(\xi) \equiv \exp(\xi K_+ - \xi^* K_-). \tag{8.21}$$

An arbitrary group transformation $g \in \mathrm{SU}(1,1)$ acting on the lowest-weight state $|k,k\rangle$ can be expressed as $g|k,k\rangle = \Omega(\xi)h|k,k\rangle = e^{i\varphi}|k,\xi\rangle$, where

$$|k,\xi\rangle \equiv \Omega(\xi)|k,k\rangle = \exp(\xi K_+ - \xi^* K_-)|k,k\rangle \tag{8.22}$$

gives the first definition of the SU(1, 1) coherent states. By using the Baker-Campbell-Hausdorff (BCH) formula, one may write the coset representative $\Omega(\xi)$ in the normal form as

$$\Omega(\xi) = \exp(\zeta K_+)\exp(\gamma K_0)\exp(-\zeta^* K_-), \tag{8.23}$$

where $\zeta \equiv \xi/|\xi| \tanh|\xi|$ and $\gamma \equiv -2\ln\cosh|\xi| = \ln(1-|\zeta|^2)$. Similarly, one may write the coset representative $\Omega(\xi)$ in the anti-normal form as

$$\Omega(\xi) = \exp(-\zeta^* K_-)\exp(-\gamma K_0)\exp(\zeta K_+). \tag{8.24}$$

As the parameters ζ and γ are independent of k, it is sufficient to verify the above formulas in the non-unitary faithful representation with $K_0 \equiv \frac{1}{2}\sigma_3$ and $K_\pm \equiv \mp\frac{1}{2}(\sigma_1 \pm i\sigma_2)$, where σ_j are the Pauli matrices. Indeed, a direct computation yields

$$K_+ = \begin{pmatrix} 0 & -1 \\ 0 & 0 \end{pmatrix}, K_- = \begin{pmatrix} 0 & 0 \\ 1 & 0 \end{pmatrix}, K_0 = \frac{1}{2}\begin{pmatrix} 1 & 0 \\ 0 & -1 \end{pmatrix}. \tag{8.25}$$

Hence one may write the coset representative in the non-unitary faithful representation as

$$\Omega(\xi) = \exp\begin{pmatrix} 0 & -\xi \\ -\xi^* & 0 \end{pmatrix} = \begin{pmatrix} \cosh|\xi| & -\frac{\xi}{|\xi|}\sinh|\xi| \\ -\frac{|\xi|}{\xi}\sinh|\xi| & \cosh|\xi| \end{pmatrix}. \tag{8.26}$$

From Eq. (8.23), one may also express the coset representative as

$$\begin{aligned} \Omega(\xi) &= \exp\begin{pmatrix} 0 & -\zeta \\ 0 & 0 \end{pmatrix}\exp\begin{pmatrix} \frac{\gamma}{2} & 0 \\ 0 & -\frac{\gamma}{2} \end{pmatrix}\exp\begin{pmatrix} 0 & 0 \\ -\zeta^* & 0 \end{pmatrix} \\ &= \begin{pmatrix} 1 & -\zeta \\ 0 & 1 \end{pmatrix}\begin{pmatrix} e^{\frac{\gamma}{2}} & 0 \\ 0 & e^{-\frac{\gamma}{2}} \end{pmatrix}\begin{pmatrix} 1 & 0 \\ -\zeta^* & 1 \end{pmatrix} \\ &= \begin{pmatrix} e^{\frac{\gamma}{2}} + |\zeta|^2 e^{-\frac{\gamma}{2}} & -\zeta e^{-\frac{\gamma}{2}} \\ -\zeta^* e^{-\frac{\gamma}{2}} & e^{-\frac{\gamma}{2}} \end{pmatrix}. \end{aligned} \tag{8.27}$$

Comparing Eqs. (8.26) and (8.27) yields the desired relationships among the parameters ξ, ζ, and γ

$$\zeta = \frac{\xi}{|\xi|}\tanh|\xi|, \gamma = -2\ln\cosh|\xi| = \ln(1-|\zeta|^2). \tag{8.28}$$

Applying $\Omega(\xi)$ to the lowest-weight state $|k,k\rangle$ and using its normal form, i.e., Eq. (8.23), one arrives at an equivalent definition for the SU(1, 1) coherent states

$$\begin{aligned} |k,\zeta\rangle &\equiv \exp(\zeta K_+)\exp(\gamma K_0)\exp(-\zeta^* K_-)|k,k\rangle \\ &= (1-|\zeta|^2)^k \exp(\zeta K_+)|k,k\rangle \\ &= (1-|\zeta|^2)^k \sum_{m=0}^{\infty}\sqrt{\frac{\Gamma(2k+m)}{\Gamma(2k)m!}}\zeta^m |k,k+m\rangle. \end{aligned} \tag{8.29}$$

As $|\zeta| \equiv \tanh|\xi| < 1$ for $|\xi| \in \mathbb{R}$, one may show that the SU(1, 1) coherent state space is simply the Poincaré disk (or Bolyai-Lobachevsky plane), which describes the two-dimensional hyperbolic geometry. To see this, one may compute the overlap between two arbitrary SU(1, 1) coherent states as

$$\langle k, \zeta'|k, \zeta\rangle = \frac{(1-|\zeta'^2|)^k(1-|\zeta^2|)^k}{(1-\zeta'^*\zeta)^{2k}}, \tag{8.30}$$

which yields

$$|\langle k, \zeta'|k, \zeta\rangle| = (\cosh\frac{d}{2})^{-2k}, \tag{8.31}$$

where d is the distance between the two points ζ and ζ' on the Poincaré disk determined by

$$\sinh\frac{d}{2} = \frac{|\zeta'-\zeta|}{\sqrt{1-|\zeta'|^2}\sqrt{1-|\zeta|^2}}, \tag{8.32a}$$

$$\cosh\frac{d}{2} = \frac{|1-\zeta'^*\zeta|}{\sqrt{1-|\zeta'|^2}\sqrt{1-|\zeta|^2}}. \tag{8.32b}$$

Hence, the Poincaré distance between two arbitrary points ζ and ζ' on the Poincaré disk is given by

$$d(\zeta', \zeta) \equiv 2\arctan\left(\frac{|\zeta'-\zeta|}{|1-\zeta'^*\zeta|}\right). \tag{8.33}$$

A direct computation yields the Poincaré metric

$$ds^2 = \frac{4d\zeta d\zeta^*}{(1-|\zeta|^2)^2} \equiv \lambda^2(\zeta, \zeta^*)d\zeta d\zeta^*, \tag{8.34}$$

where $\lambda(\zeta, \zeta^*) = 2(1-\zeta\zeta^*)^{-1}$ is a real and positive function of ζ and ζ^*. Hence, the Gaussian curvature of the Poincaré disk is given by

$$K = -\frac{4}{\lambda^2}\frac{\partial}{\partial\zeta}\frac{\partial}{\partial\zeta^*}\log\lambda = -1. \tag{8.35}$$

In other words, the SU(1, 1) coherent state space, i.e., the Poincaré disk, is a homogeneous manifold with a constant negative curvature $K = -1$.

Using Eq. (8.29), one can derive the resolution of identity in terms of the SU(1, 1) coherent states. To show this, one needs the hyperbolic polar coordinates (τ, ϕ) in the Poincaré disc, defined by $\zeta \equiv e^{i\phi}\tanh\frac{\tau}{2}$. In the hyperbolic polar coordinates, the Poincaré metric becomes

$$ds^2 = \frac{4d\zeta d\zeta^*}{(1-|\zeta|^2)^2} = d\tau^2 + \sinh^2\tau d\phi^2, \tag{8.36}$$

which corresponds to the metric tensor

$$g = \begin{pmatrix} 1 & 0 \\ 0 & \sinh^2\tau \end{pmatrix}. \tag{8.37}$$

Hence, the area element of the SU(1, 1) coherent state space has the form

$$d\Omega \equiv \sqrt{g} d\tau \wedge d\phi = \sinh\tau d\tau \wedge d\phi. \tag{8.38}$$

From Eq. (8.29), a direct computation yields

$$\begin{aligned} \int d\Omega |k,\zeta\rangle\langle k,\zeta| = &\sum_{m,n=0}^{\infty} \sqrt{\frac{\Gamma(2k+m)\Gamma(2k+n)}{\Gamma(2k)m!\cdot\Gamma(2k)n!}} \\ &\cdot \int d\Omega (1-|\zeta|^2)^{2k} (\zeta^*)^n \zeta^m \cdot |k,k+m\rangle\langle k,k+n| \end{aligned} \tag{8.39}$$

In the hyperbolic polar coordinates (τ, ϕ), the integral appearing in Eq. (8.39) can be evaluated as

$$\begin{aligned} &\int d\Omega (1-|\zeta|^2)^{2k} (\zeta^*)^n \zeta^m \\ =& \int_0^\infty \int_0^{2\pi} \operatorname{sech}^{4k}\left(\frac{\tau}{2}\right) e^{i(m-n)\phi} \tanh^{m+n}\left(\frac{\tau}{2}\right) \sinh\tau d\tau d\phi \\ =& 2\pi\delta_{mn} \int_0^\infty \operatorname{sech}^{4k}\left(\frac{\tau}{2}\right) \tanh^{2m}\left(\frac{\tau}{2}\right) \sinh\tau d\tau \\ =& 4\pi\delta_{mn} \int_0^1 (1-s)^{2k-2} s^m ds \\ =& 4\pi\delta_{mn} \frac{\Gamma(m+1)\Gamma(2k-1)}{\Gamma(2k+m)}, \end{aligned} \tag{8.40}$$

where $s \equiv \tanh^2\left(\frac{\tau}{2}\right)$ and $2k-1 \neq 0$. For $k = 1/2$, the integral diverges, as the Gamma function $\Gamma(2k-1)$ has a simple pole at $k = 1/2$. Substitution of Eq. (8.40) into Eq. (8.39) immediately yields

$$\begin{aligned}
&\int d\Omega|k,\zeta\rangle\langle k,\zeta| \qquad (8.41)\\
&=4\pi\sum_{m=0}^{\infty}\frac{\Gamma(2k+m)}{\Gamma(2k)m!}\frac{m!\Gamma(2k-1)}{\Gamma(2k+m)}|k,k+m\rangle\langle k,k+m|\\
&=\frac{4\pi}{2k-1}\sum_{m=0}^{\infty}|k,k+m\rangle\langle k,k+m|=\frac{4\pi}{2k-1}I.
\end{aligned}$$

In other words, for $k > 1/2$, the resolution of identity becomes

$$\int d\mu_k|k,\zeta\rangle\langle k,\zeta| = I, \qquad (8.42a)$$

$$d\mu_k \equiv \frac{2k-1}{4\pi}d\Omega = \frac{2k-1}{\pi}\frac{d^2\zeta}{(1-|\zeta|^2)^2}, \qquad (8.42b)$$

where $d^2\zeta \equiv d\text{Re}(\zeta)d\text{Im}(\zeta)$. Besides the resolution of identity, one may also derive the expectation value of a product of SU(1, 1) generators with respect to the coherent states $|k,\zeta\rangle$, i.e., $\langle k,\zeta|K_-^p K_0^q K_+^r|k,\zeta\rangle$. To evaluate this, one may use the following formula

$$K_+^r|k,\zeta\rangle = (1-|\zeta|^2)^k\sum_{m=0}^{\infty}\sqrt{\frac{\Gamma(2k+r+m)(r+m)!}{\Gamma(2k)(m!)^2}}\zeta^m|k,k+r+m\rangle, \qquad (8.43)$$

which can be directly obtained from Eq. (8.29). Without loss of generality, one can assume that $p \geq r$, so that $s \equiv p - r$ is a non-negative integer. Then, a direct computation yields

$$\begin{aligned}
&\langle k,\zeta|K_-^p K_0^q K_+^r|k,\zeta\rangle = (1-|\zeta|^2)^{2k}\zeta^s\cdot \qquad (8.44)\\
&\sum_{n=0}^{\infty}\frac{\Gamma(2k+p+n)(p+n)!}{\Gamma(2k)n!(s+n)!}(k+p+n)^q|\zeta|^{2n}.
\end{aligned}$$

In particular, the expectation values of the SU(1, 1) generators K_0, K_+, and K_- with respect to the coherent states $|k,\zeta\rangle$ are

$$\mathcal{K}_0 \equiv \langle k,\zeta|K_0|k,\zeta\rangle = k\frac{1+|\zeta|^2}{1-|\zeta|^2}, \qquad (8.45a)$$

$$\mathcal{K}_- \equiv \langle k,\zeta|K_-|k,\zeta\rangle = k\frac{2\zeta}{1-|\zeta|^2}, \qquad (8.45b)$$

$$\mathcal{K}_+ \equiv \langle k,\zeta|K_+|k,\zeta\rangle = k\frac{2\zeta^*}{1-|\zeta|^2}. \qquad (8.45c)$$

One can now show that the elements $\mathcal{K}_0$, $\mathcal{K}_-$, and $\mathcal{K}_+$, together with a suitable Poisson bracket on the coherent state space, would induce a phase space representation of the SU(1, 1) group. In fact, on the Poincaré disk, one may define a Poisson bracket $\{\cdot, \cdot\}$ in the form

$$\{f, g\} \equiv \frac{(1-|\zeta|^2)^2}{2ik}\left(\frac{\partial f}{\partial \zeta}\frac{\partial g}{\partial \zeta^*} - \frac{\partial f}{\partial \zeta^*}\frac{\partial g}{\partial \zeta}\right), \tag{8.46}$$

where f and g are two arbitrary functions of ζ and ζ^* on the Poincaré disk. After a straightforward calculation, one can show that the elements $\mathcal{K}_0$, $\mathcal{K}_-$ and $\mathcal{K}_+$ carry the algebraic structure

$$i\{\mathcal{K}_-, \mathcal{K}_+\} = 2\mathcal{K}_0, i\{\mathcal{K}_0, \mathcal{K}_\pm\} = \pm\mathcal{K}_\pm. \tag{8.47}$$

Such an algebraic structure relates to the original SU(1, 1) Lie algebra structure via the correspondence

$$\langle k, \zeta|[K_i, K_j]|k, \zeta\rangle = i\{\mathcal{K}_i, \mathcal{K}_j\}. \tag{8.48}$$

Instead of using the coordinates ζ and ζ^*, it is more convenient to define a set of canonical coordinates (q, p) via

$$\frac{q+ip}{\sqrt{4k}} \equiv \frac{\zeta}{\sqrt{1-|\zeta|^2}}. \tag{8.49}$$

Then the Poisson bracket $\{\cdot, \cdot\}$ in the new coordinate systems has the canonical form

$$\{f, g\} = \frac{\partial f}{\partial q}\frac{\partial g}{\partial p} - \frac{\partial f}{\partial p}\frac{\partial g}{\partial q}, \tag{8.50}$$

which induces a phase space structure on the SU(1, 1) coherent state space, i.e., the Poincaré disk, where the phase space representations of the SU(1, 1) generators are

$$\mathcal{K}_0 = k + \frac{q^2+p^2}{2}, \mathcal{K}_\mp = (q \pm ip)\sqrt{k + \frac{q^2+p^2}{4}}. \tag{8.51}$$

8.2 SU(3) Coherent States

For another example of the generalized coherent states, one may consider the coherent states associated with the Lie algebra $\mathfrak{su}(3, \mathbb{C})$ of the simply-connected simple compact Lie group SU(3, $\mathbb{C}$). In the fundamental representation, the Lie algebra $\mathfrak{su}(3,\mathbb{C})$ is generated by eight Hermitian traceless matrices $T_a \equiv \lambda_a/2$ for $1 \le a \le 8$, where λ_a are the Gell-Mann matrices

$$\lambda_1 \equiv \begin{pmatrix} 0 & 1 & 0 \\ 1 & 0 & 0 \\ 0 & 0 & 0 \end{pmatrix}, \lambda_2 \equiv \begin{pmatrix} 0 & -i & 0 \\ i & 0 & 0 \\ 0 & 0 & 0 \end{pmatrix}, \lambda_3 \equiv \begin{pmatrix} 1 & 0 & 0 \\ 0 & -1 & 0 \\ 0 & 0 & 0 \end{pmatrix}, \tag{8.52}$$

$$\lambda_4 \equiv \begin{pmatrix} 0 & 0 & 1 \\ 0 & 0 & 0 \\ 1 & 0 & 0 \end{pmatrix}, \lambda_5 \equiv \begin{pmatrix} 0 & 0 & -i \\ 0 & 0 & 0 \\ i & 0 & 0 \end{pmatrix},$$

$$\lambda_6 \equiv \begin{pmatrix} 0 & 0 & 0 \\ 0 & 0 & 1 \\ 0 & 1 & 0 \end{pmatrix}, \lambda_7 \equiv \begin{pmatrix} 0 & 0 & 0 \\ 0 & 0 & -i \\ 0 & i & 0 \end{pmatrix}, \lambda_8 \equiv \frac{1}{\sqrt{3}} \begin{pmatrix} 1 & 0 & 0 \\ 0 & 1 & 0 \\ 0 & 0 & -2 \end{pmatrix},$$

in analogy with the Pauli matrices. The Gell-Mann matrices λ_a are trace orthonormal, i.e., $\mathrm{Tr}(\lambda_a\lambda_b) = 2\delta_{ab}$, in analogy with the trace orthogonal relations that the Pauli matrices obey. Hence, the generators T_a of the $\mathfrak{su}(3,\mathbb{C})$ Lie algebra are normalized to a value of $\frac{1}{2}$, i.e., $\mathrm{Tr}(T_aT_b) = \frac{1}{2}\delta_{ab}$. Let us denote the $\mathfrak{su}(3,\mathbb{C})$ structure constants in the Gell-Mann basis as f_{abc}, defined by $[T_a, T_b] = i\sum_{c=1}^{8} f_{abc}T_c$. One can prove that the structure constants f_{abc} are *real* and *totally antisymmetric* under the interchange of any pair of indices. To show this, one needs the following relation

$$\mathrm{Tr}([T_a, T_b]T_c) = \mathrm{Tr}(i\sum_{c'} f_{abc'}T_{c'}T_c) = \frac{i}{2}\sum_{c'} f_{abc'}\delta_{cc'} = \frac{i}{2}f_{abc}. \tag{8.53}$$

Using the cyclic property of the trace, $\mathrm{Tr}(ABC) = \mathrm{Tr}(CAB)$, one obtains

$$\begin{aligned} \mathrm{Tr}([T_a, T_b]T_c) &= \mathrm{Tr}(T_aT_bT_c - T_bT_aT_c + T_aT_cT_b - T_aT_cT_b) \\ &= \mathrm{Tr}(T_a[T_b, T_c] - T_bT_aT_c + T_aT_cT_b) \\ &= \mathrm{Tr}([T_b, T_c]T_a), \end{aligned} \tag{8.54}$$

which implies immediately that $f_{abc} = f_{bca} = -f_{cba}$. Also, as the generators T_a are Hermitian matrices, one obtains $[T_a, T_b]^\dagger = [T_b^\dagger, T_a^\dagger] = -[T_a, T_b]$, which yields $-[T_a, T_b] = [T_a, T_b]^\dagger = -i\sum_{c=1}^{8} f_{abc}^* T_c$. This implies immediately that the structure constants are all real, $f_{abc} = f_{abc}^*$. The $\mathfrak{su}(3,\mathbb{C})$ Lie algebra has three independent $\mathfrak{su}(2,\mathbb{C})$ subalgebras generated by the elements $\{T_1, T_2, T_3\}$, $\{T_4, T_5, \frac{1}{2}(\sqrt{3}T_8 + T_3)\}$, and $\{T_6, T_7, \frac{1}{2}(\sqrt{3}T_8 - T_3)\}$. Hence, one obtains $f_{123} = 1$, $f_{345} = \frac{1}{2}$, $f_{458} = \frac{\sqrt{3}}{2}$, $f_{367} = -\frac{1}{2}$, and $f_{678} = \frac{\sqrt{3}}{2}$. Also, the structure constants f_{abc} vanish when the generators T_a and T_b are in the same $\mathfrak{su}(2,\mathbb{C})$ subalgebra, and the generator T_c is in another $\mathfrak{su}(2,\mathbb{C})$ subalgebra. Finally, the commutation relation $[\lambda_1, \lambda_4] = i\lambda_7$ yields $[T_1, T_4] = \frac{i}{2}T_7$ and $f_{147} = \frac{1}{2}$, and similar computations yield $f_{246} = f_{257} = \frac{1}{2}$ and $f_{156} = -\frac{1}{2}$.

In any irreducible representation of the Lie algebra $\mathfrak{su}(3, \mathbb{C})$, there exist two independent simultaneously diagonalized traceless Hermitian matrices which span the Cartan subalgebra $\mathfrak{h}$. In the fundamental representation, the two simultaneously diagonalized traceless Hermitian matrices in the Cartan subalgebra $\mathfrak{h}$ are already known, which are T_3 and T_8, respectively. Hence, one may write the two elements in the Cartan subalgebra as

$$H_1 \equiv T_3 = \frac{1}{2}\begin{pmatrix} 1 & 0 & 0 \\ 0 & -1 & 0 \\ 0 & 0 & 0 \end{pmatrix}, H_2 \equiv T_8 = \frac{1}{2\sqrt{3}}\begin{pmatrix} 1 & 0 & 0 \\ 0 & 1 & 0 \\ 0 & 0 & -2 \end{pmatrix}, \tag{8.55}$$

which automatically satisfy the commutation relation $[H_1, H_2] = 0$. For the remaining six generators which are not in the Cartan subalgebra $\mathfrak{h}$, one can guess the forms of the shift operators E_α by noting that the Gell-Mann matrices λ_1 and λ_2 are in analogy with the Pauli matrices σ_x and σ_y in the $\mathfrak{su}(2, \mathbb{C})$ subalgebras generated by $\{T_1, T_2, T_3\}$ and similarly for λ_4 and λ_5 and λ_6 and λ_7. Hence, one may infer that the shift operators E_α are proportional to $T_1 \pm iT_2$, $T_4 \pm iT_5$, and $T_6 \pm iT_7$. One can now confirm that these operators are indeed proportional to the shift operators and find the roots by evaluating the commutators with the elements in the Cartan subalgebra $\mathfrak{h}$. A direct computation yields

$$[T_3, T_1 \pm iT_2] = \pm(T_1 \pm iT_2), [T_8, T_1 \pm iT_2] = 0, \tag{8.56a}$$

$$[T_3, T_4 \pm iT_5] = \pm\frac{1}{2}(T_4 \pm iT_5), [T_8, T_4 \pm iT_5] = \pm\frac{\sqrt{3}}{2}(T_4 \pm iT_5), \tag{8.56b}$$

$$[T_3, T_6 \pm iT_7] = \mp\frac{1}{2}(T_6 \pm iT_7), [T_8, T_6 \pm iT_7] = \pm\frac{\sqrt{3}}{2}(T_6 \pm iT_7). \tag{8.56c}$$

A similar computation yields

$$[T_4 + iT_5, T_4 - iT_5] = T_3 + \sqrt{3}T_8, \tag{8.57a}$$

$$[T_6 - iT_7, T_6 + iT_7] = T_3 - \sqrt{3}T_8, \tag{8.57b}$$

$$[T_1 + iT_2, T_1 - iT_2] = 2T_3, \tag{8.57c}$$

$$[T_4 + iT_5, T_6 - iT_7] = T_1 + iT_2, \tag{8.57d}$$

$$[T_4 - iT_5, T_1 + iT_2] = T_6 - iT_7, \tag{8.57e}$$

$$[T_1 + iT_2, T_6 + iT_7] = T_4 + iT_5. \tag{8.57f}$$

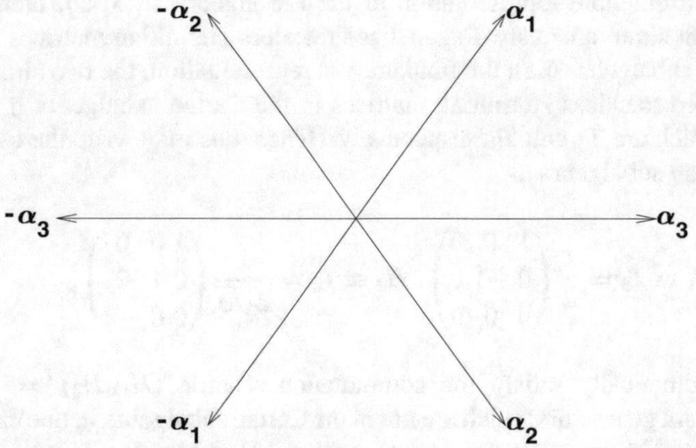

Fig. 8.1 The root space diagram for the Lie algebra $\mathfrak{su}(3, \mathbb{C})$. The six roots $\pm\boldsymbol{\alpha}_1$, $\pm\boldsymbol{\alpha}_2$, and $\pm\boldsymbol{\alpha}_3$ are of unit length and form a regular hexagon. Here $\boldsymbol{\alpha}_2$ and $\boldsymbol{\alpha}_2$ are the simple roots which form a basis of the root system

Hence, the roots $\pm\boldsymbol{\alpha}_1$, $\pm\boldsymbol{\alpha}_2$, and $\pm\boldsymbol{\alpha}_3$ of the Lie algebra $\mathfrak{su}(3, \mathbb{C})$ are determined by (see Fig. 8.1)

$$\boldsymbol{\alpha}_1 \equiv \left(\frac{1}{2}, \frac{\sqrt{3}}{2}\right), \boldsymbol{\alpha}_2 \equiv \left(\frac{1}{2}, -\frac{\sqrt{3}}{2}\right), \boldsymbol{\alpha}_3 \equiv (1, 0), \tag{8.58}$$

where $\boldsymbol{\alpha}_1$ and $\boldsymbol{\alpha}_2$ are the **simple roots** which form a basis of the root system of the Lie algebra $\mathfrak{su}(3, \mathbb{C})$, as $\boldsymbol{\alpha}_3$ is linearly dependent on $\boldsymbol{\alpha}_1$ and $\boldsymbol{\alpha}_2$, i.e., $\boldsymbol{\alpha}_3 = \boldsymbol{\alpha}_1 + \boldsymbol{\alpha}_2$. Hence, the commutation relations between the shift operators and the elements of the Cartan subalgebra $\mathfrak{h}$ become $[\mathbf{H}, E_{\pm\boldsymbol{\alpha}_1}] = \pm\boldsymbol{\alpha}_1 E_{\pm\boldsymbol{\alpha}_1}$, $[\mathbf{H}, E_{\pm\boldsymbol{\alpha}_2}] = \pm\boldsymbol{\alpha}_2 E_{\pm\boldsymbol{\alpha}_2}$, and $[\mathbf{H}, E_{\pm\boldsymbol{\alpha}_3}] = \pm\boldsymbol{\alpha}_3 E_{\pm\boldsymbol{\alpha}_3}$, and the remaining commutations between the shift operators are

$$[E_{\boldsymbol{\alpha}_1}, E_{-\boldsymbol{\alpha}_1}] = \boldsymbol{\alpha}_1 \cdot \mathbf{H}, [E_{\boldsymbol{\alpha}_2}, E_{-\boldsymbol{\alpha}_2}] = \boldsymbol{\alpha}_2 \cdot \mathbf{H}, \tag{8.59a}$$

$$[E_{\boldsymbol{\alpha}_3}, E_{-\boldsymbol{\alpha}_3}] = \boldsymbol{\alpha}_3 \cdot \mathbf{H}, [E_{\boldsymbol{\alpha}_1}, E_{\boldsymbol{\alpha}_2}] = N_{1,2} E_{\boldsymbol{\alpha}_3}, \tag{8.59b}$$

$$[E_{\boldsymbol{\alpha}_{-1}}, E_{\boldsymbol{\alpha}_3}] = N_{-1,3} E_{\boldsymbol{\alpha}_2}, [E_{\boldsymbol{\alpha}_3}, E_{\boldsymbol{\alpha}_{-2}}] = N_{3,-2} E_{\boldsymbol{\alpha}_1} \tag{8.59c}$$

where $\mathbf{H} \equiv (H_1, H_2)$, $N_{1,2} = N_{-1,3} = N_{3,-2} = \frac{1}{\sqrt{2}}$, and

$$E_{\pm\boldsymbol{\alpha}_1} \equiv \frac{1}{\sqrt{2}}(T_4 \pm i T_5), E_{\pm\boldsymbol{\alpha}_2} \equiv \frac{1}{\sqrt{2}}(T_6 \mp i T_7), E_{\pm\boldsymbol{\alpha}_3} \equiv \frac{1}{\sqrt{2}}(T_1 \pm i T_2). \tag{8.60}$$

In the fundamental representation, the six shift operators $E_{\pm\alpha_1}$, $E_{\pm\alpha_2}$, and $E_{\pm\alpha_3}$ can be written as

$$E_{\alpha_1} = \frac{1}{\sqrt{2}}\begin{pmatrix} 0 & 0 & 1 \\ 0 & 0 & 0 \\ 0 & 0 & 0 \end{pmatrix}, E_{\alpha_{-1}} = \frac{1}{\sqrt{2}}\begin{pmatrix} 0 & 0 & 0 \\ 0 & 0 & 0 \\ 1 & 0 & 0 \end{pmatrix}, \tag{8.61a}$$

$$E_{\alpha_2} = \frac{1}{\sqrt{2}}\begin{pmatrix} 0 & 0 & 0 \\ 0 & 0 & 0 \\ 0 & 1 & 0 \end{pmatrix}, E_{\alpha_{-2}} = \frac{1}{\sqrt{2}}\begin{pmatrix} 0 & 0 & 0 \\ 0 & 0 & 1 \\ 0 & 0 & 0 \end{pmatrix}, \tag{8.61b}$$

$$E_{\alpha_3} = \frac{1}{\sqrt{2}}\begin{pmatrix} 0 & 1 & 0 \\ 0 & 0 & 0 \\ 0 & 0 & 0 \end{pmatrix}, E_{\alpha_{-3}} = \frac{1}{\sqrt{2}}\begin{pmatrix} 0 & 0 & 0 \\ 1 & 0 & 0 \\ 0 & 0 & 0 \end{pmatrix}. \tag{8.61c}$$

Now we see that the Lie algebra $\mathfrak{su}(3, \mathbb{C})$ has eight elements

$$\{H_1, H_2, E_{\pm\alpha_1}, E_{\pm\alpha_2}, E_{\pm\alpha_3}\} \tag{8.62}$$

in the Cartan basis, where the two elements H_1 and H_2 commute and span the Cartan subalgebra $\mathfrak{h}$. The root system of the Lie algebra $\mathfrak{su}(3, \mathbb{C})$ is determined by

$$\boldsymbol{\alpha}_{\pm 1} \equiv \pm\left(\frac{1}{2}, \frac{\sqrt{3}}{2}\right), \boldsymbol{\alpha}_{\pm 2} \equiv \pm\left(\frac{1}{2}, -\frac{\sqrt{3}}{2}\right), \boldsymbol{\alpha}_{\pm 3} \equiv \pm(1, 0), \tag{8.63}$$

where $\boldsymbol{\alpha}_1$ and $\boldsymbol{\alpha}_2$ are two simple positive roots, i.e., every positive roots are the linear combination of them with non-negative integer coefficients, e.g., $\boldsymbol{\alpha}_3 = \boldsymbol{\alpha}_1 + \boldsymbol{\alpha}_2$. As all the six roots have the same unit length, the remaining commutation relations in the Cartan basis are $[\mathbf{H}, E_{\pm\alpha_j}] = \pm\boldsymbol{\alpha}_j E_{\pm\alpha_j}$ and

$$[E_{\alpha_j}, E_{-\alpha_j}] = \boldsymbol{\alpha}_j \cdot \boldsymbol{H}, [E_{\pm\alpha_1}, E_{\pm\alpha_2}] = \pm\frac{1}{\sqrt{2}}E_{\pm\alpha_3}, \tag{8.64a}$$

$$[E_{\pm\alpha_3}, E_{\mp\alpha_2}] = \pm\frac{1}{\sqrt{2}}E_{\pm\alpha_1}, [E_{\mp\alpha_1}, E_{\pm\alpha_3}] = \pm\frac{1}{\sqrt{2}}E_{\pm\alpha_2}, \tag{8.64b}$$

where $1 \le j \le 3$. From the commutation relations, one can readily show that the elements H_1, H_2, and $E_{\pm\alpha_2}$ span the maximal compact subalgebra $\mathfrak{k}$ of $\mathfrak{su}(3, \mathbb{C})$ and the elements $E_{\pm\alpha_1}$ and $E_{\pm\alpha_3}$ span the orthogonal complementary space $\mathfrak{p}$ of $\mathfrak{k}$. Indeed, the commutation relations between the elements H_1, H_2, and $E_{\pm\alpha_2}$ are

$$[H_1, H_2] = 0, [\mathbf{H}, E_{\pm\alpha_2}] = \pm\boldsymbol{\alpha}_2 E_{\pm\alpha_2}, [E_{\alpha_2}, E_{-\alpha_2}] = \boldsymbol{\alpha}_2 \cdot \boldsymbol{H}, \tag{8.65}$$

which implies the relation $[\mathfrak{k}, \mathfrak{k}] \subseteq \mathfrak{k}$. Similarly, the commutation relations between the elements $E_{\pm\alpha_1}$ and $E_{\pm\alpha_3}$ are

$$[E_{\pm\alpha_1}, E_{\pm\alpha_3}] = 0, [E_{\mp\alpha_1}, E_{\pm\alpha_3}] = \pm\frac{1}{\sqrt{2}}E_{\pm\alpha_2}, \tag{8.66a}$$

$$[E_{\alpha_1}, E_{-\alpha_1}] = \boldsymbol{\alpha}_1 \cdot \boldsymbol{H}, [E_{\alpha_3}, E_{-\alpha_3}] = H_1, \tag{8.66b}$$

which implies the relation $[\mathfrak{p}, \mathfrak{p}] \subseteq \mathfrak{k}$. Notice that both E_1 and E_3 and E_{-1} and E_{-3} span a commutative subalgebra of $\mathfrak{p}$. Finally, the commutation relations between the elements in $\mathfrak{k}$ and the elements in the orthogonal complementary space $\mathfrak{p}$ are $[\mathbf{H}, E_{\pm\alpha_1}] = \pm\boldsymbol{\alpha}_1 E_{\pm\alpha_1}$, $[\mathbf{H}, E_{\pm\alpha_3}] = \pm\boldsymbol{\alpha}_3 E_{\pm\alpha_3}$ and

$$[E_{\pm\alpha_2}, E_{\pm\alpha_1}] = \mp\frac{1}{\sqrt{2}}E_{\pm\alpha_3}, [E_{\pm\alpha_2}, E_{\mp\alpha_1}] = 0, \tag{8.67a}$$

$$[E_{\pm\alpha_2}, E_{\pm\alpha_3}] = 0, [E_{\pm\alpha_2}, E_{\mp\alpha_3}] = \pm\frac{1}{\sqrt{2}}E_{\mp\alpha_1}, \tag{8.67b}$$

which implies the relation $[\mathfrak{k}, \mathfrak{p}] = \mathfrak{p}$. Let us denote $U(H_i)$ and $U(E_{\boldsymbol{\alpha}})$ as the representation matrix of the generators H_i of the Cartan subalgebra and the shift operators $E_{\boldsymbol{\alpha}}$ in the fundamental representation of $\mathfrak{su}(3, \mathbb{C})$. Then, the generators of the maximal compact subalgebra $\mathfrak{k}$ in the fundamental representation of $\mathfrak{su}(3, \mathbb{C})$ have the form

$$U(H_1) = \frac{1}{2}\begin{pmatrix} 1 & 0 & 0 \\ 0 & -1 & 0 \\ 0 & 0 & 0 \end{pmatrix}, U(H_2) = \frac{1}{2\sqrt{3}}\begin{pmatrix} 1 & 0 & 0 \\ 0 & 1 & 0 \\ 0 & 0 & -2 \end{pmatrix}, \tag{8.68a}$$

$$U(E_{\alpha_2}) = \frac{1}{\sqrt{2}}\begin{pmatrix} 0 & 0 & 0 \\ 0 & 0 & 0 \\ 0 & 1 & 0 \end{pmatrix}, U(E_{\alpha_{-2}}) = \frac{1}{\sqrt{2}}\begin{pmatrix} 0 & 0 & 0 \\ 0 & 0 & 1 \\ 0 & 0 & 0 \end{pmatrix}. \tag{8.68b}$$

Similarly, the generators of the orthogonal complementary subspace $\mathfrak{p}$ in the fundamental representation of $\mathfrak{su}(3, \mathbb{C})$ are

$$U(E_{\alpha_1}) = \frac{1}{\sqrt{2}}\begin{pmatrix} 0 & 0 & 1 \\ 0 & 0 & 0 \\ 0 & 0 & 0 \end{pmatrix}, U(E_{\alpha_{-1}}) = \frac{1}{\sqrt{2}}\begin{pmatrix} 0 & 0 & 0 \\ 0 & 0 & 0 \\ 1 & 0 & 0 \end{pmatrix}, \tag{8.69a}$$

$$U(E_{\alpha_3}) = \frac{1}{\sqrt{2}}\begin{pmatrix} 0 & 1 & 0 \\ 0 & 0 & 0 \\ 0 & 0 & 0 \end{pmatrix}, U(E_{\alpha_{-3}}) = \frac{1}{\sqrt{2}}\begin{pmatrix} 0 & 0 & 0 \\ 1 & 0 & 0 \\ 0 & 0 & 0 \end{pmatrix}. \tag{8.69b}$$

As a result, a general element $\mathbf{k}$ of the maximal compact subalgebra $\mathfrak{k}$ in the fundamental representation of $\mathfrak{su}(3,\mathbb{C})$ has the form

$$U(\mathbf{k}) \equiv i\lambda_1\rho(H_1) + i\lambda_2\rho(H_2) + \eta_{\alpha_2}\rho(E_{\alpha_2}) - \eta^*_{\alpha_2}\rho(E_{-\alpha_2}) \tag{8.70}$$

$$= \left(\begin{array}{c|cc} \frac{i\lambda_1}{2} + \frac{i\lambda_2}{2\sqrt{3}} & 0 & 0 \\ \hline 0 & \frac{-i\lambda_1}{2} + \frac{i\lambda_2}{2\sqrt{3}} & \frac{-1}{\sqrt{2}}\eta^*_{\alpha_2} \\ 0 & \frac{1}{\sqrt{2}}\eta_{\alpha_2} & \frac{-i\lambda_2}{\sqrt{3}} \end{array}\right) \equiv \left(\begin{array}{c|c} \mathbf{a} & \mathbf{0} \\ \hline \mathbf{0} & \mathbf{b} \end{array}\right),$$

where $\lambda_1, \lambda_2 \in \mathbb{R}$, $\eta_{\alpha_2} \in \mathbb{C}$, and $\rho(\mathbf{k})$ are anti-Hermitian and traceless, i.e., $U(\mathbf{k})^\dagger = -U(\mathbf{k})$ and $\mathrm{Tr}(U(\mathbf{k})) = 0$. Similarly, a general element $\mathbf{p}$ of the orthogonal complementary subspace $\mathfrak{p}$ in the fundamental representation of $\mathfrak{su}(3,\mathbb{C})$ has the form

$$U(\mathbf{p}) \equiv \eta_{\alpha_1}\rho(E_{\alpha_1}) + \eta_{\alpha_3}\rho(E_{\alpha_3}) - \eta^*_{\alpha_1}\rho(E_{-\alpha_1}) - \eta^*_{\alpha_3}\rho(E_{-\alpha_3}) \tag{8.71}$$

$$= \left(\begin{array}{c|cc} 0 & \frac{1}{\sqrt{2}}\eta_{\alpha_3} & \frac{1}{\sqrt{2}}\eta_{\alpha_1} \\ \hline -\frac{1}{\sqrt{2}}\eta^*_{\alpha_3} & 0 & 0 \\ -\frac{1}{\sqrt{2}}\eta^*_{\alpha_1} & 0 & 0 \end{array}\right) \equiv \left(\begin{array}{c|c} \mathbf{0} & \boldsymbol{\eta} \\ \hline -\boldsymbol{\eta}^\dagger & \mathbf{0} \end{array}\right),$$

where $\eta_{\alpha_1}, \eta_{\alpha_3} \in \mathbb{C}$, and $U(\mathbf{p})$ are anti-Hermitian and traceless, i.e., $U(\mathbf{p})^\dagger = -U(\mathbf{p})$ and $\mathrm{Tr}(U(\mathbf{p})) = 0$.

We now discuss the weights for the fundamental representation of the Lie algebra $\mathfrak{su}(3,\mathbb{C})$. For a rank-$l$ semisimple Lie algebra $\mathfrak{g}$, the basis vectors $|\boldsymbol{\mu}\rangle$ in the representation space are chosen to be the common eigenvectors of the commuting operators H_i, $H_i|\boldsymbol{\mu}\rangle = \mu_i|\boldsymbol{\mu}\rangle$, or equivalently $\mathbf{H}|\boldsymbol{\mu}\rangle = \boldsymbol{\mu}|\boldsymbol{\mu}\rangle$. The l-dimensional vector $\boldsymbol{\mu} \equiv (\mu_1, \cdots, \mu_l)$, whose components are the eigenvalues μ_j, is called the **weight** of the basis vector $|\boldsymbol{\mu}\rangle$. For our $\mathfrak{su}(3,\mathbb{C})$ example, as the two generators H_1 and H_2 of the Cartan subalgebra are diagonal, we may choose the basic vectors as

$$|\boldsymbol{\mu}_1\rangle = \begin{pmatrix}1\\0\\0\end{pmatrix}, |\boldsymbol{\mu}_2\rangle = \begin{pmatrix}0\\1\\0\end{pmatrix}, |\boldsymbol{\mu}_3\rangle = \begin{pmatrix}0\\0\\1\end{pmatrix}. \tag{8.72}$$

Then a direct computation yields the weights of the Lie algebra $\mathfrak{su}(3,\mathbb{C})$

$$\boldsymbol{\mu}_1 = \left(\frac{1}{2}, \frac{1}{2\sqrt{3}}\right), \boldsymbol{\mu}_2 = \left(-\frac{1}{2}, \frac{1}{2\sqrt{3}}\right), \boldsymbol{\mu}_3 = \left(0, -\frac{1}{\sqrt{3}}\right), \tag{8.73}$$

which are not independent and satisfy the relation $\boldsymbol{\mu}_1 + \boldsymbol{\mu}_2 + \boldsymbol{\mu}_3 = 0$. Comparing Eqs. (8.63) and (8.73), one immediately obtains the relations between the weights

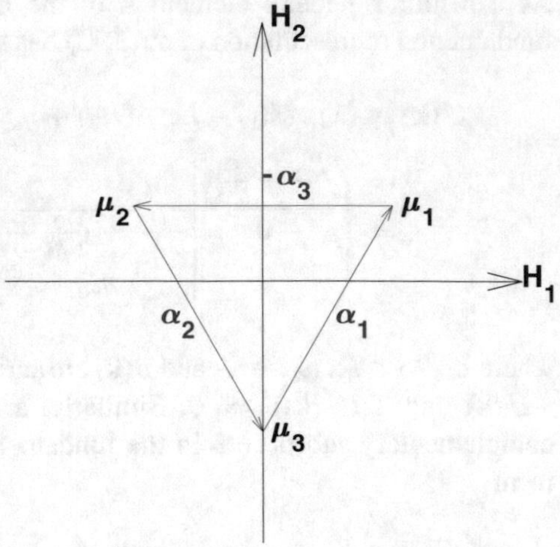

Fig. 8.2 The planar weight diagram of the fundamental representation of the Lie algebra $\mathfrak{su}(3, \mathbb{C})$ together with the three roots $\boldsymbol{\alpha}_1, \boldsymbol{\alpha}_2$, and $\boldsymbol{\alpha}_3$

and the roots of the Lie algebra $\mathfrak{su}(3, \mathbb{C})$ (see Fig. 8.2)

$$\boldsymbol{\alpha}_1 = \boldsymbol{\mu}_1 - \boldsymbol{\mu}_3, \boldsymbol{\alpha}_2 = \boldsymbol{\mu}_3 - \boldsymbol{\mu}_2, \boldsymbol{\alpha}_3 = \boldsymbol{\mu}_1 - \boldsymbol{\mu}_2. \tag{8.74}$$

Hence, the partial order on the weights of the fundamental representation of $\mathfrak{su}(3, \mathbb{C})$ is $\boldsymbol{\mu}_1 > \boldsymbol{\mu}_3 > \boldsymbol{\mu}_2$, as both $\boldsymbol{\mu}_1 - \boldsymbol{\mu}_3$ and $\boldsymbol{\mu}_3 - \boldsymbol{\mu}_2$ are positive roots. Applying the shift operators $E_{\boldsymbol{\alpha}}$ on $|\boldsymbol{\mu}_1\rangle$, $|\boldsymbol{\mu}_2\rangle$ and $|\boldsymbol{\mu}_3\rangle$ immediately yields

$$E_{-\boldsymbol{\alpha}_1}|\boldsymbol{\mu}_1\rangle = \frac{1}{\sqrt{2}}|\boldsymbol{\mu}_3\rangle, E_{-\boldsymbol{\alpha}_3}|\boldsymbol{\mu}_1\rangle = \frac{1}{\sqrt{2}}|\boldsymbol{\mu}_2\rangle, E_{\boldsymbol{\alpha}}|\boldsymbol{\mu}_1\rangle = 0, \forall \boldsymbol{\alpha} \neq -\boldsymbol{\alpha}_1, -\boldsymbol{\alpha}_3;$$

$$E_{\boldsymbol{\alpha}_2}|\boldsymbol{\mu}_2\rangle = \frac{1}{\sqrt{2}}|\boldsymbol{\mu}_3\rangle, E_{\boldsymbol{\alpha}_3}|\boldsymbol{\mu}_2\rangle = \frac{1}{\sqrt{2}}|\boldsymbol{\mu}_1\rangle, E_{\boldsymbol{\alpha}}|\boldsymbol{\mu}_2\rangle = 0, \forall \boldsymbol{\alpha} \neq \boldsymbol{\alpha}_2, \boldsymbol{\alpha}_3;$$

$$E_{\boldsymbol{\alpha}_1}|\boldsymbol{\mu}_3\rangle = \frac{1}{\sqrt{2}}|\boldsymbol{\mu}_1\rangle, E_{-\boldsymbol{\alpha}_2}|\boldsymbol{\mu}_3\rangle = \frac{1}{\sqrt{2}}|\boldsymbol{\mu}_2\rangle, E_{\boldsymbol{\alpha}}|\boldsymbol{\mu}_3\rangle = 0, \forall \boldsymbol{\alpha} \neq \boldsymbol{\alpha}_1, -\boldsymbol{\alpha}_2.$$

As a result, all the basic states in the fundamental representation of $\mathfrak{su}(3, \mathbb{C})$ can be determined by the highest weight state $|\boldsymbol{\mu}_1\rangle$ as

$$|\boldsymbol{\mu}_2\rangle = \sqrt{2}E_{-\boldsymbol{\alpha}_3}|\boldsymbol{\mu}_1\rangle, |\boldsymbol{\mu}_3\rangle = \sqrt{2}E_{-\boldsymbol{\alpha}_1}|\boldsymbol{\mu}_1\rangle. \tag{8.75}$$

Another representation of the Lie algebra $\mathfrak{su}(3, \mathbb{C})$ is obtained by taking the negative of the complex conjugate of the matrices in the fundamental representation. The reason is that by taking the complex conjugate of the commutation relation $[T_a, T_b] = i\sum_{c=1}^{8} f_{abc}T_c$ and using the fact that the structure constants are real, one can obtain $[T_a^\dagger, T_b^\dagger] = -i\sum_{c=1}^{8} f_{abc}T_c^\dagger$, which implies that $[-T_a^\dagger, -T_b^\dagger] = i\sum_{c=1}^{8} f_{abc}(-T_c^\dagger)$, i.e., the matrices $-T_a^\dagger$ satisfy the same algebraic equation as the matrices T_a themselves. A representation obtained in this way is called the **complex**

conjugate representation. In such a representation, one may still choose the basic vectors as

$$|\boldsymbol{\nu}_1\rangle = \begin{pmatrix}1\\0\\0\end{pmatrix}, |\boldsymbol{\nu}_2\rangle = \begin{pmatrix}0\\1\\0\end{pmatrix}, |\boldsymbol{\nu}_3\rangle = \begin{pmatrix}0\\0\\1\end{pmatrix}. \tag{8.76}$$

Applying the diagonal generators $-H_1^\dagger$ and $-H_2^\dagger$ on the basic vectors yields

$$-\mathbf{H}^\dagger|\boldsymbol{\nu}_1\rangle = \boldsymbol{\nu}_1|\boldsymbol{\nu}_1\rangle, -\mathbf{H}^\dagger|\boldsymbol{\nu}_2\rangle = \boldsymbol{\nu}_2|\boldsymbol{\nu}_2\rangle, -\mathbf{H}^\dagger|\boldsymbol{\nu}_3\rangle = \boldsymbol{\nu}_3|\boldsymbol{\nu}_3\rangle, \tag{8.77}$$

where we have used the fact that $H_i^\dagger = H_i$ in any unitary irreducible representation, and the weights $\boldsymbol{\nu}_1$, $\boldsymbol{\nu}_2$, and $\boldsymbol{\nu}_3$ of the Lie algebra $\mathfrak{su}(3, \mathbb{C})$ in the complex conjugate representation are given by

$$\boldsymbol{\nu}_1 = \left(-\frac{1}{2}, -\frac{1}{2\sqrt{3}}\right), \boldsymbol{\nu}_2 = \left(\frac{1}{2}, -\frac{1}{2\sqrt{3}}\right), \boldsymbol{\nu}_3 = \left(0, \frac{1}{\sqrt{3}}\right). \tag{8.78}$$

which are not independent and satisfy the relation $\boldsymbol{\nu}_1 + \boldsymbol{\nu}_2 + \boldsymbol{\nu}_3 = 0$. Notice that the weights of the complex conjugate representation can be obtained by reflecting the weights of the fundamental representation through the origin, i.e., $\boldsymbol{\nu}_k = -\boldsymbol{\mu}_k$ for $k = 1, 2$, and 3. A direct computation yields the relations between the weights and the roots in the complex conjugate representation of the Lie algebra $\mathfrak{su}(3, \mathbb{C})$ (see Fig. 8.3)

$$\boldsymbol{\alpha}_1 = \boldsymbol{\nu}_3 - \boldsymbol{\nu}_1, \boldsymbol{\alpha}_2 = \boldsymbol{\nu}_2 - \boldsymbol{\nu}_3, \boldsymbol{\alpha}_3 = \boldsymbol{\nu}_2 - \boldsymbol{\nu}_1. \tag{8.79}$$

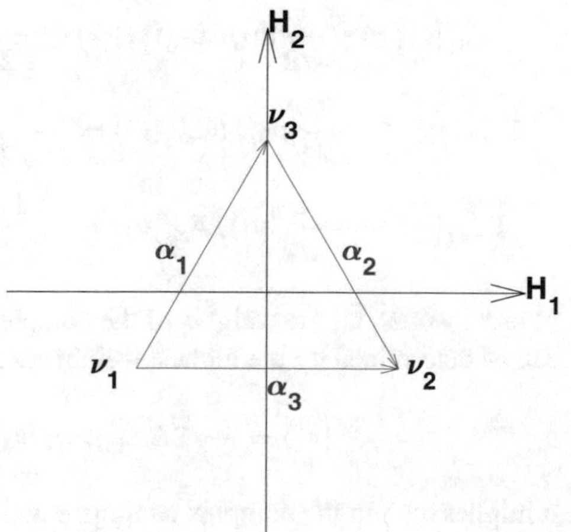

Fig. 8.3 The planar weight diagram of the complex conjugate representation of the Lie algebra $\mathfrak{su}(3, \mathbb{C})$ together with the three roots $\boldsymbol{\alpha}_1$, $\boldsymbol{\alpha}_2$, and $\boldsymbol{\alpha}_3$

Hence, the partial order on the weights of the complex conjugate representation of $\mathfrak{su}(3, \mathbb{C})$ is $\boldsymbol{\nu}_2 > \boldsymbol{\nu}_3 > \boldsymbol{\nu}_1$, as both $\boldsymbol{\nu}_2 - \boldsymbol{\nu}_3$ and $\boldsymbol{\nu}_3 - \boldsymbol{\nu}_1$ are positive roots. If one denotes the representation matrices of the shift operators $E_{\boldsymbol{\alpha}}$ in the complex conjugate representation as $\bar{U}(E_{\boldsymbol{\alpha}})$, then the representation matrix in the complex conjugate representation is related to that of the fundamental representation by $\bar{U}(E_{\boldsymbol{\alpha}}) = -U(E_{\boldsymbol{\alpha}}^{\dagger}) = -U(E_{-\boldsymbol{\alpha}})$. Hence, the representation matrices of the generators of the Cartan subalgebra and the shift operators in the complex conjugate representation are

$$\bar{U}(H_1) = \frac{1}{2}\begin{pmatrix} -1 & 0 & 0 \\ 0 & 1 & 0 \\ 0 & 0 & 0 \end{pmatrix}, \bar{U}(H_2) = \frac{1}{2\sqrt{3}}\begin{pmatrix} -1 & 0 & 0 \\ 0 & -1 & 0 \\ 0 & 0 & 2 \end{pmatrix}, \tag{8.80a}$$

$$\bar{U}(E_{\boldsymbol{\alpha}_1}) = \frac{1}{\sqrt{2}}\begin{pmatrix} 0 & 0 & 0 \\ 0 & 0 & 0 \\ -1 & 0 & 0 \end{pmatrix}, \bar{U}(E_{\boldsymbol{\alpha}_{-1}}) = \frac{1}{\sqrt{2}}\begin{pmatrix} 0 & 0 & -1 \\ 0 & 0 & 0 \\ 0 & 0 & 0 \end{pmatrix}. \tag{8.80b}$$

$$\bar{U}(E_{\boldsymbol{\alpha}_2}) = \frac{1}{\sqrt{2}}\begin{pmatrix} 0 & 0 & 0 \\ 0 & 0 & -1 \\ 0 & 0 & 0 \end{pmatrix}, \bar{U}(E_{\boldsymbol{\alpha}_{-2}}) = \frac{1}{\sqrt{2}}\begin{pmatrix} 0 & 0 & 0 \\ 0 & 0 & 0 \\ 0 & -1 & 0 \end{pmatrix}, \tag{8.80c}$$

$$\bar{U}(E_{\boldsymbol{\alpha}_3}) = \frac{1}{\sqrt{2}}\begin{pmatrix} 0 & 0 & 0 \\ -1 & 0 & 0 \\ 0 & 0 & 0 \end{pmatrix}, \bar{U}(E_{\boldsymbol{\alpha}_{-3}}) = \frac{1}{\sqrt{2}}\begin{pmatrix} 0 & -1 & 0 \\ 0 & 0 & 0 \\ 0 & 0 & 0 \end{pmatrix}. \tag{8.80d}$$

In the complex conjugate representation, the application of the shift operators $E_{\boldsymbol{\alpha}}$ on the basic vectors $|\boldsymbol{\nu}_1\rangle$, $|\boldsymbol{\nu}_2\rangle$ and $|\boldsymbol{\nu}_3\rangle$ yields

$$E_{\boldsymbol{\alpha}_1}|\boldsymbol{\nu}_1\rangle = -\frac{1}{\sqrt{2}}|\boldsymbol{\nu}_3\rangle,\ E_{\boldsymbol{\alpha}_3}|\boldsymbol{\nu}_1\rangle = -\frac{1}{\sqrt{2}}|\boldsymbol{\nu}_2\rangle, E_{\boldsymbol{\alpha}}|\boldsymbol{\nu}_1\rangle = 0, \forall \boldsymbol{\alpha} \neq \boldsymbol{\alpha}_1, \boldsymbol{\alpha}_3;$$

$$E_{-\boldsymbol{\alpha}_2}|\boldsymbol{\nu}_2\rangle = -\frac{1}{\sqrt{2}}|\boldsymbol{\nu}_3\rangle,\ E_{-\boldsymbol{\alpha}_3}|\boldsymbol{\nu}_2\rangle = -\frac{1}{\sqrt{2}}|\boldsymbol{\nu}_1\rangle, E_{\boldsymbol{\alpha}}|\boldsymbol{\nu}_2\rangle = 0, \forall \boldsymbol{\alpha} \neq -\boldsymbol{\alpha}_2, -\boldsymbol{\alpha}_3;$$

$$E_{-\boldsymbol{\alpha}_1}|\boldsymbol{\nu}_3\rangle = -\frac{1}{\sqrt{2}}|\boldsymbol{\nu}_1\rangle,\ E_{\boldsymbol{\alpha}_2}|\boldsymbol{\nu}_3\rangle = -\frac{1}{\sqrt{2}}|\boldsymbol{\nu}_2\rangle, E_{\boldsymbol{\alpha}}|\boldsymbol{\nu}_3\rangle = 0, \forall \boldsymbol{\alpha} \neq -\boldsymbol{\alpha}_1, \boldsymbol{\alpha}_2.$$

In order words, all the weights of the complex conjugate representation of $\mathfrak{su}(3, \mathbb{C})$ can be determined by the highest weight state $|\boldsymbol{\nu}_2\rangle$ via

$$|\boldsymbol{\nu}_1\rangle = -\sqrt{2}E_{-\boldsymbol{\alpha}_3}|\boldsymbol{\nu}_2\rangle, |\boldsymbol{\nu}_3\rangle = -\sqrt{2}E_{-\boldsymbol{\alpha}_2}|\boldsymbol{\nu}_2\rangle. \tag{8.81}$$

It implies that, in the complex conjugate representation of the Lie algebra $\mathfrak{su}(3, \mathbb{C})$, the elements H_1, H_2, and $E_{\pm\boldsymbol{\alpha}_1}$ span the maximal compact subalgebra $\mathfrak{k}$, whereas the elements $E_{\pm\boldsymbol{\alpha}_2}$ and $E_{\pm\boldsymbol{\alpha}_3}$ form the orthogonal complementary subspace $\mathfrak{p}$. (as $E_{\boldsymbol{\alpha}}|\boldsymbol{\nu}_2\rangle = 0$ for all positive roots and one negative root $-\boldsymbol{\alpha}_1$, except for the negative roots $-\boldsymbol{\alpha}_2$ and $-\boldsymbol{\alpha}_3$). Hence, a general element $\mathbf{k}$ of the maximal compact subalgebra $\mathfrak{k}$ in the complex conjugate representation of $\mathfrak{su}(3, \mathbb{C})$ has the form

$$\bar{U}(\mathbf{k}) \equiv i\lambda_1\bar{\rho}(H_1) + i\lambda_2\bar{\rho}(H_2) + \eta_{\alpha_1}\bar{\rho}(E_{\alpha_1}) - \eta^*_{\alpha_1}\bar{\rho}(E_{-\alpha_1}) \tag{8.82}$$
$$= \begin{pmatrix} -\frac{i\lambda_1}{2} - \frac{i\lambda_2}{2\sqrt{3}} & 0 & \frac{1}{\sqrt{2}}\eta^*_{\alpha_1} \\ 0 & \frac{i\lambda_1}{2} - \frac{i\lambda_2}{2\sqrt{3}} & 0 \\ -\frac{1}{\sqrt{2}}\eta_{\alpha_1} & 0 & \frac{i\lambda_2}{\sqrt{3}} \end{pmatrix},$$

where $\lambda_1, \lambda_2 \in \mathbb{R}$, $\eta_{\alpha_1} \in \mathbb{C}$, and $\bar{U}(\mathbf{k})$ are anti-Hermitian and traceless. Similarly, a general element $\mathbf{p}$ of the orthogonal complementary subspace $\mathfrak{p}$ in the complex conjugate representation of $\mathfrak{su}(3, \mathbb{C})$ has the form

$$\bar{U}(\mathbf{p}) \equiv \eta_{\alpha_2}\bar{U}(E_{\alpha_2}) + \eta_{\alpha_3}\bar{U}(E_{\alpha_3}) - \eta^*_{\alpha_2}\bar{U}(E_{-\alpha_2}) - \eta^*_{\alpha_3}\bar{U}(E_{-\alpha_3}) \tag{8.83}$$
$$= \frac{1}{\sqrt{2}}\begin{pmatrix} 0 & \eta^*_{\alpha_3} & 0 \\ -\eta_{\alpha_3} & 0 & -\eta_{\alpha_2} \\ 0 & \eta^*_{\alpha_2} & 0 \end{pmatrix},$$

where $\eta_{\alpha_2}, \eta_{\alpha_3} \in \mathbb{C}$, and $\bar{U}(\mathbf{p})$ are anti-Hermitian and traceless. Here, a direct computation yields

$$\bar{U}(\mathbf{p})^2 = -\frac{1}{2}\begin{pmatrix} |\eta_{\alpha_3}|^2 & 0 & \eta_{\alpha_2}\eta^*_{\alpha_3} \\ 0 & |\eta_{\alpha_3}|^2 + |\eta_{\alpha_2}|^2 & 0 \\ \eta_{\alpha_3}\eta^*_{\alpha_2} & 0 & |\eta_{\alpha_2}|^2 \end{pmatrix}, \tag{8.84}$$
$$\bar{U}(\mathbf{p})^{2k+1} = \left[-\frac{1}{2}(|\eta_{\alpha_3}|^2 + |\eta_{\alpha_2}|^2)\right]^k \bar{U}(\mathbf{p}) \equiv (-1)^k\theta^{2k}\bar{U}(\mathbf{p}).$$

Hence, if we choose $G = \mathrm{SU}(3, \mathbb{C})$ and $H \equiv \exp(\mathfrak{k})$ as its maximal isotropy subgroup, then in the complex conjugate representation of $\mathrm{SU}(3, \mathbb{C})$, a general element Ω in the coset space G/H has the form

$$\bar{U}(\Omega) \equiv \bar{U}(\exp(\mathbf{p})) = I + \frac{\sin\theta}{\theta}\bar{U}(\mathbf{p}) + \frac{1-\cos\theta}{\theta^2}\bar{U}(\mathbf{p})^2. \tag{8.85}$$

As a consequence, the $\mathrm{SU}(3, \mathbb{C})$ coherent state in the complex conjugate representation becomes

$$|\boldsymbol{v}_2, \Omega\rangle \equiv \bar{U}(\Omega)|\boldsymbol{v}_2\rangle = \begin{pmatrix} \frac{1}{\sqrt{2}}\eta^*_{\alpha_3}\frac{\sin\theta}{\theta} \\ \cos\theta \\ \frac{1}{\sqrt{2}}\eta^*_{\alpha_2}\frac{\sin\theta}{\theta} \end{pmatrix}. \tag{8.86}$$

One immediately sees that the SU(3, $\mathbb{C}$) coherent states in the complex conjugate representation are normalized to unity, i.e., $\langle \boldsymbol{\nu}_2, \Omega | \boldsymbol{\nu}_2, \Omega \rangle = 1$. Similarly, a general element $\mathbf{p}$ of the orthogonal complementary subspace $\mathfrak{p}$ in the fundamental representation of $\mathfrak{su}(3, \mathbb{C})$ has the form

$$U(\mathbf{p}) \equiv \eta_{\alpha_1} U(E_{\alpha_1}) + \eta_{\alpha_3} U(E_{\alpha_3}) - \eta^*_{\alpha_1} U(E_{-\alpha_1}) - \eta^*_{\alpha_3} U(E_{-\alpha_3}) \tag{8.87}$$
$$= \left(\begin{array}{c|cc} 0 & \frac{1}{\sqrt{2}}\eta_{\alpha_3} & \frac{1}{\sqrt{2}}\eta_{\alpha_1} \\ \hline -\frac{1}{\sqrt{2}}\eta^*_{\alpha_3} & 0 & 0 \\ -\frac{1}{\sqrt{2}}\eta^*_{\alpha_1} & 0 & 0 \end{array} \right) \equiv \left(\begin{array}{c|c} 0 & \boldsymbol{\eta} \\ \hline -\boldsymbol{\eta}^\dagger & \mathbf{0} \end{array} \right),$$

and a general element $\Omega \in \mathrm{SU}(3, \mathbb{C})/\exp(\mathfrak{k})$ in the fundamental representation of SU(3, $\mathbb{C}$) has the form

$$U(\Omega) \equiv U(\exp(\mathbf{p})) = I + \frac{\sin\theta}{\theta}\bar{U}(\mathbf{p}) + \frac{1-\cos\theta}{\theta^2}\bar{U}(\mathbf{p})^2 \tag{8.88}$$
$$= \left(\begin{array}{c|c} \cos\theta & \frac{\sin\theta}{\theta}\boldsymbol{\eta} \\ \hline -\frac{\sin\theta}{\theta}\boldsymbol{\eta}^\dagger & \boldsymbol{I}_2 - \frac{1-\cos\theta}{\theta^2}\boldsymbol{\eta}^\dagger\boldsymbol{\eta} \end{array} \right),$$
$$= \left(\begin{array}{c|c} \sqrt{1 - \boldsymbol{z}\boldsymbol{z}^\dagger} & \boldsymbol{z} \\ \hline -\boldsymbol{z}^\dagger & \sqrt{\boldsymbol{I}_2 - \boldsymbol{z}^\dagger\boldsymbol{z}} \end{array} \right),$$

where $\theta \equiv \sqrt{\boldsymbol{\eta}\boldsymbol{\eta}^\dagger}$ and $\boldsymbol{z} \equiv \frac{\sin\theta}{\theta}\boldsymbol{\eta} = \boldsymbol{\eta}\frac{\sin\sqrt{\boldsymbol{\eta}^\dagger\boldsymbol{\eta}}}{\sqrt{\boldsymbol{\eta}^\dagger\boldsymbol{\eta}}}$. Hence, the SU(3, $\mathbb{C}$) coherent state in the fundamental representation has the form

$$|\boldsymbol{\mu}_1, \Omega\rangle \equiv U(\Omega)|\boldsymbol{\mu}_1\rangle = \begin{pmatrix} \cos\theta \\ -\frac{1}{\sqrt{2}}\eta^*_{\alpha_3}\frac{\sin\theta}{\theta} \\ -\frac{1}{\sqrt{2}}\eta^*_{\alpha_1}\frac{\sin\theta}{\theta} \end{pmatrix}. \tag{8.89}$$

One immediately sees that the SU(3, $\mathbb{C}$) coherent states in the fundamental representation are normalized to unity, i.e., $\langle \boldsymbol{\mu}_1, \Omega | \boldsymbol{\mu}_1, \Omega \rangle = 1$. As a remark, both the fundamental and the complex conjugate representations of SU(3, $\mathbb{C}$) are **degenerate** in the sense that the highest weight $\boldsymbol{\Lambda}$ in such presentations is orthogonal to some root $\boldsymbol{\alpha}$, i.e., $\boldsymbol{\Lambda} \cdot \boldsymbol{\alpha} = 0$. For the former case, we have $\boldsymbol{\mu}_1 \cdot \boldsymbol{\alpha}_2 = 0$ (see Fig. 8.2), whereas for the latter case, we have $\boldsymbol{\nu}_2 \cdot \boldsymbol{\alpha}_1 = 0$ (see Fig. 8.3). For an arbitrary irreducible representation of SU(3, $\mathbb{C}$), the highest weight $\boldsymbol{\Lambda}$ can always be expressed as $\boldsymbol{\Lambda} \equiv \lambda_1\boldsymbol{\mu}_1 + \lambda_2\boldsymbol{\nu}_2$ with $\lambda_1 \neq 0$ and $\lambda_2 \neq 0$. Such a representation is non-degenerate, as

$$\boldsymbol{\Lambda} \cdot \boldsymbol{\alpha}_1 = \frac{\lambda_1}{2} \neq 0, \boldsymbol{\Lambda} \cdot \boldsymbol{\alpha}_2 = \frac{\lambda_2}{2} \neq 0, \boldsymbol{\Lambda} \cdot \boldsymbol{\alpha}_3 = \frac{\lambda_1 + \lambda_2}{2} \neq 0, \tag{8.90}$$

where we have used the fact that the highest weight $\mathbf{\Lambda} \equiv \lambda_1 \boldsymbol{\mu}_1 + \lambda_2 \boldsymbol{\nu}_2$ always located in the right-half plane such that $\frac{1}{2}(\lambda_1 + \lambda_2) > 0$. In such a non-degenerate representation, the maximal isotropy subgroup is U(1) ⊗ U(1) with generators $\{H_1, H_2\}$. Hence, the geometry of the coherent state space is *not* the coset space SU(3)/U(2) but a larger coset space SU(3)/U(1) ⊗ U(1).

Exercises

8.1. Verify that the overlap between two SU(1, 1) coherent states is

$$\langle k, \zeta' | k, \zeta \rangle = \frac{(1 - |\zeta'^2|)^k (1 - |\zeta^2|)^k}{(1 - \zeta'^* \zeta)^{2k}}.$$

8.2. Show that the Gaussian curvature of a metric $ds^2 = \lambda^2(z, z^*) dz dz^*$ has the form

$$K = -\frac{4}{\lambda^2} \frac{\partial}{\partial \zeta} \frac{\partial}{\partial \zeta^*} \log \lambda.$$

8.3. Verify the following expressions for $\mathcal{K}_0$, $\mathcal{K}_+$, and $\mathcal{K}_-$

$$\mathcal{K}_0 = k \frac{1 + |\zeta|^2}{1 - |\zeta|^2}, \mathcal{K}_- = k \frac{2\zeta}{1 - |\zeta|^2}, \mathcal{K}_+ = k \frac{2\zeta^*}{1 - |\zeta|^2}.$$

8.4. Verify the following Poisson brackets between $\mathcal{K}_0$, $\mathcal{K}_-$ and $\mathcal{K}_+$

$$i\{\mathcal{K}_-, \mathcal{K}_+\} = 2\mathcal{K}_0, i\{\mathcal{K}_0, \mathcal{K}_\pm\} = \pm \mathcal{K}_\pm.$$

8.5. Verify the following relation for arbitrary coordinates (ζ, ζ^*) and (z, z^*)

$$\frac{\partial f}{\partial \zeta} \frac{\partial g}{\partial \zeta^*} - \frac{\partial f}{\partial \zeta^*} \frac{\partial g}{\partial \zeta} = \left(\frac{\partial f}{\partial z} \frac{\partial g}{\partial z^*} - \frac{\partial f}{\partial z^*} \frac{\partial g}{\partial z} \right) \left(\frac{\partial z}{\partial \zeta} \frac{\partial z^*}{\partial \zeta^*} - \frac{\partial z}{\partial \zeta^*} \frac{\partial z^*}{\partial \zeta} \right)$$

8.6. Show that the Poisson bracket $\{\cdot, \cdot\}$ in the coordinate (z, z^*) defined via $z \equiv \sqrt{2k}\zeta / \sqrt{1 - |\zeta|^2}$ has the form

$$\{f, g\} \equiv \frac{1}{i} \left(\frac{\partial f}{\partial z} \frac{\partial g}{\partial z^*} - \frac{\partial f}{\partial z^*} \frac{\partial g}{\partial z} \right).$$

Lie Group Generalizations of Coherent States 9

9.1 Generalized Coherent States

In the previous chapters, we have extensively discussed the construction of coherent states by the utilization of group-theoretical methodologies with underlying groups H(4), SU(2), SU(1, 1), and SU(3). Of course, the breath of physics is not simply confined to these groups. In this chapter, we will discuss in detail the generalization of coherent states for arbitrary Lie groups, which is known as the Gilmore-Perelomov group-theoretic coherent state. Such a generalization will provide a unified framework for Glauber's coherent states, spin coherent states, squeezed coherent states, and all other coherent states based on Lie group theory. The group-theoretical construction of coherent states was carried out almost entirely independently by Askold M. Perelomov [8] and Robert Gilmore [9, 10] in 1972. Remarkably, the key idea behind the group-theoretical construction was proposed by J. R. Klauder nearly a decade earlier [6, 7]. In the following, we shall first discuss some generic properties of Lie group and Lie algebra, especially the structure theory, classification, and representation theory of complex semisimple Lie algebra, which are much needed for us to discuss generalized coherent states.

In order to classify structures of Lie algebras, the concepts of subalgebra and ideal are essential. Let $\mathfrak{m}, \mathfrak{n}$ be subsets of a Lie algebra $\mathfrak{g}$, and let $[\mathfrak{m}, \mathfrak{n}]$ be the vector subspace spanned by all elements in $\mathfrak{g}$ of the form $[M, N]$ ($M \in \mathfrak{m}$, $N \in \mathfrak{n}$. A vector subspace $\mathfrak{h}$ of $\mathfrak{g}$ is called a **Lie subalgebra** of $\mathfrak{g}$, if $[\mathfrak{h}, \mathfrak{h}] \subseteq \mathfrak{h}$, i.e., for arbitrary $X, Y \in \mathfrak{h}$, $[X, Y] \in \mathfrak{h}$. In other words, a Lie subalgebra $\mathfrak{h}$ of $\mathfrak{g}$ is a vector subspace of $\mathfrak{g}$ which is also closed under the Lie bracket. An **ideal** $\mathfrak{i}$ of $\mathfrak{g}$ is a Lie subalgebra of $\mathfrak{g}$ which satisfies a stronger condition: $[\mathfrak{g}, \mathfrak{i}] \subseteq \mathfrak{i}$, i.e., for arbitrary $X \in \mathfrak{g}$ and $Y \in \mathfrak{i}$, $[X, Y] \in \mathfrak{i}$. An ideal of a Lie algebra is also called a **invariant subalgebra**, as it is mapped into itself by all elements of the algebra. Evidently, the Lie algebra $\mathfrak{g}$ itself and the single element subalgebra $\{0\} \in \mathfrak{g}$ are both ideals of the Lie algebra $\mathfrak{g}$. Also, if $\mathfrak{i}$ and $\mathfrak{j}$ are ideals of a Lie algebra $\mathfrak{g}$, then $[\mathfrak{i}, \mathfrak{j}]$ is also an ideal of $\mathfrak{g}$. The reason is simple. From the Jacobi identity, one obtains

C.-F. Kam et al., *Coherent States*, Lecture Notes in Physics 1011,
https://doi.org/10.1007/978-3-031-20766-2_9

$[g, [x, y]] = -[x, [y, g]] - [y, [g, x]] \in [\mathfrak{i}, \mathfrak{j}]$ for any $a \in \mathfrak{g}$, $x \in \mathfrak{i}$ and $y \in \mathfrak{j}$. Hence, $[\mathfrak{g}, [\mathfrak{i}, \mathfrak{j}]] \subseteq [\mathfrak{i}, \mathfrak{j}]$, and thus $[\mathfrak{i}, \mathfrak{j}]$ is an ideal of $\mathfrak{g}$.

As an example, the **special linear Lie algebra** of order n, denoted as $\mathfrak{sl}(n, \mathbb{R})$, is the Lie algebra of $n \times n$ matrices with real entries and trace zero. It is a Lie subalgebra and is also an ideal of the general linear Lie algebra $\mathfrak{gl}(n, \mathbb{R})$. The reason is that for arbitrary $X, Y \in \mathfrak{gl}(n, \mathbb{R})$, $\mathrm{Tr}[X, Y] = \mathrm{Tr}(XY - YX) = 0$, and thus $[X, Y] \in \mathfrak{sl}(n, \mathbb{R})$.

We now turn our attention to the structure theory of Lie algebras. Let us begin with the concept of simple Lie algebra. A **simple Lie algebra** is a Lie algebra $\mathfrak{g}$ which is non-abelian and has no other ideals besides $\{0\}$ and $\mathfrak{g}$ itself. A basic property of a simple lie algebra $\mathfrak{g}$ is that it satisfies $[\mathfrak{g}, \mathfrak{g}] = \mathfrak{g}$. The reason is that as $[\mathfrak{g}, \mathfrak{g}]$ is an ideal of the simple lie algebra $\mathfrak{g}$, it is either $\{0\}$ or $\mathfrak{g}$ itself. But since $\mathfrak{g}$ is non-abelian, the only choice is $[\mathfrak{g}, \mathfrak{g}] = \mathfrak{g}$. In other words, the simple Lie algebras regenerate themselves under commutation. This nice property is also retained for the **semisimple Lie algebra**, which is a direct sum of simple Lie algebras. The reason is simple. For simplicity, one may let $\mathfrak{g}$ be a direct sum of two simple Lie algebras $\mathfrak{g}_1$ and $\mathfrak{g}_2$, and then one obtains $[\mathfrak{g}_1 + \mathfrak{g}_2, \mathfrak{g}_1 + \mathfrak{g}_2] = [\mathfrak{g}_1, \mathfrak{g}_1] + [\mathfrak{g}_2, \mathfrak{g}_2] = \mathfrak{g}_1 + \mathfrak{g}_2$, which means $[\mathfrak{g}, \mathfrak{g}] = \mathfrak{g}$.

With the above discussions, we are now ready to study the central result in the structure theory of simple Lie algebras over complex numbers, namely, the **root space decomposition** or **Cartan decomposition**, which is a complete classification scheme for these algebras. The key is the existence of a maximal commuting Lie subalgebra, of dimension l, called the **Cartan subalgebra**, denoted as $\mathfrak{h}$. If one denotes the bases of the Cartan subalgebra $\mathfrak{h}$ as $H_1, H_2, \cdots, H_l$, one immediately obtains the commutation relations $[H_i, H_j] = 0$ for $1 \leq i, j \leq l$. Notice that the choice of the Cartan subalgebra $\mathfrak{h}$ is not unique. However, all different choices of $\mathfrak{h}$ are isomorphic and will have the same dimension l, known as the **rank** of the Lie algebra. Then, a simple complex Lie algebra $\mathfrak{g}$ is the direct sum of the l-dimensional Cartan subalgebra $\mathfrak{h}$ and all the one-dimensional **root spaces** $\mathfrak{g}_\alpha$

$$\mathfrak{g} = \mathfrak{h} \oplus \left(\bigoplus_{\alpha \in \Phi} \mathfrak{g}_\alpha \right), \tag{9.1}$$

where $\boldsymbol{\Phi}$ is a **root system** which consists of a finite set of nonzero vectors $\boldsymbol{\alpha}$ in the l-dimensional Euclidean space $\mathbb{R}^l$, known as the **roots** of the semisimple Lie algebra $\mathfrak{g}$.

The roots are required to satisfy the following conditions: (i) the roots span the Euclidean space $\mathbb{R}^l$; (ii) the only scalar multiples of a root $\boldsymbol{\alpha} \in \boldsymbol{\Phi}$ that belong to $\boldsymbol{\Phi}$ are $\boldsymbol{\alpha}$ itself and $-\boldsymbol{\alpha}$; (iii) for every root $\boldsymbol{\alpha} \in \boldsymbol{\Phi}$, the set $\boldsymbol{\Phi}$ is closed under reflection through the hyperplane perpendicular to $\boldsymbol{\alpha}$; and (iv) if $\boldsymbol{\alpha}$ and $\boldsymbol{\beta}$ are roots in $\boldsymbol{\Phi}$, then the projection of $\boldsymbol{\beta}$ onto the line through $\boldsymbol{\alpha}$ is an integer or half-integer multiple of $\boldsymbol{\alpha}$. In other words, for any two roots $\boldsymbol{\alpha}$ and $\boldsymbol{\beta}$, the number $\langle \boldsymbol{\beta}, \boldsymbol{\alpha} \rangle \equiv 2(\boldsymbol{\beta} \cdot \boldsymbol{\alpha})/(\boldsymbol{\alpha} \cdot \boldsymbol{\alpha})$ is an integer. For a given root system $\boldsymbol{\Phi}$, one may always choose a set of **positive roots**, which is a subset $\boldsymbol{\Phi}^+ \in \boldsymbol{\Phi}$ such that (i) for every root $\boldsymbol{\alpha} \in \boldsymbol{\Phi}^+$, the opposite root

$-\boldsymbol{\alpha} \notin \boldsymbol{\Phi}^+$; (ii) for any two roots $\boldsymbol{\alpha}, \boldsymbol{\beta} \in \boldsymbol{\Phi}^+$, if $\boldsymbol{\alpha}+\boldsymbol{\beta} \in \boldsymbol{\Phi}$, then $\boldsymbol{\alpha}+\boldsymbol{\beta} \in \boldsymbol{\Phi}^+$. If a set of positive roots is chosen, the subset $\boldsymbol{\Phi}^- \in \boldsymbol{\Phi}$ which consists of all opposite roots $-\boldsymbol{\alpha}$ of the roots $\boldsymbol{\alpha} \in \boldsymbol{\Phi}^+$ is called the set of **negative roots**. A set of positive roots can be constructed by choosing a hyperplane not containing any root and setting Φ^+ to be the set of all the roots lying on the same side of the hyperplane. An element of $\boldsymbol{\Phi}^+$ is called a **simple positive root** if it cannot be written as the sum of two positive roots $\boldsymbol{\alpha}, \boldsymbol{\beta} \in \boldsymbol{\Phi}^+$. It follows that the set of positive simple roots is a basis of the Euclidean space $\mathbb{R}^l$ which satisfies the condition: every positive root is a linear combination of simple positive roots with non-negative integer coefficients.

Let us denote $\boldsymbol{\alpha} \equiv (\alpha_1, \alpha_2, \cdots, \alpha_l)$. Then for any nonzero roots $\boldsymbol{\alpha} \in \boldsymbol{\Phi}$, there is a unique element $E_{\boldsymbol{\alpha}} \in \mathfrak{g}_\alpha$ such that $[H_i, E_{\boldsymbol{\alpha}}] = \alpha_i E_{\boldsymbol{\alpha}}$ for any $H_i \in \mathfrak{h}$. Moreover, if $\boldsymbol{\alpha}+\boldsymbol{\beta}$ is a nonzero root, it follows that $[E_{\boldsymbol{\alpha}}, E_{\boldsymbol{\beta}}]$ is a multiple of $E_{\boldsymbol{\alpha}+\boldsymbol{\beta}}$, while if $\boldsymbol{\alpha}+\boldsymbol{\beta}$ is not a root, then $[E_{\boldsymbol{\alpha}}, E_{\boldsymbol{\beta}}] = 0$. Finally, for the case $\boldsymbol{\alpha}+\boldsymbol{\beta} = 0$, the commutator of $E_{\boldsymbol{\alpha}}$ and $E_{\boldsymbol{\beta}}$ is an element in $\mathfrak{h}$, i.e., $[E_{\boldsymbol{\alpha}}, E_{-\boldsymbol{\alpha}}] = \boldsymbol{\alpha} \cdot \mathbf{H}$ with $\mathbf{H} \equiv (H_1, H_2, \cdots, H_l)$. As such, the canonical commutation relations between the elements of a Lie algebra $\mathfrak{g}$ in the Cartan basis become

$$[H_i, H_j] = 0, \tag{9.2a}$$

$$[H_i, E_\alpha] = \alpha_i E_\alpha, \tag{9.2b}$$

$$[E_{\boldsymbol{\alpha}}, E_{-\boldsymbol{\alpha}}] = \boldsymbol{\alpha} \cdot \mathbf{H}, \tag{9.2c}$$

$$[E_{\boldsymbol{\alpha}}, E_{\boldsymbol{\beta}}] = N_{\boldsymbol{\alpha},\boldsymbol{\beta}} E_{\boldsymbol{\alpha}+\boldsymbol{\beta}}, \ (\boldsymbol{\alpha}+\boldsymbol{\beta} \neq 0). \tag{9.2d}$$

Before we start the detailed discussion of the generalized coherent states, we shall recapitulate Glauber's coherent state, $|\alpha\rangle \equiv \hat{D}(\alpha)|0\rangle$, which is obtained by applying a displacement operator $\hat{D}(\alpha)$ on the vacuum state $|0\rangle$ of the harmonic oscillator. In fact, the displacement operator $\hat{D}(\alpha) \equiv \exp\{\alpha\hat{a}^\dagger - \alpha^*\hat{a}\}$ is simply a representation of the **Heisenberg-Weyl group** in the Fock space. Here, a Heisenberg-Weyl group is a Lie group whose Lie algebra is the **Heisenberg-Weyl algebra** generated from four elements $\{\hat{n}, \hat{a}^\dagger, \hat{a}, \hat{I}\}$, subject to the nontrivial Lie brackets $[\hat{a}, \hat{a}^\dagger] = \hat{I}$, $[\hat{n}, \hat{a}] = \hat{a}^\dagger$, and $[\hat{n}, \hat{a}] = -\hat{a}$. A general element of the Heisenberg-Weyl group (denoted as H_4) may be written as

$$g = \exp\{\alpha\hat{a}^\dagger - \alpha\hat{a}\}\exp\{i(\delta\hat{n} + \varphi\hat{I})\} = \hat{D}(\alpha)\exp\{i(\delta\hat{n} + \varphi\hat{I})\}. \tag{9.3}$$

Hence, any group transformation g of H_4 acting on the vacuum state $|0\rangle$ can be identified with Glauber's coherent state, up to a phase factor $e^{i\varphi}$

$$g|0\rangle = \hat{D}(\alpha)\exp\{i(\delta\hat{n} + \varphi\hat{I})\}|0\rangle = e^{i\varphi}\hat{D}(\alpha)|0\rangle = e^{i\varphi}|\alpha\rangle. \tag{9.4}$$

Accordingly, the group-theoretical formulation of coherent states consists of three inputs: a Lie group, its unitary representation on a Hilbert space, and a fixed reference state. With this group-theoretical approach, one can readily generalize the concept of coherent states to arbitrary dynamical systems whose Hamiltonians are constructed by the elements of a Lie algebra.

In the most general setting, the group-theoretical algorithm for constructing coherent states requires three inputs: (a) a **topological group** G; (b) a **continuous unitary irreducible representation** V of G by unitary operators $\hat{U}$ on a Hilbert space $\mathcal{H}$; and (c) a **reference state** $|\Phi_0\rangle$ within the Hilbert space. With these three inputs, one might naïvely define the coherent states as $\hat{U}(g)|\Psi_0\rangle$ for those $g \in G$. However, one cannot parametrize the generalized coherent states in this simple manner, as two group elements may correspond to the same state. In quantum mechanics, two vectors $|\psi_1\rangle$ and $|\psi_2\rangle$ in $\mathcal{H}$ correspond to the same state if they only differ by a phase factor, $|\psi_1\rangle = e^{i\varphi}|\psi_2\rangle$. Hence, one sees that two vectors $\hat{U}(g_1)|\Phi_0\rangle$ and $\hat{U}(g_2)|\Phi_0\rangle$ in $\mathcal{H}$ correspond to the same state if and only if $\hat{U}(g_2^{-1}g_1)|\Phi_0\rangle = e^{i\varphi}|\Phi_0\rangle$. It immediately leads to the important concept of **stabilizer subgroup** (also called the **isotropy subgroup**). An isotropy subgroup H of G is a subgroup of G such that for all $h \in H$, $\hat{U}(h)|\Phi_0\rangle = e^{i\varphi}|\Phi_0\rangle$. In other words, two vectors $\hat{U}(g_1)|\Phi_0\rangle$ and $\hat{U}(g_2)|\Phi_0\rangle$ in $\mathcal{H}$ correspond to the same state if and only if two group elements g_1 and g_2 belong to the same left coset, i.e., $g_1 H = g_2 H$, or equivalently $g_2^{-1} g_1 \in H$. In this regard, the generalized coherent states can be parametrized by elements in the **coset space** $G/H \equiv \{gH | g \in G\}$, i.e., the set of equivalence classes of elements of G where two are regarded as equivalent if they differ by right multiplication with an element in H.

The following are several remarks about the group-theoretical algorithm. First, as we shall see below, the irreducibility of the representation V of G is crucial, as it is one of the necessary conditions for the over-completeness properties of the generalized coherent states. Second, the coset space G/H does not necessarily form a group, unless H is a normal subgroup of G which satisfies $gH = Hg$ for every group element $g \in G$. But the coset space G/H forms a **homogeneous space**, a space in which "all points are the same" instead. In particular, if G is a Lie group and H is a closed Lie subgroup of G such that the coset space G/H is connected, then the coset space is a smooth manifold referred to as the space of **Klein geometry**, which generalizes the usual Euclidean, spherical, projective, and hyperbolic geometries.

More precisely, the output of the group-theoretical algorithm gives the coherent states, which is done in three steps: (a) a **maximal isotropy subgroup** H of G which is a subgroup of G that consists of all the group elements $h \in G$ that will leave the reference state $|\Phi_0\rangle$ unchanged up to a phase factor, $h|\Phi_0\rangle = e^{i\varphi}|\Phi_0\rangle$, and is not a subgroup of any of the other isotropy subgroups; (b) a **coset space** G/H such that for every group element $g \in G$, there is a unique left coset decomposition of g in H—$g = \Omega h$ for $g \in G$, $h \in H$, and $\Omega \in G/H$; and (c) the **coherent states** $|\Omega\rangle \equiv \hat{U}(\Omega)|\Phi_0\rangle$ which are parametrized by elements in the coset space G/H. In other words, the **coherent state space** is in one-to-one correspondence with the set space G/H. Moreover, since $\hat{U}(\Omega)$ are unitary operators, the generalized coherent states are normalized to unity

$$\langle\Omega|\Omega\rangle = \langle\Phi_0|\hat{U}(\Omega^{-1})\hat{U}(\Omega)|\Phi_0\rangle = \langle\Phi_0|\Phi_0\rangle = 1. \tag{9.5}$$

The following are several remarks on the constructions of generalized coherent states. The coherent states constructed in the group-theoretical algorithm depend on the choice of the group G, its representation on $\mathcal{H}$, and the reference state $|\Phi_0\rangle$. Hence, there are various possible constructions of the coherent states utilizing the group-theoretical algorithm which possesses different properties. First, the group G may be an arbitrary topological group in general, as in Gilmore's construction. One may impose additional structure on G, i.e., a smooth manifold structure, to make it a Lie group, as in Perelomov's construction. Perelomov's more restrictive choice for G allows always for the construction of BCH formulas. One may also impose additional structures on G, i.e., a compact manifold structure in the sense of a "manifold without boundary," to make it a compact Lie group. Second, the unitary irreducible representation V of G may be arbitrary, as in Perelomov's construction. But as in Gilmore's construction, one may also require that G be a locally compact topological group and demand that the representation V of G to be **square-integrable**, i.e., there exists nonzero vectors $|\phi\rangle, |\psi\rangle \in \mathcal{H}$ such that

$$\int_G |\langle\phi|\hat{U}(g)|\psi\rangle|^2 d\mu_G(g) < \infty, \tag{9.6}$$

where μ_G is the **left Haar measure** on G, which assigns an invariant volume to subsets of G, in the sense that μ_G is a left translation invariant on G such that $\mu_G(gS) = \mu_G(S)$ for all $g \in G$ and all measurable subsets $S \in G$. The square integrability assumption adopted here is crucial as it guarantees the existence of the over-completeness relations for the generalized coherent states. Third, the reference state $|\Phi_0\rangle \in \mathcal{H}$ may be an arbitrary state, as in Perelomov's construction. However, one may also require that $|\Phi_0\rangle$ is the eigenstate of an unperturbed Hamiltonian. In particular, if G is a simple Lie group, then one may choose $|\Phi_0\rangle$ to be an **extremal state**, e.g., the highest-weight state or the lowest-weight state in the square-integrable irreducible highest weight representation $V^{\mathbf{\Lambda}}$. Here, the highest-weight state $|\mathbf{\Lambda}, \mathbf{\Lambda}\rangle$ of the square-integrable irreducible irreducible representation $V^{\mathbf{\Lambda}}$ is defined by $E_{\boldsymbol{\alpha}}|\mathbf{\Lambda}, \mathbf{\Lambda}\rangle = 0$ for all positive roots $\boldsymbol{\alpha} \in \mathbf{\Phi}^+$ in the root space.

To summarize, if one were to demand that the generalized coherent states are normalized to unity, over-complete, and allow for the applications of BCH formulas, one needs to consider a Lie group G, and a square-integrable unitary irreducible representation $V^{\mathbf{\Lambda}}$ of G in the group-theoretical algorithm for constructing coherent states. Besides, it is useful to choose the reference state $|\Phi_0\rangle$ to be a ground state of some unperturbed Hamiltonians, which becomes the extremal states of the irreducible representation $V^{\mathbf{\Lambda}}$ of G when the Hamiltonian is linear in the elements of G in the absence of interaction. In this regard, the coset space G/H that the generalized coherent states belong to is constructed by the exponential map of the shift-down operators and their conjugate operators, where the shift-down operators $E_{-\boldsymbol{\alpha}}$ obey $E_{-\boldsymbol{\alpha}}|\mathbf{\Lambda}, \mathbf{\Lambda}\rangle \neq 0$ when acting on the highest-weight state.

To be specific, let us consider a Lie group G corresponding to a finite-dimensional semisimple Lie algebra $\mathfrak{g}$. In the Cartan basis, there are two types of operators H_i and $E_{\boldsymbol{\alpha}}$, where the operators H_i are diagonal in any unitary irreducible representation and $E_{\boldsymbol{\alpha}}$ are the "shift operators" which obey $H_i^\dagger = H_i$ and $E_{\boldsymbol{\alpha}}^\dagger = E_{-\boldsymbol{\alpha}}$. Every group element $g \in G$ can be written as the exponential of anti-Hermitian matrix which is a complex linear combination of H_i and $E_{\boldsymbol{\alpha}}$. Let us denote $\boldsymbol{\Lambda}$ to be the highest weight of an irreducible unitary representation $V^{\boldsymbol{\Lambda}}$ of G; the highest-weight state $|\boldsymbol{\Lambda}, \boldsymbol{\Lambda}\rangle$, which is often the ground state of the unperturbed Hamiltonian, satisfies the following properties: (i) $|\boldsymbol{\Lambda}, \boldsymbol{\Lambda}\rangle$ is annihilated by all the shift-up operators $E_{\boldsymbol{\alpha}}$ with $\boldsymbol{\alpha}$ being a positive root, $E_{\boldsymbol{\alpha}}|\boldsymbol{\Lambda}, \boldsymbol{\Lambda}\rangle = 0, \forall \boldsymbol{\alpha} \in \boldsymbol{\Phi}^+$; (ii) $|\boldsymbol{\Lambda}, \boldsymbol{\Lambda}\rangle$ is the eigenstate of H_i with an eigenvalue Λ_i, $H_i|\boldsymbol{\Lambda}, \boldsymbol{\Lambda}\rangle = \Lambda_i|\boldsymbol{\Lambda}, \boldsymbol{\Lambda}\rangle$; and (iii) $|\boldsymbol{\Lambda}, \boldsymbol{\Lambda}\rangle$ is annihilated by some shift-down operators $E_{\boldsymbol{\alpha}}$ with $\boldsymbol{\alpha} \in \boldsymbol{\Phi}^-$ and is mapped to the states $|\boldsymbol{\Lambda}, \boldsymbol{\Lambda} + \boldsymbol{\beta}\rangle$ with a lower weight for other $E_{\boldsymbol{\beta}}$ with $\boldsymbol{\beta} \in \boldsymbol{\Phi}^-$:

$$E_{\boldsymbol{\alpha}}|\boldsymbol{\Lambda}, \boldsymbol{\Lambda}\rangle = 0, \qquad \text{for some } \boldsymbol{\alpha} \in \boldsymbol{\Phi}^-, \tag{9.7a}$$

$$E_{\boldsymbol{\beta}}|\boldsymbol{\Lambda}, \boldsymbol{\Lambda}\rangle = |\boldsymbol{\Lambda}, \boldsymbol{\Lambda} + \boldsymbol{\beta}\rangle \times \text{factor}, \; \text{for other } \boldsymbol{\beta} \in \boldsymbol{\Phi}^-. \tag{9.7b}$$

Then the generalized coherent states can be explicitly written as $|\boldsymbol{\Lambda}, \Omega\rangle \equiv \Omega|\boldsymbol{\Lambda}, \boldsymbol{\Lambda}\rangle$, where Ω is a generalized "displacement operator" which has one-to-one correspondence with the elements in the coset space G/H

$$\Omega = \exp \sum_{\boldsymbol{\beta}} (\eta_{\boldsymbol{\beta}} E_{\boldsymbol{\beta}} - \eta_{\boldsymbol{\beta}}^* E_{-\boldsymbol{\beta}}), \tag{9.8}$$

where $\eta_{\boldsymbol{\beta}}$ are some complex parameters and the summation is restricted to those negative roots $\boldsymbol{\beta} \in \boldsymbol{\Phi}^-$ such that $E_{\boldsymbol{\beta}}|\boldsymbol{\Lambda}, \boldsymbol{\Lambda}\rangle \neq 0$.

9.2 General Properties of Coherent States

To begin with, we consider the geometric properties of the generalized coherent states. As the generalized coherent states $|\boldsymbol{\Lambda}, \Omega\rangle$ are in one-to-one correspondence with the elements of the coset space G/H, the space of coherent states and the coset space are topologically equivalent. More importantly, the coset space admits three different geometric structures: (i) a **complex structure** along with an atlas of charts to an open disk in $\mathbb{C}^n$ such that the transition maps are holomorphic; (ii) a **Riemannian structure** equipped with a Riemannian metric which measures distance, area and volume in the space; and (iii) a **symplectic structure** equipped with a non-degenerate closed 2-form, which provides a way of measuring area in the space with a changing shape.

To discuss the complex structure of the coset space, let us recall that the generalized coherent states $|\boldsymbol{\Lambda}, \Omega\rangle$ for a semisimple Lie group G can be explicitly expressed as

$$|\boldsymbol{\Lambda}, \Omega\rangle = \exp \sum_{\boldsymbol{\beta}} (\eta_{\boldsymbol{\beta}} E_{\boldsymbol{\beta}} - \eta_{\boldsymbol{\beta}}^* E_{-\boldsymbol{\beta}})|\boldsymbol{\Lambda}, \boldsymbol{\Lambda}\rangle, \tag{9.9}$$

where the summation is restricted to those roots which satisfy $E_\beta|\mathbf{\Lambda}, \mathbf{\Lambda}\rangle \neq 0$. It shows that the generalized coherent states are in one-to-one correspondence with the elements of the coset space G/H. As the Lie algebra $\mathfrak{g}$ of the Lie group G is semisimple, it has the Cartan decomposition $\mathfrak{g} = \mathfrak{k} \oplus \mathfrak{p}$, where $\mathfrak{k}$ is the Lie algebra of the maximal isotropy subgroup H which forms a subalgebra of $\mathfrak{g}$, also known as the **maximal compact subalgebra** when $\mathfrak{g}$ is a compact semisimple Lie algebra, and $\mathfrak{p}$ is the Lie algebra of the coset space G/H which forms the **orthogonal complementary subspace** of $\mathfrak{k}$. Here, $\mathfrak{p} = \sum_\beta(\eta_\beta E_\beta - \eta_\beta^* E_{-\beta})$ is spanned by the shift operators E_β and $E_{-\beta}$ with $\boldsymbol{\beta} \in \mathbf{\Phi}^-$ and $E_\beta|\mathbf{\Lambda}, \mathbf{\Lambda}\rangle \neq 0$. Precisely, the Lie subalgebra $\mathfrak{k}$ of $\mathfrak{g}$ and its orthogonal complementary subspace $\mathfrak{p}$ satisfy the relations

$$[\mathfrak{k}, \mathfrak{k}] \subseteq \mathfrak{k}, [\mathfrak{k}, \mathfrak{p}] \subseteq \mathfrak{p}, [\mathfrak{p}, \mathfrak{p}] \subseteq \mathfrak{k}, \tag{9.10}$$

such that $\mathfrak{k}$ is a Lie subalgebra and any subalgebra of $\mathfrak{p}$ is commutative.

We now go back to the discussion of the geometric properties of the generalized coherent states. Let G be a compact semisimple Lie group and $\mathfrak{g}$ be its Lie algebra, and then in the fundamental representation of $\mathfrak{g}$, the maximal compact subalgebra $\mathfrak{k}$ and its orthogonal complementary subspace $\mathfrak{p}$ which satisfy $[\mathfrak{k}, \mathfrak{k}] \subseteq \mathfrak{k}$, $[\mathfrak{k}, \mathfrak{p}] \subseteq \mathfrak{p}$ and $[\mathfrak{p}, \mathfrak{p}] \subseteq \mathfrak{k}$ have the form

$$U(\mathbf{k}) \equiv \left(\begin{array}{c|c} \mathbf{a} & \mathbf{0} \\ \hline \mathbf{0} & \mathbf{b} \end{array}\right), \ U(\mathbf{p}) \equiv \left(\begin{array}{c|c} \mathbf{0} & \eta \\ \hline -\eta^\dagger & \mathbf{0} \end{array}\right), \ \forall \mathbf{k} \in \mathfrak{k}, \mathbf{p} \in \mathfrak{p}, \tag{9.11}$$

where $\mathbf{a}$ is a $m \times m$ matrix, $\mathbf{b}$ is an $n \times n$ matrix, and $\boldsymbol{\eta}$ is a $m \times n$ matrix. Here, n and m are the numbers of the shift operators $E_{-\alpha}$ with $\boldsymbol{\alpha} \in \mathbf{\Phi}^+$ which satisfy $E_{-\alpha}|\mathbf{\Lambda}, \mathbf{\Lambda}\rangle \neq 0$ or $E_{-\alpha}|\mathbf{\Lambda}, \mathbf{\Lambda}\rangle = 0$, respectively. Let H be the maximal isotropy subgroup of G whose Lie algebra is $\mathfrak{k}$, and then the coset space G/H in the fundamental representation can be written in a matrix form as

$$U(\Omega) = \left(\begin{array}{c|c} \sqrt{I_m - zz^\dagger} & z \\ \hline -z^\dagger & \sqrt{I_n - z^\dagger z} \end{array}\right), \ \Omega \in G/H, \tag{9.12}$$

where

$$z \equiv \frac{\sin\sqrt{\eta\eta^\dagger}}{\sqrt{\eta\eta^\dagger}}\eta = \eta\frac{\sin\sqrt{\eta^\dagger\eta}}{\sqrt{\eta^\dagger\eta}}. \tag{9.13}$$

Similarly, for a non-compact semisimple Lie group G and its Lie algebra $\mathfrak{g}$, the maximal compact subalgebra $\mathfrak{k}$ of $\mathfrak{g}$ and its orthogonal complementary subspace $\mathfrak{p}$ in the fundamental representation have the form

$$U(\mathbf{k}) \equiv \left(\begin{array}{c|c} \mathbf{a} & \mathbf{0} \\ \hline \mathbf{0} & \mathbf{b} \end{array}\right), \ U(\mathbf{p}) \equiv \left(\begin{array}{c|c} \mathbf{0} & \eta \\ \hline \eta^\dagger & \mathbf{0} \end{array}\right), \ \forall \mathbf{k} \in \mathfrak{k}, \mathbf{p} \in \mathfrak{p}, \tag{9.14}$$

so that the coset space G/H in the fundamental representation can be written in a matrix form as

$$U(\Omega) = \left(\begin{array}{c|c} \sqrt{\boldsymbol{I}_m + \boldsymbol{z}\boldsymbol{z}^\dagger} & \boldsymbol{z} \\ \hline \boldsymbol{z}^\dagger & \sqrt{\boldsymbol{I}_n + \boldsymbol{z}^\dagger \boldsymbol{z}} \end{array} \right), \Omega \in G/H, \tag{9.15}$$

where

$$\boldsymbol{z} \equiv \frac{\sinh \sqrt{\boldsymbol{\eta}\boldsymbol{\eta}^\dagger}}{\sqrt{\boldsymbol{\eta}\boldsymbol{\eta}^\dagger}} \boldsymbol{\eta} = \boldsymbol{\eta} \frac{\sinh \sqrt{\boldsymbol{\eta}^\dagger \boldsymbol{\eta}}}{\sqrt{\boldsymbol{\eta}^\dagger \boldsymbol{\eta}}}. \tag{9.16}$$

The difference in the geometries of compact and non-compact groups are manifested in the trigonometric and hyperbolic functions in Eqs. (9.13) and (9.16), respectively. Besides the complex coordinates z, one may also introduce the complex projective coordinates of G/H via

$$\boldsymbol{\tau}(\boldsymbol{z}) \equiv \frac{\boldsymbol{z}}{\sqrt{\boldsymbol{I}_n \mp \boldsymbol{z}^\dagger \boldsymbol{z}}}, \tag{9.17}$$

where the minus and plus signs are for compact and non-compact groups, respectively. Hence, the inverse mapping $\boldsymbol{z}(\boldsymbol{\tau})$ has the form

$$\boldsymbol{z}(\boldsymbol{\tau}) = \frac{\boldsymbol{\tau}}{\sqrt{\boldsymbol{I}_n \pm \boldsymbol{\tau}^\dagger \boldsymbol{\tau}}}, \tag{9.18}$$

where the plus and minus signs are for compact and non-compact groups, respectively. Using the complex projective coordinates, any group transformations g acting on the coset space G/H is a **holomorphic mapping** of G/H onto itself,

$$g\boldsymbol{\tau} = \frac{\boldsymbol{A}\boldsymbol{\tau} + \boldsymbol{B}}{\boldsymbol{C}\boldsymbol{\tau} + \boldsymbol{D}}, g \equiv \left(\begin{array}{c|c} \boldsymbol{A} & \boldsymbol{B} \\ \hline \boldsymbol{C} & \boldsymbol{D} \end{array} \right) \in G. \tag{9.19}$$

Besides the complex structure discussed above, there exists another geometric structure on the coset manifold G/H, namely, a **Riemannian structure**. On the coset manifold G/H, a Hermitian metric h can be expressed in terms of the local coordinates by a Hermitian symmetric tensor as

$$h \equiv \sum_{j,k} h_{j\bar{k}} d\tau_j \otimes d\tau_k^*, h_{j\bar{k}} \equiv \frac{\partial^2 \ln K(\boldsymbol{\tau}, \boldsymbol{\tau}^*)}{\partial \tau_j \partial \tau_k^*}, \tag{9.20}$$

where $h_{j\bar{k}}$ are the components of a positive-definite Hermitian matrix $(h_{j\bar{k}})$, i.e., $(h_{j\bar{k}})^\dagger = (h_{j\bar{k}})$ and $\boldsymbol{w}^\dagger (h_{j\bar{k}}) \boldsymbol{w} > 0$ for all $\boldsymbol{w} \in \mathbb{C}^n$, and the function $K(\boldsymbol{\tau}, \boldsymbol{\tau}^*)$ is called the **kernel function** of G/H. Here, j and $\bar{k}$ are summed over all the matrix elements of $\boldsymbol{\tau}$ and $\boldsymbol{\tau}^\dagger$ given by Eq. (9.17), where $\boldsymbol{\tau}$ is an $m \times n$ matrix. A

Hermitian metric h on the coset manifold G/H induces a Riemannian metric g on the underlying manifold as the real part of h, i.e., $g \equiv \frac{1}{2}(h + h^*)$, which can be expressed in terms of the local coordinates as

$$g = \frac{1}{2}\sum_{j,k} h_{j\bar{k}}(d\tau_j \otimes d\tau_k^* + d\tau_k^* \otimes d\tau_j), \tag{9.21}$$

where $h_{\bar{k}j} = h_{j\bar{k}}$. The Riemannian metric g is called a **Kähler metric** on the coset manifold G/H, which has the properties

$$\Gamma^{\bar{l}}_{jk} = \Gamma^{l}_{\bar{j}k} = \Gamma^{\bar{l}}_{j\bar{k}} = \Gamma^{l}_{\bar{j}\bar{k}} = 0, \tag{9.22}$$

where Γ^l_{jk} are the Christoffel symbols. The kernel function is obtained from the unnormalized form of the coherent states $||\mathbf{\Lambda}, \boldsymbol{\tau}\rangle$ on the coset manifold G/H by using the Baker-Campbell-Hausdorff (BCH) formula

$$\begin{aligned} |\mathbf{\Lambda}, \Omega\rangle &\equiv \exp\sum_{\alpha}(\eta_\alpha E_\alpha - \eta_\alpha^* E_{-\alpha})|\mathbf{\Lambda}, \mathbf{\Lambda}\rangle \\ &= \frac{1}{\sqrt{K(\boldsymbol{\tau}, \boldsymbol{\tau}^*)}} \exp\sum_{\alpha}(\tau_\alpha E_\alpha)|\mathbf{\Lambda}, \mathbf{\Lambda}\rangle \\ &= \frac{1}{\sqrt{K(\boldsymbol{\tau}, \boldsymbol{\tau}^*)}}||\mathbf{\Lambda}, \boldsymbol{\tau}\rangle, \end{aligned} \tag{9.23}$$

where the summation is restricted to those roots which obey $E_{-\alpha}|\mathbf{\Lambda}, \mathbf{\Lambda}\rangle \neq 0$. In the fundamental representation of a semisimple Lie group, the explicit expression of the kernel function is

$$K(\boldsymbol{\tau}, \boldsymbol{\tau}^*) = \det(\boldsymbol{I}_n \pm \boldsymbol{\tau}^\dagger \boldsymbol{\tau})^{\pm 1}, \tag{9.24}$$

where the plus and minus signs are for compact and non-compact groups, respectively. Finally, a group-invariant measure of the coset manifold G/H is given by

$$d\mu \equiv \text{const} \times \det(h_{j\bar{k}}) \prod_i \frac{dz_i dz_i^*}{\pi}. \tag{9.25}$$

As an example, one may consider the coherent states of a harmonic oscillator which possesses the Heisenberg-Weyl algebra $\mathfrak{h}_4$ with four generators $\{\hat{n}, \hat{a}^\dagger, \hat{a}, \hat{I}\}$. The coherent state space is then constructed by applying the exponential mapping of the subspace $\mathfrak{p}$ spanned by $\{\hat{a}^\dagger, \hat{a}\}$ of $\mathfrak{h}_4$

$$\Omega(\alpha) \equiv \exp(\alpha\hat{a}^\dagger - \alpha^*\hat{a}) \in H_4/U(1) \otimes U(1), \tag{9.26}$$

where $\alpha \in \mathbb{C}$ such that the coset space $H_4/U(1) \otimes U(1)$ is isomorphic to the ordinary complex plane, and $U(1) \otimes U(1)$ is the maximal isotropy subgroup of H_4 with respect to the vacuum state $|0\rangle$, which possesses the infinitesimal generators $\hat{a}^\dagger \hat{a}$ and I. A group unitary transformation $g = \exp\{\beta\hat{a}^\dagger - \beta^*\hat{a} + i\eta\hat{n} + i\delta\hat{I})\}$ with $\beta \in \mathbb{C}$ and $\eta, \delta \in \mathbb{R}$ acting on the coset space $H_4/U(1) \otimes U(1)$ is determined by

$$g\Omega(\alpha) = \Omega(\alpha')h, h \in U(1) \otimes U(1), \tag{9.27a}$$

$$\alpha' = \alpha e^{i\eta} + \frac{\beta}{i\eta}(e^{i\eta} - 1). \tag{9.27b}$$

When we restrict $g \in H_4/U(1) \otimes U(1)$, Eq. (9.27b) describes a translation on the complex plane via $\alpha' = \alpha + \beta$. The geometric structure of the coset space $H_4/U(1) \otimes U(1)$ can also be understood via the coherent state of the Heisenberg-Weyl group H_4, which is defined by $\Omega(\alpha)$ acting on the vacuum state $|0\rangle$

$$|\alpha\rangle \equiv \Omega(\alpha)|0\rangle = \exp(\alpha a^\dagger - \alpha^* a)|0\rangle = e^{-\frac{1}{2}\alpha\alpha^*} e^{\alpha\hat{a}^\dagger}|0\rangle \equiv e^{-\frac{1}{2}\alpha\alpha^*}||\alpha\rangle, \tag{9.28}$$

where the normalization constant of the unnormalized form of the coherent states gives the kernel function, $K(\alpha, \alpha^*) \equiv \langle\alpha||\alpha\rangle = e^{\alpha\alpha^*}$. Hence, a direct computation yields the standard Riemannian metric on the complex plane

$$g = \frac{1}{2}(d\alpha \otimes d\alpha^* + d\alpha^* \otimes d\alpha) = dx \otimes dx + dy \otimes dy, \tag{9.29}$$

where $\alpha \equiv x + iy$.

Using the Hermitian metric h, one may define another important geometric structure on the coset manifold G/H, namely, a **symplectic structure** $(G/H, \omega)$ with a differential 2-form ω which is closed $d\omega = 0$ and of maximal rank $\omega^n \neq 0$. In terms of the Hermitian metric h, one may explicitly define the closed non-degenerate 2-form as minus the imaginary part of h, i.e., $\omega \equiv \frac{i}{2}(h - \bar{h})$, which can be written in terms of the local coordinates as

$$\omega = i\sum_{j,k} h_{j\bar{k}} d\tau_j \wedge d\tau_k^*, \tag{9.30}$$

where the wedge product $d\tau_j \wedge d\tau_k^*$ of the vectors $d\tau_j$ and $d\tau_k^*$ is an element of the space of antisymmetric tensors defined via the embedding $d\tau_j \wedge d\tau_k^* \mapsto \frac{1}{2}(d\tau_j \otimes d\tau_k^* - d\tau_k^* \otimes d\tau_j)$. As the coset manifold has a symplectic structure $(G/H, \omega)$, one may equip it with a natural Poisson bracket $\{\cdot, \cdot\}_\omega$ via

$$\{f, g\}_\omega \equiv \frac{1}{i}\sum_{j,k} h^{j\bar{k}}\left(\frac{\partial f}{\partial\tau_j}\frac{\partial g}{\partial\tau_k^*} - \frac{\partial g}{\partial\tau_j}\frac{\partial f}{\partial\tau_k^*}\right), \tag{9.31}$$

where f and g are two functions defined on the coset manifold and $h^{j\bar{k}}$ are the components of the inverse of the matrix $(h_{j\bar{k}})$. One may verify that by performing the coordinate transformation $z(\boldsymbol{\tau})$ in Eq. (9.18), one obtains a diagonal non-degenerate closed 2-form

$$\omega = i \sum_{j=1}^{nm} dz_j \wedge dz_j^*, \tag{9.32}$$

which induces a diagonal Poisson bracket $\{\cdot, \cdot\}_\omega$ in the form

$$\{f, g\}_\omega = \frac{1}{i} \sum_{j=1}^{nm} \left(\frac{\partial f}{\partial z_j} \frac{\partial g}{\partial z_j^*} - \frac{\partial g}{\partial z_j} \frac{\partial f}{\partial z_j^*} \right). \tag{9.33}$$

If one further performs a coordinate transformation $z_j \equiv (q_j + ip_j)/\sqrt{2}$ and $z_j^* \equiv (q_j - ip_j)/\sqrt{2}$, one obtains the symplectic 2-form $\omega = \sum_{j=1}^{nm} dq_j \wedge dp_k$, and the standard Poisson bracket used in classical mechanics

$$\{f, g\}_\omega = \sum_{j=1}^{nm} \left(\frac{\partial f}{\partial q_j} \frac{\partial g}{\partial p_j} - \frac{\partial g}{\partial q_j} \frac{\partial f}{\partial p_j} \right). \tag{9.34}$$

Besides the geometric properties mentioned above, there are some other important Hilbert-space properties that the generalized coherent states have in common, namely, the **non-orthogonality** and the **over-completeness**. In the most general setting, one may consider a general Lie group G and its maximal isotropy subgroup $H \subset G$. Then in an arbitrary unitary irreducible representation $V^{\boldsymbol{\Lambda}}$ of G, the generalized coherent state $|\boldsymbol{\Lambda}, \Omega\rangle \equiv \Omega|\Phi_0\rangle$ is in one-to-one correspondence with the elements Ω in the coset space G/H, where $|\Phi_0\rangle$ is an arbitrary reference state. As such, the generalized coherent states $|\boldsymbol{\Lambda}, \Omega\rangle$ are generally non-orthogonal, expect for a set of measure zero. The reason is that when applying arbitrary elements $g, g' \in G$ on the reference state $|\Phi_0\rangle$, one obtains

$$\begin{aligned} g|\Phi_0\rangle &= \Omega h|\Phi_0\rangle = e^{i\varphi(h)}|\boldsymbol{\Lambda}, \Omega\rangle, \\ g'|\Phi_0\rangle &= \Omega' h'|\Phi_0\rangle = e^{i\varphi(h')}|\boldsymbol{\Lambda}, \Omega'\rangle, \end{aligned} \tag{9.35}$$

where $h, h' \in H, \Omega, \Omega' \in G/H$. Hence, the overlap between the two generalized coherent states $|\boldsymbol{\Lambda}, \Omega\rangle$ and $|\boldsymbol{\Lambda}, \Omega'\rangle$ becomes

$$\begin{aligned} \langle \boldsymbol{\Lambda}, \Omega | \boldsymbol{\Lambda}, \Omega' \rangle &= e^{i(\varphi(h') - \varphi(h))} \langle \Phi_0 | g^{-1} g' | \Phi_0 \rangle \\ &= e^{i(\varphi(h') - \varphi(h))} \langle \Phi_0 | g'' | \Phi_0 \rangle \neq 0, \end{aligned} \tag{9.36}$$

where we have used the fact that $g^{-1} \in G$ and $g^{-1}g' \equiv g'' \in G$. From the above discussion, one immediately sees that the generalized coherent states are normalized to unity: $\langle \boldsymbol{\Lambda}, \Omega | \boldsymbol{\Lambda}, \Omega \rangle = \langle \Phi_0 | g^{-1} g | \Phi_0 \rangle = \langle \Phi_0 | \Phi_0 \rangle = 1$. Moreover, applying the unitary representation $\hat{U}(g_1)$ of an arbitrary element $g_1 \in G$ on the coherent state $|\mathbf{\Lambda}, \Omega\rangle$, one simply obtains another coherent state, up to a phase factor

$$\begin{aligned}\hat{U}(g_1)|\mathbf{\Lambda}, \Omega\rangle &= e^{-i\varphi(h)}\hat{U}(g_1)\hat{U}(g)|\Phi_0\rangle \\ &= e^{-i\varphi(h)}\hat{U}(g_1 g)|\Phi_0\rangle \\ &= e^{i(\varphi(h')-\varphi(h))}|\mathbf{\Lambda}, \Omega'\rangle,\end{aligned} \tag{9.37}$$

where $g' \equiv g_1 g = \Omega' h'$. Turning to the problem of completeness, one may notice that there exists an invariant measure $d\mu_G$ on the group G, which induces an invariant measure $d\mu_\Omega$ on the coset space G/H. One may now consider the operator

$$\hat{O} \equiv \int d\mu_\Omega |\mathbf{\Lambda}, \Omega\rangle\langle\mathbf{\Lambda}, \Omega|. \tag{9.38}$$

One can readily show that the operator $\hat{O}$ is invariant under the transformation of G

$$\begin{aligned}\hat{U}(g_1)\hat{O}\hat{U}(g_1)^{-1} &= \int d\mu_\Omega \hat{U}(g_1)|\mathbf{\Lambda}, \Omega\rangle\langle\mathbf{\Lambda}, \Omega|\hat{U}(g_1)^{-1} \\ &= \int d\mu_\Omega e^{i(\varphi(h')-\varphi(h))}|\mathbf{\Lambda}, \Omega'\rangle\langle\mathbf{\Lambda}, \Omega'|e^{-i(\varphi(h')-\varphi(h))} \\ &= \int d\mu_{\Omega'}|\mathbf{\Lambda}, \Omega'\rangle\langle\mathbf{\Lambda}, \Omega'|,\end{aligned} \tag{9.39}$$

where we used the fact that $d\mu_\Omega = d\mu_{\Omega'}$ is an invariant measure of the coset space G/H. As the operator $\hat{O}$ commutes with all the operators $\hat{U}(g_1)$ and thus due to the irreducibility of the representation $\hat{U}$ of G, the operator $\hat{O}$ must be a multiple of the identity operator, according to **Schur's lemma** in representation theory. With an appropriately normalized measure $d\mu_\Omega$ on the square-integrable function space $L^2(G/H)$, one obtains

$$\int d\mu_\Omega |\mathbf{\Lambda}, \Omega\rangle\langle\mathbf{\Lambda}, \Omega| = \hat{I}. \tag{9.40}$$

Combining with the fact that the coherent states are non-orthogonal in general, $\langle\mathbf{\Lambda}, \Omega|\mathbf{\Lambda}, \Omega'\rangle \neq 0$, one concludes that the system of the coherent states $|\mathbf{\Lambda}, \Omega\rangle$ is over-complete, i.e., it contains subsystems of coherent states which are complete.

Making use of the over-completeness properties, one may expand an arbitrary state $|\Psi\rangle \in V^{\mathbf{\Lambda}}$ in terms of the coherent states

$$|\Psi\rangle = \int d\mu_\Omega f_{\mathbf{\Lambda}}(\Omega)|\mathbf{\Lambda}, \Omega\rangle,\ f_{\mathbf{\Lambda}}(\Omega) \equiv \langle\mathbf{\Lambda}, \Omega|\Psi\rangle, \tag{9.41}$$

which immediately yields

$$\langle \Psi | \Psi \rangle = \int d\mu_{\Omega'} |f_{\mathbf{\Lambda}}(\Omega)|^2. \tag{9.42}$$

The function $f_{\mathbf{\Lambda}}(\Omega)$ defined on the coset space G/H is not arbitrary but needs to be continuous in Ω which satisfy the relation

$$f_{\mathbf{\Lambda}}(\Omega) = \int d\mu_{\Omega'} \langle \mathbf{\Lambda}, \Omega | \mathbf{\Lambda}, \Omega' \rangle f_{\mathbf{\Lambda}}(\Omega') \equiv \int d\mu_{\Omega'} K_{\mathbf{\Lambda}}(\Omega, \Omega') f_{\mathbf{\Lambda}}(\Omega'). \tag{9.43}$$

Hence, the kernel $K_{\mathbf{\Lambda}}(\Omega, \Omega') \equiv \langle \mathbf{\Lambda}, \Omega | \mathbf{\Lambda}, \Omega' \rangle$ is a reproducing one

$$K_{\mathbf{\Lambda}}(\Omega, \Omega'') = \int d\mu_{\Omega'} K_{\mathbf{\Lambda}}(\Omega, \Omega') K_{\mathbf{\Lambda}}(\Omega', \Omega''). \tag{9.44}$$

From Eq. (9.41), one can readily see that there are linear dependences between the coherent states

$$|\mathbf{\Lambda}, \Omega\rangle = \int d\mu_{\Omega'} K_{\mathbf{\Lambda}}(\Omega', \Omega) |\mathbf{\Lambda}, \Omega'\rangle \tag{9.45}$$

Finally, the scalar product between two arbitrary states $|\Psi\rangle$ and $|\Psi'\rangle$ can also be expressed as a coherent state integral over $L^2(G/H)$

$$\begin{aligned} \langle \Psi | \Psi' \rangle &= \iint d\mu_{\Omega} d\mu_{\Omega'} f^*_{\mathbf{\Lambda}}(\Omega) f_{\mathbf{\Lambda}}(\Omega') K(\Omega, \Omega') \\ &= \int d\mu_{\Omega} f^*_{\mathbf{\Lambda}}(\Omega) f_{\mathbf{\Lambda}}(\Omega), \end{aligned} \tag{9.46}$$

where we have used Eq. (9.43) in the last step. Conversely, the coherent states can be expressed in terms of arbitrary diagonal states. Let us consider a complete set of orthogonal states of $V^{\mathbf{\Lambda}}$, denoted as $|\mathbf{\Lambda}, \boldsymbol{\lambda}\rangle$, where

$$\sum_{\boldsymbol{\lambda}} |\mathbf{\Lambda}, \boldsymbol{\lambda}\rangle \langle \mathbf{\Lambda}, \boldsymbol{\lambda}| = I, \tag{9.47}$$

and $\boldsymbol{\lambda}$ is the usual **Gelfand-Tsetlin pattern** which indexes all the basic states in $V^{\boldsymbol{\lambda}}$, including weight and multiplicity. Thus the coherent states $|\mathbf{\Lambda}, \Omega\rangle$ can be expressed in terms of this basis of $V^{\mathbf{\Lambda}}$ via

$$|\mathbf{\Lambda}, \Omega\rangle = N^{-1/2}(\Omega) \sum_{\boldsymbol{\lambda}} f_{\mathbf{\Lambda},\boldsymbol{\lambda}}(\Omega) |\mathbf{\Lambda}, \boldsymbol{\lambda}\rangle, \tag{9.48}$$

where $f_{\mathbf{\Lambda},\boldsymbol{\lambda}}(\Omega) \equiv N^{1/2}(\Omega)\langle\mathbf{\Lambda}, \boldsymbol{\lambda}|\mathbf{\Lambda}, \Omega\rangle = \langle\mathbf{\Lambda}, \boldsymbol{\lambda}||\mathbf{\Lambda}, \boldsymbol{\tau}\rangle$ is an entire function on the coset space G/H.

Similar to the Hilbert-space expansions of arbitrary states in terms of coherent states, one may expand an arbitrary operator A that acts on $V^{\mathbf{\Lambda}}$ in terms of coherent states

$$A = \iint d\mu_{\Omega} d\mu_{\Omega'} |\mathbf{\Lambda}, \Omega\rangle\langle\mathbf{\Lambda}, \Omega|A|\mathbf{\Lambda}, \Omega'\rangle\langle\mathbf{\Lambda}, \Omega'|. \tag{9.49}$$

Such an expansion induces three special representations, namely, the P, Q, and W representations, which are analogous to their counterparts for bosonic coherent states. Thus the coherent states provide a natural phase-space structure and three useful phase-space distributions for a quantum system, which are discussed below. The P- and Q-representations for an operator A that maps $V^{\mathbf{\Lambda}}$ onto itself are defined over the coset space G/H as

$$A \equiv \int d\mu_{\Omega} A_P(\mathbf{\Lambda}, \Omega)|\mathbf{\Lambda}, \Omega\rangle\langle\mathbf{\Lambda}, \Omega|, \tag{9.50a}$$

$$A_Q(\mathbf{\Lambda}, \Omega) \equiv \langle\mathbf{\Lambda}, \Omega|A|\mathbf{\Lambda}, \Omega\rangle = \int d\mu_{\Omega'} |K_{\mathbf{\Lambda}}(\Omega, \Omega')|^2 A_P(\mathbf{\Lambda}, \Omega'). \tag{9.50b}$$

Similar to its counterpart for bosonic coherent states, the density operator ρ for a pure coherent state $|\mathbf{\Lambda}, \Omega\rangle$ is the projection operator $\rho \equiv |\mathbf{\Lambda}, \Omega\rangle\langle\mathbf{\Lambda}, \Omega|$. But for the statistical mixtures of the pure coherent states, the density operator is a superposition of the projection operator $|\mathbf{\Lambda}, \Omega\rangle\langle\mathbf{\Lambda}, \Omega|$

$$\rho = \int d\mu_{\Omega} P(\mathbf{\Lambda}, \Omega)|\mathbf{\Lambda}, \Omega\rangle\langle\mathbf{\Lambda}, \Omega|, \tag{9.51}$$

where the function $P(\mathbf{\Lambda}, \Omega)$, known as the P-representation of the density operator, can be regarded as a probability distribution function of Ω over the coset manifold G/H. Similarly, the Q-representation of the density operator has the form

$$Q(\mathbf{\Lambda}, \Omega) \equiv \langle\mathbf{\Lambda}, \Omega|\rho|\mathbf{\Lambda}, \Omega\rangle = \int d\mu_{\Omega} P(\mathbf{\Lambda}, \Omega). \tag{9.52}$$

Correspondingly, the statistical average of an operator A is given by

$$\mathrm{Tr}(\rho A) = \int d\mu_{\Omega} Q(\mathbf{\Lambda}, \Omega) A_P(\mathbf{\Lambda}, \Omega) = \int d\mu_{\Omega} P(\mathbf{\Lambda}, \Omega) A_Q(\mathbf{\Lambda}, \Omega). \tag{9.53}$$

Based on the P- and Q-distributions, the W-distribution for an operator A is defined through an integral kernel $\chi_{\mathbf{\Lambda}}(\Omega, \Omega')$ via the relations

$$A_W(\mathbf{\Lambda}, \Omega) \equiv \int d\mu_{\Omega'} \chi_{\mathbf{\Lambda}}(\Omega, \Omega') A_P(\mathbf{\Lambda}, \Omega'), \tag{9.54a}$$

$$A_Q(\mathbf{\Lambda}, \Omega) \equiv \int d\mu_{\Omega'} \chi_{\mathbf{\Lambda}}(\Omega, \Omega') A_W(\mathbf{\Lambda}, \Omega'), \tag{9.54b}$$

where the integral kernel $\chi_{\mathbf{\Lambda}}(\Omega, \Omega')$ obeys $\chi^*_{\mathbf{\Lambda}}(\Omega, \Omega') = \chi_{\mathbf{\Lambda}}(\Omega', \Omega)$. Substituting Eq. (9.57a) into Eq. (9.57b) yields

$$A_Q(\mathbf{\Lambda}, \Omega) = \iint d\mu_{\Omega'} d\mu_{\Omega''} \chi_{\mathbf{\Lambda}}(\Omega, \Omega') \chi_{\mathbf{\Lambda}}(\Omega', \Omega'') A_P(\mathbf{\Lambda}, \Omega''). \tag{9.55}$$

Comparing Eqs. (9.55) and (9.50b) immediately yields

$$\int d\mu_{\Omega'} \chi_{\mathbf{\Lambda}}(\Omega, \Omega') \chi_{\mathbf{\Lambda}}(\Omega', \Omega'') = |K_{\mathbf{\Lambda}}(\Omega, \Omega'')|^2. \tag{9.56}$$

Hence, the P-, Q-, and W-distributions for the density operator ρ satisfy the relations

$$W(\mathbf{\Lambda}, \Omega) \equiv \int d\mu_{\Omega'} \chi_{\mathbf{\Lambda}}(\Omega, \Omega') P(\mathbf{\Lambda}, \Omega'), \tag{9.57a}$$

$$Q(\mathbf{\Lambda}, \Omega) \equiv \int d\mu_{\Omega'} \chi_{\mathbf{\Lambda}}(\Omega, \Omega') W(\mathbf{\Lambda}, \Omega'), \tag{9.57b}$$

$$Q(\mathbf{\Lambda}, \Omega) \equiv \int d\mu_{\Omega'} |K_{\mathbf{\Lambda}}(\Omega, \Omega')|^2 P(\mathbf{\Lambda}, \Omega'). \tag{9.57c}$$

As a first example, for the bosonic coherent states $|\alpha\rangle$ defined on the complex-α plane, the integral kernel $\chi(\alpha, \beta)$ for the W-distribution and the reproducing kernel $K(\alpha, \beta) \equiv \langle\alpha|\beta\rangle$ have the form $\chi(\alpha, \beta) = 2e^{-2|\alpha-\beta|^2}$ and $K(\alpha, \beta) \equiv e^{-|\alpha-\beta|^2}$, so that

$$W(\alpha) \equiv 2 \int \frac{d^2\beta}{\pi} e^{-2|\alpha-\beta|^2} P(\beta), \tag{9.58a}$$

$$Q(\alpha) \equiv 2 \int \frac{d^2\beta}{\pi} e^{-2|\alpha-\beta|^2} W(\beta), \tag{9.58b}$$

$$Q(\alpha) \equiv \int \frac{d^2\beta}{\pi} e^{-|\alpha-\beta|^2} P(\beta). \tag{9.58c}$$

Finally, one may show that the inner product between two arbitrary operators A and B on $V^{\mathbf{\Lambda}}$ can also be expressed in terms of the W-distribution

$$\begin{aligned}(A, B) \equiv \mathrm{Tr}(A^{\dagger}B) &= \int d\mu_{\Omega} A_Q^*(\mathbf{\Lambda}, \Omega) B_P(\mathbf{\Lambda}, \Omega) \\ &= \iint d\mu_{\Omega} d\mu_{\Omega'} \chi_{\mathbf{\Lambda}}^*(\Omega, \Omega') A_W^*(\mathbf{\Lambda}, \Omega') B_P(\mathbf{\Lambda}, \Omega) \\ &= \int d\mu_{\Omega'} A_W^*(\mathbf{\Lambda}, \Omega') B_W(\mathbf{\Lambda}, \Omega').\end{aligned} \tag{9.59}$$

Besides the geometric properties of the coset space and the Hilbert-space properties of the coherent states, there are some additional group-theoretic formulas that are useful in applications and thus worth discussing. For example, one may consider the Baker-Campbell-Hausdorff (BCH) formulas, which are the analytic isomorphism connecting exponentials of the Lie algebraic elements and the products of exponentials. A first example is the familiar BCH formula for the bosonic coherent states

$$e^{\alpha a^{\dagger} - \alpha^* a} = e^{-\frac{1}{2}\alpha\alpha^*} e^{\alpha a^{\dagger}} e^{-\alpha^* a} = e^{\frac{1}{2}\alpha\alpha^*} e^{-\alpha^* a} e^{\alpha a^{\dagger}}. \tag{9.60}$$

In general, employing the Cartan decomposition of the semisimple Lie algebra $\mathfrak{g}$, i.e., $\mathfrak{g} = \mathfrak{k} \oplus \mathfrak{p}$, the exponential of a general element $k \in \mathfrak{p}$ onto the coset space G/H can be disentangled as the following product

$$\begin{aligned}\Omega = \exp \sum_{\alpha} (\eta_{\alpha} E_{\alpha} - \eta_{\alpha}^* E_{-\alpha}) &\\ = (\exp \sum_{\alpha} \tau_{\alpha} E_{\alpha})(\exp \sum_{i} \gamma_i H_i)(\exp \sum_{\beta} -\tau_{\beta}^* E_{-\beta}) &\\ = (\exp \sum_{\beta} -\tau_{\beta}^* E_{-\beta})(\exp \sum_{i} -\gamma_i H_i)(\exp \sum_{\alpha} \tau_{\alpha} E_{\alpha}), &\end{aligned} \tag{9.61}$$

where the relation between τ_{α} and η_{α} can be derived from the matrix representation of G. For example, in the fundamental representation of G, the coset space G/H is a symmetric space, whose representation matrix can be disentangled in the form

$$\begin{aligned}U(\Omega) &= \left(\begin{array}{c|c} \sqrt{\boldsymbol{I}_m \pm z z^{\dagger}} & z \\ \hline \pm z^{\dagger} & \sqrt{\boldsymbol{I}_n \pm z^{\dagger} z} \end{array}\right) \\ &= \left(\begin{array}{c|c} \boldsymbol{I}_m & \boldsymbol{\tau} \\ \hline 0 & \boldsymbol{I}_n \end{array}\right) \left(\begin{array}{c|c} \exp \boldsymbol{\gamma}_m & 0 \\ \hline 0 & \exp \boldsymbol{\gamma}_n \end{array}\right) \left(\begin{array}{c|c} \boldsymbol{I}_m & 0 \\ \hline \pm \boldsymbol{\tau}^{\dagger} & \boldsymbol{I}_n \end{array}\right),\end{aligned} \tag{9.62}$$

where the plus and minus signs are for compact and non-compact Lie groups, respectively. A direct computation yields

$$\boldsymbol{\tau} = \frac{\boldsymbol{z}}{\sqrt{\boldsymbol{I}_n \pm \boldsymbol{z}^\dagger \boldsymbol{z}}}, \exp \boldsymbol{\gamma}_m = \frac{\boldsymbol{I}_m}{\sqrt{\boldsymbol{I}_m \pm \boldsymbol{z}\boldsymbol{z}^\dagger}}, \exp \boldsymbol{\gamma}_n = \sqrt{\boldsymbol{I}_n \pm \boldsymbol{z}^\dagger \boldsymbol{z}}. \tag{9.63}$$

Exercises

9.1. For an arbitrary $m \times n$ matrix $\boldsymbol{\eta}$, verify the relations

$$\frac{\sin\sqrt{\boldsymbol{\eta}\boldsymbol{\eta}^\dagger}}{\sqrt{\boldsymbol{\eta}\boldsymbol{\eta}^\dagger}}\boldsymbol{\eta} = \boldsymbol{\eta}\frac{\sin\sqrt{\boldsymbol{\eta}^\dagger\boldsymbol{\eta}}}{\sqrt{\boldsymbol{\eta}^\dagger\boldsymbol{\eta}}} \quad \text{and} \quad \frac{\sinh\sqrt{\boldsymbol{\eta}\boldsymbol{\eta}^\dagger}}{\sqrt{\boldsymbol{\eta}\boldsymbol{\eta}^\dagger}}\boldsymbol{\eta} = \boldsymbol{\eta}\frac{\sinh\sqrt{\boldsymbol{\eta}^\dagger\boldsymbol{\eta}}}{\sqrt{\boldsymbol{\eta}^\dagger\boldsymbol{\eta}}}.$$

9.2. Verify Eqs. (9.27a) and (9.27b) for the Heisenberg-Weyl algebra $\mathfrak{h}_4$.

10 Quantum Many-Body Systems

10.1 Mean-Field Approach with Coherent States

In nature, there are fundamentally two types of particles, namely, bosons and fermions, for which the basic operators obey different commutation relations and the corresponding states have distinct statistical properties. Bosons, which carry intrinsic property of integer spins, such as photons or pairs of electrons, obey the standard canonical commutation relations and Boson-Einstein statistics, whereas fermions, which carry intrinsic property of half-integer spins, such as electrons, obey anti-commutation relations and Fermi-Dirac statistics. The differences of bosons and fermions are manifested in their different group structures as well as their associated Hilbert space structures. Indeed, a large part of condensed matter theory, such as band theory, Fermi liquid theory, superfluidity, superconductivity, and quantum Hall effect, to name a few, are consequences of this fact [127]. In practice, quantum systems usually couple these two types of particles. As the algebras of bosons and fermions commute, their associated coherent states can be separately constructed. In the following, we focus on the coherent states for fermions.

The archetypical fermion systems are atoms, molecules, and nuclei. However, unlike the hydrogen atom and deuteron with specific interactions which can be solved exactly, many-body systems usually consist of an immense number of interacting particles, i.e., of the order of 10^2 in nuclei [17], or of the order of one valence electron per atom in crystal lattices like solids. The estimation of number of particles in solids comes from the adiabatic approximation [128], where the nuclei are considered fixed at their empirically known equilibrium positions, and also one can divide the atomic electrons into those belonging to closed shells whose excitation potentials are high, and the relatively loosely bound valence electrons, such that the effect of closed shells upon the valence electrons can be regarded as only a static potential. It is exactly such an immense number of particles that makes exact solutions of many-body systems practically impossible. In this regard,

C.-F. Kam et al., *Coherent States*, Lecture Notes in Physics 1011,
https://doi.org/10.1007/978-3-031-20766-2_10

the development of many-body theory has been, since the genesis of quantum mechanics, primarily a search for better approximation schemes. One such scheme is the variational principle, whose input is a proper choice of good trial wave function. The main criteria of a good trial wave function are that it can maximize quantum correlations between particles and that it is simple enough to use. The coherent states, directly constructed from the dynamical group structure of a system, are no doubt one of the best candidates as trial wave functions for the many-body systems.

To start with, let us discuss the algebra of a single-fermion system, of which the Hamiltonian can be written as a function of the fermionic creation and annihilation operators $c^\dagger$ and c of a single fermion that obey the standard anti-commutation relations

$$\{c, c^\dagger\} = 1, \{c, c\} = \{c^\dagger, c^\dagger\} = 0. \tag{10.1}$$

Interestingly, such a system can be described by the familiar SU(2) dynamical group [21], whose generators are c, $c^\dagger$, and $c^\dagger c - \frac{1}{2}$. The generators obey the following commutation relations:

$$[c^\dagger, c] = 2(c^\dagger c - \frac{1}{2}), [c^\dagger c - \frac{1}{2}, c] = -c, [c^\dagger c - \frac{1}{2}, c^\dagger] = c^\dagger. \tag{10.2}$$

Evidently, the operators $c^\dagger$, c, and $c^\dagger c - \frac{1}{2}$ are in one-to-one correspondence with the quantum angular momentum operators, i.e., $c^\dagger \Leftrightarrow J_+$, $c \Leftrightarrow J_-$, and $c^\dagger c - \frac{1}{2} \Leftrightarrow J_0$. Hence, the coherent states of a single-fermion system can be constructed by following the general algorithm for the SU(2) dynamical group.

As the Hilbert space of a single fermion contains only two states, it can be realized by the simplest irreducible representation of SU(2), i.e., spin-1/2 representation, whose basis vectors are $|\frac{1}{2}, \frac{1}{2}\rangle$ and $|\frac{1}{2}, -\frac{1}{2}\rangle$. The fermion coherent states are then constructed as follows:

$$\begin{aligned} |\frac{1}{2}, \xi\rangle &\equiv \exp(\xi c^\dagger - \xi^* c)|\frac{1}{2}, -\frac{1}{2}\rangle \\ &= \sin(\frac{\theta}{2})e^{-i\varphi}|\frac{1}{2}, \frac{1}{2}\rangle + \cos(\frac{\theta}{2})|\frac{1}{2}, -\frac{1}{2}\rangle, \end{aligned} \tag{10.3}$$

where $\xi \equiv \frac{\theta}{2}e^{-i\varphi}$ and $|\frac{1}{2}, \frac{1}{2}\rangle \equiv c^\dagger|\frac{1}{2}, -\frac{1}{2}\rangle$. Equation (10.3) shows that the group definition of the fermion coherent states has a natural topological space, i.e., SU(2)/U(1)$\simeq S^2$. Hence, all the results for spin-1/2 coherent states are applicable to the fermion coherent states. For example, the completeness of the fermion coherent states is given by

$$\frac{1}{2\pi}\int d\Omega|\frac{1}{2}, \xi\rangle\langle\frac{1}{2}, \xi| = I_2, \tag{10.4}$$

where $d\Omega \equiv \sin\theta d\theta d\varphi$. Similarly, the BCH formula for fermionic algebra can be expressed as

$$\exp(\xi c^\dagger - \xi^* c) = \exp(\tau c^\dagger)\exp[\ln(1+\tau^*\tau)(c^\dagger c - \frac{1}{2})]\exp(-\tau^* c) \tag{10.5}$$
$$= \exp(-\tau^* c)\exp[-\ln(1+\tau^*\tau)(c^\dagger c - \frac{1}{2})]\exp(\tau c^\dagger),$$

where $\tau \equiv \tan\frac{\theta}{2}e^{-i\varphi}$.

We now discuss a system with r fermions, where the creation and annihilation operators for the constituting fermions, e.g., the electrons in atomic systems or the nucleons in a nucleus, obey the anti-commutation relations

$$\{c_i, c_j^\dagger\} = \delta_{ij}, \{c_i, c_j\} = \{c_i^\dagger, c_j^\dagger\} = 0 \tag{10.6}$$

for $1 \le i, j \le r$. One can easily construct several distinct algebras from these operators, which implies that those coherent states constructed from the associated algebra have different algebraic and geometric properties.

The simplest algebra which can be constructed from the r^2 fermionic pairing operators $c_i^\dagger c_j$ with $1 \le i, j \le r$ is the Lie algebra $\mathfrak{u}(r)$, whose commutation relations are given by

$$[c_i^\dagger c_j, c_k^\dagger c_l] = \delta_{jk}c_i^\dagger c_l - \delta_{il}c_k^\dagger c_j. \tag{10.7}$$

Let $H_i \equiv c_i^\dagger c_i$; then the set $\{H_i | i = 1, 2, \cdots, r\}$ constitutes the maximal Abelian Cartan subalgebra of $\mathfrak{u}(r)$, whose commutation relations are simply given by

$$[H_i, c_j^\dagger c_k] = (\delta_{ij} - \delta_{ik})c_j^\dagger c_k. \tag{10.8}$$

Hence, the roots $\mathbf{e}_i - \mathbf{e}_j$ with $1 \le i, j \le r$ span the root space A_{r-1} of the multi-fermion $\mathfrak{u}(r)$ algebra. Although not directly related to the fermion irreducible representation of $\mathfrak{u}(r)$, the faithful matrix representation of $\mathfrak{u}(r)$ is useful. In the faithful matrix representation of $\mathfrak{u}(r)$, each generator corresponds to an $r \times r$ matrix, i.e., $c_i^\dagger c_j \Leftrightarrow E_{ij}$, where E_{ij} is an $r \times r$ matrix with 1 in the ith row and jth column and zero otherwise.

In the fully antisymmetric representation of $\mathfrak{u}(r)$ labeled by the highest weight $\mathbf{\Lambda} \equiv (\Lambda_1, \Lambda_2, \cdots, \Lambda_r) = (1, \cdots, 1, 0, \cdots, 0) = (\mathbf{1}_k, \mathbf{0}_{r-k})$, the basic states in the Hilbert space are the set $\{|n_1, n_2, \cdots, n_r\rangle\}$, with $n_i = 0, 1$ and $\sum n_i = k$. Clearly, there are $C_r^k = r!/k!(r-k)!$ basic states in the Hilbert space. Hence, one immediately obtains

$$c_i^\dagger c_j |\mathbf{\Lambda}, \mathbf{\Lambda}\rangle \equiv c_i^\dagger c_j |1, \cdots, 1, 0, \cdots, 0\rangle = 0, \tag{10.9}$$

where $1 \leq i \neq j \leq k$ or $k+1 \leq i \neq j \leq r$. One may verify that the operators $c_i^\dagger c_j$ with $1 \leq i \leq k$ and $k+1 \leq j \leq r$ span a subalgebra $\mathfrak{u}(k) \oplus \mathfrak{u}(r-k)$ of $\mathfrak{u}(r)$, where the corresponding Lie group $\mathrm{U}(k) \otimes \mathrm{U}(r-k)$ is the isotropy subgroup of $\mathrm{U}(r)$ which keeps the extremal state $|\boldsymbol{\Lambda}\rangle$ unchanged. Hence, the coherent states for the multi-fermion $\mathrm{U}(r)$ group can be defined as

$$|\boldsymbol{\Lambda}, \Omega\rangle \equiv \exp \sum_{i,j} (\eta_{ij} c_i^\dagger c_j - \eta_{ij}^* c_j^\dagger c_i) |\boldsymbol{\Lambda}, \boldsymbol{\Lambda}\rangle, \tag{10.10}$$

where $k+1 \leq i \leq r$, $1 \leq j \leq k$, and η_{ij} are some complex parameters. Under such a definition, one may readily show that the coherent states for the multi-fermion $\mathrm{U}(r)$ group have a natural topological coset space $\mathrm{U}(r)/\mathrm{U}(k) \otimes \mathrm{U}(r-k)$. In the following, we discuss in detail the completeness relation, the symplectic structure, and the BCH formula for the multi-fermion $\mathrm{U}(r)$ coherent states.

First of all, according to the general theorem of completeness, the coherent states $|\boldsymbol{\Lambda}, \Omega\rangle$ satisfy the completeness relation

$$\int d\mu_\Omega |\boldsymbol{\Lambda}, \Omega\rangle\langle\boldsymbol{\Lambda}, \Omega| = \boldsymbol{I}. \tag{10.11}$$

In our case, the measure $d\mu_\Omega(\boldsymbol{\tau}, \boldsymbol{\tau}^*)$ is given by

$$d\mu_\Omega(\tau_\alpha, \tau_\alpha^*) \equiv \frac{\dim V^{\boldsymbol{\Lambda}}}{\mathrm{Vol}[\mathrm{U}(r)/\mathrm{U}(r-k)]} \det(I_k + \boldsymbol{\tau}^\dagger \boldsymbol{\tau})^{-r} \prod_\alpha d\tau_\alpha d\tau_\alpha^*, \tag{10.12}$$

where $\boldsymbol{\tau}$ is a $(r-k) \times k$ matrix with complex entries given by

$$\boldsymbol{\tau} \equiv \frac{z}{\sqrt{\boldsymbol{I}_k - \boldsymbol{\tau}^\dagger \boldsymbol{\tau}}}, z \equiv \boldsymbol{\eta} \frac{\sin\sqrt{\boldsymbol{\eta}^\dagger \boldsymbol{\eta}}}{\sqrt{\boldsymbol{\eta}^\dagger \boldsymbol{\eta}}}, (\boldsymbol{\eta})_{ij} \equiv \eta_{ij}. \tag{10.13}$$

Hence, any states $|\psi\rangle$ in the Hilbert space of the fully antisymmetric representation of the $\mathrm{U}(r)$ group of r fermions can be expanded in terms of the coherent states $|\boldsymbol{\Lambda}, \Omega\rangle$ as

$$|\psi\rangle = \int d\mu_\Omega N^{-1}(\tau_\alpha, \tau_\alpha^*) \langle\boldsymbol{\Lambda}, \boldsymbol{\mu}|\psi\rangle |\boldsymbol{\Lambda}, \boldsymbol{\mu}\rangle, \tag{10.14}$$

where $|\boldsymbol{\Lambda}, \Omega\rangle \equiv N^{-1/2}(\tau_\alpha, \tau_\alpha^*)|\boldsymbol{\Lambda}, \boldsymbol{\tau}\rangle$ and $N(\tau_\alpha, \tau_\alpha^*)$ is a normalization factor determined by

$$N(\tau_\alpha, \tau_\alpha^*) = \det(\boldsymbol{I}_k + \boldsymbol{\tau}^\dagger \boldsymbol{\tau}). \tag{10.15}$$

To display the symplectic structure of the coset space $\mathrm{U}(r)/\mathrm{U}(k) \otimes \mathrm{U}(r-k)$ of the fermion $\mathrm{U}(r)$ coherent states, one only needs to assign a metric $\boldsymbol{g} \equiv (g_{\alpha\beta})$ as follows:

$$ds^2 \equiv \sum_{\alpha,\beta} g_{\alpha\beta} d\tau_\alpha d\tau_\beta^* = \sum_{\alpha,\beta} \frac{\partial^2 F}{\partial\tau_\alpha \partial\tau_\beta^*} d\tau_\alpha d\tau_\beta^*, \tag{10.16}$$

where $\tau_\alpha = \tau_{ij}$ and $F(\tau_\alpha, \tau_\alpha^*) \equiv \ln\det(\boldsymbol{I}_k + \boldsymbol{\tau}^\dagger \boldsymbol{\tau})$. Using the metric $\boldsymbol{g}$, one can explicitly construct a symplectic structure with a non-degenerate closed 2-form

$$\omega \equiv \frac{i}{2} \sum_{\alpha,\beta} g_{\alpha,\beta} d\tau_\alpha \wedge d\tau_\beta^*, \tag{10.17}$$

where the function $F(\tau_\alpha, \tau_\alpha^*)$ is called the Kähler potential of the coset manifold.

Finally, the BCH formula of the U(r) group of r fermions can be obtained by using the faithful matrix representation of its group elements. As an example, the following BCH formula of U(r) group is useful in the application of the fermion U(r) coherent states:

$$\begin{aligned} &\exp \sum_{i,j} (\eta_{ij} c_i^\dagger c_j - \eta_{ij}^* c_i c_j^\dagger) \\ &= \exp \sum_{i,j} (\tau_{ij} c_i^\dagger c_j) \exp \sum_{ij} (\lambda_{ij} c_i^\dagger c_j) \exp \sum_{i,j} (-\tau_{ij}^* c_i^\dagger c_j), \end{aligned} \tag{10.18}$$

where the relation between τ_{ij}, λ_{ij}, and η_{ij} can be found explicitly in the following matrix representation:

$$\begin{aligned} &\exp \sum_{i,j} (\tau_{ij} c_i^\dagger c_j) \exp \sum_{ij} (\lambda_{ij} c_i^\dagger c_j) \exp \sum_{i,j} (-\tau_{ij}^* c_i^\dagger c_j) \\ &= \left(\begin{array}{c|c} \boldsymbol{I}_{r-k} & \boldsymbol{\tau} \\ \hline \boldsymbol{0} & \boldsymbol{I}_k \end{array} \right) \left(\begin{array}{c|c} \exp \boldsymbol{\lambda}_1 & \boldsymbol{0} \\ \hline \boldsymbol{0} & \exp \boldsymbol{\lambda}_2 \end{array} \right) \left(\begin{array}{c|c} \boldsymbol{I}_{r-k} & \boldsymbol{0} \\ \hline -\boldsymbol{\tau}^\dagger & \boldsymbol{I}_k \end{array} \right) \\ &= \left(\begin{array}{c|c} \exp \boldsymbol{\lambda}_1 - \boldsymbol{\tau} \exp \boldsymbol{\lambda}_2 \boldsymbol{\tau}^\dagger & \boldsymbol{\tau} \exp \boldsymbol{\lambda}_2 \\ \hline -\exp \boldsymbol{\lambda}_2 \boldsymbol{\tau}^\dagger & \exp \boldsymbol{\lambda}_2. \end{array} \right) \\ &= \left(\begin{array}{c|c} \sqrt{\boldsymbol{I}_{r-k} - z z^\dagger} & z \\ \hline -z^\dagger & \sqrt{\boldsymbol{I}_k - z^\dagger z} \end{array} \right) \end{aligned} \tag{10.19}$$

Hence, a direct comparison yields

$$\boldsymbol{\tau} = \frac{z}{\sqrt{\boldsymbol{I}_k - z^\dagger z}}, \exp \boldsymbol{\lambda}_1 = \frac{\boldsymbol{I}_{r-k}}{\sqrt{\boldsymbol{I}_{r-k} - z z^\dagger}}, \exp \boldsymbol{\lambda}_2 = \sqrt{\boldsymbol{I}_k - z^\dagger z}. \tag{10.20}$$

To continue, we shall consider a fermion system consisted of r single fermions with pairing correlations. Let c_i and $c_i^\dagger$ with $1 \le i \le r$ be the annihilation and creation operators of the fermions. In the Hartree-Fock-Bogoliubov mean-field approach to many-fermion systems, the Hamiltonian can be expressed as a linear combination of $r(2r-1)$ fermion pair operators $E^i_j \equiv c_i^\dagger c_j - \frac{1}{2}\delta_{ij}$ with $1 \le i, j \le r$, $E_{ij} \equiv c_i c_j$, and $E^{ij} \equiv c_i^\dagger c_j^\dagger$ with $1 \le i < j \le r$, which obey the properties

$$(E^i_j)^\dagger = E^j_i, (E_{ij})^\dagger = E^{ji}, (E^{ij})^\dagger = E_{ji}, E_{ij} = -E_{ji}, E^{ij} = -E^{ji}. \tag{10.21}$$

One may show that the $r(2r-1)$ fermion pair operators form the Lie algebra $\mathfrak{so}(2r)$ of the special orthogonal group SO($2r$). The commutation relations for the fermion pair operators in the $\mathfrak{so}(2r)$ Lie algebra are

$$[E^i_j, E^k_l] = \delta_{jk} E^i_l - \delta_{il} E^k_j, \tag{10.22a}$$

$$[E^i_j, E^{kl}] = \delta_{jk} E^{il} - \delta_{jl} E^{ik}, \tag{10.22b}$$

$$[E^i_j, E_{kl}] = \delta_{ik} E_{lj} - \delta_{il} E_{kj}, \tag{10.22c}$$

$$[E_{ij}, E^{kl}] = \delta_{ik} E^l_j - \delta_{jk} E^l_i + \delta_{jl} E^k_i - \delta_{il} E^k_j, \tag{10.22d}$$

$$[E_{ij}, E_{kl}] = [E^{ij}, E^{kl}] = 0. \tag{10.22e}$$

From Eq. (10.22a), one notices that the r^2 operators E^i_j form a $\mathfrak{u}(r)$ subalgebra of $\mathfrak{so}(2r)$. Let us denote $H_i \equiv E^i_i$ as the elements of the Cartan subalgebra. Then from Eqs. (10.22a)–(10.22c), one obtains the following commutation relations in the Cartan basis:

$$[H_i, E^j_k] = (\delta_{ij} - \delta_{jk}) E^j_k, \tag{10.23a}$$

$$[H_i, E^{jk}] = (\delta_{ij} + \delta_{ik}) E^{jk}, \tag{10.23b}$$

$$[H_i, E_{jk}] = -(\delta_{ij} + \delta_{ik}) E_{jk}, \tag{10.23c}$$

Equations (10.23a)–(10.23c) show that the multi-fermion $\mathfrak{so}(2r)$ algebra has $2r(r-1)$ roots given by $\pm \boldsymbol{e}_i \pm \boldsymbol{e}_j$ with $1 \le i < j \le r$. Similar to the multi-fermion $\mathfrak{u}(r)$ case, the faithful matrix representation of the multi-fermion $\mathfrak{so}(2r)$ algebra is useful in practical applications, which are given by

$$E^i_j \equiv c_i^\dagger c_j - \frac{1}{2}\delta_{ij} \Leftrightarrow E_{i,j} - E_{r+j,r+i}, \tag{10.24}$$

$$E^{ij} \equiv c_i^\dagger c_j^\dagger \Leftrightarrow E_{i,r+j} - E_{j,r+i}, \tag{10.25}$$

$$E_{ij} \equiv c_i c_j \Leftrightarrow E_{r+i,j} - E_{r+j,i}, \tag{10.26}$$

where $E_{i,j}$ is a $2r \times 2r$ matrix with 1 in the ith column and jth row and zeros elsewhere. Clearly, the set of diagonal matrices $E_{i,i} - E_{r+i,r+i}$ forms the Cartan subalgebra of $\mathfrak{so}(2r)$.

The basic states in the Hilbert space for the multi-fermion $\mathfrak{so}(2r)$ algebra are $|n_1, n_2, \cdots, n_r\rangle$ with $n_k = 0$ or 1. There are two different spinor representations $(\frac{1}{2}, \frac{1}{2}, \cdots, \pm\frac{1}{2})$, where the plus and minus signs correspond to the even and odd total particle number cases. Hence, there are 2^{r-1} states in each spinor representation of $(\frac{1}{2}, \frac{1}{2}, \cdots, \pm\frac{1}{2})$. For the two distinct spinor representations of the multi-fermion Spin$(2r)$ group, where Spin$(2r)$ is the covering group of SO$(2r)$, the extremal states can be taken as

$$|\mathbf{0}\rangle \equiv |0, 0, \cdots, 0\rangle \text{ for } (\frac{1}{2}, \frac{1}{2}, \cdots, \frac{1}{2}), \tag{10.27}$$
$$|\mathbf{1}\rangle \equiv |1, 0, \cdots, 0\rangle \text{ for } (\frac{1}{2}, \frac{1}{2}, \cdots, -\frac{1}{2}).$$

One then obtains the following relations $c_i^\dagger c_j|\mathbf{0}\rangle = 0$ for $1 \le i, j \le r$ and

$$c_1^\dagger c_1|\mathbf{1}\rangle = |\mathbf{1}\rangle, c_i^\dagger c_j|\mathbf{1}\rangle = 0, (2 \le i, j \le r), \tag{10.28}$$
$$c_1^\dagger c_i^\dagger|\mathbf{1}\rangle = c_i c_1|\mathbf{1}\rangle = 0, (2 \le i \le r).$$

Clearly, the operators $E^i_j \equiv c_i^\dagger c_j - \frac{1}{2}\delta_{ij}$ with $1 \le i, j \le r$ form a subalgebra $\mathfrak{u}(r)$ of $\mathfrak{so}(2r)$. Besides, one may show that the operators $E^i_j \equiv c_i^\dagger c_j - \frac{1}{2}\delta_{ij}$ with $2 \le i, j \le r$, $E^1_1 \equiv c_1^\dagger c_1 - \frac{1}{2}$, $E^{1i} \equiv c_1^\dagger c_i^\dagger$, and $E_{j1} \equiv c_j c_1$ with $2 \le i \le r$ also form a subalgebra $\mathfrak{u}(r)$ of $\mathfrak{so}(2r)$. The corresponding Lie groups U(r) are the stability subgroup of Spin$(2r)$, which keep the extremal state $|\mathbf{0}\rangle$ of the irreducible representation $(\frac{1}{2}, \frac{1}{2}, \cdots, \frac{1}{2})$ unchanged, or the extremal state $|\mathbf{1}\rangle$ of the irreducible representation $(\frac{1}{2}, \frac{1}{2}, \cdots, -\frac{1}{2})$ unchanged. One may then construct the coherent states for the multi-fermion $\mathfrak{so}(2r)$ according to the general algorithm. For simplicity, in the following, we shall discuss only the coherent states for the irreducible representation $(\frac{1}{2}, \frac{1}{2}, \cdots, \frac{1}{2})$ which corresponds to the case of even total particle number. The same procedure can also be applied to the irreducible representation $(\frac{1}{2}, \frac{1}{2}, \cdots, -\frac{1}{2})$ which corresponds to the case of odd total particle number.

Since the isotropy subgroup of Spin$(2r)$ in the irreducible representation $(\frac{1}{2}, \frac{1}{2}, \cdots, \frac{1}{2})$ is U(r) generated by the operators $E^i_j \equiv c_i^\dagger c_j - \frac{1}{2}\delta_{ij}$, the coherent state spaces are then isomorphic to the coset space Spin$(2r)/$U(r), and the coherent states are generated by

$$|\mathbf{\Lambda}, \Omega\rangle \equiv \Omega|\mathbf{\Lambda}, \mathbf{\Lambda}\rangle \equiv \exp \sum_{1\le i<j\le r} (\eta_{ij} c_i^\dagger c_j^\dagger - \eta_{ij}^* c_j c_i)|\mathbf{0}\rangle, \tag{10.29}$$

where Ω is the general form of the coset representative of Spin$(2r)/\mathrm{U}(r)$. Hence, the coherent states so defined naturally have the topological structure of Spin$(2r)/\mathrm{U}(r)$, from which one may derive more geometric properties.

One of the key features of the topological space Spin$(2r)/\mathrm{U}(r)$ of the multi-fermion Spin$(2r)$ coherent states is a symmetric space, i.e., a $r \times r$-dimensional complex manifold. As the coset space Spin$(2r)/\mathrm{U}(r)$ is isomorphic to SO$(2r)/\mathrm{U}(r)$, the symplectic structure of Spin$(2r)/\mathrm{U}(r)$ can be found from that of SO$(2r)/\mathrm{U}(r)$. In the faithful matrix representation of SO$(2r)$, the coset representative of SO$(2r)/\mathrm{U}(r)$ is

$$\Omega = \exp\left(\begin{array}{c|c} \mathbf{0} & \boldsymbol{\eta} \\ \hline -\boldsymbol{\eta}^\dagger & \mathbf{0} \end{array}\right) = \left(\begin{array}{c|c} \sqrt{\boldsymbol{I}_r - \boldsymbol{z}\boldsymbol{z}^\dagger} & \boldsymbol{z} \\ \hline -\boldsymbol{z}^\dagger & \sqrt{\boldsymbol{I}_r - \boldsymbol{z}^\dagger \boldsymbol{z}} \end{array}\right), \tag{10.30}$$

where $\boldsymbol{\eta}$ and $\boldsymbol{z}$ are $r \times r$ antisymmetric complex matrices given by

$$\boldsymbol{\eta} \equiv (\eta_{ij}),\, \boldsymbol{z} \equiv \boldsymbol{\eta} \frac{\sin\sqrt{\boldsymbol{\eta}^\dagger \boldsymbol{\eta}}}{\sqrt{\boldsymbol{\eta}^\dagger \boldsymbol{\eta}}}. \tag{10.31}$$

By introducing the projected coset representation via

$$\boldsymbol{\tau} \equiv \frac{\boldsymbol{z}}{\sqrt{\boldsymbol{I}_r - \boldsymbol{z}^\dagger \boldsymbol{z}}} = \boldsymbol{\eta} \frac{\tan\sqrt{\boldsymbol{\eta}^\dagger \boldsymbol{\eta}}}{\sqrt{\boldsymbol{\eta}^\dagger \boldsymbol{\eta}}}, \tag{10.32}$$

one can explicitly express a group transformation

$$g = \left(\begin{array}{c|c} \boldsymbol{A} & \boldsymbol{B} \\ \hline \boldsymbol{C} & \boldsymbol{D} \end{array}\right) \in \mathrm{SO}(2r), \tag{10.33}$$

on the coset space SO$(2r)/\mathrm{U}(r)$ as

$$\boldsymbol{\tau}' = \frac{\boldsymbol{A}\boldsymbol{\tau} + \boldsymbol{B}}{\boldsymbol{C}\boldsymbol{\tau} + \boldsymbol{D}}. \tag{10.34}$$

The Riemann metric of the coset space can be obtained from the non-normalized form $|\boldsymbol{\tau}\rangle$ of the coherent states

$$|\boldsymbol{\Lambda}, \Omega\rangle = \frac{1}{\sqrt{N(\tau, \tau^*)}} |\boldsymbol{\Lambda}, \boldsymbol{\tau}\rangle \equiv \frac{1}{\sqrt{N(\tau, \tau^*)}} \exp \sum_{1 \le i \ne j \le r} (\eta_{ij} c_i^\dagger c_j^\dagger) |\mathbf{0}\rangle \tag{10.35}$$

via the relation

$$ds^2 = \sum_{\alpha,\beta} g_{\alpha\beta} d\tau_\alpha d\tau_\beta^* = \sum_{\alpha,\beta} \frac{\partial^2 F}{\partial \tau_\alpha \partial \tau_\beta^*} d\tau_\alpha d\tau_\beta^*, \tag{10.36}$$

where $N(\tau, \tau^*) = \sqrt{\det(\boldsymbol{I}_r + \boldsymbol{\tau}^\dagger \boldsymbol{\tau})}$ and $F \equiv \ln N(\tau, \tau^*) = \frac{1}{2} \ln \det(\boldsymbol{I}_r + \boldsymbol{\tau}^\dagger \boldsymbol{\tau})$. Finally, the coset space Spin$(2r)$/U(r) is also a Kähler manifold which possesses a symplectic structure with a closed non-degenerate 2-form

$$\omega \equiv \frac{i}{2} \sum_{\alpha,\beta} g_{\alpha,\beta} d\tau_\alpha \wedge d\tau_\beta^*. \tag{10.37}$$

Using the faithful matrix representation of SO$(2r)$, one may obtain the following BCH formula of SO$(2r)$:

$$\begin{aligned} \Omega &\equiv \exp \sum_{i<j} (\eta_{ij} c_i^\dagger c_j^\dagger - \eta_{ij}^* c_j c_i) \\ &= \exp \sum_{i<j} (\tau_{ij} c_i^\dagger c_j^\dagger) \exp \sum_{i<j} [\lambda_{ij} (c_i^\dagger c_j - \frac{1}{2} \delta_{ij})] \exp \sum_{i<j} (-\tau_{ij}^* c_i c_j), \end{aligned} \tag{10.38}$$

where the right-hand side of Eq. (10.39) can be written in the faithful matrix representation of SO$(2r)$ as

$$\begin{aligned} &\left(\begin{array}{c|c} \boldsymbol{I}_r & \boldsymbol{\tau} \\ \hline \boldsymbol{0} & \boldsymbol{I}_r \end{array}\right) \left(\begin{array}{c|c} \exp \boldsymbol{\lambda} & \boldsymbol{0} \\ \hline \boldsymbol{0} & \exp -\boldsymbol{\lambda}^\top \end{array}\right) \left(\begin{array}{c|c} \boldsymbol{I}_r & \boldsymbol{0} \\ \hline -\boldsymbol{\tau}^\dagger & \boldsymbol{I}_r \end{array}\right) \\ &= \left(\begin{array}{c|c} e^{\boldsymbol{\lambda}} - \boldsymbol{\tau} e^{-\boldsymbol{\lambda}^\top} \boldsymbol{\tau}^\dagger & \boldsymbol{\tau} e^{-\boldsymbol{\lambda}^\top} \\ \hline -e^{-\boldsymbol{\lambda}^\top} \boldsymbol{\tau}^\dagger & e^{-\boldsymbol{\lambda}^\top} \end{array}\right), \end{aligned} \tag{10.39}$$

where $\boldsymbol{\tau} \equiv (\tau_{ij})$ and $\boldsymbol{\lambda} \equiv (\lambda_{ij})$. Comparing Eqs. (10.30) and (10.39), one immediately obtains

$$\begin{aligned} \boldsymbol{\tau} &= \frac{\boldsymbol{z}}{\sqrt{\boldsymbol{I}_r - \boldsymbol{z}^\dagger \boldsymbol{z}}} = \boldsymbol{\eta} \frac{\tan \sqrt{\boldsymbol{\eta}^\dagger \boldsymbol{\eta}}}{\sqrt{\boldsymbol{\eta}^\dagger \boldsymbol{\eta}}}, \\ e^{\boldsymbol{\lambda}} &= \frac{\boldsymbol{I}_r}{\sqrt{\boldsymbol{I}_r - \boldsymbol{z}\boldsymbol{z}^\dagger}} = \frac{\boldsymbol{I}_r}{\cos \sqrt{\boldsymbol{\eta}\boldsymbol{\eta}^\dagger}}. \end{aligned} \tag{10.40}$$

However, Eq. (10.40) is only the BCH formula for SO$(2r)$. In order to obtain the BCH formula for Spin$(2r)$, one notes that the subgroup U(r) of Spin$(2r)$ is double-valued in the SO$(2r)$ representation. Hence, $\boldsymbol{\lambda}$ must be replaced by $\boldsymbol{\lambda}/2$ in the case of Spin$(2r)$, which yields the following relationship between $\boldsymbol{z}$ and $\boldsymbol{\lambda}$ and $\boldsymbol{\eta}$ for the Spin$(2r)$:

$$\boldsymbol{\tau} = \frac{\boldsymbol{z}}{\sqrt{\boldsymbol{I}_r - \boldsymbol{z}^\dagger \boldsymbol{z}}}, \quad \exp \frac{\boldsymbol{\lambda}}{\boldsymbol{2}} = \frac{\boldsymbol{I}_r}{\sqrt{\boldsymbol{I}_r - \boldsymbol{z}^\dagger \boldsymbol{z}}}. \tag{10.41}$$

Finally, from the general theorem of completeness, the coherent states $|\mathbf{\Lambda}, \Omega\rangle \equiv \Omega|\mathbf{0}\rangle$ obey the completeness relation

$$\int d\mu_{\Omega}(\boldsymbol{\tau}, \boldsymbol{\tau}^*)|\mathbf{\Lambda}, \Omega\rangle\langle\mathbf{\Lambda}, \Omega| = \boldsymbol{I}_r, \tag{10.42}$$

where the measure of the coset space can be expressed in terms of the projected coset representation as

$$d\mu_{\Omega}(\tau_\alpha, \tau_\alpha^*) = \frac{\dim V^{\mathbf{\Lambda}}}{\mathrm{Vol}(\mathrm{Spin}(2r)/\mathrm{U}(r))}[\det((\boldsymbol{I}_r + \boldsymbol{\tau}^\dagger\boldsymbol{\tau}))]^{-2r}\prod_\alpha d\tau_\alpha d\tau_\alpha^*. \tag{10.43}$$

Hence, any state $|\psi\rangle$ in the irreducible representation $(\frac{1}{2}, \frac{1}{2}, \cdots, \frac{1}{2})$ can be expanded in terms of the coherent states as

$$|\psi\rangle = \int d\mu_{\Omega}(\boldsymbol{\tau}, \boldsymbol{\tau}^*)N^{-1}(\tau_\alpha, \tau_\alpha^*)f(\boldsymbol{\tau})|\boldsymbol{\tau}\rangle,\ f(\boldsymbol{\tau}) \equiv \langle\boldsymbol{\tau}|\psi\rangle. \tag{10.44}$$

Finally, when one adds the single-particle creation and annihilation operators $c_i^\dagger$ and c_j into the set of generators of SO$(2r)$, i.e., $\{E^i_j \equiv c_i^\dagger c_j - \frac{1}{2}\delta_{ij}, E_{ij} \equiv c_i c_j, E^{ij} \equiv c_i^\dagger c_j^\dagger\}$, the enlarged set of operator still forms a Lie algebra $\mathfrak{so}(2r+1)$, the maximum dynamical Lie algebra of a system with r fermions. The additional commutation relations besides those for $\mathfrak{so}(2r)$ are

$$\begin{aligned}
&[c_i^\dagger, c_j] = 2E^i_j, [c_i, c_j] = 2E_{ij}, [c_i^\dagger, c_j^\dagger] = 2E^{ij}, \\
&[c_i, E^j_k] = \delta_{ij} f_k, [c_i^\dagger, E^j_k] = -\delta_{ik}c_j^\dagger, [c_i, E_{jk}] = [c_i^\dagger, E^{jk}] = 0, \\
&[c_i, E^{jk}] = \delta_{ij}c_k^\dagger - \delta_{ik}c_j^\dagger, [c_i^\dagger, E_{jk}] = \delta_{ij}c_k - \delta_{ik}c_j.
\end{aligned} \tag{10.45}$$

Evidently, the sets of Fermion operators $\{E^i_j\}$ and $\{E^i_j, E_{ij}, E^{ij}\}$ form the $\mathfrak{u}(r)$ and $\mathfrak{so}(2r)$ subalgebras of $\mathfrak{so}(2r+1)$, respectively. Similar to the $\mathfrak{so}(2r)$ case, the r operators $H_i \equiv E^i_i$ with $1 \leq i \leq r$ form the maximal commutating Cartan subalgebra of $\mathfrak{so}(2r+1)$. Then from Eq. (10.45), one obtains the following commutating relations in the Cartan basis:

$$\begin{aligned}
&[H_i, c_j] = -\delta_{ij}c_j, [H_i, c_j^\dagger] = \delta_{ij}c_i^\dagger, [H_i, E^j_k] = (\delta_{ij} - \delta_{jk})E^j_k, \\
&[H_i, E^{jk}] = (\delta_{ij} + \delta_{ik})E^{jk}, [H_i, E_{jk}] = -(\delta_{ij} + \delta_{ik})E_{jk}.
\end{aligned} \tag{10.46}$$

Equation (10.46) shows that the multi-fermion $\mathfrak{so}(2r+1)$ algebra has $2r^2$ roots given by $\pm\boldsymbol{e}_i$ and $\pm\boldsymbol{e}_i \pm \boldsymbol{e}_j$ with $1 \leq i < j \leq r$. Similar to the multi-fermion $\mathfrak{u}(r)$ and $\mathfrak{so}(2r)$ cases, there exists a faithful matrix representation of the multi-fermion

$\mathfrak{so}(2r+1)$ algebra, which is given by

$$c_i^\dagger \Leftrightarrow E_{i,0} - E_{0,r+i}, \tag{10.47a}$$

$$c_i \Leftrightarrow E_{0,i} - E_{r+i,0}, \tag{10.47b}$$

$$E^i_j \equiv c_i^\dagger c_j - \frac{1}{2}\delta_{ij} \Leftrightarrow E_{i,j} - E_{r+j,r+i}, \tag{10.47c}$$

$$E^{ij} \equiv c_i^\dagger c_j^\dagger \Leftrightarrow E_{i,r+j} - E_{j,r+i}, \tag{10.47d}$$

$$E_{ij} \equiv c_i c_j \Leftrightarrow E_{r+i,j} - E_{r+j,i}, \tag{10.47e}$$

where $E_{i,j}$ is a $(2r+1)\times(2r+1)$ matrix with 1 in the ith row and jth column and zeros elsewhere. Similar to the multi-fermion $\mathfrak{u}(r)$ and $\mathfrak{so}(2r)$ cases, the diagonal matrices $H_i \equiv E_{i,i} - E_{r+i,r+i}$ form the Cartan subalgebra of $\mathfrak{so}(2r+1)$.

For the irreducible spinor representation $\mathbf{\Lambda} \equiv (\frac{1}{2}, \frac{1}{2}, \cdots, \frac{1}{2})$ of the multi-fermion $\mathfrak{so}(2r+1)$ algebra, the extremal state can be taken as

$$|\mathbf{\Lambda}, \mathbf{\Lambda}\rangle \equiv |\mathbf{0}\rangle = |0, 0, \cdots, 0\rangle. \tag{10.48}$$

Hence, the isotropy subgroup of $\mathrm{Spin}(2r+1)$ in the irreducible spinor representation $(\frac{1}{2}, \frac{1}{2}, \cdots, \frac{1}{2})$ which leaves the extremal state $|\mathbf{0}\rangle$ invariant is $\mathrm{U}(r)$ generated by the operators $E^i_j \equiv c_i^\dagger c_j - \frac{1}{2}\delta_{ij}$, and the coherent state space is then isomorphic to the coset space $\mathrm{Spin}(2r+1)/\mathrm{U}(r)$. Following the same procedure as in the $\mathfrak{u}(r)$ and $\mathfrak{so}(2r)$ cases, the coherent states for the multi-fermion $\mathfrak{so}(2r+1)$ algebra are given by $|\mathbf{\Lambda}, \Omega\rangle \equiv \Omega|\mathbf{\Lambda}, \mathbf{\Lambda}\rangle$, where

$$\Omega \equiv \exp\left\{ \sum_{1\le i<j\le r} (\eta_{0i} c_i^\dagger - \eta_{0i}^* c_i + \eta_{ij} c_i^\dagger c_j^\dagger - \eta_{ij}^* c_j c_i) \right\} \tag{10.49}$$

is the general form of the coset representative of the coset space $\mathrm{Spin}(2r+1)/\mathrm{U}(r)$.

10.2 Bogoliubov Transformations

In many-body quantum theory, the concept of **quasi-particles**, coined by the physicist Lev D. Landau, is a way to describe complicated collective behaviors of real particles in many-body systems in terms of imagined quasi-particles, which behave more like noninteracting particles. The basic idea is to represent the ground state of a many-body system as the vacuum state with respect to the quasi-particles. In principle, the is no simple relationship between the Landau quasi-particles and the bare particles of a general physical system. However, for the so-called Bogoliubov quasi-particles, coined by the physicist Nikolay N. Bogoliubov, the quasi-particles are indeed related to the bare particles by a simple linear unitary transformation. The simple mathematical relationship, which is essence to the **Hartree-Fock-Bogoliubov theory**, comes at the expense of the fact that the Bogoliubov quasi-particle vacuum states and the corresponding one-quasi-particle states are only approximations of the exact eigenstates of the many-body Hamiltonian. Remarkably, both the concept of Bogoliubov quasi-particles and the Hartree-Fock-Bogoliubov theory can be best understood in the context of coherent states, which we shall discuss below.

When the Hamiltonian of a many-body system is constrained by a dynamical group, its associated Hilbert space must be reduced to an irreducible representation of the group, and thus the general Bogoliubov quasi-particle transformation has to be restricted to a unitary transformation within the group. More precisely, let G be the dynamical symmetry group of the system, and let g be an element of G, i.e., a general unitary transformation of G. Then the Bogoliubov transformation of the bare vacuum state $|0\rangle$ to the **quasi-particle vacuum state** $|\Phi\rangle$ has the form

$$|\Phi\rangle = g|0\rangle = \Omega h|0\rangle = \Omega|0\rangle e^{i\varphi(h)}, \tag{10.50}$$

where $h \in H$, $\Omega \in G/H$, and H is the maximum isotropy subgroup of G which keeps the bare vacuum state $|0\rangle$ invariant up to a phase factor $e^{i\varphi(h)}$. According to the general algorithm for constructing coherent states, the Bogoliubov quasi-particle vacuum state $|\Phi\rangle$ is precisely the coherent state on the coset space G/H.

As an example, for a finite many-fermion system of r single-fermion states, in the even fermion spinor irreducible representation of SO$(2r)$, the corresponding Bogoliubov quasi-particle vacuum state is

$$|\Omega\rangle \equiv \Omega|\mathbf{0}\rangle = \exp \sum_{1\le i<j\le r} (\eta_{ij} c_i^\dagger c_j^\dagger - \eta_{ij}^* c_j c_i)|\mathbf{0}\rangle, \tag{10.51}$$

where $|\Omega\rangle$ is a general representative of the coset space SO$(2r)$/U(r), c_i and $c_i^\dagger$ are the single-particle fermionic annihilation and creation operators which obey the anti-commutation relations $\{c_i, c_j^\dagger\} = \delta_{ij}$, $\{c_i, c_j\} = \{c_i^\dagger, c_j^\dagger\} = 0$, and $|\mathbf{0}\rangle$ is the no-particle bare vacuum state annihilated by all the single-particle fermion annihilation operators c_i. Equation (10.51) clearly shows that the Bogoliubov quasi-particle vacuum state is nothing but a coherent superposition of fermionic pairing states. In such a case, the Bogoliubov transformation from the single-particle operators c_i and $c_i^\dagger$ to the quasi-particle operators b_i and $b_i^\dagger$ is given by [129]

$$b_j \equiv \Omega c_j \Omega^{-1} \equiv \sum_{1\le i\le r} (u_{ij}^* c_i + v_{ij}^* c_i^\dagger), \tag{10.52a}$$

$$b_j^\dagger \equiv \Omega c_j^\dagger \Omega^{-1} \equiv \sum_{1\le i\le r} (u_{ij} c_i^\dagger + v_{ij} c_i). \tag{10.52b}$$

As such, the Bogoliubov quasi-particle annihilation and creation operators b_i and $b_i^\dagger$ are forced to obey the same fermionic anti-commutation relations

$$\{b_i, b_j^\dagger\} = \{\Omega c_i \Omega^{-1}, \Omega c_j^\dagger \Omega^{-1}\} = \Omega\{c_i, c_j^\dagger\}\Omega^{-1} = \delta_{ij}, \tag{10.53a}$$

$$\{b_i, b_j\} = \{\Omega c_i \Omega^{-1}, \Omega c_j \Omega^{-1}\} = \Omega\{c_i, c_j\}\Omega^{-1} = 0, \tag{10.53b}$$

$$\{b_i^\dagger, b_j^\dagger\} = \{\Omega c_i^\dagger \Omega^{-1}, \Omega c_j^\dagger \Omega^{-1}\} = \Omega\{c_i^\dagger, c_j^\dagger\}\Omega^{-1} = 0, \tag{10.53c}$$

and thus the coefficients u_{ij} and v_{ij} appearing in the Bogoliubov transformation, Eqs. (10.52a) and (10.52b), are not completely arbitrary, but are restricted to obey the condition

$$\boldsymbol{U}^\dagger \boldsymbol{U} + \boldsymbol{V}^\dagger \boldsymbol{V} = \boldsymbol{I}_r, \boldsymbol{U}^\dagger \boldsymbol{V}^* + \boldsymbol{V}^\dagger \boldsymbol{U}^* = \boldsymbol{0}. \tag{10.54}$$

where $\boldsymbol{U} \equiv (u_{ij})$ and $\boldsymbol{V} \equiv (v_{ij})$. Using the two matrices $\boldsymbol{U}$ and $\boldsymbol{V}$, the Bogoliubov transformation from the single-particle annihilation and creation operators $(\boldsymbol{c}, \boldsymbol{c}^\dagger)^\top \equiv (c_1, \cdots, c_r, c_1^\dagger, \cdots, c_r^\dagger)^\top$ to the quasi-particle annihilation and creation operators $(\boldsymbol{b}, \boldsymbol{b}^\dagger)^\top \equiv (b_1, \cdots, b_r, b_1^\dagger, \cdots, b_r^\dagger)^\top$, which acts in a $2r$-dimensional space, can be expressed as

$$\begin{pmatrix} \boldsymbol{b} \\ \boldsymbol{b}^\dagger \end{pmatrix} = \mathbb{T}^\dagger \begin{pmatrix} \boldsymbol{c} \\ \boldsymbol{c}^\dagger \end{pmatrix}, \mathbb{T}^\dagger \equiv \begin{pmatrix} \boldsymbol{U}^\dagger & \boldsymbol{V}^\dagger \\ \boldsymbol{V}^\top & \boldsymbol{U}^\top \end{pmatrix}, \tag{10.55}$$

where the inverse Bogoliubov transformation is determined by

$$\begin{pmatrix} \boldsymbol{c} \\ \boldsymbol{c}^\dagger \end{pmatrix} = (\mathbb{T}^\dagger)^{-1} \begin{pmatrix} \boldsymbol{b} \\ \boldsymbol{b}^\dagger \end{pmatrix}, (\mathbb{T}^\dagger)^{-1} = \begin{pmatrix} \boldsymbol{U} & \boldsymbol{V}^* \\ \boldsymbol{V} & \boldsymbol{U}^* \end{pmatrix}. \tag{10.56}$$

Hence, the explicit expression of the inverse Bogoliubov transformation from the quasi-particle operators b_i and $b_i^\dagger$ to the single-particle operators c_i and $c_i^\dagger$ is

$$c_i = \sum_{1\le j\le r} (u_{ij} b_j + v_{ij}^* b_j^\dagger), c_i^\dagger = \sum_{1\le j\le r} (v_{ij} b_j + u_{ij}^* b_j^\dagger). \tag{10.57}$$

As a coherent state, the Bogoliubov quasi-particle vacuum state $|\Omega\rangle$ is automatically annihilated by all the quasi-particle annihilation operators b_i

$$b_i|\Omega\rangle = \Omega c_i \Omega^{-1}|\Omega\rangle = \Omega c_i \Omega^{-1}\Omega|\boldsymbol{0}\rangle = \Omega c_i|\boldsymbol{0}\rangle = 0. \tag{10.58}$$

Using Eq. (10.52a), one may readily show that the Bogoliubov quasi-particle vacuum state is equivalently annihilated by the following combinations of fermionic creation and annihilation operators

$$(c_k + \sum_{1\leq i\leq r} (\boldsymbol{V}^*\boldsymbol{U}^{*-1})_{i,k} c_i^\dagger)|\boldsymbol{0}\rangle = 0. \tag{10.59}$$

In order to find out the mathematical relationship between the parameters u_{ij} and v_{ij} appearing in the Bogoliubov transformation from the single-particle operators c_i and $c_i^\dagger$ to the quasi-particle operators b_i and $b_i^\dagger$, and the parameters η_{ij} appearing in the Bogoliubov quasi-particle vacuum state $|\Omega\rangle$, one may employ the following identity:

$$\begin{aligned}
&\exp\left\{\sum_{i<j} \tau_{ij} c_i^\dagger c_j^\dagger\right\} c_k \exp\left\{-\sum_{i<j} \tau_{ij} c_i^\dagger c_j^\dagger\right\} \qquad (10.60)\\
&= c_k + \sum_{i<j} \tau_{ij} [c_i^\dagger c_j^\dagger, c_k]\\
&= c_k + \sum_{i<j} \tau_{ij} (\delta_{kj} c_i^\dagger - \delta_{ki} c_j^\dagger)\\
&= c_k + \sum_{i<k} \tau_{ik} c_i^\dagger - \sum_{j>k} \tau_{kj} c_j^\dagger,
\end{aligned}$$

where we have used the commutation relations $[c_k, c_i^\dagger c_j^\dagger] = \delta_{ki} c_j^\dagger - \delta_{kj} c_i^\dagger$ and $[c_k^\dagger, c_i^\dagger c_j^\dagger] = 0$. If one regards τ_{ij} as the elements of an antisymmetric matrix $\boldsymbol{\tau} \equiv (\tau_{ij})$, which obey the relations $\tau_{ji} = -\tau_{ij}$, then one obtains

$$\exp\left\{\sum_{i<j} \tau_{ij} c_i^\dagger c_j^\dagger\right\} c_k \exp\left\{-\sum_{i<j} \tau_{ij} c_i^\dagger c_j^\dagger\right\} = c_k + \sum_{1\leq i\leq r} \tau_{ik} c_i^\dagger. \tag{10.61}$$

Moreover, one also needs the following identity:

$$\begin{aligned}
&\exp\left\{\sum_{i<j} \lambda_{ij} (c_i^\dagger c_j - \frac{1}{2}\delta_{ij})\right\} c_k \exp\left\{-\sum_{i<j} \lambda_{ij} (c_i^\dagger c_j - \frac{1}{2}\delta_{ij})\right\} \qquad (10.62)\\
&= c_k - \sum_j \lambda_{kj} c_j + \frac{1}{2!} \sum_{j,l} \lambda_{kj}\lambda_{jl} c_l - \frac{1}{3!} \sum_{j,l,m} \lambda_{kj}\lambda_{jl}\lambda_{lm} c_m + \cdots\\
&= c_k - \sum_j \boldsymbol{\lambda}_{k,j} c_j + \frac{1}{2!} \sum_l (\boldsymbol{\lambda}^2)_{k,l} c_l - \frac{1}{3!} \sum_m (\boldsymbol{\lambda}^3)_{k,m} c_m + \cdots\\
&= \sum_{1\leq j\leq r} (e^{-\boldsymbol{\lambda}})_{k,j} c_j,
\end{aligned}$$

where we have used the commutation relation $[c_k, E^i_j] = \delta_{ki} c_j$ with $E^i_j \equiv c_i^\dagger c_j - \frac{1}{2}\delta_{ij}$. Using the BCH formula for the coset representative

$$\Omega \equiv \exp \sum_{i<j} (\eta_{ij} c_i^\dagger c_j^\dagger - \eta_{ij}^* c_i c_j) \tag{10.63}$$

$$= \exp \sum_{i<j} (\tau_{ij} c_i^\dagger c_j^\dagger) \exp \sum_{i<j} [\lambda_{ij} (c_i^\dagger c_j - \frac{1}{2}\delta_{ij})] \exp \sum_{i<j} (-\tau_{ij}^* c_i c_j),$$

and the identities Eqs. (10.61) and (10.62), one will obtain

$$\Omega c_j \Omega^{-1} = \exp\left\{\sum_{i<j} \tau_{ij} c_i^\dagger c_j^\dagger\right\} \sum_k (e^{-\boldsymbol{\lambda}})_{j,k} c_k \exp\left\{-\sum_{i<j} \tau_{ij} c_i^\dagger c_j^\dagger\right\} \tag{10.64}$$

$$= \sum_{1 \le k \le r} (e^{-\boldsymbol{\lambda}})_{j,k} (c_k + \sum_{1 \le i \le r} \tau_{ik} c_i^\dagger)$$

$$= \sum_{1 \le k \le r} (e^{-\boldsymbol{\lambda}^\top})_{k,j} c_k + \sum_{1 \le i,k \le r} (\boldsymbol{\tau})_{i,k} (e^{-\boldsymbol{\lambda}^\top})_{k,j} c_i^\dagger$$

$$= \sum_{1 \le i \le r} [(e^{-\boldsymbol{\lambda}^\top})_{i,j} c_i + (\boldsymbol{\tau} e^{-\boldsymbol{\lambda}^\top})_{i,j} c_i^\dagger],$$

where we have used the commutation relation $[c_k, c_i c_j] = 0$. A direct comparison between Eqs. (10.52a) and (10.64) yields

$$u_{ij}^* = (e^{-\boldsymbol{\lambda}^\top})_{i,j},\ v_{ij}^* = (\boldsymbol{\tau} e^{-\boldsymbol{\lambda}^\top})_{i,j}, \tag{10.65}$$

which is equivalent to the following expression:

$$\boldsymbol{U}^* = e^{-\boldsymbol{\lambda}^\top} = \cos\sqrt{\boldsymbol{\eta}^\dagger \boldsymbol{\eta}},\ \boldsymbol{V}^* = \boldsymbol{\tau} e^{-\boldsymbol{\lambda}^\top} = \boldsymbol{\eta} \frac{\sin\sqrt{\boldsymbol{\eta}^\dagger \boldsymbol{\eta}}}{\sqrt{\boldsymbol{\eta}^\dagger \boldsymbol{\eta}}}. \tag{10.66}$$

From Eq. (10.66), one can immediately derive the following identities:

$$\boldsymbol{U}^\dagger = e^{-\boldsymbol{\lambda}} = \cos\sqrt{\boldsymbol{\eta}\boldsymbol{\eta}^\dagger} = \boldsymbol{U}, \tag{10.67a}$$

$$\boldsymbol{V}^\dagger = -\frac{\sin\sqrt{\boldsymbol{\eta}\boldsymbol{\eta}^\dagger}}{\sqrt{\boldsymbol{\eta}\boldsymbol{\eta}^\dagger}} \boldsymbol{\eta} = -\boldsymbol{\eta} \frac{\sin\sqrt{\boldsymbol{\eta}^\dagger \boldsymbol{\eta}}}{\sqrt{\boldsymbol{\eta}^\dagger \boldsymbol{\eta}}} = -\boldsymbol{V}^*, \tag{10.67b}$$

where we used the relations $\boldsymbol{\eta}^\top = -\boldsymbol{\eta}$ and $(\boldsymbol{\eta}^\dagger)^\top = -\boldsymbol{\eta}^\dagger$ for antisymmetric matrices $\boldsymbol{\eta}$ and $\boldsymbol{\eta}^\dagger$. Substituting Eq. (10.66) into Eq. (10.59), one immediately obtains

$$(c_k + \sum_{1\le i\le r} \tau_{ik} c_i^\dagger)|\mathbf{0}\rangle = 0, \boldsymbol{V}^*\boldsymbol{U}^{*-1} = \boldsymbol{\tau} = \boldsymbol{\eta}\frac{\tan\sqrt{\boldsymbol{\eta}^\dagger\boldsymbol{\eta}}}{\sqrt{\boldsymbol{\eta}^\dagger\boldsymbol{\eta}}}. \tag{10.68}$$

Substitution of Eqs. (10.66) and (10.67) into Eq. (10.56) immediately yields

$$\begin{pmatrix} \boldsymbol{b} \\ \boldsymbol{b}^\dagger \end{pmatrix} = \mathbb{T}^\dagger \begin{pmatrix} \boldsymbol{c} \\ \boldsymbol{c}^\dagger \end{pmatrix}, \mathbb{T}^\dagger \equiv \begin{pmatrix} \cos\sqrt{\boldsymbol{\eta}\boldsymbol{\eta}^\dagger} & -\boldsymbol{\eta}\frac{\sin\sqrt{\boldsymbol{\eta}^\dagger\boldsymbol{\eta}}}{\sqrt{\boldsymbol{\eta}^\dagger\boldsymbol{\eta}}} \\ \frac{\sin\sqrt{\boldsymbol{\eta}^\dagger\boldsymbol{\eta}}}{\sqrt{\boldsymbol{\eta}^\dagger\boldsymbol{\eta}}}\boldsymbol{\eta}^\dagger & \cos\sqrt{\boldsymbol{\eta}^\dagger\boldsymbol{\eta}} \end{pmatrix}. \tag{10.69}$$

A direct computation yields the orthogonal conditions for the coefficient matrices $\mathbb{T}^\dagger$ and $\mathbb{T}$

$$\mathbb{T}^\dagger\mathbb{T} = \exp\begin{pmatrix} \mathbf{0} & -\boldsymbol{\eta} \\ \boldsymbol{\eta}^\dagger & \mathbf{0} \end{pmatrix} \exp\begin{pmatrix} \mathbf{0} & \boldsymbol{\eta} \\ -\boldsymbol{\eta}^\dagger & \mathbf{0} \end{pmatrix} = \boldsymbol{I}_{2r}, \tag{10.70a}$$

$$\mathbb{T}\mathbb{T}^\dagger = \exp\begin{pmatrix} \mathbf{0} & \boldsymbol{\eta} \\ -\boldsymbol{\eta}^\dagger & \mathbf{0} \end{pmatrix} \exp\begin{pmatrix} \mathbf{0} & -\boldsymbol{\eta} \\ \boldsymbol{\eta}^\dagger & \mathbf{0} \end{pmatrix} = \boldsymbol{I}_{2r}, \tag{10.70b}$$

from which one recovers the orthogonal conditions for the coefficient matrices $\boldsymbol{U}$ and $\boldsymbol{V}$

$$\boldsymbol{U}^\dagger\boldsymbol{U} + \boldsymbol{V}^\dagger\boldsymbol{V} = \boldsymbol{I}_r, \boldsymbol{U}^\dagger\boldsymbol{V}^* + \boldsymbol{V}^\dagger\boldsymbol{U}^* = \mathbf{0}. \tag{10.71}$$

From Eq. (10.71), one also recovers that $\boldsymbol{\tau}$ is antisymmetric, which satisfies

$$\boldsymbol{\tau}^* = \boldsymbol{V}^*\boldsymbol{U}^{*-1} = -\boldsymbol{U}^{\dagger-1}\boldsymbol{V}^\dagger = -(\boldsymbol{V}\boldsymbol{U}^{-1})^\dagger = -\boldsymbol{\tau}^\dagger, \tag{10.72}$$

or equivalently $\boldsymbol{\tau}^\top = -\boldsymbol{\tau}$. As $\mathbb{T}^\dagger\mathbb{T} = \mathbb{T}\mathbb{T}^\dagger = \boldsymbol{I}_{2r}$, one may simply invert the matrix $\mathbb{T}^\dagger$ and obtain single-particle operators c_i and $c_i^\dagger$ from the quasi-particle operators b_i and $b_i^\dagger$

$$\begin{pmatrix} \boldsymbol{c} \\ \boldsymbol{c}^\dagger \end{pmatrix} = \mathbb{T} \begin{pmatrix} \boldsymbol{b} \\ \boldsymbol{b}^\dagger \end{pmatrix}, \mathbb{T} \equiv \begin{pmatrix} \cos\sqrt{\boldsymbol{\eta}\boldsymbol{\eta}^\dagger} & \boldsymbol{\eta}\frac{\sin\sqrt{\boldsymbol{\eta}^\dagger\boldsymbol{\eta}}}{\sqrt{\boldsymbol{\eta}^\dagger\boldsymbol{\eta}}} \\ -\frac{\sin\sqrt{\boldsymbol{\eta}^\dagger\boldsymbol{\eta}}}{\sqrt{\boldsymbol{\eta}^\dagger\boldsymbol{\eta}}}\boldsymbol{\eta}^\dagger & \cos\sqrt{\boldsymbol{\eta}^\dagger\boldsymbol{\eta}} \end{pmatrix}. \tag{10.73}$$

Remarkably, the coefficient matrix $\mathbb{T}$, which brings the quasi-particle operators b_i and $b_i^\dagger$ to the single-particle operators c_i and $c_i^\dagger$, is nothing but the faithful representation of the coset representative $\boldsymbol{\Omega}$. This fact reveals the real power of the coherent state method in many-body theory. The usage of coherent states greatly simplifies the computation of the Bogoliubov transformation where a large dynamical group is involved to a point that it cannot be further simplified, as the faithful representation of a group has the smallest dimension among all the other matrix representations.

Up to now, we have only investigated the formal mathematical structure of the Hartree-Fock-Bogoliubov theory and the corresponding Bogoliubov quasi-particle vacuum state $|\Omega\rangle$ using the coherent state method. In the following, we will derive an equation for the coefficients u_{ij} and v_{ij} which defines the Bogoliubov quasi-particles and the Bogoliubov quasi-particle vacuum state $|\Omega\rangle$.

We assume that the Bogoliubov quasi-particle vacuum state $|\Omega\rangle$ is a good variational ansatz for the exact ground state of the many-body Hamiltonian

$$\hat{H} = \sum_{ij} \epsilon_{ij} c_i^\dagger c_j + \frac{1}{4} \sum_{ijkl} g_{ijkl} c_i^\dagger c_j^\dagger c_l c_k, \tag{10.74}$$

and will derive the Hartree-Fock-Bogoliubov equations using the variation principle. However, one has to notice that the variation has to be restricted by the subsidiary condition that the expectation value of the particle number has the value N, i.e., $\langle\Omega|\hat{N}|\Omega\rangle = N$, which can be achieved by adding the term $-\lambda\hat{N}$ to the variational Hamiltonian $\hat{H}' = \hat{H} - \lambda\hat{N}$. Here, the Lagrange multiplier λ is also called the **chemical potential** or the **Fermi energy**, as it represents the increase of the energy $E = \langle\Omega|\hat{H}|\Omega\rangle$ for a change in the particle number, i.e., $\lambda = dE/dN$.

From the variational principle, $\delta\langle\Omega|\hat{H}'|\Omega\rangle = 0$, one can investigate small variations $|\delta\Omega\rangle$ in the vicinity of the variational ansatz $|\Omega\rangle$. As the first step, one needs **Thouless's theorem**, which states that any quasi-particle state $|\tilde{\Omega}\rangle \equiv |\Omega\rangle + |\delta\Omega\rangle$ of the Hartree-Fock-Bogoliubov type, which is not orthogonal to $|\Omega\rangle$, can be expressed in the form

$$|\tilde{\Omega}\rangle = \langle\Omega|\tilde{\Omega}\rangle \exp\left(\sum_{1\le i<j\le r} Z_{ij} b_i^\dagger b_j^\dagger \right) |\Omega\rangle. \tag{10.75}$$

In order to prove Thouless's theorem, one starts with two sets of Bogoliubov quasi-particle annihilation and creation operators, $\{b_j, b_j^\dagger\}$ and $\{\tilde{b}_j, \tilde{b}_j^\dagger\}$, which are associated to the two quasi-particle states $|\Omega\rangle$ and $|\tilde{\Omega}\rangle$, where

$$b_j^\dagger \equiv \sum_{1\le i\le r} (u_{ij} c_i^\dagger + v_{ij} c_i), \tilde{b}_j^\dagger \equiv \sum_{1\le i\le r} (\tilde{u}_{ij} c_i^\dagger + \tilde{v}_{ij} c_i). \tag{10.76}$$

If one expresses $\tilde{b}_j$ and $\tilde{b}_j^\dagger$ in terms of b_j and $b_j^\dagger$ as

$$\tilde{b}_j^\dagger = \sum_{1\le i\le r} (u'_{ij} b_i^\dagger + v'_{ij} b_i), \tilde{b}_j = \sum_{1\le i\le r} (u'^*_{ij} b_i + v'^*_{ij} b_i^\dagger), \tag{10.77}$$

one immediately obtains

$$\begin{aligned}\tilde{b}_j^\dagger &= \sum_{1\le i,k\le r} [u'_{ij}(u_{ki}c_k^\dagger + v_{ki}c_k) + v'_{ij}(u_{ki}^* c_k + v_{ki}^* c_k^\dagger)] \\ &= \sum_{1\le i,k\le r} [(u_{ki}u'_{ij} + v_{ki}^* v'_{ij})c_k^\dagger + (v_{ki}u'_{ij} + u_{ki}^* v'_{ij})c_k].\end{aligned} \tag{10.78}$$

A direct comparison between Eqs. (10.78) and (10.76) yields

$$\tilde{u}_{ij} = \sum_i (u_{ki}u'_{ij} + v_{ki}^* v'_{ij}),\ \tilde{v}_{ij} = \sum_i (v_{ki}u'_{ij} + u_{ki}^* v'_{ij}), \tag{10.79}$$

or equivalently

$$\tilde{\boldsymbol{U}} = \boldsymbol{U}\boldsymbol{U}' + \boldsymbol{V}^*\boldsymbol{V}',\ \tilde{\boldsymbol{V}} = \boldsymbol{V}\boldsymbol{U}' + \boldsymbol{U}^*\boldsymbol{V}', \tag{10.80}$$

where $\tilde{\boldsymbol{U}} \equiv (\tilde{u}_{ij})$, $\boldsymbol{U}' \equiv (u'_{ij})$, $\tilde{\boldsymbol{V}} \equiv (\tilde{v}_{ij})$, and $\boldsymbol{V}' \equiv (v'_{ij})$. Using Eqs. (10.71) and (10.80), one immediately obtains

$$\boldsymbol{U}' = \boldsymbol{U}^\dagger\tilde{\boldsymbol{U}} + \boldsymbol{V}^\dagger\tilde{\boldsymbol{V}},\ \boldsymbol{V}' = \boldsymbol{V}^\top\tilde{\boldsymbol{U}} + \boldsymbol{U}^\top\tilde{\boldsymbol{V}}. \tag{10.81}$$

As both sets of quasi-particle operators, $\{b_j, b_j^\dagger\}$ and $\{\tilde{b}_j, \tilde{b}_j^\dagger\}$, obey the anti-commutation relations, the transformation matrices $\boldsymbol{U}'$ and $\boldsymbol{V}'$ have to obey

$$\boldsymbol{U}'^\dagger\boldsymbol{U}' + \boldsymbol{V}'^\dagger\boldsymbol{V}' = \boldsymbol{I}_r,\ \boldsymbol{U}'^\dagger\boldsymbol{V}'^* + \boldsymbol{V}'^\dagger\boldsymbol{U}'^* = \boldsymbol{0}. \tag{10.82}$$

From the fact that the two quasi-particle states $|\tilde{\Omega}\rangle \equiv |\Omega\rangle + |\delta\Omega\rangle$ and $|\Omega\rangle$ are non-orthogonal, one may invert the matrix $\boldsymbol{U}'$ and obtain

$$\gamma_k^\dagger \equiv \sum_{1\le j\le r} U'^{-1}_{j,k}\tilde{b}_j^\dagger = b_k^\dagger + \sum_{1\le i\le r} (\boldsymbol{V}'\boldsymbol{U}'^{-1})_{i,k} b_i. \tag{10.83}$$

Let $Z_{ij}^* \equiv (\boldsymbol{V}'\boldsymbol{U}'^{-1})_{ij}$; Eq. (10.83) immediately yields

$$\gamma_k^\dagger = b_k^\dagger + \sum_{1\le i\le r} Z_{ik}^* b_i,\ \gamma_k = b_k + \sum_{1\le i\le r} Z_{ik} b_i^\dagger. \tag{10.84}$$

From Eq. (10.82), one can readily show that the coefficient matrix $\boldsymbol{Z} \equiv (Z_{ij})$ is antisymmetric, i.e., $\boldsymbol{Z}^\top = -\boldsymbol{Z}$. As the transformation equation (10.83) involves only the quasi-particle creation operators $\tilde{b}_j^\dagger$, the state $|\tilde{\Omega}\rangle$ is still the quasi-particle vacuum state with respect to the quasi-particle operators γ_k, i.e., $\gamma_k|\tilde{\Omega}\rangle = 0$ for all k. This condition will determine the state $|\tilde{\Omega}\rangle$ up to a normalization constant. Hence, one only needs to prove that

$$\gamma_k \exp(\sum_{i<j} Z_{ij} b_i^\dagger b_j^\dagger)|\Omega\rangle = 0, \tag{10.85}$$

which can be validated by a direct computation:

$$\begin{aligned}
&\exp(-\sum_{i<j} Z_{ij} b_i^\dagger b_j^\dagger)\gamma_k \exp(\sum_{i<j} Z_{ij} b_i^\dagger b_j^\dagger)|\Omega\rangle \\
&= \exp(-\sum_{i<j} Z_{ij} b_i^\dagger b_j^\dagger)(b_k + \sum_i Z_{ik} b_i^\dagger)\exp(\sum_{i<j} Z_{ij} b_i^\dagger b_j^\dagger)|\Omega\rangle \\
&= (b_k - \sum_i Z_{ik} b_i^\dagger + \sum_i Z_{ik} b_i^\dagger)|\Omega\rangle = b_k|\Omega\rangle = 0.
\end{aligned} \tag{10.86}$$

As the second step, one needs to express the many-body Hamiltonian equation (10.74) in terms of the quasi-particle operators b_i and $b_i^\dagger$. Using the inverse Bogoliubov transformation equation (10.57), the one-body terms become

$$\begin{aligned}
\sum_{ij} \epsilon_{ij} c_i^\dagger c_j &= \sum_{ijlm} \epsilon_{ij} (v_{il} b_l + u_{il}^* b_l^\dagger)(u_{jm} b_m + v_{jm}^* b_m^\dagger) \\
&= \mathrm{Tr}(\boldsymbol{\epsilon\rho}) + \sum_{lm} [\xi_{l,m} b_l^\dagger b_m + \frac{1}{2}(\zeta_{l,m} b_l^\dagger b_m^\dagger + \zeta_{l,m}^\dagger b_m b_l)],
\end{aligned} \tag{10.87}$$

where $\boldsymbol{\epsilon} \equiv (\epsilon_{ij})$ is a Hermitian matrix, i.e., $\boldsymbol{\epsilon}^\dagger = \boldsymbol{\epsilon}$, $\boldsymbol{\rho} \equiv \boldsymbol{V}^* \boldsymbol{V}^\top$, $\boldsymbol{\xi} \equiv \boldsymbol{U}^\dagger \boldsymbol{\epsilon} \boldsymbol{U} - \boldsymbol{V}^\dagger \boldsymbol{\epsilon}^\top \boldsymbol{V}$, and $\boldsymbol{\zeta} \equiv \boldsymbol{U}^\dagger \boldsymbol{\epsilon} \boldsymbol{V}^* - \boldsymbol{V}^\dagger \boldsymbol{\epsilon}^\top \boldsymbol{U}^*$. By definition, $\boldsymbol{\rho}$ is also a Hermitian matrix, i.e., $\boldsymbol{\rho}^\dagger = \boldsymbol{\rho}$.

In the many-body Hamiltonian equation (10.74), for the two-body terms with three quasi-particle creation operators $b_i^\dagger$ and one quasi-particle annihilation operator b_i, one obtains

$$\begin{aligned}
&\frac{1}{4}\sum_{IJ} g_{ijkl}(v_{im} u_{jn}^* v_{lp}^* v_{kq}^* b_m b_n^\dagger b_p^\dagger b_q^\dagger + u_{im}^* v_{jn} v_{lp}^* v_{kq}^* b_m^\dagger b_n b_p^\dagger b_q^\dagger) \\
&= \frac{1}{4}\sum_{IJ} (g_{ijkl} + g_{jilk}) v_{im} u_{jn}^* v_{lp}^* v_{kq}^* b_m b_n^\dagger b_p^\dagger b_q^\dagger - \frac{1}{2}\sum_{klpq} \Delta_{k,l}^* v_{lp}^* v_{kq}^* b_p^\dagger b_q^\dagger \\
&= \frac{1}{2}\sum_{IJ} g_{ijkl} v_{im} u_{jn}^* v_{lp}^* v_{kq}^* b_m b_n^\dagger b_p^\dagger b_q^\dagger + \frac{1}{2}\sum_{pq} (\boldsymbol{V}^\dagger \boldsymbol{\Delta} \boldsymbol{V}^*)_{q,p} b_q^\dagger b_p^\dagger,
\end{aligned} \tag{10.88}$$

where I and J are the index sets $\{1 \le i, j, k, l \le r\}$ and $\{1 \le m, n, p, q \le r\}$, respectively, and we have used the condition $g_{ijkl} = g_{jilk}$ in the last step. Here, the coefficient matrix $\boldsymbol{\Delta}$ is defined by

$$\boldsymbol{\Delta}_{i,j} \equiv \frac{1}{2}\sum_{kl} g_{ijkl}\boldsymbol{\kappa}_{k,l}, \tag{10.89}$$

and $\boldsymbol{\kappa} \equiv \boldsymbol{V}^*\boldsymbol{U}^\top$ is an antisymmetric matrix, i.e., $\boldsymbol{\kappa}^\top = -\boldsymbol{\kappa}$, which can be validated as follows:

$$(\boldsymbol{V}^*\boldsymbol{U}^\top)^\top = \boldsymbol{U}\boldsymbol{V}^\dagger = -\boldsymbol{U}^\dagger\boldsymbol{V}^* = \boldsymbol{V}^\dagger\boldsymbol{U}^* = -\boldsymbol{V}^*\boldsymbol{U}^\top, \tag{10.90}$$

where we have used the orthogonal condition $\boldsymbol{U}^\dagger\boldsymbol{V}^* + \boldsymbol{V}^\dagger\boldsymbol{U}^* = 0$ and the condition that the coefficient matrices $\boldsymbol{U}$ and $\boldsymbol{V}$ are Hermitian and antisymmetric, respectively, i.e., $\boldsymbol{U}^\dagger = \boldsymbol{U}$ and $\boldsymbol{V}^\top = -\boldsymbol{V}$. Equation (10.88) can also be expressed as

$$\begin{aligned}
&\frac{1}{4}\sum_{IJ} g_{ijkl}(v_{im}u^*_{jn}v^*_{lp}v^*_{kq}b_m b^\dagger_n b^\dagger_p b^\dagger_q + u^*_{im}v_{jn}v^*_{lp}v^*_{kq}b^\dagger_m b_n b^\dagger_p b^\dagger_q) \\
&= \frac{1}{4}\sum_{IJ}(g_{ijkl} - g_{jikl})u^*_{im}v_{jn}v^*_{lp}v^*_{kq}b^\dagger_m b_n b^\dagger_p b^\dagger_q + \frac{1}{2}\sum_{klpq}\boldsymbol{\Delta}^*_{k,l}v^*_{lp}v^*_{kq}b^\dagger_p b^\dagger_q \\
&= \frac{1}{2}\sum_{IJ} g_{ijkl}u^*_{im}v_{jn}v^*_{lp}v^*_{kq}b^\dagger_m b_n b^\dagger_p b^\dagger_q - \frac{1}{2}\sum_{pq}(\boldsymbol{V}^\dagger\boldsymbol{\Delta}^*\boldsymbol{V}^*)_{q,p}b^\dagger_q b^\dagger_p,
\end{aligned} \tag{10.91}$$

where we have used the antisymmetric condition $g_{jikl} = -g_{ijkl}$ in the last step. Using the commutation relation $[b^\dagger_i b_j, b^\dagger_k b^\dagger_l] = \delta_{jk}b^\dagger_i b^\dagger_l - \delta_{jl}b^\dagger_i b^\dagger_k$, one can express the two-body terms in Eq. (10.91) in the normal ordered form as

$$\begin{aligned}
&\sum_{IJ} g_{ijkl}u^*_{im}v_{jn}v^*_{lp}v^*_{kq}b^\dagger_m b_n b^\dagger_p b^\dagger_q \\
&= \sum_{IJ} g_{ijkl}u^*_{im}v_{jn}v^*_{lp}v^*_{kq}b^\dagger_p b^\dagger_q b^\dagger_m b_n + 2\sum_{ikmq}\boldsymbol{\Gamma}_{i,k}u^*_{im}v^*_{kq}b^\dagger_m b^\dagger_q \\
&= \sum_{IJ} g_{ijkl}u^*_{ip}v_{jq}v^*_{lm}v^*_{kn}b^\dagger_m b^\dagger_n b^\dagger_p b_q + \sum_{mq}(\boldsymbol{U}^\dagger\boldsymbol{\Gamma}\boldsymbol{V}^* - \boldsymbol{V}^\dagger\boldsymbol{\Gamma}^\top\boldsymbol{U}^*)_{m,q}b^\dagger_m b^\dagger_q,
\end{aligned} \tag{10.92}$$

where we have used the antisymmetric condition $g_{jilk} = -g_{ijkl}$ in the last second step. Here, the coefficient matrix $\boldsymbol{\Gamma}$ is defined by

$$\boldsymbol{\Gamma}_{i,k} \equiv \sum_{jl} g_{ijkl}(\boldsymbol{V}\boldsymbol{V}^\dagger)_{j,l}. \tag{10.93}$$

For the remaining two-body terms with three quasi-particle creation operators $b_i^\dagger$ and one quasi-particle annihilation operator b_i, one obtains

$$\frac{1}{4}\sum_{IJ} g_{ijkl}(u^*_{im}u^*_{jn}v^*_{lp}u_{kq}b^\dagger_m b^\dagger_n b^\dagger_p b_q + u^*_{im}u^*_{jn}u_{lp}v^*_{kq}b^\dagger_m b^\dagger_n b_p b^\dagger_q) \tag{10.94}$$

$$= \frac{1}{4}\sum_{IJ}(g_{ijkl} - g_{ijlk})u^*_{im}u^*_{jn}v^*_{lp}u_{kq}b^\dagger_m b^\dagger_n b^\dagger_p b_q + \frac{1}{2}\sum_{ijmn} \boldsymbol{\Delta}_{i,j}u^*_{im}u^*_{jn}b^\dagger_m b^\dagger_n$$

$$= \frac{1}{2}\sum_{IJ} g_{ijkl}u^*_{im}u^*_{jn}v^*_{lp}u_{kq}b^\dagger_m b^\dagger_n b^\dagger_p b_q + \frac{1}{2}\sum_{mn}(\boldsymbol{U}^\dagger \boldsymbol{\Delta} \boldsymbol{U}^*)_{m,n}b^\dagger_m b^\dagger_n,$$

where we have used the condition $g_{jilk} = -g_{ijkl}$ in the last step. Combining equations (10.91), (10.92), and (10.94), the four two-body terms with three quasi-particle creation operators $b_i^\dagger$ and one quasi-particle annihilation operator b_i can be expressed in the normal ordered form as

$$\frac{1}{4}\sum_{IJ} g_{ijkl}(u^*_{im}u^*_{jn}v^*_{lp}u_{kq}b^\dagger_m b^\dagger_n b^\dagger_p b_q + u^*_{im}u^*_{jn}u_{lp}v^*_{kq}b^\dagger_m b^\dagger_n b_p b^\dagger_q \tag{10.95}$$

$$+ v_{im}u^*_{jn}v^*_{lp}v^*_{kq}b_m b^\dagger_n b^\dagger_p b^\dagger_q + u^*_{im}v_{jn}v^*_{lp}v^*_{kq}b^\dagger_m b_n b^\dagger_p b^\dagger_q)$$

$$= \frac{1}{2}\sum_{IJ} g_{ijkl}(u^*_{im}u^*_{jn}v^*_{lp}u_{kq} + u^*_{im}v_{jn}v^*_{lp}v^*_{kq})b^\dagger_m b^\dagger_n b^\dagger_p b_q$$

$$+ \frac{1}{2}\sum_{mn}(\boldsymbol{U}^\dagger\boldsymbol{\Gamma}\boldsymbol{V}^* - \boldsymbol{V}^\dagger\boldsymbol{\Gamma}^\top\boldsymbol{U}^* + \boldsymbol{U}^\dagger\boldsymbol{\Delta}\boldsymbol{U}^* - \boldsymbol{V}^\dagger\boldsymbol{\Delta}^*\boldsymbol{V}^*)_{m,n}b^\dagger_m b^\dagger_n.$$

Similarly, for the first pair of two-body terms with two quasi-particle creation operators b_i and two quasi-particle annihilation operators $b_i^\dagger$, one obtains

$$\sum_{IJ} g_{ijkl}(u^*_{im}u^*_{jn}u_{lp}u_{kq}b^\dagger_m b^\dagger_n b_p b_q + v_{im}v_{jn}v^*_{lp}v^*_{kq}b_m b_n b^\dagger_p b^\dagger_q) \tag{10.96}$$

$$= \sum_{IJ} g_{ijkl}(u^*_{im}u^*_{jn}u_{lp}u_{kq}b^\dagger_m b^\dagger_n b_p b_q + v_{im}v_{jn}v^*_{lp}v^*_{kq}b^\dagger_p b^\dagger_q b_m b_n)$$

$$+ \sum_{ijklnp}(g_{ijlk} + g_{jikl} - g_{jilk} - g_{ijkl})(\boldsymbol{V}\boldsymbol{V}^\dagger)_{ik}v_{jn}v^*_{lp}b^\dagger_p b_n + 2(\boldsymbol{\rho}, \boldsymbol{\rho})$$

$$= \sum_{IJ} g_{ijkl}(u^*_{im}u^*_{jn}u_{lp}u_{kq} + v^*_{lm}v^*_{kn}v_{ip}v_{jq})b^\dagger_m b^\dagger_n b_p b_q$$

$$- 4\sum_{np}(\boldsymbol{V}^\dagger\boldsymbol{\Gamma}^\top\boldsymbol{V})_{p,n}b^\dagger_p b_n + 2(\boldsymbol{\rho}, \boldsymbol{\rho}),$$

where $(\boldsymbol{\rho}, \boldsymbol{\rho}) \equiv \sum_I g_{ijkl}\rho_{ik}\rho_{jl}$. Here we have used the $\mathfrak{so}(2r)$ commutation relation in the last second step and the antisymmetric conditions $g_{jikl} = g_{ijlk} = -g_{jilk}$ in the last step. For the second pair of the two-body terms with two quasi-particle creation operators b_i and two quasi-particle annihilation operators $b_i^\dagger$, one obtains

$$\begin{aligned}
&\sum_{IJ} g_{ijkl}(u^*_{im}v_{jn}v^*_{lp}u_{kq}b^\dagger_m b_n b^\dagger_p b_q + u^*_{im}v_{jn}u_{lp}v^*_{kq}b^\dagger_m b_n b_p b^\dagger_q) \qquad (10.97)\\
&= \sum_{IJ} g_{ijkl}(u^*_{im}v_{jn}u_{lp}v^*_{kq}b^\dagger_m b^\dagger_q b_n b_p - u^*_{im}v_{jn}v^*_{lp}u_{kq}b^\dagger_m b^\dagger_p b_n b_q)\\
&\quad + \sum_{ijklmn}[(g_{ijkl} - g_{ijlk})(\boldsymbol{V}\boldsymbol{V}^\dagger)_{jl}u^*_{im}u_{kn} + g_{ijkl}(\boldsymbol{V}^*\boldsymbol{U}^\top)_{kl}u^*_{im}v_{jn}]b^\dagger_m b_n\\
&= \sum_{IJ} g_{ijkl}(u^*_{im}v^*_{kn}v_{jp}u_{lq} - u^*_{im}v^*_{ln}v_{jp}u_{kq})b^\dagger_m b^\dagger_n b_p b_q\\
&\quad + 2\sum_{mn}(\boldsymbol{U}^\dagger\boldsymbol{\Gamma}\boldsymbol{U} + \boldsymbol{U}^\dagger\boldsymbol{\Delta}\boldsymbol{V})_{m,n}b^\dagger_m b_n,
\end{aligned}$$

where we have used the fermionic $\mathfrak{so}(2r)$ commutation relation $[E_{np}, b_q^\dagger] = \delta_{pq}b_n - \delta_{nq}b_p$ with $E_{np} \equiv b_n b_p$ in the last second step and the antisymmetric condition $g_{ijlk} = -g_{ijkl}$ in both the second last step and the last step. For the last pair of the two-body terms with two quasi-particle creation operators b_i and two quasi-particle annihilation operators $b_i^\dagger$, one obtains

$$\begin{aligned}
&\sum_{IJ} g_{ijkl}(v_{im}u^*_{jn}v^*_{lp}u_{kq}b_m b^\dagger_n b^\dagger_p b_q + v_{im}u^*_{jn}u_{lp}v^*_{kq}b_m b^\dagger_n b_p b^\dagger_q) \qquad (10.98)\\
&= \sum_{IJ} g_{ijkl}(u^*_{jm}v^*_{ln}v_{ip}u_{kq} - u^*_{jm}v^*_{kn}v_{ip}u_{lq})b^\dagger_m b^\dagger_n b_p b_q\\
&\quad + 2\sum_{mn}(\boldsymbol{U}^\dagger\boldsymbol{\Gamma}\boldsymbol{U} + \boldsymbol{U}^\dagger\boldsymbol{\Delta}\boldsymbol{V} - 2\boldsymbol{V}^\dagger\boldsymbol{\Delta}^*\boldsymbol{U})_{m,n}b^\dagger_m b_n + \sum_I g_{ijkl}\kappa^*_{ij}\kappa_{kl},
\end{aligned}$$

Combining equations (10.96), (10.97), and (10.98), the two-body terms with two quasi-particle creation operators and two quasi-particle annihilation operators can be expressed in the normal ordered form as

$$\begin{aligned}
&\sum_{IJ} g_{ijkl}(u^*_{im}u^*_{jn}u_{lp}u_{kq} + u^*_{im}v^*_{kn}v_{jp}u_{lq} - u^*_{im}v^*_{ln}v_{jp}u_{kq} \qquad (10.99)\\
&\quad + u^*_{jm}v^*_{ln}v_{ip}u_{kq} - u^*_{jm}v^*_{kn}v_{ip}u_{lq} + v^*_{lm}v^*_{kn}v_{ip}v_{jq})b^\dagger_m b^\dagger_n b_p b_q\\
&\quad + 4\sum_{mn}(\boldsymbol{U}^\dagger\boldsymbol{\Gamma}\boldsymbol{U} - \boldsymbol{V}^\dagger\boldsymbol{\Gamma}^\top\boldsymbol{V} + \boldsymbol{U}^\dagger\boldsymbol{\Delta}\boldsymbol{V} - \boldsymbol{V}^\dagger\boldsymbol{\Delta}^*\boldsymbol{U})_{m,n}b^\dagger_m b_n + (\boldsymbol{\kappa}, \boldsymbol{\kappa}),
\end{aligned}$$

where $(\boldsymbol{\kappa}, \boldsymbol{\kappa}) \equiv \sum_I \kappa^*_{ij} g_{ijkl} \kappa_{kl}$. Notice that $g^*_{ijkl} = g_{klij} = g_{lkji}$ as a result of the Hermiticity of the Hamiltonian equation (10.74) and the anti-commutation relations between the fermionic operator c_i and $c^\dagger_j$. Combining equations (10.87), (10.95), and (10.99), the many-body Hamiltonian equation (10.74) can be expressed in the normal ordered form in terms of the quasi-particle operators

$$\begin{aligned}\hat{H} &= \mathrm{Tr}(\boldsymbol{\epsilon}\boldsymbol{\rho}) + \frac{1}{2}(\boldsymbol{\rho}, \boldsymbol{\rho}) + \frac{1}{4}(\boldsymbol{\kappa}, \boldsymbol{\kappa}) \\ &+ \sum_{mn} [\xi'_{m,n} b^\dagger_m b_n + \frac{1}{2}(\zeta'_{m,n} b^\dagger_m b^\dagger_n + \text{h.c.})] \\ &+ \sum_{mnpq} (H^{22}_{mnpq} b^\dagger_m b^\dagger_n b_p b_q + H^{31}_{mnpq} b^\dagger_m b^\dagger_n b^\dagger_p b_q + H^{40}_{mnpq} b^\dagger_m b^\dagger_n b^\dagger_p b^\dagger_q + \text{h.c.}),\end{aligned} \tag{10.100}$$

where $\boldsymbol{\xi}' \equiv \boldsymbol{U}^\dagger \boldsymbol{\Gamma} \boldsymbol{U} - \boldsymbol{V}^\dagger \boldsymbol{\Gamma}^\top \boldsymbol{V} + \boldsymbol{U}^\dagger \boldsymbol{\Delta} \boldsymbol{V} - \boldsymbol{V}^\dagger \boldsymbol{\Delta}^* \boldsymbol{U} + \boldsymbol{\xi}$, $\boldsymbol{\zeta}' \equiv \boldsymbol{U}^\dagger \boldsymbol{\Gamma} \boldsymbol{V}^* - \boldsymbol{V}^\dagger \boldsymbol{\Gamma}^\top \boldsymbol{U}^* + \boldsymbol{U}^\dagger \boldsymbol{\Delta} \boldsymbol{U}^* - \boldsymbol{V}^\dagger \boldsymbol{\Delta} \boldsymbol{V}^* + \boldsymbol{\zeta}$, and

$$\begin{aligned}H^{22}_{mnpq} &\equiv \frac{1}{4}\sum_I g_{ijkl}(u^*_{im}u^*_{jn}u_{lp}u_{kq} + u^*_{im}v^*_{kn}v_{jp}u_{lq} - u^*_{im}v^*_{ln}v_{jp}u_{kq} \\ &+ u^*_{jm}v^*_{ln}v_{ip}u_{kq} - u^*_{jm}v^*_{kn}v_{ip}u_{lq} + v^*_{lm}v^*_{kn}v_{ip}v_{jq}), \\ H^{31}_{mnpq} &\equiv \frac{1}{2}\sum_I g_{ijkl}(u^*_{im}u^*_{jn}v^*_{lp}u_{kq} + u^*_{im}v_{jn}v^*_{lp}v^*_{kq}), \\ H^{40}_{mnpq} &\equiv \frac{1}{4}\sum_I g_{ijkl}u^*_{im}u^*_{jn}v^*_{lp}v^*_{kq}.\end{aligned} \tag{10.101}$$

As the last step, one can now apply the variational principle to the expectation of the many-body Hamiltonian with respect to the trial quasi-particle state $|\tilde{\Omega}\rangle$. For infinitesimal variations, one can expand the exponential functions in Eq. (10.75) up to the second order in Z_{ij} and obtain

$$\begin{aligned}\frac{\langle\tilde{\Omega}|\hat{H}|\tilde{\Omega}\rangle}{\langle\tilde{\Omega}|\tilde{\Omega}\rangle} &= \langle\Omega| \exp\left(\sum_{i<j} Z^*_{ij} b_j b_i\right) \hat{H} \exp\left(\sum_{i<j} Z_{ij} b^\dagger_i b^\dagger_j\right) |\Omega\rangle \\ &= \langle\Omega|\hat{H} + \sum_{i<j}(Z^*_{ij} b_j b_i \hat{H} + Z_{ij}\hat{H} b^\dagger_i b^\dagger_j) + \frac{1}{2}\sum_{i<j,l<m} Z^*_{ij} Z^*_{lm} b_j b_i b_m b_l \hat{H} \\ &+ \sum_{i<j,l<m} (Z^*_{ij} Z_{lm} b_j b_i \hat{H} b^\dagger_l b^\dagger_m + \frac{1}{2} Z_{ij} Z_{lm} \hat{H} b^\dagger_i b^\dagger_j b^\dagger_l b^\dagger_m)|\Omega\rangle.\end{aligned} \tag{10.102}$$

Here, in accordance to the Hartree-Fock-Bogoliubov theory, $\hat{H}$ is chosen to be the truncated many-body Hamiltonian of Eq. (10.100), of which all the higher than second-order terms expressed in the normal ordered form in terms of the quasi-particle operators b_i and $b_i^\dagger$ are neglected:

$$\hat{H} \equiv H^0 + \sum_{\alpha\beta} \left(H^{11}_{\alpha\beta} b^\dagger_\alpha b_\beta + \frac{1}{2} H^{20}_{\alpha\beta} b^\dagger_\alpha b^\dagger_\beta + \frac{1}{2} H^{02}_{\alpha\beta} b_\beta b_\alpha \right), \tag{10.103}$$

where $H^{02}_{\alpha\beta} = H^{20*}_{\beta\alpha}$ and the coefficient matrices $\boldsymbol{H}^{11} \equiv (H^{11}_{\alpha\beta})$ and $\boldsymbol{H}^{20} \equiv (H^{20}_{\alpha\beta})$ are determined by

$$\begin{pmatrix} \boldsymbol{H}^{11} & \boldsymbol{H}^{20} \\ -\boldsymbol{H}^{20*} & -\boldsymbol{H}^{11*} \end{pmatrix} \equiv \mathbb{T}^\dagger \begin{pmatrix} \boldsymbol{h} & \boldsymbol{\Delta} \\ -\boldsymbol{\Delta}^* & -\boldsymbol{h}^* \end{pmatrix} \mathbb{T}, \tag{10.104a}$$

$$\mathbb{T}^\dagger \equiv \begin{pmatrix} \boldsymbol{U}^\dagger & \boldsymbol{V}^\dagger \\ \boldsymbol{V}^\top & \boldsymbol{U}^\top \end{pmatrix} = \exp \begin{pmatrix} \boldsymbol{0} & -\boldsymbol{\eta} \\ \boldsymbol{\eta}^\dagger & \boldsymbol{0} \end{pmatrix}, \tag{10.104b}$$

$$\mathbb{T} \equiv \begin{pmatrix} \boldsymbol{U} & \boldsymbol{V}^* \\ \boldsymbol{V} & \boldsymbol{U}^* \end{pmatrix} = \exp \begin{pmatrix} \boldsymbol{0} & \boldsymbol{\eta} \\ -\boldsymbol{\eta}^\dagger & \boldsymbol{0} \end{pmatrix}. \tag{10.104c}$$

As $|\Omega\rangle$ is the quasi-particle vacuum state, which is annihilated by all the quasi-particle annihilation operators b_i, i.e., $b_i|\Omega\rangle = 0$, the nonzero contribution from the term $\langle\Omega|b_j b_i b_m b_l \hat{H}|\Omega\rangle$ should come from

$$\begin{aligned} \sum_{\alpha\beta} H^{20}_{\alpha\beta} \langle\Omega|b_j b_i b_m b_l b^\dagger_\alpha b^\dagger_\beta|\Omega\rangle &= \sum_{\alpha\beta} H^{20}_{\alpha\beta} \langle\Omega|b_j b_i b^\dagger_\alpha b^\dagger_\beta b_m b_l|\Omega\rangle \\ &+ \sum_{\alpha\beta} H^{20}_{\alpha\beta} \langle\Omega|b_j b_i (\delta_{m\alpha} E^\beta_l - \delta_{l\alpha} E^\beta_m + \delta_{l\beta} E^\alpha_m - \delta_{m\beta} E^\alpha_l)|\Omega\rangle \\ &= -\sum_{\alpha\beta} H^{20}_{\alpha\beta} (\delta_{m\alpha}\delta_{\beta l} - \delta_{l\alpha}\delta_{\beta m}) \langle\Omega|b_j b_i|\Omega\rangle = 0, \end{aligned} \tag{10.105}$$

where we have used the fermionic $\mathfrak{so}(2r)$ commutation relation $[E_{\beta\alpha}, E^{ij}] = \delta_{\beta i} E^j_\alpha - \delta_{\alpha i} E^j_\beta + \delta_{\alpha j} E^i_\beta - \delta_{\beta j} E^i_\alpha$ with $E_{\beta\alpha} \equiv b_\beta b_\alpha$, $E^{ij} \equiv b^\dagger_i b^\dagger_j$, and $E^i_\alpha \equiv b^\dagger_i b_\alpha - \frac{1}{2}\delta_{i\alpha}$ in the last second step. Similarly, the term $\langle\Omega|\hat{H} b^\dagger_i b^\dagger_j b^\dagger_l b^\dagger_m|\Omega\rangle$ that appeared in the expectation of the truncated Hamiltonian with respect to the trial quasi-particle state $|\tilde{\Omega}\rangle$ vanishes identically. Notice that without the truncation of the full Hamiltonian, both the terms $\langle\Omega|b_j b_i b_m b_l \hat{H}|\Omega\rangle$ and $\langle\Omega|\hat{H} b^\dagger_i b^\dagger_j b^\dagger_l b^\dagger_m|\Omega\rangle$ are nonvanishing and have to be taken into account in evaluating the expectation $\langle\tilde{\Omega}|\hat{H}|\tilde{\Omega}\rangle$. For the other terms appearing in the expectation of the truncated Hamiltonian with respect to the trial quasi-particle state $|\tilde{\Omega}\rangle$, a direct computation yields

$$\langle\Omega|\hat{H}b_i^\dagger b_j^\dagger|\Omega\rangle = \sum_{\alpha\beta} H_{\alpha\beta}^{02}\langle\Omega|b_\beta b_\alpha b_i^\dagger b_j^\dagger|\Omega\rangle \tag{10.106}$$

$$= \sum_{\alpha\beta} H_{\alpha\beta}^{02}\langle\Omega|b_i^\dagger b_j^\dagger b_\beta b_\alpha + \delta_{\beta i}E_\alpha^j - \delta_{\alpha i}E_\beta^j + \delta_{\alpha j}E_\beta^i - \delta_{\beta j}E_\alpha^i|\Omega\rangle$$

$$= \sum_{\alpha\beta} H_{\alpha\beta}^{02}(\delta_{i\alpha}\delta_{j\beta} - \delta_{i\beta}\delta_{j\alpha}),$$

where we have used the fermionic $\mathfrak{so}(2r)$ commutation relation $[E_{\beta\alpha}, E^{ij}] = \delta_{\beta i}E_\alpha^j - \delta_{\alpha i}E_\beta^j + \delta_{\alpha j}E_\beta^i - \delta_{\beta j}E_\alpha^i$ again in the last second step. Similarly, a direct computation yields

$$\langle\Omega|b_j b_i \hat{H}|\Omega\rangle = \sum_{\alpha\beta} H_{\alpha\beta}^{20}(\delta_{i\alpha}\delta_{j\beta} - \delta_{i\beta}\delta_{j\alpha}), \tag{10.107}$$

and

$$\langle\Omega|b_j b_i \hat{H} b_l^\dagger b_m^\dagger|\Omega\rangle = \langle\Omega|H^0 b_j b_i b_l^\dagger b_m^\dagger + \sum_{\alpha\beta} H_{\alpha\beta}^{11} b_j b_i b_\alpha^\dagger b_\beta b_l^\dagger b_m^\dagger|\Omega\rangle \tag{10.108}$$

$$= \langle\Omega|H^0 b_j b_i b_l^\dagger b_m^\dagger + \sum_{\alpha\beta} H_{\alpha\beta}^{11}(\delta_{i\alpha}b_j - \delta_{j\alpha}b_i)(\delta_{l\beta}b_m^\dagger - \delta_{m\beta}b_l^\dagger)|\Omega\rangle$$

$$= H^0(\delta_{il}\delta_{jm} - \delta_{jl}\delta_{im}) + \sum_{\alpha\beta} H_{\alpha\beta}^{11}[(\delta_{i\alpha}\delta_{jm} - \delta_{j\alpha}\delta_{im})\delta_{l\beta} - (l \leftrightarrow m)],$$

where we have used the relation $\langle\Omega|b_j b_i b_\alpha^\dagger b_\beta^\dagger b_l^\dagger b_m^\dagger|\Omega\rangle = \langle\Omega|b_j b_i b_\beta b_\alpha b_l^\dagger b_m^\dagger|\Omega\rangle = 0$ in the first step and have used two other fermionic $\mathfrak{so}(2r)$ commutation relations $[E_{ji}, b_\alpha^\dagger] = \delta_{i\alpha}b_j - \delta_{j\alpha}b_i$ and $[b_\beta, E^{lm}] = \delta_{l\beta}b_m^\dagger - \delta_{m\beta}b_l^\dagger$ in the last second step. Combining equations (10.106), (10.107), and (10.108), the expectation of the many-body Hamiltonian $\hat{H}' \equiv \hat{H} - \lambda\hat{N}$ with respect to the trial quasi-particle state $|\tilde{\Omega}\rangle$, up to the second order in Z_{ij}, has the form

$$\frac{\langle\tilde{\Omega}|\hat{H}'|\tilde{\Omega}\rangle}{\langle\tilde{\Omega}|\tilde{\Omega}\rangle} = H^0 + \sum_{\alpha\beta}(H_{\alpha\beta}^{02}Z_{\alpha\beta} + H_{\alpha\beta}^{20}Z_{\alpha\beta}^*) \tag{10.109}$$

$$+ \left(\frac{H^0}{2} - \lambda\right)\sum_{\alpha\beta} Z_{\alpha\beta}^* Z_{\alpha\beta} - \sum_{\alpha\beta\gamma} Z_{\gamma\alpha}^* H_{\alpha\beta}^{11} Z_{\beta\gamma}.$$

Applying the variation principle, the variation of Eq. (10.109) with respect to the independent variables $Z^*_{\alpha\beta}$ and $Z_{\alpha\beta}$ yields

$$\frac{\partial}{\partial Z^*_{\alpha\beta}} \frac{\langle\tilde{\Omega}|\hat{H}'|\tilde{\Omega}\rangle}{\langle\tilde{\Omega}|\tilde{\Omega}\rangle}\bigg|_{Z_{\alpha\beta}=0} = H^{20}_{\alpha\beta} = 0, \tag{10.110a}$$

$$\frac{\partial}{\partial Z_{\alpha\beta}} \frac{\langle\tilde{\Omega}|\hat{H}'|\tilde{\Omega}\rangle}{\langle\tilde{\Omega}|\tilde{\Omega}\rangle}\bigg|_{Z^*_{\alpha\beta}=0} = H^{02}_{\alpha\beta} = 0, \tag{10.110b}$$

which implies that all the linear terms in Eq. (10.109) vanish at the stationary point. However, since the variational equations $H^{20}_{\alpha\beta} = H^{02}_{\alpha\beta} = 0$ themselves do not completely fix the Bogoliubov transformation coordinates $\boldsymbol{U}$ and $\boldsymbol{V}$, one can require, in addition to the variational equations $H^{20}_{\alpha\beta} = H^{02}_{\alpha\beta} = 0$, that the $\boldsymbol{H}^{11}$ matrix involved in the truncated many-body Hamiltonian equation (10.103) is diagonalized. Hence, using Eqs. (10.104a) and (10.104c), one can readily obtain

$$\mathbb{T}\begin{pmatrix} \boldsymbol{E}' & 0 \\ 0 & -\boldsymbol{E}' \end{pmatrix} = \begin{pmatrix} \boldsymbol{h} & \boldsymbol{\Delta} \\ -\boldsymbol{\Delta}^* & -\boldsymbol{h}^* \end{pmatrix}\mathbb{T}, \tag{10.111}$$

where $\boldsymbol{E}' \equiv \boldsymbol{E} - \lambda \boldsymbol{I}_r$ and $\boldsymbol{E}_{\alpha,\beta} \equiv E_\alpha \delta_{\alpha\beta}$. Equation (10.111) is equivalent to

$$\begin{pmatrix} \boldsymbol{U}\boldsymbol{E}' & -\boldsymbol{V}^*\boldsymbol{E}' \\ \boldsymbol{V}\boldsymbol{E}' & -\boldsymbol{U}^*\boldsymbol{E}' \end{pmatrix} = \begin{pmatrix} \boldsymbol{h} & \boldsymbol{\Delta} \\ -\boldsymbol{\Delta}^* & -\boldsymbol{h}^* \end{pmatrix}\begin{pmatrix} \boldsymbol{U} & \boldsymbol{V}^* \\ \boldsymbol{V} & \boldsymbol{U}^* \end{pmatrix}, \tag{10.112}$$

which will lead to the Hartree-Fock-Bogoliubov equation for the Bogoliubov transformation coordinates $\boldsymbol{U}$ and $\boldsymbol{V}$

$$\begin{pmatrix} \boldsymbol{h} & \boldsymbol{\Delta} \\ -\boldsymbol{\Delta}^* & -\boldsymbol{h}^* \end{pmatrix}\begin{pmatrix} \boldsymbol{U} \\ \boldsymbol{V} \end{pmatrix} = \begin{pmatrix} \boldsymbol{U} \\ \boldsymbol{V} \end{pmatrix}\boldsymbol{E}'. \tag{10.113}$$

where $\boldsymbol{h} \equiv \boldsymbol{\epsilon} + \boldsymbol{\Gamma}$. As both $\boldsymbol{\epsilon}$ and $\boldsymbol{\Gamma}$ are Hermitian matrices, i.e., $\boldsymbol{\epsilon}^\dagger = \boldsymbol{\epsilon}$ and $\boldsymbol{\Gamma}^\dagger = \boldsymbol{\Gamma}$, $\boldsymbol{h}$ is also Hermitian. Here, the matrices $\boldsymbol{\Gamma}$ and $\boldsymbol{\Delta}$ are explicitly determined by

$$\boldsymbol{\Gamma}_{i,k} \equiv \sum_{jl} g_{ijkl}(\boldsymbol{V}\boldsymbol{V}^\dagger)_{j,l}, \; \boldsymbol{\Delta}_{i,j} \equiv \frac{1}{2}\sum_{kl} g_{ijkl}(\boldsymbol{V}^*\boldsymbol{U}^\top)_{k,l}, \tag{10.114}$$

where $\boldsymbol{\Delta}$ is an antisymmetric matrix, i.e., $\boldsymbol{\Delta}^\top = -\boldsymbol{\Delta}$.

As an immediate application, one can write down the variational ansatz for the ground state of a superconductor, in which only particular fermionic pairs, i.e., Cooper pairs $c^\dagger_{k\uparrow}c^\dagger_{-k\downarrow}$, are considered

$$|\mathrm{BCS}\rangle = \prod_{k=1}^{r} \frac{1}{\sqrt{1+|\tau_k|^2}} \exp\left\{\sum_{k=1}^{r} \tau_k c^\dagger_{k\uparrow}c^\dagger_{-k\downarrow}\right\}|\boldsymbol{0}\rangle. \tag{10.115}$$

If one introduces the coordinates $u_k \equiv 1/\sqrt{1+|\tau_k|^2}$ and $v_k \equiv \tau_k/\sqrt{1+|\tau_k|^2}$, one immediately obtains the original form of the variational ground state proposed by Bardeen, Cooper, and Schrieffer [130]

$$|\mathrm{BCS}\rangle = \prod_{k>0}(u_k + v_k c^\dagger_{k\uparrow} c^\dagger_{-k\downarrow})|\mathbf{0}\rangle. \tag{10.116}$$

At first sight, it seems that the variational ansatz equation (10.115) of the superconducting ground state is quite arbitrary. But one can readily show that the fermionic pairing operators $\{c^\dagger_{k\uparrow}c^\dagger_{-k\downarrow}, c_{-k\downarrow}c_{k\uparrow}, c^\dagger_{k\uparrow}c_{k\uparrow} + c^\dagger_{-k\downarrow}c_{-k\downarrow}\}$ span a $\mathfrak{su}(2)$ algebra, where the pairing operators are mapped to the $\mathfrak{su}(2)$ operators via

$$c^\dagger_{k\uparrow}c^\dagger_{-k\downarrow} \Leftrightarrow J^k_+, c_{-k\downarrow}c_{k\uparrow} \Leftrightarrow J^k_-, c^\dagger_{k\uparrow}c_{k\uparrow} + c^\dagger_{-k\downarrow}c_{-k\downarrow} \Leftrightarrow 2J^k_z + 1, \tag{10.117}$$

and $J^k_\pm$ and J^k_z obey the standard $\mathfrak{su}(2)$ commutation relations $[J^k_+, J^k_-] = 2J^k_z$ and $[J^k_z, J^k_\pm] = \pm J^k_\pm$. Hence, the Bardeen-Cooper-Schrieffer (BCS) variational ground state, Eq. (10.115), is simply a product of $\mathfrak{su}(2)$ coherent states defined on the coset space $\mathrm{SU}(2)^{\otimes r}/\mathrm{U}(1)^{\otimes r}$. In other words, the BCS ground state of a superconductor is a special case of the Bogoliubov quasi-particle vacuum state defined on the coset space $\mathrm{SO}(2r)/\mathrm{U}(r)$.

Exercises

10.1. For a system with a single fermion, use the relation $f^\dagger f|n\rangle = n|n\rangle$ to verify the results

$$f|0\rangle = 0,\ f|1\rangle = |0\rangle,\ f^\dagger|0\rangle = |1\rangle,\ f^\dagger|1\rangle = 0.$$

10.2. Verify Eqs. (10.2) and (10.3).

10.3. Verify Eq. (10.5) by using the spin-1/2 representation for a single fermion.

10.4. Verity Eqs. (10.7) and (10.8).

10.5. Verify that the operators $f^\dagger_i f_j$ with $1 \le i \le k$ and $k+1 \le j \le r$ span a subalgebra $\mathfrak{u}(k) \oplus \mathfrak{u}(r-k)$ of $\mathfrak{u}(r)$.

10.6. Verify by direct computation that the faithful matrix representation of $E^i_{\ j}$, E^{ij}, and E_{ij} forms the Lie algebra $\mathfrak{so}(2r)$, where

$$E^i_{\ j} \Leftrightarrow E_{i,j} - E_{r+j,r+i},$$

$$E^{ij} \Leftrightarrow E_{i,r+j} - E_{j,r+i},$$

$$E_{ij} \Leftrightarrow E_{r+i,j} - E_{r+j,i}.$$

10.7. Verify Eq. (10.87).

10.8. Verify by direct computation that

$$\langle\Omega|b_j b_i \hat{H}|\Omega\rangle = \sum_{\alpha\beta} H^{20}_{\alpha\beta}(\delta_{i\alpha}\delta_{j\beta} - \delta_{i\beta}\delta_{j\alpha}),$$

10.9. Using Eqs. (10.106) and (10.107), verify the following relations:

$$\langle\Omega|\sum_{i<j} Z_{ij}\hat{H}b_i^\dagger b_j^\dagger|\Omega\rangle = \sum_{\alpha\beta} H^{02}_{\alpha\beta}Z_{\alpha\beta},$$

$$\langle\Omega|\sum_{i<j} Z^*_{ij}b_j b_i \hat{H}|\Omega\rangle = \sum_{\alpha\beta} H^{20}_{\alpha\beta}Z^*_{\alpha\beta}.$$

10.10. Using Eq. (10.108), verify the following relation:

$$\langle\Omega| \sum_{i<j,l<m} Z^*_{ij}Z_{lm}b_j b_i \hat{H}b_l^\dagger b_m^\dagger|\Omega\rangle = \frac{H^0}{2}\sum_{\alpha\beta} Z^*_{\alpha\beta}Z_{\alpha\beta} - \sum_{\alpha\beta\gamma} Z^*_{\gamma\alpha}H^{11}_{\alpha\beta}Z_{\beta\gamma}.$$

10.11. Verify by direct computation that

$$\lambda\langle\Omega| \sum_{i<j,l<m} Z^*_{ij}Z_{lm}b_j b_i \hat{N}b_l^\dagger b_m^\dagger|\Omega\rangle = \lambda\sum_{\alpha\beta} Z^*_{\alpha\beta}Z_{\alpha\beta},$$

where $\hat{N} \equiv \sum_\alpha b_\alpha^\dagger b_\alpha$.

10.12. Using Eq. (10.115) to verify Eq. (10.116).

Quantum Phase Transitions 11

11.1 Lieb-Berezin Inequality and Large N Limit

Traditionally, since Glauber's groundbreaking work on coherent states and optical coherence theory, the concept of coherent states has been tightly connected with quantum optics. But in the last chapter, we have shown that the Gilmore-Perelomov group-theoretic coherent states could play a fundamental role in many-body physics, which are essential to the variational calculations in the Hartree-Fock-Bogoliubov theory in quantum chemistry and the theory of superconductivity. Furthermore, in equilibrium statistical mechanics, it is well known that the partition function is the fundamental physical quantity, in which once the partition function is known, one can immediately obtain all the thermodynamic quantities. To this end, one may wonder whether the Gilmore-Perelomov group-theoretic coherent states may also be useful in evaluating quantum partition function.

Any direct evaluation of the partition function is known to be exceedingly difficult, if not possible. However, fortunately and interestingly, instead of directly evaluating the quantum partition functions, Elliott H. Lieb had pioneered in 1973 that one can rigorously obtain the upper and lower bounds of the quantum partition functions for spin systems by using the spin coherent states [38]. Remarkably, a year earlier, Felix A. Berezin was able to derive the same inequality by using the covariant and contravariant symbols related to coherent states [39]. Barry Simon was able to generalize Lieb's ingenious result [131] in 1980 by using the Gilmore-Perelomov group-theoretic coherent states to arbitrary quantum systems possessing a compact dynamical symmetry group. In this regard, as it is in general exceedingly difficult to solve the exact quantum partition function, the advantage of using coherent states in statistical mechanics is manifested by such an approach. In fact, in this context, one can obtain not only reasonable but also rigorous estimate of the quantum partition function in the thermodynamic limit.

Besides the usage in the Lieb-Berezin inequalities, the coherent states are also indispensable in the imaginary time path integrals formalism of quantum statistical

C.-F. Kam et al., *Coherent States*, Lecture Notes in Physics 1011,
https://doi.org/10.1007/978-3-031-20766-2_11

mechanics, as the Gilmore-Perelomov group-theoretic coherent states provide an over-complete family of states in a Hilbert space. Thus, the coherent states can be used as a basis for a simpler representation of the quantum partition function of a given many-body Hamiltonian as an integral over the coset manifold [132].

Moreover, the Gilmore-Perelomov group-theoretic coherent state is also valuable in studying quantum phase transitions. Quantum phase transitions are associated with abrupt change in the physical properties of a many-body quantum system caused by the variation of the Hamiltonian parameters. It has been extensively studied for various many-body systems in recent years [133, 134]. Remarkably, the earliest studies on quantum phase transitions can be dated back to the late 1970s. In 1976, Hertz studied the magnetic quantum phase transitions in three dimensions, in which the coherent states path integral is used to describe the low-energy collective excitations of the fermionic quasi-particles [135]. In 1978, Gilmore and Feng studied the shape phase transitions in atomic nuclei in the context of the Lipkin-Meshkov-Glick model by using the spin coherent states [136]. Subsequently, in 1980, Feng, Gilmore, and Deans showed that the expectation of the many-body Hamiltonian in the interacting boson model of the atomic nuclei with respect to the Gilmore-Perelomov group-theoretic coherent states can demonstrate shape phase transitions of first and second order [40]. In the following, we will discuss the Lieb-Berezin inequalities, the imaginary time path integrals formalism of quantum statistical mechanics, and the shape phase transitions in atomic nuclei in a unified group-theoretical viewpoint.

As is well known, the statistical properties of a physical system associated with a Hamiltonian H can either be determined from the partition function $Z(\beta)$, or the from the free energy $F(\beta)$ [137]

$$e^{-\beta F(\beta)} \equiv Z(\beta) \equiv \mathrm{Tr}\, e^{-\beta H}, \tag{11.1}$$

where $\beta \equiv 1/k_B T$, k_B is the Boltzmann constant, and the trace is taken over the entire Hilbert space. For our purpose, we consider primary a Hamiltonian H which is constructed from the generators of a Lie algebra $\mathfrak{g}$ such that $H = H(T_i)$ for $T_i \in \mathfrak{g}$, while in most applications of quantum mechanics, the Hamiltonians are either linearly spanned by the generators T_i of the Lie algebra $\mathfrak{g}$

$$H = \sum_i c_i T_i, \tag{11.2}$$

or simplified into the generic form of superposition of linear and quadratic functions of the generators T_i under suitable mean-field approximations

$$H = \sum_i c_i T_i + \sum_{i,j} c_{ij} T_i T_j. \tag{11.3}$$

With these approximations, the whole Hilbert space is decomposed into $\mathfrak{g}$-invariant subspaces $V^{\mathbf{\Lambda}}$, where each subspace is represented by a weight $\mathbf{\Lambda}$ with degeneracy

$Y_{\mathbf{\Lambda}}$. As the generators T_i have nonzero matrix elements only within an irreducible subspace, the trace in Eq. (11.1) becomes

$$Z(\beta) = \mathrm{Tr}\, e^{-\beta H} = \sum_{\mathbf{\Lambda}} Y_{\mathbf{\Lambda}}\, \mathrm{Tr}_{\mathbf{\Lambda}}\, e^{-\beta H}, \tag{11.4}$$

where $\mathrm{Tr}_{\mathbf{\Lambda}}$ implies that the trace is restricted to an invariant subspace with a weight $\mathbf{\Lambda}$. In the coherent state representation, as a result of the resolution of identity, the trace within each invariant subspace $V^{\mathbf{\Lambda}}$ can be expressed as a coherent state integral [38, 132]

$$\begin{aligned} \mathrm{Tr}_{\mathbf{\Lambda}}\, e^{-\beta H} &= \mathrm{Tr}_{\mathbf{\Lambda}} \int d\mu_{\Omega} |\mathbf{\Lambda}, \Omega\rangle\langle\mathbf{\Lambda}, \Omega| e^{-\beta H} \\ &= \int d\mu_{\Omega} \langle\mathbf{\Lambda}, \Omega| e^{-\beta H} |\mathbf{\Lambda}, \Omega\rangle. \end{aligned} \tag{11.5}$$

In general, except for a few special cases, the above coherent state integral cannot be computed exactly. However, in 1973, in studying the classical limit of quantum partition functions, Lieb ingeniously derived an ingenious thermodynamics inequality using spin coherent states as a useful approximation of the partition function for quantum spin systems [38]. Interestingly, 1 year before this study, in 1972, for the purpose of studying the quantization of classical systems, Felix A. Berezin also had derived the same inequality through the usage of the covariant and contravariant symbols related to an over-complete family of coherent states [39]. These thermodynamic inequalities, now known as Lieb-Berezin inequalities, give both the upper and lower bounds of the quantum free energy $F(\beta)$ in terms of two classical free energies. In 1980, Barry Simon generalized the Lieb-Berezin inequalities to a large class of dynamical systems using generalized coherent states for arbitrary compact Lie groups [131].

The lower bound of the quantum partition function $Z(\beta)$ can be obtained from the so-called Peierls-Bogoliubov inequality, $\langle\Phi|e^X|\Phi\rangle \geq \exp\langle\Phi|X|\Phi\rangle$, for any normalized states $|\Phi\rangle \in V^{\mathbf{\Lambda}}$ and self-adjoint operator X, which yields

$$\begin{aligned} \mathrm{Tr}_{\mathbf{\Lambda}}\, e^{-\beta H} &= \int d\mu_{\Omega} \langle\mathbf{\Lambda}, \Omega| e^{-\beta H} |\mathbf{\Lambda}, \Omega\rangle \\ &\geq \int d\mu_{\Omega} \exp(-\beta\langle\mathbf{\Lambda}, \Omega| H |\mathbf{\Lambda}, \Omega\rangle) \\ &= \int d\mu_{\Omega} e^{-\beta H_Q(\mathbf{\Lambda}, \Omega)}, \end{aligned} \tag{11.6}$$

where $H_Q(\mathbf{\Lambda}, \Omega)$ is the Q-representation of the Hamiltonian H. In fact, the Peierls-Bogoliubov inequality turns out to be a special case of the Jensen inequality for convex functions and self-adjoint operators. To see this, we need the Jensen inequality on a bounded interval $[m, M]$ on $\mathbb{R}$ [138]: Let $f(x)$ be a convex function on an interval $[m, M]$; then for every $x_1, x_2, \cdots, x_k \in [m, M]$ and every positive real numbers $c_1, c_2, \cdots, c_k$ with $\sum_{j=1}^{k} c_k = 1$, one has

$$f\left(\sum_{j=1}^{k} c_j x_j\right) \leq \sum_{j=1}^{k} c_j f(x_j). \tag{11.7}$$

The Jensen inequality for two points reflects a geometric fact that the secant line of a convex function lies above the graph of the function, where $f(c_1x_1 + c_2x_2)$ represents the value of the convex function on the point $c_1x_1 + c_2x_2$ which lies between x_1 and x_2 and $c_1 f(x_1) + c_2 f(x_2)$ represents the associated value on the secant line which passes though the two points x_1 and x_2.

To show that the Peierls-Bogoliubov inequality is a special case of the Jensen inequality, one needs to know that if X is a self-adjoint operator, or a Hermitian $k \times k$ matrix for a finite-dimensional Hilbert space, then there exists a spectral decomposition of X such that $X = U^\dagger DU$, where $D = \text{diag}(x_1, \cdots, x_k)$ is a diagonal matrix whose entries are the eigenvalues $\lambda_i \in \mathbb{R}$ of X, and U is a unitary matrix which satisfies $U^\dagger U = UU^\dagger = I_k$. As $U|\Phi\rangle$ is normalized to unity, one may write $U|\Phi\rangle = (\sqrt{c_1}e^{i\varphi_1}, \cdots, \sqrt{c_k}e^{i\varphi_k})^\mathsf{T}$. Hence, applying the Jensen inequality to the exponential function, one immediately obtains

$$\exp\langle\Phi|X|\Phi\rangle = \exp\left(\sum_{j=1}^{k} c_j x_j\right) \leq \sum_{j=1}^{k} c_j e^{x_j} = \langle\Phi|e^X|\Phi\rangle. \tag{11.8}$$

One may also estimate an upper bound of the quantum partition function $Z(\beta)$ by using the Jensen inequality. Let $|\varphi_j\rangle$ be a complete orthonormal set of eigenfunctions of $X \equiv -\beta H$ with $X|\varphi_j\rangle = x_j|\varphi_j\rangle$ in the $\mathfrak{g}$-invariant subspace $V^{\mathbf{\Lambda}}$, so that $U|\varphi_1\rangle = (1, 0, \cdots, 0)^\mathsf{T}$, $\cdots$, $U|\varphi_k\rangle = (0, 0, \cdots, 1)^\mathsf{T}$, and $|\langle\varphi_j|\Phi\rangle|^2 = |\langle\varphi_j|U^\dagger U|\Phi\rangle|^2 = c_j$. Let $|\Phi\rangle \equiv |\mathbf{\Lambda}, \Omega\rangle$ be a generalized coherent state and $\mathcal{P}(\mathbf{\Lambda}, \Omega) \equiv |\mathbf{\Lambda}, \Omega\rangle\langle\mathbf{\Lambda}, \Omega|$ be a projection operator; one then has $\langle\varphi_j|\mathcal{P}(\mathbf{\Lambda}, \Omega)|\varphi_j\rangle = c_j(\mathbf{\Lambda}, \Omega)$ and

$$\begin{aligned}
\langle\varphi_j|\exp(-\beta H)|\varphi_j\rangle &= \exp(-\beta\langle\varphi_j|H|\varphi_j\rangle) \\
&= \exp(-\beta\langle\varphi_j|\int d\mu_\Omega H_P(\mathbf{\Lambda}, \Omega)\mathcal{P}(\mathbf{\Lambda}, \Omega)|\varphi_j\rangle) \\
&= \exp(-\beta\int d\mu_\Omega c_j(\mathbf{\Lambda}, \Omega)H_P(\mathbf{\Lambda}, \Omega)) \\
&\leq \int d\mu_\Omega c_j(\mathbf{\Lambda}, \Omega)e^{-\beta H_p(\mathbf{\Lambda}, \Omega)},
\end{aligned} \tag{11.9}$$

where $H_P(\mathbf{\Lambda}, \Omega)$ is the P-representation of the Hamiltonian H, and in the last step, we used the integral form of the Jensen inequality by using the fact

$\int d\mu_\Omega c_j(\mathbf{\Lambda}, \Omega) = 1$. Finally, in Eq. (11.9), summation over j will yield

$$\mathrm{Tr}_{\mathbf{\Lambda}}\, e^{-\beta H} \leq \int d\mu_\Omega e^{-\beta H_P(\mathbf{\Lambda},\Omega)}. \tag{11.10}$$

It is worth understanding here that since only the Jensen inequality is used, the resulting inequalities can be extended to any convex functions, and are not limited to the exponential function. This fact was first noted by Berezin [39] and was later pointed out by Simon [131]. Combining Eqs. (11.6) and (11.9), one immediately obtains the Lieb-Berezin thermodynamic inequalities

$$\int d\mu_\Omega e^{-\beta H_Q(\mathbf{\Lambda},\Omega)} \leq \mathrm{Tr}_{\mathbf{\Lambda}}\, e^{-\beta H} \leq \int d\mu_\Omega e^{-\beta H_P(\mathbf{\Lambda},\Omega)}. \tag{11.11}$$

Using the Lieb-Berezin inequalities, one is able to approximately describe quantum dynamical systems via the coherent state representation. In particular, the upper and lower bounds of the quantum free energy can be used to study the qualitative statistical behavior of the systems as well as thermodynamic phase transitions. From Eq. (11.11), the upper and lower bounds of the partition function as well as the free energy are given by

$$\sum_{\mathbf{\Lambda}} Y_{\mathbf{\Lambda}} \int d\mu_\Omega e^{-\beta H_Q(\mathbf{\Lambda},\Omega)} \leq e^{-F(\beta)} \leq \sum_{\mathbf{\Lambda}} Y_{\mathbf{\Lambda}} \int d\mu_\Omega e^{-\beta H_P(\mathbf{\Lambda},\Omega)}. \tag{11.12}$$

For a quantum system with a dynamical symmetry group G, the classical limit $\hbar \to 0$ is in fact equivalent to the large N limit, where N is some measure of the number of dynamical variables. One may prove this equivalence within the group-theoretical framework of coherent states [131, 139]. For example, the classical limit of a quantum spin system with a dynamical symmetry group SU(2) is known to be $j \to \infty$, where j is the total angular momentum quantum number and $N = 2j$ [38].

In particular, when the Hamiltonian of the spin system is linear in the spin operators of each spin, i.e., one allows multiple site interactions of arbitrary complexity such as $\hat{S}_x^1 \hat{S}_y^2 \hat{S}_y^3 \hat{S}_z^4$, but does not allow monomials such as $(\hat{S}_x^1)^2$ or $\hat{S}_x^1 \hat{S}_y^1$. In such cases, the Q- and P-representations of the Hamiltonian can be easily obtained. The Q- and P-representations for some operators commonly appearing in quantum spin Hamiltonians are listed in the following table: [38] (Table 11.1).

One can readily see that the Q-representation of a Hamiltonian which is linear in the spin operators of each spin is precisely the classical partition function in which each spin operator is replaced by j^i times a classical unit vector, i.e., $\hat{\mathbf{S}}^i \to j^i(\sin\theta^i \cos\varphi^i, \sin\theta^i \sin\varphi^i, \cos\theta^i)$. Similarly, the P-representation of the same Hamiltonian is precisely the classical partition function in which each spin operator is replaced by $(j^i + 1)$ times the same unit vector, i.e., $\hat{\mathbf{S}}^i \to (j+1)^i(\sin\theta^i \cos\varphi^i, \sin\theta^i \sin\varphi^i, \cos\theta^i)$. Hence, applying the Lieb-Berezin

Table 11.1 Table of the Q- and P-representations of some operators that commonly appear in quantum spin Hamiltonians, where the spin-j coherent state $|\theta, \varphi\rangle$ is used

Operator	Q-representation	P-representation
$\hat{S}_z$	$j\cos\theta$	$(j+1)\cos\theta$
$\hat{S}_x$	$j\sin\theta\cos\varphi$	$(j+1)\sin\theta\cos\varphi$
$\hat{S}_y$	$j\sin\theta\sin\varphi$	$(j+1)\sin\theta\sin\varphi$
$\hat{S}_z^2$	$j(j-\frac{1}{2})(\cos\theta)^2+\frac{j}{2}$	$(j+1)(j+\frac{3}{2})(\cos\theta)^2-\frac{j+1}{2}$
$\hat{S}_x^2$	$j(j-\frac{1}{2})(\sin\theta\cos\varphi)^2+\frac{j}{2}$	$(j+1)(j+\frac{3}{2})(\sin\theta\cos\varphi)^2-\frac{j+1}{2}$
$\hat{S}_y^2$	$j(j-\frac{1}{2})(\sin\theta\sin\varphi)^2+\frac{j}{2}$	$(j+1)(j+\frac{3}{2})(\sin\theta\sin\varphi)^2-\frac{j+1}{2}$

inequalities to a quantum spin system whose Hamiltonian is linear in the spin operators will yield [38]

$$Z^{cl}(\beta, j^1, \cdots, j^N) \le Z(\beta) \le Z^{cl}(\beta, j^1+1, \cdots, j^N+1). \tag{11.13}$$

The Lieb-Berezin inequalities show that as j increases, the quantum and classical free energies form two decreasing, interlacing sequences.

Using the Lieb-Berezin inequalities, one may readily show that in the thermodynamic limit $N \to \infty$, the Q- and P-representations of the Hamiltonian per particle as well as the quantum free energy per particle are reduced to their classical forms. Let H_N be a quantum Hamiltonian of N spins. Without loss of generality, the j values for different spins are assumed to be the same, i.e., $j^1 = j^2 = \cdots = j^N = j$. One may replace each spin operator $\hat{\mathbf{S}}^i$ by $\hat{\mathbf{S}}^i/j$ and denotes the resulting Hamiltonian and the partition function as $H_N(j)$ and $Z_N(\beta, j)$, respectively. Then the free energy per spin is determined by $f_N(\beta, j) = -(N\beta)^{-1}\ln Z_N(\beta, j)$. From the Lieb-Berezin inequalities, one immediately obtains the upper and lower bounds for the quantum free energy per spin [38]

$$f_N^{cl}(\beta) \ge f_N(\beta, j) \ge f_N^{cl}(\beta, \delta_j), \tag{11.14}$$

where $f_N^{cl}(\beta)$ is the classical free energy per spin in which each spin operator $\hat{\mathbf{S}}^i$ is replaced by a classical unit vector $(\sin\theta^i\cos\varphi^i, \sin\theta^i\sin\varphi^i, \cos\theta^i)$ and $f_N^{cl}(\beta, \delta_j)$ is the classical free energy per spin in which each vector is multiplied by $\delta_j \equiv (j+1)/j$.

If we regard δ_j as a variable, then the classical Hamiltonian $H_N^{cl}(\delta_j)$ is a continuous function of δ_j on $\mathcal{S}_N \equiv S^2 \times \cdots \times S^2$, i.e., the Cartesian product of N copies of the unit sphere S^2. If one assumes that the thermodynamic limit of H_N exists, then the sequence of continuous functions $h_N^{cl}(\delta_j) \equiv N^{-1}H_N^{cl}(\delta_j)$ is *bounded* and *equicontinuous*, i.e., for any given $\epsilon > 0$, it is possible to find a $\delta > 0$, such that $\|h_N^{cl}(\delta_j + x) - h_N^{cl}(\delta_j)\| < \epsilon$ for $|x| < \delta$, independent of N, where $\|h_N^{cl}\|$ is the supremum norm on $\mathcal{S}_N$, i.e., $\|h_N^{cl}\| \equiv \sup|h_N^{cl}(\boldsymbol{\theta}, \boldsymbol{\varphi})|$ for $(\boldsymbol{\theta}, \boldsymbol{\varphi}) \equiv (\theta^1, \varphi^1, \cdots, \theta^N, \varphi^N) \in \mathcal{S}_N$. According to the Arzelà-Ascoli theorem [140–142], as $h_N^{cl}(\delta_j)$ is bounded and equicontinuous sequence on a compact metric space $\mathcal{S}_N$, there exists a subsequence of it that converges uniformly. As a result of the uniform limit theorem [143], the uniform limit function $h^{cl}(\delta_j) \equiv \lim_{N\to\infty} h_N^{cl}(\delta_j)$ is continuous in δ_j. Hence, the uniform limit function

$$f^{cl}(\beta, \delta_j) \equiv \lim_{N\to\infty} f_N^{cl}(\beta, \delta_j) \tag{11.15}$$

is also continuous in δ_j. By the squeeze theorem, as both the left and right sequences of Eq. (11.14) have the same limit when $N \to \infty$ (thermodynamic limit) and $j \to \infty$ (classical limit) and both sequences converge uniformly

$$\lim_{j\to\infty} \lim_{N\to\infty} f_N^{cl}(\beta, \delta_j) = \lim_{j\to\infty} f^{cl}(\beta, \delta_j) = f^{cl}(\beta) \equiv \lim_{N\to\infty} f_N^{cl}(\beta), \tag{11.16}$$

the sequence of quantum free energy per spin approaches to the same limit

$$\lim_{j\to\infty} \lim_{N\to\infty} f_N(\beta, j) = f^{cl}(\beta) \equiv \lim_{N\to\infty} f_N^{cl}(\beta). \tag{11.17}$$

The above conclusion can be extended to arbitrary quantum systems with a compact dynamical symmetry group, or those with a non-compact dynamical symmetry group with a square-integrable Hilbert space [131, 139, 144].

In the following, we will derive the classical limit of the quantum partition function for a general system with an arbitrary compact symmetry group G. Similar to the SU(2) case, we assume that the quantum Hamiltonian is a sum of monomials which are products of commuting operators. Let $V^{\boldsymbol{\Lambda}}$ be the Hilbert space which carries a unitary irreducible representation of the compact group G, where $\boldsymbol{\Lambda}$ is the highest weight in the representation space $V^{\boldsymbol{\Lambda}}$. Without loss of generality, the highest weight in $V^{\boldsymbol{\Lambda}}$ is assumed to be a multiple of a single fundamental weight $\boldsymbol{f}$, i.e., $\boldsymbol{\Lambda} = L\boldsymbol{f}$. Let $d = \dim V^{\boldsymbol{\Lambda}}$ and T_i be the generators of the group G. Similar to the SU(2) case, one may construct the quantum partition function

$$Z(\beta, L) \equiv d^{-|\boldsymbol{\Lambda}|} \operatorname{Tr}_{\boldsymbol{\Lambda}} \exp[-\beta H(T_i/L)]. \tag{11.18}$$

Using the Lieb-Berezin inequalities, one can derive the following upper and lower bounds for the quantum partition function

$$Z^{cl}(\beta) \leq Z(\beta, L) \leq Z^{cl}(\beta, \delta_L), \tag{11.19}$$

where $\delta_L = 1 + 2(\boldsymbol{\Lambda}, \boldsymbol{\delta})/(\boldsymbol{\Lambda}, \boldsymbol{\Lambda}) = 1 + 2(\boldsymbol{f}, \boldsymbol{\delta})/(\boldsymbol{f}, \boldsymbol{f}) \cdot L^{-1}$ is a scale factor multiplied on each generator T_i, and $\boldsymbol{\delta} \equiv \frac{1}{2}\sum_{\boldsymbol{\alpha}\in\boldsymbol{\Phi}^+} \boldsymbol{\alpha}$ is the half-sum of the positive roots, namely, the **Weyl vector**. As a first example, the single positive root of SU(2) is $\boldsymbol{\alpha} = 1$ and thus $\boldsymbol{\delta} = 1/2$. In the irreducible representation with highest weight $\boldsymbol{\Lambda} = j$, i.e., a spin-j system, one obtains $\boldsymbol{f} = \boldsymbol{\alpha} = 1$ and $L \equiv j$, and thus $\delta_L \equiv 1 + 1/j$. As another example, the three positive roots of SU(3) are $\boldsymbol{\alpha}_1 \equiv (\frac{1}{2}, \frac{\sqrt{3}}{2})$, $\boldsymbol{\alpha}_2 \equiv (\frac{1}{2}, \frac{-\sqrt{3}}{2})$, and $\boldsymbol{\alpha}_3 \equiv (1, 0)$, respectively, and thus $\boldsymbol{\delta} = (1, 0)$. In the fundamental representation **3** and its complex conjugate representation $\bar{\mathbf{3}}$

of SU(3), the fundamental weights are $\boldsymbol{f}_1 \equiv (\frac{1}{2}, \frac{1}{2\sqrt{3}})$ and $\boldsymbol{f}_2 \equiv (\frac{1}{2}, -\frac{1}{2\sqrt{3}})$, respectively. Hence, in the irreducible representations of SU(3) with highest weight $\boldsymbol{\Lambda} = L\boldsymbol{f}_1$ or $\boldsymbol{\Lambda} = L\boldsymbol{f}_2$, the scale factor $\delta_L = 1 + 3/L$. Finally, from Eq. (11.19), the classical limit of the quantum partition function is

$$\lim_{L\to\infty} Z(\beta, L) = Z^{cl}(\beta) = \lim_{L\to\infty} Z^{cl}(\beta, \delta_L). \tag{11.20}$$

11.2 Coherent State Representation of Partition Function

The quantum partition function of a single particle governed by a Hamiltonian $H(\hat{p}, \hat{x})$ can be written as

$$Z(\beta) \equiv \mathrm{Tr}\, e^{-\beta\hat{H}} = \int dx \langle x|e^{-\beta\hat{H}}|x\rangle, \tag{11.21}$$

where $\beta \equiv 1/(k_B T)$, k_B is the Boltzmann constant, and $|x\rangle$ is the position eigenstate. Hence, in order to evaluate the quantum partition function $Z(\beta)$, one only needs to evaluate the transition amplitude of the imaginary time evolution operator from a space-time point $(x_a, -i\tau_a)$ to another space-time point $(x_b, -i\tau_b)$ for an interval $\tau_b - \tau_a = \beta\hbar$

$$K(x_b, \tau_b; x_a, \tau_a) \equiv \langle x_b|e^{-(\tau_b-\tau_a)\hat{H}/\hbar}|x_a\rangle. \tag{11.22}$$

Following all the steps in the derivation of the real-time path integral, one obtains the transition amplitude for the case of imaginary time

$$\begin{aligned}
K(x_b, \tau_b; x_a, \tau_a) &= \lim_{N\to\infty} \prod_{n=1}^{N} \int_{-\infty}^{\infty} dx_n \prod_{n=1}^{N+1} \langle x_n|e^{-\epsilon\hat{H}/\hbar}|x_{n-1}\rangle \qquad (11.23)\\
&= \lim_{N\to\infty} \prod_{n=1}^{N} \int_{-\infty}^{\infty} dx_n \prod_{n=1}^{N+1} \int_{-\infty}^{\infty} \frac{dp_n}{2\pi\hbar} e^{\frac{1}{\hbar}\sum_{n=1}^{N+1}[ip_n(x_n-x_{n-1})-\epsilon H(p_n,x_n)]}\\
&= \lim_{N\to\infty} \prod_{n=1}^{N} \int_{-\infty}^{\infty} dx_n \left(\frac{m}{2\pi\hbar\epsilon}\right)^{\frac{N+1}{2}} e^{-\frac{\epsilon}{\hbar}\sum_{i=1}^{N+1}[\frac{m}{2}\frac{(x_n-x_{n-1})^2}{\epsilon^2}+V(x_n)]}\\
&= \int_{(x_a,\tau_a)}^{(x_b,\tau_b)} \mathcal{D}[x(\tau)] e^{\frac{-1}{\hbar}\int_{\tau_a}^{\tau_b} d\tau\left[\frac{m}{2}(\frac{dx(\tau)}{d\tau})^2+V(x(\tau))\right]}\\
&= \int_{(x_a,\tau_a)}^{(x_b,\tau_b)} \mathcal{D}[x(\tau)] e^{\frac{-1}{\hbar}\int_{\tau_a}^{\tau_b} d\tau H(x(\tau))},
\end{aligned}$$

where $\epsilon \equiv (\tau_b - \tau_a)/(N+1)$. Hence, one sees that the imaginary time path integral is nothing but a sum over the trajectories which start at the point (x_a, τ_a) and terminate on (x_b, τ_b), where the exponential of the action is modified by a sign change in the kinetic energy, which yields a Hamiltonian instead of a Lagrangian in the exponential. As a remark, the measure appearing in the imaginary time path integral is equivalent to the measure defined in the study of continuous stochastic processes by Wiener in 1932, which is named as the Wiener process thereafter. Using the imaginary time path integral, the quantum partition function can be expressed as

$$Z(\beta) = \int dx \int_{x(0)=x}^{x(\beta\hbar)=x} \mathcal{D}[x(\tau)] e^{\frac{-1}{\hbar}\int_0^{\beta\hbar} d\tau \left[\frac{m}{2}(\frac{dx(\tau)}{d\tau})^2 + V(x(\tau))\right]}. \tag{11.24}$$

Hence, the quantum partition function is a sum over all periodic trajectories of period $\beta\hbar$.

For a general many-body Hamiltonian expressed in a second quantized form, instead of using the position and momentum eigenstates, one may use the coherent states to establish a functional integral representation of the many-body evolution operator. As before, one needs to evaluate the matrix element of the evolution operator between an initial coherent state and a finial coherent state, i.e., $U(\alpha^*_{k,b}, t_b; \alpha_{k,a}, t_a) \equiv \langle \boldsymbol{\alpha}_b | e^{-i(t_b - t_a)\hat{H}/\hbar} | \boldsymbol{\alpha}_a \rangle$, where

$$|\boldsymbol{\alpha}\rangle \equiv \exp\left(-\frac{1}{2}\sum_k |\alpha_k|^2\right) \exp\left(\sum_k \alpha_k a_k^\dagger\right) |\mathbf{0}\rangle, \tag{11.25}$$

$$\langle\boldsymbol{\alpha}| \equiv \langle\mathbf{0}| \exp\left(-\frac{1}{2}\sum_k |\alpha_k|^2\right) \exp\left(\sum_k \alpha_k^* a_k\right),$$

and $|\mathbf{0}\rangle$ is the vacuum state. As usual, one may break the time interval $[t_a, t_b]$ into $N+1$ time steps of size $\epsilon \equiv (t_b - t_a)/(N+1)$, and inserting the over-completeness relations for the coherent states at the nth time step

$$I = \prod_k \frac{d^2\alpha_{k,n}}{\pi} |\alpha_{k,n}\rangle\langle\alpha_{k,n}|, \tag{11.26}$$

then the matrix element of the evolution operator can be written as

$$U(\alpha^*_{k,b}, t_b; \alpha_{k,a}, t_a) = \lim_{N\to\infty} \int \prod_{n=1}^{N} \prod_k \frac{d^2\alpha_{k,n}}{\pi} e^{-\frac{i}{\hbar}S(\boldsymbol{\alpha}^*, \boldsymbol{\alpha})}, \tag{11.27}$$

$$\frac{-i}{\hbar} S(\boldsymbol{\alpha}^*, \boldsymbol{\alpha}) \equiv \sum_{n=1}^{N+1} \left(\sum_k \alpha^*_{k,n}(\alpha_{k,n-1} - \alpha_{k,n}) - \frac{i\epsilon}{\hbar} H(\alpha^*_{k,n}, \alpha_{k,n-1}) \right).$$

In real time, the term $e^{-i\epsilon H/\hbar}$ is oscillatory, and the convergence of the integrals involved is ensured by the factor $e^{-\alpha^*_{k,n}\alpha_{k,n}}$ arising from the measure, whereas in imaginary time, the term $e^{-\epsilon H/\hbar}$ is bounded, as a physical Hamiltonian is bound from below, which implies that the Hamiltonian function $H(\alpha^*_{k,n}, \alpha_{k,n-1}) \equiv \langle\alpha_{k,n}|\hat{H}|\alpha_{k,n-1}\rangle/\langle\alpha_{k,n}|\alpha_{k,n-1}\rangle$ is also bound from below. Hence, the convergence of the integrals is again ensured by the Gaussian factor arising from the measure.

Using the imaginary time path integral for the transition amplitude, the partition function for a bosonic many-body system can be expressed as

$$
\begin{aligned}
Z(\beta) \equiv \mathrm{Tr}\, e^{-\beta\hat{H}} &= \prod_k \frac{d^2\alpha_k}{\pi}\langle\alpha_k|e^{-\beta\hat{H}}|\alpha_k\rangle \\
&= \lim_{N\to\infty}\int\prod_{n=1}^{N+1}\prod_k \frac{d^2\alpha_{k,n}}{\pi} e^{-\frac{1}{\hbar}S(\boldsymbol{\alpha}^*,\boldsymbol{\alpha})},
\end{aligned}
\tag{11.28}
$$

where the periodic boundary conditions $\alpha_{k,0} = \alpha_k$ and $\alpha_{k,N+1} = \alpha^*_k$ are imposed, and the Euclidean action is given by

$$
\begin{aligned}
\frac{1}{\hbar}S(\boldsymbol{\alpha}^*,\boldsymbol{\alpha}) \equiv{}& \alpha^*_{k,1}(\alpha_{k,1} - \alpha_{k,N+1}) + \frac{\epsilon}{\hbar}H(\alpha^*_{k,1}, \alpha_{k,N+1}) \\
&+ \sum_{n=2}^{N+1}\left[\sum_k \alpha^*_{k,n}(\alpha_{k,n} - \alpha_{k,n-1}) + \frac{\epsilon}{\hbar}H(\alpha^*_{k,n}, \alpha_{k,n-1})\right].
\end{aligned}
\tag{11.29}
$$

Using the Wiener measure, the continuous limit of the above coherent states path integral can be written as

$$
Z(\beta) = \int_{\alpha_k(\beta\hbar)=\alpha_k(0)} \mathcal{D}(\alpha^*_k, \alpha_k) e^{\frac{-1}{\hbar}\int_0^{\beta\hbar} d\tau\{\sum_k \alpha^*_k\partial_\tau\alpha_k + H(\alpha^*_k,\alpha_k)\}}. \tag{11.30}
$$

As a first example, one may evaluate the quantum partition function for a system of noninteracting particles which are described by the Hamiltonian

$$
H = \sum_k \epsilon_k \hat{a}^\dagger_k \hat{a}_k. \tag{11.31}
$$

Using Eqs. (11.28) and (11.29), one immediately obtains

$$
\begin{aligned}
Z(\beta) &= \lim_{N\to\infty}\prod_k\prod_{n=1}^{N+1}\int\frac{d^2\alpha_{k,n}}{\pi}\exp\left\{-\sum_{i,j=1}^{N+1}\alpha^*_{k;i}S^{(k)}_{ij}(\beta)\alpha_{k,j}\right\} \\
&= \lim_{N\to\infty}\prod_k[\det S^{(k)}(\beta)]^{-1}
\end{aligned}
\tag{11.32}
$$

where $S_{ij}^{(k)}(\beta)$ are the elements of a matrix $S^{(k)}(\beta)$ given by

$$S^{(k)}(\beta) \equiv \begin{pmatrix} 1 & 0 & \cdots & \cdots & 0 & -s_k \\ -s_k & 1 & 0 & \cdots & 0 & 0 \\ 0 & -s_k & 1 & \cdots & 0 & 0 \\ \vdots & \vdots & \vdots & \vdots & \vdots & \vdots \\ 0 & 0 & \cdots & -s_k & 1 & 0 \\ 0 & 0 & \cdots & \cdots & -s_k & 1 \end{pmatrix}, \tag{11.33}$$

and $s_k \equiv 1 - \beta\epsilon_k/(N+1)$. Expanding by minors of the first row, one obtains

$$\begin{aligned} \lim_{N\to\infty} \det S^{(k)}(\beta) &= \lim_{N\to\infty} (1 + (-1)^{N+2}(-s_k)^{N+1}) \\ &= \lim_{N\to\infty} \det S^{(k)}(\beta) \left[1 - (1 - \frac{\beta\epsilon_k}{N+1})^{N+1} \right] \\ &= 1 - e^{-\beta\epsilon_k}. \end{aligned} \tag{11.34}$$

Hence, one immediately obtains the familiar result for noninteracting bosons

$$Z(\beta) = \prod_k (1 - e^{-\beta\epsilon_k})^{-1}. \tag{11.35}$$

11.3 Quantum Phase Transitions

When classical phase transition is mentioned, one refers it to an abrupt change in the physical properties of a system caused by a change of temperature. The reason for such a phenomenon is the change of symmetry of the phases involved, which is driven by thermal fluctuations. As a result, classical phase transition cannot happen when the temperature drops to zero, as the thermal fluctuations are absented. However, quantum phase transitions driven by a change of control parameters controlling an interaction strength in the system's Hamiltonian persist at zero temperature, as quantum fluctuations always exist at zero temperature. In this regard, quantum phase transitions are qualitative changes in the structure of quantum systems induced by a change in the parameters of the Hamiltonian [145, 146] (e.g., varying coupling constants).

The term quantum phase transitions, or *quantum critical phenomena*, was originated from condensed matter physics, which was first introduced for ordered-disordered phase transitions, i.e., ferromagnetic-paramagnetic transitions, in spin systems at zero temperature by John Hertz [135]. He evaluated the quantum partition function using coherent states path integrals for fermions and investigated the critical behaviors of the spin systems at finite temperatures, as well as the zero-temperature limit by using the momentum-space renormalization group approach.

Nearly at the same period, the research of quantum phase transitions was conducted by a group of nuclear physicists [136, 144] in the context of Lipkin-Meshkov-Glick model [147] under the name of *ground state phase transitions* and was subsequently further developed [40, 148, 149] in even-even nuclei in the context of the interacting boson model [150] and in the fermion dynamical symmetry model for high- and low-spin nuclear collective states [151, 152, 152] in the 1980s. Interestingly, since then, the Lipkin-Meshkov-Glick model has been extensively used to study the relationship between quantum entanglement and quantum phase transitions [153–156]. With the aid of the generalized coherent states, they are sufficient to build up a conceptual framework to explain quantum phase transitions.

The interacting boson model is formulated in terms of Lie algebras, but the shape of the nucleus is based on geometry. In order to extract geometry from the algebra of the interacting boson model, one needs the theory of generalized coherent states, as well as the associated geometric coset spaces [9, 10]. In the interacting boson model, the collective excitations of nuclei are described by bosons. In particular, the low-lying collective states of nuclei are described by a monopole s-boson and a quadrupole d-boson. In the language of second quantization, the boson creation and annihilation operators are denoted as $\hat{s}^\dagger$, $\hat{d}^\dagger_\mu$, $\hat{s}$, and $\hat{d}_\mu$, respectively, where $\mu = 0, \pm 1, \pm 2$. The operators satisfy the bosonic commutation relations

$$
\begin{aligned}
&[\hat{s}, \hat{s}^\dagger] = 1, [\hat{s}, \hat{s}] = [\hat{s}^\dagger, \hat{s}^\dagger] = 0, \\
&[\hat{d}_\mu, \hat{d}^\dagger_{\mu'}] = \delta_{\mu\mu'}, [\hat{d}_\mu, \hat{d}_{\mu'}] = [\hat{d}^\dagger_\mu, \hat{d}^\dagger_{\mu'}] = 0, \\
&[\hat{s}, \hat{d}^\dagger_\mu] = [\hat{s}, \hat{d}_\mu] = [\hat{s}^\dagger, \hat{d}^\dagger_\mu] = [\hat{s}^\dagger, \hat{d}_\mu] = 0.
\end{aligned}
\tag{11.36}
$$

One may use a more compact notation for the boson operators, i.e., $\hat{b}^\dagger_\alpha$ and $\hat{b}_\alpha$ with $\alpha = 1, \cdots, 6$ and $\hat{b}_1 \equiv \hat{s}$, $\hat{b}_2 \equiv \hat{d}_2$, $\hat{b}_3 \equiv \hat{d}_1$, $\hat{b}_4 \equiv \hat{d}_0$, $\hat{b}_5 \equiv \hat{d}_{-1}$, and $\hat{b}_6 \equiv \hat{d}_{-2}$. Then the commutation relations become the standard ones

$$
[\hat{b}_\alpha, \hat{b}^\dagger_{\alpha'}] = \delta_{\alpha\alpha'}, [\hat{b}_\alpha, \hat{b}_{\alpha'}] = [\hat{b}^\dagger_\alpha, \hat{b}^\dagger_{\alpha'}] = 0. \tag{11.37}
$$

A direct computation shows that the set of bilinear products of boson creation and annihilation operators satisfies the canonical commutation relations

$$
[\hat{b}^\dagger_\alpha \hat{b}_\beta, \hat{b}^\dagger_\gamma \hat{b}_\delta] = \hat{b}^\dagger_\alpha \hat{b}_\delta \delta_{\beta\gamma} - \hat{b}^\dagger_\gamma \hat{b}_\beta \delta_{\delta\alpha}. \tag{11.38}
$$

which are nothing but the standard commutation relations of $\mathfrak{u}(6)$, the unitary algebra in six dimensions. In other words, the 36 operators $\hat{b}^\dagger_\alpha \hat{b}_\beta$ span the Lie algebra $\mathfrak{u}(6)$. From Eq. (11.38), one immediately obtains $[\hat{b}^\dagger_i \hat{b}_i, \hat{b}^\dagger_j \hat{b}_j] = 0$. Hence, the set of operators $\hat{b}^\dagger_i \hat{b}_i$ $(i = 1, \cdots, 6)$ spans the maximal commutative subalgebra, namely, the Cartan subalgebra of $\mathfrak{u}(6)$. Also, from Eq. (11.38), one obtains

$$
[\hat{b}^\dagger_i \hat{b}_i, \hat{b}^\dagger_j \hat{b}_k] = (\delta_{ij} - \delta_{ik}) \hat{b}^\dagger_j \hat{b}_k. \tag{11.39}
$$

Let $E_{ij} \equiv \hat{b}_i^\dagger \hat{b}_j$ with $i \neq j$, $H_i \equiv \hat{b}_i^\dagger \hat{b}_i$ be an element of the Cartan subalgebra, and $\mathbf{H} \equiv (H_1, \cdots, H_6)$. From Eq. (11.39), one can readily show that

$$\begin{aligned}
&[\mathbf{H}, E_{12}] = (1, -1, 0, 0, 0, 0)E_{12}, [\mathbf{H}, E_{21}] = (-1, 1, 0, 0, 0, 0)E_{21}, \\
&[\mathbf{H}, E_{13}] = (1, 0, -1, 0, 0, 0)E_{13}, [\mathbf{H}, E_{31}] = (-1, 0, 1, 0, 0, 0)E_{31}, \\
&\vdots \\
&[\mathbf{H}, E_{56}] = (0, 0, 0, 0, 1, -1)E_{56}, [\mathbf{H}, E_{65}] = (0, 0, 0, 0, -1, 1)E_{65}.
\end{aligned} \tag{11.40}$$

Let $\mathbf{e}_i$ $(i = 1, \cdots, 6)$ be the basis vectors of a six-dimensional Euclidean space. Then Eq. (11.40) yields $[\mathbf{H}, E_{\boldsymbol{\alpha}}] = \boldsymbol{\alpha} E_{\boldsymbol{\alpha}}$, which is the standard commutation relations between the elements of the Cartan subalgebra and the shift operators, where $E_{\mathbf{e}_i - \mathbf{e}_j} \equiv E_{ij} = \hat{b}_i^\dagger \hat{b}_j$. Hence, the Lie algebra $\mathfrak{u}(6)$ is classified by the 30 roots $\mathbf{e}_i - \mathbf{e}_j$ with $i \neq j$. One may choose the 15 roots $\mathbf{e}_i - \mathbf{e}_j$ with $i < j$ as a set of positive roots. Let $\boldsymbol{\alpha}_i \equiv \mathbf{e}_i - \mathbf{e}_{i+1}$. Among the positive roots, one can show that $\boldsymbol{\alpha}_1, \cdots, \boldsymbol{\alpha}_5$ are the simple roots.

In accordance with the above choice of positive roots, the highest weight state in the fully symmetric representation of $\mathfrak{u}(6)$ should be

$$|\boldsymbol{\Lambda}, \boldsymbol{\Lambda}\rangle = |N, 0, 0, 0, 0, 0\rangle = \frac{(\hat{b}_1^\dagger)^N}{\sqrt{N!}} |0, 0, 0, 0, 0, 0\rangle, \tag{11.41}$$

where $|0, 0, 0, 0, 0, 0\rangle$ is the vacuum state, and the highest weight state is annihilated by all shift operators labeled by positive roots, i.e., $E_{\mathbf{e}_i - \mathbf{e}_j} |\boldsymbol{\Lambda}, \boldsymbol{\Lambda}\rangle = \hat{b}_i^\dagger \hat{b}_j |\boldsymbol{\Lambda}, \boldsymbol{\Lambda}\rangle = 0$ for $i < j$. The highest weight $\boldsymbol{\Lambda}$ is determined by the eigenvalues of the elements of the Cartan subalgebra

$$\begin{aligned}
H_1|\boldsymbol{\Lambda}, \boldsymbol{\Lambda}\rangle &= \hat{b}_1^\dagger \hat{b}_1 |N, 0, 0, 0, 0, 0\rangle = N|\boldsymbol{\Lambda}, \boldsymbol{\Lambda}\rangle, \\
H_i|\boldsymbol{\Lambda}, \boldsymbol{\Lambda}\rangle &= \hat{b}_i^\dagger \hat{b}_i |N, 0, 0, 0, 0, 0\rangle = 0 \text{ for } i > 1,
\end{aligned} \tag{11.42}$$

or equivalently $\mathbf{H}|\boldsymbol{\Lambda}, \boldsymbol{\Lambda}\rangle = \boldsymbol{\Lambda}|\boldsymbol{\Lambda}, \boldsymbol{\Lambda}\rangle$, which yields $\boldsymbol{\Lambda} = (N, 0, 0, 0, 0, 0)$. For those shift operators labeled by negative roots, i.e., $E_{e_i - e_j} = \hat{b}_i^\dagger \hat{b}_j$ with $i > j$, one obtains

$$E_{\mathbf{e}_i - \mathbf{e}_j} |\boldsymbol{\Lambda}, \boldsymbol{\Lambda}\rangle = \hat{b}_i^\dagger \hat{b}_j |N, 0, 0, 0, 0, 0\rangle = 0 \text{ for } i > j > 1, \tag{11.43a}$$

$$E_{\mathbf{e}_i - \mathbf{e}_1} |\boldsymbol{\Lambda}, \boldsymbol{\Lambda}\rangle = \hat{b}_i^\dagger \hat{b}_1 |N, 0, 0, 0, 0, 0\rangle = \sqrt{N} |\boldsymbol{\Lambda} + \mathbf{e}_i - \mathbf{e}_1\rangle, \tag{11.43b}$$

where $\mathbf{\Lambda} + \boldsymbol{e}_2 - \boldsymbol{e}_1 = (N-1, 1, 0, 0, 0, 0)$, $\mathbf{\Lambda} + \boldsymbol{e}_3 - \boldsymbol{e}_1 = (N-1, 0, 1, 0, 0, 0)$, etc. Hence, the $\mathfrak{su}(6)$ coherent states $|\mathbf{\Lambda}, \Omega\rangle$ are given by

$$|\mathbf{\Lambda}, \Omega\rangle \equiv \Omega|\mathbf{\Lambda}, \mathbf{\Lambda}\rangle = \exp[\sum_i (\eta_i E_{\boldsymbol{e}_i - \boldsymbol{e}_1} - \eta_i^* E_{\boldsymbol{e}_1 - \boldsymbol{e}_i})]|\mathbf{\Lambda}, \mathbf{\Lambda}\rangle \quad (11.44)$$

$$\equiv \exp[\sum_i (\eta_i \hat{b}_i^\dagger \hat{b}_1 - \eta_i^* \hat{b}_1^\dagger \hat{b}_i)]|\mathbf{\Lambda}, \mathbf{\Lambda}\rangle.$$

Using the original notation, the $\mathfrak{su}(6)$ coherent states can be written as

$$|N; \eta_\mu\rangle \equiv \exp[\sum_\mu (\eta_\mu \hat{d}_\mu^\dagger \hat{s} - \eta_\mu^* \hat{s}^\dagger \hat{d}_\mu)] \frac{\hat{s}^{\dagger N}}{\sqrt{N!}}|0\rangle, \quad (11.45)$$

where $|0\rangle$ is a shorthand of the vacuum state $|0, 0, 0, 0, 0, 0\rangle$. One can show that the $\mathfrak{su}(6)$ coherent states can also be written as

$$|N; \zeta_\mu\rangle = \frac{1}{\sqrt{N!}} \left[\sqrt{1 - |\boldsymbol{\zeta}|^2} \hat{s}^\dagger + \sum_\mu \zeta_\mu \hat{d}_\mu^\dagger \right]^N |0\rangle, \quad (11.46)$$

where $\zeta_\mu = (\sin|\boldsymbol{\eta}|/|\boldsymbol{\eta}|)\eta_\mu$, $|\boldsymbol{\zeta}|^2 \equiv \sum_\mu \zeta_\mu^* \zeta_\mu$, and $|\boldsymbol{\eta}|^2 \equiv \sum_\mu \eta_\mu^* \eta_\mu$. Consequently, the $\mathfrak{su}(6)$ coherent state $|N; \zeta_\mu\rangle$ represents a condensate of N bosons which are in the same single-particle state and is obtained by applying to the vacuum state the condensate creation operator N times. Here, the condensate creation operator $\hat{b}_c^\dagger \equiv \sqrt{1 - |\boldsymbol{\zeta}|^2} \hat{s}^\dagger + \sum_\mu \zeta_\mu \hat{d}_\mu^\dagger$ is a specific superposition of the s and d bosons.

One may also introduce five complex projective coordinates α_μ via $\alpha_\mu \equiv \zeta_\mu / \sqrt{1 - |\boldsymbol{\zeta}|^2}$, so that the $\mathfrak{su}(6)$ coherent state becomes

$$|N; \alpha_\mu\rangle = \frac{1}{\sqrt{N!}} \left[\frac{1}{\sqrt{1 + |\boldsymbol{\alpha}|^2}} (\hat{s}^\dagger + \sum_\mu \alpha_\mu \hat{d}_\mu^\dagger) \right]^N |0\rangle, \quad (11.47)$$

where $|\boldsymbol{\alpha}|^2 \equiv \sum_\mu \alpha_\mu^* \alpha_\mu = |\boldsymbol{\zeta}|^2/(1 - |\boldsymbol{\zeta}|^2) = \tan^2|\boldsymbol{\eta}|$. The five variables α_μ are the collective surface variables in the laboratory frame, which define the quadrupole moments of a nucleus. Reality of the surface yields the relation $\alpha_\mu^* = (-1)^\mu \alpha_{-\mu}$. Hence, the coordinate space of the collective model, when restricted to quadrupole degrees of freedom, is a real five-dimensional Euclidian space $\mathbb{R}^5$. As the shape of the surface is independent of its orientation, the collective surface variables α_μ transform like the covariant components of irreducible spherical tensor of rank 2, i.e., under arbitrary rotation of the coordinate system described by the Euler angles $\Omega \equiv (\theta_1, \theta_2, \theta_3)$, the collective surface variables α_μ transform in accordance with [157]

$$\hat{D}(\Omega)\alpha_\mu = \sum_\nu \alpha_\nu D^2_{\nu\mu}(\Omega), \tag{11.48}$$

where $\hat{D}(\Omega) \equiv e^{-i\theta_1 \hat{J}_z} e^{-i\theta_2 \hat{J}_y} e^{-i\theta_3 \hat{J}_z}$ is the rotation operator in the $z - y' - z'$ convention and $D^2_{\nu\mu}(\Omega) \equiv \langle 2, \nu|\hat{D}(\Omega)|2, \mu\rangle$ are the elements of the Wigner D-matrix. One may transform the laboratory collective variables α_μ to the intrinsic variables a_μ in the rotating frame of reference, so that

$$a_\mu = \sum_\nu \alpha_\nu D^2_{\nu\mu}(\Omega) \text{ or } \alpha_\lambda = \sum_\mu a_\mu D^2_{\lambda\mu}(\Omega)^*. \tag{11.49}$$

The intrinsic collective variables a_μ are fixed by the conditions $a_\mu \in \mathbb{R}$ and $a^*_\mu = (-1)^\mu a_{-\mu}$, which yields $a_0 = \beta \cos\gamma$, $a_{\pm 1} = 0$, and $a_{\pm 2} = \frac{1}{\sqrt{2}}\beta \sin\gamma$. Hence, in the rotating frame of reference, the $\mathfrak{su}(6)$ coherent state becomes

$$|N; \beta, \gamma\rangle \equiv \frac{1}{\sqrt{N!}}[\hat{b}^\dagger_c(\beta, \gamma)]^N |0\rangle, \tag{11.50}$$

where

$$\hat{b}^\dagger_c(\beta, \gamma) \equiv \frac{1}{\sqrt{1+\beta^2}}(s^\dagger + \beta \cos\gamma\, \hat{d}^\dagger_0 + \frac{\beta}{\sqrt{2}} \sin\gamma\, (\hat{d}^\dagger_2 + \hat{d}^\dagger_{-2})) \tag{11.51}$$

is the condensate creation operator for the collective excitations of nuclei, where β and γ are two intrinsic surface variables, and $\beta^2 \equiv \sum_\mu |\alpha_\mu|^2$ is a rotational invariant obtained from the scalar product of the collective variables α_μ by themselves. The radius β quantifies the degree of deformation, where $\beta = 0$ corresponds to spherical nuclei and $\beta > 0$ corresponds to deformed shapes. The angular variable γ quantifies the type and orientation of the deformed shape, where values of γ equal to multiples of $\pi/3$ correspond to prolate spheroid (American football) or oblate spheroid (M&M's candy) with different symmetry axes, while intermediate values of γ are associated with triaxial ellipsoid without axial symmetry.

In order to study quantum phase transitions in the shapes of nuclei, one may consider the following Hamiltonian:

$$\hat{H} = \epsilon_0[(1-\zeta)\hat{n}_d - \frac{\zeta}{4N}\hat{Q}_\chi \cdot \hat{Q}_\chi], \tag{11.52}$$

where $\hat{T} \cdot \hat{T} \equiv \sum_\mu (-1)^\mu \hat{T}_\mu \hat{T}_{-\mu} = \sum_\mu \hat{T}^\dagger_\mu \hat{T}_\mu$ is the square of a tensor operator $\hat{T}$, $\hat{n}_d \equiv \hat{d}^\dagger \cdot \tilde{d} = \sum_\mu \hat{d}^\dagger_\mu \hat{d}_\mu$ is the d boson number operator, $\tilde{d}_\mu \equiv (-1)^\mu \hat{d}_{-\mu}$ are the adjoint operators of $\hat{d}_\mu$, and $\hat{Q}_\chi \equiv (\hat{d}^\dagger \times \hat{s} + \hat{s}^\dagger \times \tilde{d})^{(2)} + \chi(\hat{d}^\dagger \times \tilde{d})^{(2)}$ is the quadrupole operator. Here, $\hat{T}^{(j_1)} \times \hat{T}^{(j_2)}$ is the tensor product of two spherical tensor operators $\hat{T}^{(j_1)}$ and $\hat{T}^{(j_2)}$ defined by

$$[\hat{T}^{(j_1)} \times \hat{T}^{(j_2)}]^j_m \equiv \sum_{m_1,m_2} \langle j_1, m_1; j_2, m_2 | j, m\rangle \hat{T}^{(j_1)}_{m_1} \hat{T}^{(j_2)}_{m_2}, \tag{11.53}$$

where $\langle j_1, m_1; j_2, m_2 | j, m\rangle$ denotes the standard Clebsch-Gordan coefficients. A direct computation yields the following expectation values with respect to the $\mathfrak{su}(6)$ coherent states $|N; \beta, \gamma\rangle$

$$\langle N; \beta, \gamma | \hat{n}_d | N; \beta, \gamma\rangle = \frac{N\beta^2}{1+\beta^2}, \tag{11.54a}$$

$$\langle N; \beta, \gamma | \hat{Q}_\chi \cdot \hat{Q}_\chi | N; \beta, \gamma\rangle = \frac{N}{1+\beta^2}\left[\frac{4(N-1)\beta^2}{1+\beta^2} + 5 + \beta^2\right] \tag{11.54b}$$
$$-4\sqrt{\frac{2}{7}}\frac{\chi N(N-1)\beta^3}{(1+\beta^2)^2}\cos 3\gamma + \frac{2}{7}\frac{\chi^2 N(N-1)\beta^4}{(1+\beta^2)^2} + \frac{\chi^2 N\beta^2}{1+\beta^2}.$$

Hence, the expectation value of the boson Hamiltonian, Eq. (11.52), with respect to the $\mathfrak{su}(6)$ coherent state $|N; \beta, \gamma\rangle$ is given by

$$E(N; \beta, \gamma) \equiv \langle N; \beta, \gamma | \hat{H} | N; \beta, \gamma\rangle \tag{11.55}$$
$$= \epsilon_0\left[(1-\zeta)N - (1+\chi^2)\frac{\zeta}{4}\right]\left(\frac{\beta^2}{1+\beta^2}\right) - \frac{5\epsilon_0\zeta}{4}\left(\frac{1}{1+\beta^2}\right)$$
$$-\frac{\epsilon_0\zeta}{4}\frac{N-1}{(1+\beta^2)^2}\left(4\beta^2 - 4\sqrt{\frac{2}{7}}\chi\beta^3\cos 3\gamma + \frac{2}{7}\chi^2\beta^4\right).$$

which yields the following energy function:

$$E(N; \beta, \gamma) = \mathcal{E}_0 + \frac{A\beta^2 + B\beta^3\cos 3\gamma + C\beta^4}{(1+\beta^2)^2}, \tag{11.56}$$

where the parameters $\mathcal{E}_0$, A, B, and C are related to N, ϵ_0, ζ, and χ by $\mathcal{E}_0 = -\frac{5}{4}\epsilon_0\zeta$ and

$$A = \epsilon_0[(1-\zeta)N + \zeta(1 - \frac{1}{4}\chi^2 - (N-1))], \tag{11.57}$$

$$B = \epsilon_0\zeta(N-1)\sqrt{\frac{2}{7}}\chi,$$

$$C = \epsilon_0[(1-\zeta)N + \zeta(1 - \frac{1}{4}\chi^2 - \frac{(N-1)\chi^2}{14})].$$

The ground state energy of the interacting boson model Hamiltonian is obtained by minimizing $E(N; \beta, \gamma)$ with respect to β and γ, i.e., $E_0(N) = \min E(N; \beta, \gamma)$. Here, the local extrema of the energy function $E(N; \beta, \gamma)$ are determined by

$$\frac{\partial E}{\partial \beta} = \frac{\beta}{(1+\beta^2)^2}(2A + 3B\beta\cos 3\gamma + 4C\beta^2) - \frac{4\beta^3}{(1+\beta^2)^3}(A + B\beta\cos 3\gamma + C\beta^2) = 0, \tag{11.58a}$$

$$\frac{\partial E}{\partial \gamma} = \frac{-3B\beta^3}{(1+\beta^2)^2}\sin 3\gamma = 0. \tag{11.58b}$$

From the above equations, one immediately sees that the point $\beta = 0$ is always the local extrema of the energy function $E(N; \beta, \gamma)$. For $\beta \neq 0$, Eq. (11.58b) yields $\sin 3\gamma = 0$, or equivalently $\gamma = 0, \pi/3, 2\pi/3, \pi, 4\pi/3$ or $5\pi/3$, which corresponds to axially symmetric quadrupole shapes, i.e., prolate spheroid or oblate spheroid; Eq. (11.58a) yields

$$2A \pm 3B\beta + (4C - 2A)\beta^2 \mp B\beta^3 = 0, \tag{11.59}$$

where the plus and minus signs correspond to $\gamma = 0, 2\pi$ or 4π, and $\gamma = \pi, 3\pi$ or 5π, respectively. Without loss of generality, we will assume that $\gamma = 0$ in the following discussion: For $C > 0$ and $A < B^2/4C$, the energy function $E(N; \beta, \gamma)$ has a double root at $\beta = 0$ and two real roots at $\beta = \beta_\pm$, where

$$E(N; \beta, \gamma)|_{\gamma=0} = \mathcal{E}_0 + \frac{\beta^2(\beta - \beta_+)(\beta - \beta_-)}{(1+\beta^2)^2}, \tag{11.60}$$

$$\beta_\pm \equiv \frac{-B \pm \sqrt{B^2 - 4AC}}{2C}.$$

Hence, the energy function $E(N; \beta, \gamma)$ has a global minimum at $\beta > 0$ (or $\beta < 0$, depending on the sign of B), which corresponds to a deformed axially symmetric shape. On the contrary, for $C > 0$ and $A > B^2/4C$, the energy function $E(N; \beta, \gamma)$ for $\gamma = 0$ only has a double root at $\beta = 0$, and hence the energy function has a single global minimum at $\beta = 0$, which corresponds to a spherical shape. At the critical value $A = B^2/4C$, the minimum at $\beta > 0$ (or $\beta < 0$) and the minimum at $\beta = 0$ swap.

As the ground state energy $E_0(N)$ of the interacting boson model Hamiltonian is obtained by minimizing the energy function $E(N; \beta, \gamma)$ with respect to β and γ, one concludes that the ground state energy $E_0(N)$ is an increasing function of the parameter A when $A < B^2/4C$ and becomes a constant when $A > B^2/4C$. Hence, the parabola $B^2 = 4AC$ in the parameter space constitutes a quantum phase transition in the shape of atomic nuclei from deformed axially symmetric shape to spherical shape. From Eq. (11.56), one immediately sees that near the critical parabola $B^2 = 4AC$ in the parameter space, the ground state energy $E_0(N) = \min E(N; \beta, \gamma)$ as a function of the parameter A has the form

$$\begin{cases} E_0(N) = \mathcal{E}_0 + \dfrac{\beta_c^2(A + B\beta_c + C\beta_c^2)}{(1+\beta_c^2)^2}, & \text{for } A \lessapprox B^2/4C, \\ E_0(N) = \mathcal{E}_0, & \text{for } A > B^2/4C, \end{cases} \tag{11.61}$$

where β_c is the smallest root of the cubic equation $\beta^3 + (\eta - 2/\eta)\beta^2 - 3\beta - \eta = 0$ with $\eta \equiv B/2C$. As an example, for $B = 2$ and $C = 1$, one gets $\beta_c = -1$, and thus the ground state energy for $A \lessapprox 1$ has the form $E_0(N) \approx \mathcal{E}_0 + \frac{1}{4}(A - 1)$. From Eq. (11.61), one immediately obtains

$$\begin{cases} \dfrac{\partial E_0(N)}{\partial A} = \dfrac{\beta_c^2}{(1+\beta_c^2)^2}, & \text{for } A \lessapprox B^2/4C, \\ \dfrac{\partial E_0(N)}{\partial A} = 0, & \text{for } A > B^2/4C. \end{cases} \tag{11.62}$$

It shows that the ground state energy itself is continuous across the critical phase parabola $B^2 = 4AC$, but the slope of the ground state energy with respect to the parameter A has a jump, which corresponds to a first-order phase transition in the Ehrenfest classification.

We now focus on the case with $C > 0$ and $A < 0$. In such a case, the energy function $E(N; \beta, \gamma)$ always has a double root at $\beta = 0$ and two real roots β_+ and β_-. For $B > 0$, from the relation $\beta_+ + \beta_- = -B/C$, one sees that either both roots are negative or $|\beta_-| > \beta_+$, when β_+ is positive. For both cases, the global minimum of the energy function $E(N; \beta, \gamma)$ is located at $\beta < 0$. On the contrary, for $B < 0$, from the relation $\beta_+ + \beta_- = -B/C$, one sees that either both roots are positive or $\beta_+ > |\beta_-|$, when β_+ is positive. For both cases, the global minimum of the energy function $E(N; \beta, \gamma)$ is located at $\beta > 0$. In particular, when $B \approx 0$, from Eq. (11.59), one obtains $2A + (4C - 2A)\beta^2 \approx 0$, and thus the global minimum of the energy function $E(N; \beta, \gamma)$ is located at $\beta_c \approx \pm\sqrt{|A|/(2C - A)}$, depending the sign of B. From Eq. (11.56), one immediately sees that near the critical line $B = 0$ for $A < 0$, the ground state energy $E_0(N) = \min E(N; \beta, \gamma)$ as a function of the parameter B has the form

$$\begin{cases} E_0(N) = \mathcal{E}_0 + \dfrac{\beta_c^2(A + B|\beta_c| + C\beta_c^2)}{(1+\beta_c^2)^2}, & \text{for } B \lessapprox 0, \\ E_0(N) = \mathcal{E}_0 + \dfrac{\beta_c^2(A - B|\beta_c| + C\beta_c^2)}{(1+\beta_c^2)^2}, & \text{for } B \gtrapprox 0. \end{cases} \tag{11.63}$$

As an example, for $C = -A = 1$, one obtains $\beta_c = \pm\sqrt{1/3}$ and $E_0(N) = \mathcal{E}_0 + \frac{\sqrt{3}}{16}(B - \frac{2}{\sqrt{3}})$ for $B \lessapprox 0$, and $E_0(N) = \mathcal{E}_0 - \frac{\sqrt{3}}{16}(B + \frac{2}{\sqrt{3}})$ for $B \gtrapprox 0$. From

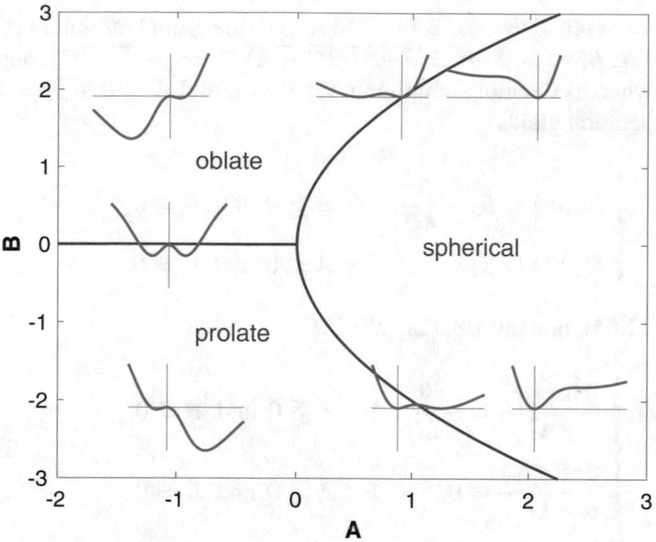

Fig. 11.1 Schematic of the quantum phase diagram for shape transitions in atomic nuclei. The critical line $B = 0$ for $A < 0$ which separates the two phases of oblate and prolate shapes and the critical hyperbola $B^2 = 4CA$ which separates the two phases of spherical and deformed shapes are shown in red. Here, the parameter C is fixed at $C = 1$. The variations of the energy function $E(N; \beta, \gamma)$ with respect to parameters A and B are shown in blue, where we have neglected the constant energy $\mathcal{E}_0$. The top three insets are for $(A, B) = (-1, 2)$, $(1, 2)$, and $(2, 2)$, respectively, the middle inset is for $(A, B) = (-1, 0)$, and the bottom three insets are for $(A, B) = (-1, -2)$, $(1, -2)$, and $(2, -2)$, respectively

Eq. (11.63), one immediately obtains

$$\begin{cases} \dfrac{\partial E_0(N)}{\partial B} = \dfrac{\beta_c^2|\beta_c|}{(1+\beta_c^2)^2}, & \text{for } B \lessapprox 0, \\ \dfrac{\partial E_0(N)}{\partial B} = \dfrac{-\beta_c^2|\beta_c|}{(1+\beta_c^2)^2}, & \text{for } B \gtrapprox 0, \end{cases} \tag{11.64}$$

where $|\beta_c| \equiv \sqrt{|A|/(2C - A)}$. It shows that the ground state energy itself is continuous across the critical line $B = 0$ for $A < 0$, but the slope of the ground state energy with respect to the parameter B has a jump, which also corresponds to a first-order phase transition in the Ehrenfest classification (Fig. 11.1).

To summarize, the critical parabola $B^2 = 4CA$ constitutes a quantum phase transition between the spherical and deformed shapes of the nucleus, whereas the critical line $B = 0$ for $A < 0$ constitutes a quantum phase transition between the prolate and oblate shapes of the nucleus. All the spherical-prolate, spherical-oblate, and prolate-oblate phase transitions are of the first order in the sense of the Ehrenfest classification. However, there is one exception at the triple point $(A, B) = (0, 0)$ at which the three phases coexist. One may show that the quantum phase transition at the triple point is of the second order in the sense of the Ehrenfest classification. To show this, let us recall that for $B \approx 0$ and $A \lessapprox 0$, the global minimum of the energy function $E(N; \beta, \gamma)$ is $\beta_c \approx \pm\sqrt{-A/(2C-A)} \approx \pm\sqrt{-A/(2C)}$, depending the sign of B, whereas the global minimum for $B = 0$ and $A \gtrapprox 0$ is $\beta_c = 0$. Hence, a direct computation yields

$$\begin{cases} E_0(N) = \mathcal{E}_0 - \dfrac{A^2}{4C}, & \text{for } A \lessapprox 0 \text{ and } B = 0, \\ E_0(N) = \mathcal{E}_0, & \text{for } A \gtrapprox 0 \text{ and } B = 0. \end{cases} \tag{11.65}$$

From Eq. (11.65), one immediately obtains

$$\begin{cases} \dfrac{\partial^2 E_0(N)}{\partial A^2} = -\dfrac{1}{2C}, & \text{for } A \lessapprox 0 \text{ and } B = 0, \\ \dfrac{\partial^2 E_0(N)}{\partial A^2} = 0, & \text{for } A \gtrapprox 0 \text{ and } B = 0. \end{cases} \tag{11.66}$$

It shows that both the ground state energy and its first-order derivative with respect to the parameter A are continuous at the triple point $(A, B) = (0, 0)$. But the second-order derivative of the ground state energy with respect to the parameter A has a jump at the triple point, which constitutes a second-order phase transition in the Ehrenfest classification.

Exercises

11.1. For a semisimple Lie algebra with rank r, show that the Casimir operator C_2 in the Cartan basis is given by

$$C_2 \equiv \sum_{i=1}^{r} H_i^2 + \sum_{\boldsymbol{\alpha}\in\boldsymbol{\Phi}^+} (E_{\boldsymbol{\alpha}} E_{-\boldsymbol{\alpha}} + E_{-\boldsymbol{\alpha}} E_{\boldsymbol{\alpha}}) = (\boldsymbol{\Lambda} + 2\boldsymbol{\delta}, \boldsymbol{\Lambda}) I,$$

where I is the unit matrix in the associated irreducible representation and $\delta \equiv \frac{1}{2}\sum_{\boldsymbol{\alpha}\in\boldsymbol{\Phi}^+} \alpha$ is the corresponding Weyl vector, i.e., the half-sum of the positive roots.

11.2. For $n = 1, 2$, verify that the $\mathfrak{su}(6)$ coherent state can be written in two different ways by direct computation

$$|N; \eta_\mu\rangle \equiv \exp[\sum_\mu (\eta_\mu \hat{d}_\mu^\dagger \hat{s} - \eta_\mu^* \hat{s}^\dagger \hat{d}_\mu)] \frac{\hat{s}^{\dagger N}}{\sqrt{N!}} |0\rangle$$

$$= \frac{1}{\sqrt{N!}} \left[\sqrt{1 - |\zeta|^2} \hat{s}^\dagger + \sum_\mu \zeta_\mu \hat{d}_\mu^\dagger \right]^N |0\rangle,$$

where $\zeta_\mu = (\sin|\boldsymbol{\eta}|/|\boldsymbol{\eta}|)\eta_\mu$, $|\boldsymbol{\zeta}|^2 \equiv \sum_\mu \zeta_\mu^* \zeta_\mu$, and $|\boldsymbol{\eta}|^2 \equiv \sum_\mu \eta_\mu^* \eta_\mu$.

11.3. Verify that the boson condensate with particle number N

$$|N;\beta,\gamma\rangle = \frac{1}{\sqrt{N!}}[\hat{b}_c^\dagger(\beta,\gamma)]^N|0\rangle$$

satisfies the normalization condition $\langle N;\beta,\gamma|N;\beta,\gamma\rangle = 1$.

11.4. Verify the following relations:

$$\hat{s}|N;\beta,\gamma\rangle = \sqrt{\frac{N}{1+\beta^2}}|N-1;\beta,\gamma\rangle,$$

$$\hat{d}_\mu|N;\beta,\gamma\rangle = \sqrt{\frac{N}{1+\beta^2}}a_\mu|N-1;\beta,\gamma\rangle.$$

11.5. Show that $[\hat{d}^\dagger \times \tilde{s}]_M^{(J)} = \delta_{J,2}\hat{d}_M^\dagger s$ and $[\tilde{s}^\dagger \times \tilde{d}]_M^{(J)} = \delta_{J,2}\hat{s}^\dagger \hat{d}_M$.

11.6. Verify the following relations:

$$[\hat{d}^\dagger \times \tilde{d}]_2^{(2)} = \sqrt{\frac{2}{7}}(\hat{d}_2^\dagger \hat{d}_0 + \hat{d}_0^\dagger \hat{d}_{-2}) + \sqrt{\frac{3}{7}}\hat{d}_1^\dagger \hat{d}_{-1},$$

$$[\hat{d}^\dagger \times \tilde{d}]_1^{(2)} = \sqrt{\frac{3}{7}}(\hat{d}_{-1}^\dagger \hat{d}_{-2} - \hat{d}_2^\dagger \hat{d}_1) - \sqrt{\frac{1}{14}}(\hat{d}_1^\dagger \hat{d}_0 - \hat{d}_0^\dagger \hat{d}_{-1}),$$

$$[\hat{d}^\dagger \times \tilde{d}]_0^{(2)} = \sqrt{\frac{2}{7}}(\hat{d}_2^\dagger \hat{d}_2 - \hat{d}_0^\dagger \hat{d}_0 + \hat{d}_{-2}^\dagger \hat{d}_{-2}) - \sqrt{\frac{1}{14}}(\hat{d}_1^\dagger \hat{d}_1 - \hat{d}_{-1}^\dagger \hat{d}_{-1}),$$

$$[\hat{d}^\dagger \times \tilde{d}]_{-1}^{(2)} = \sqrt{\frac{3}{7}}(\hat{d}_1^\dagger \hat{d}_2 - \hat{d}_{-2}^\dagger \hat{d}_{-1}) - \sqrt{\frac{1}{14}}(\hat{d}_{-1}^\dagger \hat{d}_0 - \hat{d}_0^\dagger \hat{d}_1),$$

$$[\hat{d}^\dagger \times \tilde{d}]_{-2}^{(2)} = \sqrt{\frac{2}{7}}(\hat{d}_{-2}^\dagger \hat{d}_0 + \hat{d}_0^\dagger \hat{d}_2) + \sqrt{\frac{3}{7}}\hat{d}_{-1}^\dagger \hat{d}_1.$$

11.7. Verify the following relations:

$$\langle N;\beta,\gamma|(\hat{d}^\dagger \times \tilde{d})_{\pm 2}^{(2)\dagger}(\hat{d}^\dagger \times \tilde{d})_{\pm 2}^{(2)}|N;\beta,\gamma\rangle = \frac{N}{7}[\frac{(N-1)\beta^4 \sin^2 2\gamma}{(1+\beta^2)^2} + \frac{\beta^2(1+\cos^2\gamma)}{1+\beta^2}],$$

$$\langle N;\beta,\gamma|(\hat{d}^\dagger \times \tilde{d})_{\pm 1}^{(2)\dagger}(\hat{d}^\dagger \times \tilde{d})_{\pm 1}^{(2)}|N;\beta,\gamma\rangle = \frac{N}{7}\frac{\beta^2}{1+\beta^2}(\sin^2\gamma + \frac{1}{2}),$$

$$\langle N;\beta,\gamma|(\hat{d}^\dagger\times\tilde{d})_0^{(2)\dagger}(\hat{d}^\dagger\times\tilde{d})_0^{(2)}|N;\beta,\gamma\rangle=\frac{2N}{7}[\frac{(N-1)\beta^4\cos^2 2\gamma}{(1+\beta^2)^2}+\frac{\beta^2}{1+\beta^2}].$$

11.8. Verify the following expected values:

$$\langle N;\beta,\gamma|\hat{n}_d|N;\beta,\gamma\rangle=\frac{N\beta^2}{1+\beta^2},$$

$$\langle N;\beta,\gamma|\hat{Q}_0\cdot\hat{Q}_0|N;\beta,\gamma\rangle=\frac{N}{1+\beta^2}\left(\frac{4(N-1)\beta^2}{1+\beta^2}+5+\beta^2\right),$$

$$\langle N;\beta,\gamma|(\hat{Q}_\chi-\hat{Q}_0)\cdot(\hat{Q}_\chi-\hat{Q}_0)|N;\beta,\gamma\rangle=\frac{2}{7}\frac{\chi^2N(N-1)\beta^4}{(1+\beta^2)^2}+\frac{\chi^2N\beta^2}{1+\beta^2},$$

$$\langle N;\beta,\gamma|\hat{Q}_0\cdot(\hat{Q}_\chi-\hat{Q}_0)|N;\beta,\gamma\rangle=-2\sqrt{\frac{2}{7}}\frac{\chi N(N-1)\beta^3}{(1+\beta^2)^2}\cos 3\gamma.$$

where $\hat{n}_d\equiv\hat{d}^\dagger\cdot\tilde{d}=\sum_\mu\hat{d}^\dagger_\mu\hat{d}_\mu$, $\tilde{d}_\mu\equiv(-1)^\mu\hat{d}_{-\mu}$, $\hat{Q}_0\equiv(\hat{d}^\dagger\times\hat{s}+\hat{s}^\dagger\times\tilde{d})^{(2)}$, and $\hat{Q}_\chi\equiv\hat{Q}_0+\chi(\hat{d}^\dagger\times\tilde{d})^{(2)}$.

Quantum Chaos 12

12.1 Overview

Since the dawn of science began with the prediction of solar eclipse by Thales of Miletus in the sixth century B.C., the attempt to predict future events based on past causes is one of the primary goals of science. It is no wonder that in the rational era of Kepler, Newton, and Laplace, the central dogma of natural philosophy, or what is now called classical mechanics, was to provide a complete description of a dynamical system, e.g., future positions of a collection of particles, based on the properties of the present physical state. In fact, the causal determinism à la Laplace was the dominant philosophy among scientists and philosophers for more than 300 years. Amidst such an overwhelming deterministic thinking, it was Henri Poincaré who discovered what is now known as **deterministic chaos** through his studies of restricted three-body problem in the 1880s.

In the middle of 1885, to honor the 60th birthday of King Oscar II of Sweden and Norway, an international prize competition was officially announced in Volume 7 of the newly established journal *Acta Mathematica*. In order to attract the most brilliant mathematicians in the realm of mathematical analysis, the prize jury which consisted of Mittag-Leffler in Stockholm, Karl Weierstrass in Berlin, and Charles Hermite in Paris agreed on four problems. One of the proposed problem was related to the study of Fuchsian function, which was invented by Poincaré himself and made him famous 5 years earlier. But, in correspondence with Mittag-Leffler, Poincaré decided to tackle the first and the most difficult problem, proposed by Weierstrass, namely, the n-body problem: "For a system of arbitrarily many mass points that attract each other according to Newton's law, assuming that no two points ever collide, find a series expansion of the coordinates of each point in known functions of time converging uniformly for any period of time." Throughout the eighteenth and nineteenth centuries, even the greatest mathematicians and astronomers held the idea that the solar system had to be stable. For this, Weierstrass had had a great hope on receiving an analytical method of solution, and with which one could exclude that

C.-F. Kam et al., *Coherent States*, Lecture Notes in Physics 1011,
https://doi.org/10.1007/978-3-031-20766-2_12

the orbit of solar planets varies drastically over a long period of time or even that one of the planets can be rejected from the system.

In May 1888, Poincaré submitted his memoir for the Prize of King Oscar II entitled *Sur le problème des trois corps et les équations de la dynamique* (*The Three-Body Problem and the Equations of Dynamics*) after enduring hard work. In his memoir, Poincaré initiated a veritable conceptual revolution. He abandoned the transitional approach to dynamical differential equations based on infinite trigonometric series. Instead, he studied the flows in phase space by using ingenious geometric and topological methods to extract the qualitative aspects of the entire dynamical system. Having these innovative tools on hand, Poincaré is sufficient to prove his *stability theorem à la Poisson*, or what is now known as the *Poincaré recurrence theorem*: "almost every trajectory of the restricted three-body system passes arbitrarily close to its initial position infinitely often."

Although the Poincaré recurrence theorem is profound for sure, it is definitely too abstract to be useful in practice. To this end, Poincaré continued his exploration of stability in the restricted three-body systems with a body A of mass $1 - \mu$, a body B of mass μ, and a body C of negligible mass. He sought to understand how and in what conditions periodic orbits exist in the restricted three-body systems. Remarkably, Poincaré was able to prove that as long as the mass μ is sufficiently small, the restricted three-body system always possesses periodic orbits. In particular, the number of periodic trajectories tends to infinity as μ tends to 0. Bolstered by his new results about periodic orbits, Poincaré found and obtained new types of solutions, namely, *asymptotic solutions*, which asymptotically approached an unstable periodic orbit when moving forward and backward in time. Using Poincaré's terminology, one may denote S_T^s as the set of initial positions of all trajectories asymptotic to T in the future and S_T^u as the set of initial positions of all trajectories asymptotic to T in the past. Poincaré proved that both S_T^s and S_T^u are surfaces, namely, *asymptotic surfaces* associated to T, from which he elaborated on the second stability result a theorem on the coincidence of asymptotic surfaces: "As long as the mass μ is sufficiently small, the asymptotic surfaces S_T^s and S_T^u associated to the unstable periodic orbit T coincide." This theorem clearly explains the stability à la Poincaré: the trajectories that had nearly periodic motion in the past, but whose motion was disturbed later, will finally recover their initial nearly periodic motion. As such, after almost 300 handwritten pages of manuscripts full of novel and sophisticated geometric arguments, Poincaré announced a claim of stability for the restricted three-body systems. In March 1888, Poincaré received a sum of 2500 kronor together with a gold medal from the hands of the Swedish ambassador in Paris. It seems that the problem of solar stability had been settled, and the causal determinism has been recovered. But in hindsight it is, merely, the end of the beginning.

On the last day of November 1889, the prize jury Mittag-Leffler received an ominous telegram from Paris, in which Poincaré requested to stop the presses of his memoir for the Prize of King Oscar II. He had found a serious error. Soon, Mittag-Leffler discovered that the error was graver than what Poincaré described in the letter arrived the next day. In a sleepless night, Mittag-Leffler wrote disappointedly: “It is even not true that asymptotic surfaces are closed.”

The major fault that Poincaré had made was to implicitly assume that the two asymptotic surfaces S_T^s and S_T^u, which formed by the associated solutions attracted or repelled from the periodic orbit, smoothly glued together to comprise a single closed surface sheet, and no intersections occurred between them. After an intense work for more than 1 month, Poincaré was able to submit a substantially revised memoir on January 5, 1890. After altering some of those implicit assumptions that he made previously, Poincaré arrived at the inevitable conclusion that deterministic chaos was unavoidable in restricted three-body problem. Poincaré found to his astonishment that the asymptotic solution curves could in fact intersect each other and would have to intersect at an infinity of points! A sequence of new intersection points are generated by iteration from the original intersection point. Poincaré called these intersections *homoclinic points*, namely, they asymptotically both leave and return to the same unstable periodic orbit. Still more astonishing to him is he soon discovered that, due to the stretching and folding of the asymptotic surfaces, an infinity of new intersections from distinct homoclinic trajectories appears between the original ones. All these intersections added together form a seemingly endless maze-like intricate patterns made of infinitely fine web, or a *homoclinic tangle*, using Poincaré’s terminology. Even Poincaré himself found it difficult to understand the stunning implications of his discovery of homoclinic points and orbits: in a tiny volume of phase space, an infinite number of small slices of asymptotic surfaces enclose delicate stretched-and-folded regions which comprise different kinds of orbits. This makes any predictions of the fate of an orbit, if not completely impossible, practically meaningless, as it is inherently sensitive to the initial conditions.

Eventually the door of causal determinism à la Laplace was closed, but the door to the new wonderland of deterministic chaos just opened. After Poincaré’s pioneering works, many great names contributed to the qualitative studies of dynamical systems through which various aspects of deterministic chaos were manifested. Those great names included George Birkhoff from America, Aleksandr Lyapunov from Russia, and Jacques Hadamard from France, to name a few. But it was not until 1963 that mankind’s view of nature had been profoundly and fundamentally changed. At that particular year, an article by the meteorologist Edward Lorenz at MIT was published, in which a strange attractor with fractal structure, namely, the Lorenz attractor, was discovered. Unlike ordinary attractors which are a collection of points or limit cycles, a strange attractor is a dynamical equilibrium which is composed of those trajectories that pass through the entire phase space. More importantly, in studying the return maps à la Poincaré for strange attractors, iterated maps emerged, which led to the celebrated discovery of the universality of chaos.

But it was not until the mid-1970s that different train of thoughts about iterated maps started to converge. In 1975, dynamical systems theorist James A. Yorke and his PhD student Tien-Yien Li at the University of Maryland published the paper entitled “Period Three Implies Chaos” [158], which proved that in an iterated map, the existence of a cycle of period three will inevitably lead to the existence of uncountable infinitude of points that never map to any cycle. Remarkably, a slightly weaker result that a cycle of period three implies cycles of arbitrary periods had already been proved in 1964 by Oleksandr M. Sharkovsky, a Ukrainian mathematician who worked in the Ukrainian school of dynamical systems research. Meanwhile, in 1976, the mathematically oriented population biologist Robert May, who was then a biology professor at Princeton University and the chief scientific adviser to the UK Government, published a paper in *Nature* entitled "Simple Mathematical Models with Very Complicated Dynamics" [159], in which the so-called logistic map was studied in detail and popularized by himself. But the real discovery of the quantitatively universal properties of chaos had to await Mitchell J. Feigenbaum and his discovery of Feigenbaum constant $\delta \approx 4.6692016$.

Mitchell Feigenbaum was born in Philadelphia in 1944 and raised in Brooklyn. His parents are Jewish emigrants from Poland and Ukraine. In 1960, at the age of 16, Feigenbaum enrolled in the City College of New York, studying electrical engineering. After graduation he had to make a decision on whether to become a money-making electrical engineer or to become a scientist who can explore the physical world as a career option. He decided to choose the latter. In 1964, he went to MIT. He initially enrolled in a PhD program in electrical engineering, but he quickly switched to physics. In the 1960s, particle physics entered its golden age. But Feigenbaum has not taken a strong interest in this trend of development, though he finished his PhD thesis on a topic in particle physics set by his advisor Francis E. Low. In 1970, when he was 26, he found a postdoc position at Cornell. But he still struggled with motivation. During his 2 years at Cornell, Feigenbaum did not produce any visible academic output. But it was in Cornell that he first met Peter A. Carruthers who became his "Bo Le" several years later.

After Cornell, Feigenbaum left for his second postdoc position at Virginia Polytechnic Institute, at which he performed even worse than what he did in Cornell. He didn't interact with his colleagues very well. After 2 more years, he ended up with only a three-page paper. In 1974, he was 30 years old, and it still wasn't clear what was going to happen. With only one paper, his prospects after Virginia Polytechnic Institute were bleak. However, his luck was on the turn. Peter A. Carruthers had been hired to establish the theory division of Los Alamos National Laboratory and given carte blanche to hire and fire staff without legal obstacles. He recognized Feigenbaum's talent and had a strong feeling that Feigenbaum had the potential to make impactful discoveries. Thus, he brought Feigenbaum to Los Alamos, despite other people's doubt and negativity. In 1974, there was breaking news that Kenneth G. Wilson had solved the long-standing Kondo problem using his new technique called renormalization group. Carruthers suggested Feigenbaum to try to apply this method to study fluid turbulence.

Before Feigenbaum started to get involved in the research of turbulence, David P. Ruelle and Floris Takens predicted in a paper in 1971 that fluid turbulence could develop through strange attractors in the nonlinear dynamical Navier-Stokes equations. In fact, it was exactly this paper that coined the phrase "strange attractor." In their approach, one of the characteristics of fluid turbulence was the onset of bifurcation cascade. A single bifurcation is a sudden alteration of an orbit when external parameters are only slightly changed. In contrast, a bifurcation cascade was a succession of bifurcations that doubled the period of an orbit each time, until chaos fully emerged. Hence, Feigenbaum tried to understand bifurcation cascade in iterated map, a simplified version of differential equation by making time discrete. He was particularly interested in the bifurcation cascade of the logistic map: $x_{i+1} = rx_i(1 - x_i)$, where r represents the growth rate and x_i represents the ratio of existing population to the maximum population. Feigenbaum found a strong similarity between the bifurcation cascade of the iterated map and the concept of real-space renormalization, where both of which involved scale transformations from smaller to larger structures. Feigenbaum thus studied how the bifurcations of the logistic map depended on the growth rate. He computed the threshold values r_n at the n-th period-doubling bifurcation and then studied the ratios of the intervals $(r_{n+2} - r_{n+1})/(r_n - r_{n-1})$. He found to his astonishment that the ratio approached to a limiting value of $\delta \approx 4.6692016$. Feigenbaum had soon discovered that δ seemed to be a universal constant as long as the iterated map had a single quadratic maximum. In 1976, Feigenbaum tried to submit a manuscript to the prestigious journal *Advances in Mathematics* to announce his results. He waited 6 months, but his manuscript finally got rejected. He tried again, sending the manuscript to the *SIAM Journal on Applied Mathematics*. It got rejected again. Eventually, in late 1977, the editor of the *Journal of Statistical Physics*, Joel Lebowitz, agreed to publish his manuscript. Thus, Feigenbaum's results were officially announced in a paper entitled "Quantitative Universality for a Class of Nonlinear Transformations" in 1978 [160]. Remarkably, in 1979, Albert J. Libchaber in Paris reported his results on an experiment in liquid helium that period doubling is observed in the transition to turbulence, with the same exponent δ that was theoretically predicted by Feigenbaum. Feigenbaum and Libchaber immediately became famous. They shared the Wolf Prize in Physics in 1986. Since then, the theory of deterministic chaos was promoted to all corners of science and technologies and soon even became a common occurrence in the public media.

However, nearly one century ago, mankind witnessed the genesis of a new fundamental theory of physics, namely, quantum mechanics, which provides a description of the physical properties of matter at the microscopic level. As such, it is logically impossible that the strange behaviors of deterministic chaos have no relationship with quantum mechanics at all, as classical mechanics is merely a limiting theory for an aggregate of microscopic particles. But, due to the inherent different mathematical structures attached to quantum and classical mechanics, it has been a huge challenge for generations of physicists to establish a fundamental theory of deterministic chaos which is completely based on quantum mechanics. In the modern theory of deterministic chaos, a dynamical system is said to be chaotic, if it has a Lorenz attractor-like subset S of the phase space, which is invariant in the sense that every evolution starting in the subset will stay in it, and the subset S fulfills the following properties: (1) Orbits in the subset are sensitive to initial conditions, i.e., the phase space trajectories exiting from nearly points in S diverge exponentially. (2) No evolution starting in S is periodic or quasi-periodic. (3) No evolution in S tends to a periodic or quasi-periodic evolution when time tends to infinity. From the above crucial properties of chaotic systems, one clearly sees that deterministic chaos only happens in nonlinear systems. As such, the existence of deterministic chaos deeply contradicts with the linear evolution in quantum mechanics. What is even worse is that due to the Heisenberg uncertainty principle, the concept of trajectory simply does not exist in quantum mechanics, not to mention the infinitely fine web of homoclinic tangle, or strange attractor with fractal structure. Hence, it is apparently unclear how classical chaotic dynamics, comprised of trees of cycles of increasing lengths and self-similar structures, which is utterly a nonlinear phenomenon, can manifest itself at the quantum level.

In fact, since the genesis of quantum mechanics, the question of how classical mechanics can be a limiting case of quantum mechanics, besides a few examples in simple regular systems, was not rigorously answered. This was largely due to the empirical successes of quantum mechanics in explaining and predicting microscopic physical behaviors, so that a microscopic description of the discovery of Poincaré has no imminent requirement. Indeed, it was not until the latter half of the twentieth century when the research in nonlinear science has attracted an intense focus due to the increasing of computing power; the question of microscopic origin of the observed deterministic chaos became an important subject of theoretical physics.

During the last several decades, attempts to establish relationship between the classical deterministic chaos and the quantum mechanics have triggered a new field called **quantum chaos**, which is a field still full of conundrums and opportunities, rather than well-posed problems. Up to date, the research of seeking relationship between classical deterministic chaos and quantum mechanics falls into two main categories: one studies the energy spectrum, and the other studies the wave functions. For the former, the energy-level spacing distribution for regular quantum systems obeys Poisson statistics, while for generic fully chaotic systems, according to the Bohigas-Giannoni-Schmit conjecture proposed nearly 40 years ago [161], highly excited energy levels obey universal spectral statistics. The analysis of highly excited energy-level statistics-based large random matrices was developed in the late 1950s and early 1960s by Eugene P. Wigner [162], Freeman J. Dyson [163], and Madan L. Mehta [164], respectively. In the absence of geometric symmetries, the prediction of random matrix theory only depends on whether the system is not time-reversal ($\mathcal{T}$) invariant, described by Gaussian unitary ensemble (GUE), or is time-reversal invariant with either $\mathcal{T}^2 = 1$ or -1, described by Gaussian orthogonal ensemble (GOE) or Gaussian symplectic ensemble (GSE), respectively. These results have been verified numerically for many dynamical systems, but a rigorous proof for them, unfortunately, has not yet been put forward. For the latter, it was discovered by Eric J. Heller in 1984 in his study of configuration space distribution of eigenstates that many "scar" phenomena associated with the classical unstable periodic orbits are exhibited in the projection of wave function on either configuration space or phase space [165]. It implies that the classical trajectories and the phase space distribution of quantum density for classical regular motion are actually closely correlated. Interestingly, scars have recently been re-explored in a new light after experimentalists at Harvard surprisingly demonstrated periodic orbits in their quantum simulator [166]. It was suggested that these periodic orbits could be due to a many-body version of quantum scar [167].

Though the abovementioned approaches to quantum chaos have shed invaluable light on the signatures of classical deterministic chaos in quantum systems, we shall focus our attention on the time-honored question of quantum-classical correspondence. Here, quantum-classical correspondence means the search for an unambiguous classical limit, starting purely from quantum mechanics. This would be crucial to understanding quantum chaos, partly because the essential concepts for analyzing classical deterministic chaos, such as orbits, attractors, or even strange attractors, are either absent or meaningless in quantum mechanics and also because classical regularity is based on the criterion of integrability, a concept which is not yet fully developed in quantum mechanics.

Historically, the standard quantization techniques were developed in integrable systems. Hence, there is no compelling reason to apply these methods to a classical chaotic system. In this regard, the direction going from classical to quantum mechanics seems to not have the necessary theoretical underpinning. As such, it may be more fruitful to study how quantum mechanics can be used to describe nonlinear phenomena in the classical world. In other words, one has to answer the question of how the nonlinear dynamics of classical objects can be understood from quantum mechanics. To this end, a thorough understanding of deterministic chaos in the framework of quantum mechanics requires a rigorous formulation of quantum-classical correspondence, from which one can reveal how classical mechanics is hidden in the quantum world and how deterministic chaos can naturally emerge from the quantum mechanics.

In conventional quantum mechanics literatures, the fundamental problem of quantum-classical correspondence is either it has to be entirely omitted or implemented through the action $\hbar \to 0$ which is logically possible but empirically impossible. The reduced Planck constant is $\hbar = 6.58212 \times 10^{-16}\text{eV} \cdot \text{s}$ and—mathematically allowed to vary it notwithstanding—it is a constant by definition. In fact, the action of letting the universal reduced Planck constant $\hbar$ to vanish is simply impossible in nature. Mathematically, this is an ill-defined limit as there is no known physical operator which is expressible in terms of $\hbar$.

This becomes strikingly evident when the relativistic versus nonrelativistic mechanics are contrasted. In such a case, the speed of light c is a universal constant, and $v/c \to 0$ is a well-defined nonrelativistic limit. In the same vein, the reduction from quantum to classical mechanics should also depend on the ratio of a physical quantity and $\hbar$. For a one-dimensional system, the action $S[q(t)]$ given by the time integral of the Lagrangian for an input evolution $q(t)$ is such a unique quantity. However, beyond the simplest one-dimensional case, the situation becomes murky for an obvious reason. An n-dimensional integrable system can always be regarded as a combination of n one-dimensional systems. Hence, even if the action of the entire system is large, it does not automatically imply that the corresponding action for every subsystem must also be large. To this end, it is conceivable that even with a large action, a multidimensional system can still be very much quantum mechanical. For a non-integrable system, the precise meaning of $\hbar \to 0$ is even more ambiguous. This can be seen in the path integral formulation in which the action is obtained by an integration over an appropriate time interval which for a chaotic system could be infinitely large. Thus it is not all clear that a large action can be the determining condition to physically realize the limit of $\hbar \to 0$. As such, the exploration of the chaotic phenomena by varying the reduced Planck constant $\hbar$ without fully examining all its ramifications could be hazardous.

From this point of view, the basic question one must be faced with in exploring the onset of chaos from fundamental quantum theory is that what is quantum non-integrability in finite systems. Quantum mechanics is built on a framework that completely differs from classical mechanics. It stands on a very solid mathematical foundation and does not require any classical input. Hence, a crucial issue for studying quantum non-integrability is to see whether the related concept of quantum integrability can be properly defined. Overall, the integrability of the physical equations of motion, albeit quantum or classical, may be well defined. For instance, the linear Schrödinger equation is integrable. However, the story hardly terminates here as quantum dynamics describes the various correlated distributions of physical objects via the wave functions, rather than the wave functions themselves. Moreover, for any realistic quantum system, its wave function state space is an infinite-dimensional space. Hence, the linearity of the Schrödinger equation or, more precisely, the quantum mechanical hypothesis of linear superposition of wave functions by no means implies the linearity of quantum phenomena or the lack of chaos.

After all, integrability must be a rigorous mathematical concept. In order to describe the global properties of nonlinear phenomena, it has to be precisely defined. In classical mechanics, the precise definition of integrability dated back to the nineteenth century and was due to the France mathematician Joseph Liouville. According to the Liouville-Arnold theorem, a system with n degrees of freedom described by the Hamiltonian H and the Poisson bracket $\{,\}$ is said to be **Liouville integrable** if there exists n conversed quantities I_i in involution, i.e., with vanishing Poisson brackets $\{I_i, I_j\} = 0$ for arbitrary $i, j = 1, \cdots, n$. In quantum mechanics, although the Schrödinger equation is linear, or more precisely, the wave functions are linear vectors in the Hilbert space, the basic algebraic structure to determine quantum dynamics is encoded in the commutators, or Lie brackets. Mathematically both the Poisson brackets in classical mechanics and the Lie brackets in quantum mechanics lie at the same level, namely, they define the algebraic structures of group. Hence, it is natural to ask whether one can establish the concept of quantum integrability in finite systems based on the algebraic structures of group, so that the notion of classical integrability becomes a limiting case of the general concept of quantum integrability. Fortunately, the answer should be positive. In fact, it is exactly the group representation theory that provides the mathematical basis of quantum mechanics. Starting from the group theory, whether a system can be analytically solved can be completely determined and can be defined as a criterion of integrability [16, 168].

Based on the group theoretical approach, one can develop a semi-quantal approach to the problem of quantum-classical correspondence as well as quantum non-integrability. Here the term semi-quantal is used to distinguish its difference from the familiar semiclassical approach to path integrals formalism. The semiclassical approach, which is based on the stationary phase approximation of the Feynman path integral, is essentially to use the classical mechanics to approximately extract the quantum phenomena. In the last few decades, there are numerous attempts proposed to improve the semiclassical approach by including quantum corrections to the stationary phase approximation, so that the quantum dynamics can be treated approximately from the classical mechanics plus quantum corrections as a perturbative effect in $\hbar$. One can of course surmise from various integrable examples that the semiclassical approach can provide a good, and at times exact, description of the quantum system for integrable systems. But there is no theoretical foundation that these results lend confidence for such approaches for chaotic systems and from which one may still have a reasonable tool to explore the quantum chaotic phenomena.

However, it is hard to imagine that the semiclassical approaches can serve as the starting point in the study of quantum-classical correspondence, especially in addressing the time-honored problem of quantum chaos, as we argued in the above. The semiclassical methods can provide a good description of quantum dynamics only for those systems whose quantum correlations do not significantly alter their corresponding classical phase space structures. A chaotic system has an unstable phase space structure as it is susceptible to non-perturbative alteration by the slightest change in its dynamics. The quantum correlation of a microscopic system is unavoidable. For example, the leading quantum correlation, in the mean-field description, is equivalent to an effective potential. When a system is chaotic, such an effective potential will probably destroy the original classical phase space structure, and thus one is unable to expect the reproduction, even approximately, of the quantum dynamics of a chaotic system with semiclassical approaches. In contrast, although the phase space structure of an integrable system may also be altered by quantum correlation, such alteration hardly changes the topology of the phase space structure due to the constraint of integrability.

It is worth noting that the semi-quantal approach which we shall discuss in the following is still a stationary phase approximation. However, it does not trace its origin to the classical mechanics and is based on the generalized phase space path integral, which is derived from the quantum evolutionary equation using the generalized coherent state theory. One of the most notable properties of the generalized phase space path integral is that the Lagrangian or the action itself in this formalism is solely derived from quantum mechanics. As we shall see later on, this effective action includes the leading order of quantum correlations. In fact, only in certain limits, the effective action can be reduced to the classical action. When this happens, the semi-quantal approach is exactly the same as the semiclassical approach.

In addressing the problems of quantum-classical correspondence and quantum chaos, one has to go beyond the classical limit. The semi-quantal method thus provides a description of the dynamics with the influence of the leading quantum correlation on the phase structure of dynamics. On the one hand, it provides a link between quantum and classical mechanics, while on the other hand, it provides a basis for studying chaos by exploring the effect of quantum correlation in the chaotic phase space structure. For this reason, one may regard this approach suitable for examining the quantum-classical correspondence with the goal of generating the classical physics from the quantum mechanics. Of course, one may also incorporate higher-order quantum correlations in the stationary phase approximation within the semi-quantal approach, similar to that within the semiclassical approach, though it will not be treated in the following:

12.2 Geometry in Quantum Systems

For a classical system, dynamical observables are differential analytic functions defined on a phase space. The phase space is coordinated by the physically independent degrees of freedom. A point in the phase space represents a physical state. For a given initial point, the time evolution of a system is uniquely characterized by a trajectory in the phase space. Chaotic dynamics is the study of *topological* characteristics of the trajectories as a whole, i.e., the phase space structures.

On the other hand, the dynamics properties of a quantum system are described within a Hilbert space, denoted as $\mathcal{H}$. The dynamical observables are represented by self-adjoint operators acting on this space. A physical state of the system is described by a vector, or more precisely, a **ray** in $\mathcal{H}$, the equivalent class of all vectors in $\mathcal{H}$ which differs by a phase factor. Hence, the Hilbert space of a system plays a role analogous to that of classical phase space. Unfortunately, a Hilbert space cannot be directly defined as a quantum phase space since its dimensions cannot be interpreted as physical degrees of freedom, nor can it be directly reduced to a classical phase space in the classical limit. Moreover, the wave functions are not observables themselves. Hence, for quantum dynamics, there is no a priori phase space description. Yet, the study of quantum-classical correspondence demands knowledge of the associated geometry for a given Hilbert space, which quantum mechanics can be embedded in and from which classical mechanics can be reproduced.

In 1966, F. Strocchi attempted to construct a phase space description of quantum mechanics [169]. He demonstrated that for a given Hilbert space $\mathfrak{H}$ with basis $|\alpha_i\rangle$, $i = 1, 2, \cdots, N$, any state $|\psi\rangle \in \mathfrak{H}$ is expressible as

$$|\psi\rangle = \sum_{i=1}^{N} u_i |\alpha_i\rangle, \tag{12.1}$$

where $u_i (i = 1, \cdots, N)$ are complex coefficients which completely determine $|\psi\rangle$. As a remark, N can be in principle infinite. For the convenience of discussion, we shall consider N as finite. By setting

$$u_i \equiv \frac{1}{\sqrt{2}}(q_i + ip_i), u_i^* \equiv \frac{1}{\sqrt{2}}(q_i - ip_i) \tag{12.2}$$

and substituting Eq. (12.1) into the time-dependent Schrödinger equation

$$i\frac{\partial|\psi\rangle}{\partial t} = H|\psi\rangle, \tag{12.3}$$

one can readily find that the Schrödinger equation can be revised as the Hamiltonian equations

$$\frac{dq_i}{dt} = \frac{\partial\mathcal{H}(q,p)}{\partial p_i}, \frac{dp_i}{dt} = -\frac{\partial\mathcal{H}(q,p)}{\partial q_i}, \tag{12.4}$$

where $\mathcal{H}(q, p)$ is the matrix element of the Hamiltonian operator for the state $|\psi\rangle$

$$\mathcal{H}(q,p) \equiv \langle\psi|H|\psi\rangle = \sum_{ik}\langle\alpha_i|H|\alpha_k\rangle u_i^* u_k \tag{12.5}$$

$$= \frac{1}{2}\sum_{ik} H_{ik}(p_i p_k + q_i q_k + ip_k q_i - ip_i q_k).$$

Moreover, in this formulation, the commutation relationship $[A, B]$ of any two physical observables A and B can be expressed as a Poisson bracket

$$\langle\psi|[A,B]|\psi\rangle = \{\mathcal{A},\mathcal{B}\}, \tag{12.6}$$

where

$$\mathcal{A}(q,p) \equiv \langle\psi|A|\psi\rangle = \sum_{ik}\langle\alpha_i|A|\alpha_k\rangle u_i^* u_k \tag{12.7}$$

$$= \frac{1}{2}\sum_{ik} A_{ik}(p_i p_k + q_i q_k + ip_k q_i - ip_i q_k).$$

The implication of the Hamiltonian Eq. (12.4) and the Poisson bracket Eq. (12.6) is that quantum mechanics can formally be embedded into a classical mechanics which is defined in a $2N$-dimensional phase space. From this formalism, it seems that one may superficially conclude that a quantum system is always equivalent to an N-coupled classical harmonic oscillator system. More importantly, as the Hamiltonian function is quadratic, the Hamiltonian equation (Eq. (12.4)) is, by definition, integrable. One immediate consequence is that a quantum system is always integrable and thus "quantum chaos" simply does not exist.

However, the above conclusion is false. According to the probabilistic interpretation of quantum mechanics, one requires that the relevant Hilbert space is square-integrable,

$$\int ||\psi\rangle|^2 du < \infty, \tag{12.8}$$

and $|\psi\rangle$ can always be multiplied by a constant c, i.e., $|\psi\rangle \to |\tilde{\psi}\rangle \equiv c|\psi\rangle$, such that $\langle\tilde{\psi}|\tilde{\psi}\rangle = 1$. This condition imposes a constraint on the $2N$-dimensional phase space,

$$\sum_{i=1}^{N} |\tilde{u}_i|^2 = 1, \tag{12.9}$$

where $\tilde{u}_i \equiv cu_i$. As a result, the $2N$-dimensional phase space described by Eq. (12.4) is reduced to a **complex projected space**, mathematically denoted as $\mathrm{CP}(N-1)$. In this manifold, the Hamiltonian function is not an N-coupled harmonic oscillator in quadrature. Moreover, since $\mathrm{CP}(N-1)$ is compact, its topological structure is different from the usual classical phase space. In this regard, before the quantum equations can be expressed in terms of Hamilton-type equations, one needs to verify whether a symplectic structure still exists. Thus, although Eq. (12.4) is defined on the $2N$-dimensional phase space, it does not represent a formulation of a quantum system without the constraint Eq. (12.9).

Hilbert spaces are still the basic framework of quantum mechanics. In the group representation theory, the Hilbert space is realizable by a unitary irreducible representation space of an algebra. Moreover, for a given algebra, there is an associated group which carries a natural geometrical manifold. The above $\mathrm{CP}(N-1)$ space is in fact a subspace of the parametrized manifold of $\mathrm{SU}(N)$, where $\mathrm{SU}(N)$ is a group of the N-level system. In the following, we will show that all the Hilbert spaces of quantum systems are indeed associated with a subspace of their group geometric manifold, and from which the physical phase space can be firmly established [16, 168].

To establish the geometry for quantum systems, one needs a dynamical group structure. According to the axioms of quantum mechanics, a quantum system is a self-adjoint operator algebra, denoted by $\mathfrak{g}$ and spanned by the basic physical observables. The algebraic structure is given by the commutation relations. Explicitly, the systems' Hamiltonian H and the transition operators $\{A\}$ can both be expressed as functions of a set of the basic operators $\{T_i\}$ with $i = 1, \cdots, n$ of $\mathfrak{g}$

$$H = H(T_i), A = A(T_i). \tag{12.10}$$

This set of operators satisfies a closed commutation relationship, $[T_i, T_j] = \sum_{k=1}^{n} C_{ij}^k T_k$, namely, it spans an algebra where the structure constants of $\mathfrak{g}$ are C_{ij}^k.

We then introduce the concept of dynamical group. A **dynamical group**, denoted by G, is defined by a unitary exponential mapping of $\mathfrak{g}$

$$g \to G \equiv \left\{ \exp\left(i \sum_i \alpha_i T_i \right) \right\}, \tag{12.11}$$

where unitarity is required by quantum mechanics. Mathematically, G is called the **covering group** of $\mathfrak{g}$. According to the group representation theory, the state space in quantum mechanics $\mathfrak{H}$ is a representation space of $\mathfrak{g}$. Group theory ensures that $\mathfrak{H}$ can always be decomposed into a direct sum of the various unitary irreducible carrier spaces $\mathfrak{H}_\Lambda$ of $\mathfrak{g}$

$$\mathfrak{H} = \sum_\Lambda \oplus Y_\Lambda \mathfrak{H}_\Lambda, \tag{12.12}$$

where $\mathbf{\Lambda}$ is a vector which characterizes an irreducible representation of $\mathfrak{g}$. When $\mathfrak{g}$ is a simple Lie algebra, $\mathbf{\Lambda}$ is the highest weight of this irreducible representation, and Y_Λ is the degeneracy of $\mathbf{\Lambda}$ in $\mathfrak{H}$. Different irreducible representations $\mathfrak{H}_\Lambda$ do not mix. Hence, without loss of generality, we can confine our discussion to a relevant irreducible subspace $\mathfrak{H}_\Lambda$ of $\mathfrak{H}$.

Once a dynamical group structure is given, the system's Hilbert space $\mathfrak{H}$ can be completely specified. In order to investigate the dynamical properties, it is convenient to choose a basis for $\mathfrak{H}$. This is analogous to defining a coordinate system for the classical phase space. In classical mechanics, the integrability of a system depends on the number of degrees of freedom, which is specified by a choice of a set of coordinates for the phase space. For an integrable system, there is always a suitable coordinate system, namely, action and angle coordinates, in which the integrability is manifested. In order to study quantum dynamics and its integrability which contains the classical integrability as a limit, it is imperative to develop the concept of degrees of freedom in quantum mechanics. However, in the development of quantum mechanics, this problem did not receive sufficient attention, and consequently the concept of quantum integrability has not yet been defined.

A specification of the dynamical degrees of freedom is essential to establish the basis of a physical state space. In quantum mechanics, selecting a basis of $\mathfrak{H}$ is mostly a matter of taste and experience. However, in group representation theory, there exists a rule to construct all possible bases of a given $\mathfrak{H}$; see, for example, Ref. [170]. Such a rule is merely on the essence of the quantum mechanical degrees of freedom and quantum integrability, which can be used as a starting point for our discussion. Such a discussion requires in-depth details of group representation theory. But since the group representation theory is the mathematical underpinning of quantum mechanics, such discussions are unavoidable.

For most quantum systems, their operator algebras $\mathfrak{g}$ are Lie algebras. It is known that for a Lie algebra $\mathfrak{g}$ with rank l and dimension n, there are a large although finite number λ of subalgebra chains $\{\mathfrak{g}^1, \mathfrak{g}^2, \cdots, \mathfrak{g}^\lambda\}$. For each group chain $\mathfrak{g} \subset \mathfrak{g}^\alpha$, $\alpha = 1, \cdots, \lambda$, there is a complete set of commuting operators, $\{Q_i^\alpha\}$, introduced by Dirac [51], such that

$$[Q_i^\alpha, Q_j^\alpha] = 0, i, j = 1, \cdots, d_\alpha, \tag{12.13}$$

where d_α is the number of the operators in the complete set of commuting operators for a specific chain, which is chain independent, and is given by

$$d_\alpha = l + \frac{n-l}{2} = d. \tag{12.14}$$

The chain independent of d_α means that for a given $\mathfrak{g}$, this is a universal number. Moreover, $\mathfrak{H}$ can be completely specified by a set of quantum numbers related to the eigenvalues of the complete set of commuting operators

$$Q_i^\alpha |\gamma_i^\alpha\rangle = \gamma_i^\alpha |\gamma_i^\alpha\rangle, i = 1, \cdots, d. \tag{12.15}$$

Generally speaking, the complete set of commuting operators has two classes of quantum numbers: a fully degenerate set and a non-fully degenerate set. The former satisfies the identity that

$$Q_i^\alpha |\Psi\rangle = c|\Psi\rangle, \tag{12.16}$$

where c are constants for any states $|\Psi\rangle \in \mathfrak{H}$. The operators in the complete set of commuting operators which do not satisfy the above equation are called the non-fully degenerate operators. The basis of $\mathfrak{H}$ can only be specified by non-fully degenerate quantum numbers.

It must be emphasized that the number of non-fully degenerate quantum numbers depends only on the details of $\mathfrak{H}$ and does not depend on the subgroup chains. For example, if $\mathfrak{H}$ is the carrier space of the non-degenerate irreducible representation of $\mathfrak{g}$, then the non-fully degenerate operators in the complete set of commuting operators contain l Casimir operators of $\mathfrak{g}$, and the number of such operators is

$$M = d - l = \frac{n-l}{2}. \tag{12.17}$$

Obviously, this number is independent of the subgroup chains. For the degenerate irreducible representations of $\mathfrak{g}$, although $M < (n-l)/2$, one can still show that it is independent of the subgroup chains. Hence, the number M of non-fully degenerate operators is unique for a quantum system in a given $\mathfrak{g}$ and can be determined from any one of the subalgebra chains of $\mathfrak{g}$. In this regard, one may conclude that the non-fully degenerate operators in the complete set of commuting operators can completely determine the structure of $\mathfrak{g}$. The number of the non-fully degenerate operators, defined by M, plays the role of quantum dynamical degrees of freedom [168].

A crucial point is that the quantum dynamical degrees of freedom, when defined in this way, would include both the internal and intrinsic quantum degrees of freedom. It appears to be a feasible definition because of the following reasons: (i) *Consistency*. For a classical Hamiltonian system with M degrees of freedom, it is required that the canonically quantized system should have M quantum numbers to specify the basis of its Hilbert space $\mathfrak{H}$. Based on our definition, this is exactly the case. For example, for a system with n structureless particles, the corresponding degrees of freedom are $3n$. It is well known that its Hilbert space, $\mathfrak{H}$, can be spanned by the simultaneous eigenstate set $\{\psi(r_1, r_2, \cdots, r_n)\}$ of the $3n$ position observables

$$(X_i\psi)(r_1, r_2, \cdots, r_n) = x_i\psi(r_1, r_2, \cdots, r_n), \tag{12.18}$$

$$(Y_i\psi)(r_1, r_2, \cdots, r_n) = y_i\psi(r_1, r_2, \cdots, r_n), \tag{12.19}$$

$$(Z_i\psi)(r_1, r_2, \cdots, r_n) = z_i\psi(r_1, r_2, \cdots, r_n), \tag{12.20}$$

for $i = 1, \cdots, n$. Then the complete set of commuting operators here is constructed by the $3n$ observables, X_i, Y_i, and Z_i, with $i = 1, \cdots, n$, which are all the non-fully degenerate operators for the associated Hilbert space. Equation (12.18) shows that there exist $3n$ quantum numbers x_i, y_i, and z_i to specify a basis of the Hilbert space for this system with $3n$ degrees of freedom. (ii) *Universality*. Since this definition of quantum dynamical degrees of freedom does not require, explicitly or implicitly, the assumption that the system must have a classical counterpart, it is applicable to any quantum system with both classical-like and additional internal degrees of freedom. (iii) *Uniqueness*. Since the complete set of commuting operators is provided by the algebraic structure of the basic physical observables without specifying a priori the Hamiltonian operator, it depends only on the *global* structure of the system's dynamical algebra $\mathfrak{g}$. Hence, the number of quantum dynamical degrees of freedom in this definition is unique for a specific Hilbert space $\mathfrak{H}$.

Notice that the number of quantum dynamical degrees of freedom does not include the number of fully degenerate operators in the complete set of commuting operators. This is because these operators are equivalent to constant multiples of the identity operator and are important only for determining the irreducibility of $\mathfrak{H}$. The expectation values of the fully degenerate operators themselves are dynamical independent and cannot be relevant to the degrees of freedom.

Once the number of quantum dynamical degrees of freedom is determined, any state $|\psi\rangle \in \mathfrak{H}$ can be obtained by a series of repeated action of elementary excitation operators on a fixed state $|\psi_0\rangle \in \mathfrak{H}$. Explicitly, the fixed state $|\psi_0\rangle$ can be defined uniquely as follows: if G is compact, the state $|\psi_0\rangle$ is the lowest or highest weight state of $\mathfrak{H}$; if G is non-compact, it is merely the lowest bound state of $\mathfrak{H}$. After the fixed state $|\psi_0\rangle$ is specified, operators $X_i^\dagger \in \mathfrak{g}$ are elementary excitation operators if and only if

$$X_i^\dagger|\psi_0\rangle \neq 0 \text{ or } |\psi_0\rangle. \tag{12.21}$$

Hence, all the states $|\psi\rangle$ of the system can be generated as follows:

$$|\psi\rangle = F(X_i^\dagger)|\psi_0\rangle, \forall|\psi\rangle \in \mathfrak{H}_\Lambda, \tag{12.22}$$

where $F(X_i^\dagger)$ is a polynomial of $\{X_i^\dagger\}$. It can be proven that the number of independent elementary excitation operators $\{X_i^\dagger\}$ is identical to the number of non-fully degenerated commuting observables of the complete set of commuting operators, i.e., the number of quantum dynamical degrees of freedom. Let us now discuss some examples.

Example 1. The harmonic oscillator with an $\mathfrak{h}_4$ algebra and a Fock space V^F as its Hilbert space. This space can be determined by the complete set of commuting operators of $\mathfrak{h}_4$, namely, the Casimir operators of the following sub-algebraic chain:

$$\mathfrak{h}_4 \supset \mathfrak{u}(1) \otimes \mathfrak{u}(1). \tag{12.23}$$

Although $\mathfrak{h}_4$ is a rank-2 non-semisimple Lie algebra, the Casimir operators of $\mathfrak{h}_4$ and $\mathfrak{u}(1)$ are nevertheless proportional to the identity operator I. Hence, there is only one irreducible representation for $\mathfrak{h}_4$. From the algebraic chain Eq. (12.23), the basis of V^F is given by the eigenstates of $\hat{n}$: $\{|n\rangle, n = 1, 2, \cdots\}$, where $\hat{n}|n\rangle = n|n\rangle$. It is evident that $\hat{n}$ is the only non-fully degenerate operator in the complete set of commuting operators of V^F. Hence, as in the classical case, the number of quantum dynamical degrees of freedom is one. In addition, the fixed state of V^F is the ground state $|0\rangle$. The elementary excitation operator is the particle creation operator $a^\dagger$, and any state $|\psi\rangle \equiv \sum_n c_n|n\rangle \in V^F$ has the form $|\psi\rangle = F(a^\dagger)|0\rangle$, where

$$F(a^\dagger) = \sum_n c_n \frac{(a^\dagger)^n}{\sqrt{n!}}. \tag{12.24}$$

Example 2. The spin system with an $\mathfrak{su}(2)$ as its dynamical algebra and an irreducible representation space V^{2j+1} of $\mathfrak{su}(2)$. This space can be specified by the complete set of commuting operators of the following sub-algebraic chain:

$$\mathfrak{su}(2) \supset \mathfrak{u}(1), \tag{12.25}$$

where $J^2 \equiv J_0^2 + (J_+J_- + J_-J_+)$ is the Casimir operator of $\mathfrak{su}(2)$. Since $\mathfrak{su}(2)$ is rank 1, there is only one class of irreducible representation, and it is uniquely determined by the total spin j. The basis $\{|jm\rangle, m = -j, \cdots, j\}$ of V^{2j+1} is specified by J_0:

$$J^2|jm\rangle = j(j+1)|jm\rangle,\ J_0|jm\rangle = m|jm\rangle. \tag{12.26}$$

Evidently, the only non-fully degenerate operator in the complete set of commuting operators is J_0, and thus the number of quantum dynamical degrees of freedom for the spin system is one. Since SU(2), the covering group of $\mathfrak{su}(2)$, is compact, the fixed state can be taken as the lowest weight state of V^{2j+1}: $|\psi_0\rangle \equiv |j-j\rangle$. Then the elementary excitation operator of spin systems is J_+. One can generate any state $|\psi\rangle \equiv \sum_{m=-j}^{j} c_m|jm\rangle \in V^{2j+1}$ by the action of a polynomial of J_+ on $|j-j\rangle$ as $|\psi\rangle = F(J_+)|j-j\rangle$, where

$$F(J_+) \equiv \sum_{m=-j}^{j} c_m \frac{1}{(j+m)!}\left(C_{2j}^{j+m}\right)^{-1/2}(J_+)^{j+m}. \tag{12.27}$$

Example 3. The radial motion of a particle in a central potential has a dynamical algebra $\mathfrak{su}(1,1)$. Its discrete irreducible representation $D^+(k)$ has a basis $\{kn\rangle, n = 0, 1, 2, \cdots\}$ which can be specified by the complete set of commuting operators of the following sub-algebraic chain:

$$\mathfrak{su}(1,1) \supset \mathfrak{u}(1), \tag{12.28}$$

where $K^2 \equiv K_2^2 - K_1^2 - K_2^2$ is the Casimir operator of $\mathfrak{su}(1,1)$ and

$$K^2|kn\rangle = k(k-1)|kn\rangle,\ K_0|kn\rangle = (k+n)|kn\rangle. \tag{12.29}$$

The non-fully degenerate operator is K_0. Hence, the number of quantum dynamical degrees of freedom for $\mathfrak{su}(1,1)$ is one. In this irreducible representation of $D^+(k)$, the lowest bound state $|k0\rangle$ can be chosen as the fixed state. Any state $|\psi\rangle \in\equiv \sum_{m=0}^{2j} c_n|kn\rangle \in D^+(k)$ can be generated by a polynomial of K_+ acting on $|k0\rangle$ as $|\psi\rangle = F(K_+)|k0\rangle$, where $K_\pm \equiv K_1 \pm iK_2$ and

$$F(K_+) \equiv \sum_{n=0}^{2j} c_n \sqrt{\frac{\Gamma(2k)}{n!\Gamma(2k+n)}}(K_+)^n. \tag{12.30}$$

Hence, the elementary excitation operator of $\mathfrak{su}(1,1)$ is only the raising operator K_+.

In the above three examples, the corresponding irreducible representations are non-degenerate and are technically known as single irreducible representations. In such cases, the number of the quantum dynamical degrees can be obtained directly from Eq. (12.17). In more complicated cases, the corresponding quantum dynamical degrees of freedom will depend sensitively on the structure of the irreducible spaces.

Since the classical trajectories are defined in a manifold endowed with a symplectic structure, i.e., a phase space, the classical limit can occur only if such a structure can emerge from the associated quantum system. In order to study quantum non-integrability and quantum chaos which relates to classical chaos, one must construct a geometry which originates from the Hilbert space with the necessary symplectic structure.

Once the quantum dynamical degrees of freedom are defined, a geometrical manifold with a symplectic structure for the quantum system can be constructed for $\mathfrak{H}$. The result is the following: for a quantum system, namely, a structure $(\mathfrak{g}, \mathfrak{H})$ with M quantum dynamical degrees of freedom, there exists a $2M$-dimensional symplectic manifold $\wp$ which is isomorphic to the coset space G/H, where G is the covering group of $\mathfrak{g}$ and H is the maximal isotropy subgroup of G with respect to the fixed state $|\psi_0\rangle$.

One can explicitly demonstrate why such a manifold is a symplectic manifold. To this end, we begin with the space $\mathfrak{p}$ whose operators are the elementary excitation operators and their conjugates, $\{X_i^\dagger, X_i\}$ with $i = 1, \cdots, M$. Clearly, $\mathfrak{p}$ is a subspace of $\mathfrak{g}$. The manifold is obtained by the exponential mapping. One may carry out a unitary exponential mapping for $\mathfrak{p}$,

$$\sum_{i=1}^{M}(\eta^i X_i^\dagger - \text{h.c.}) \to \Omega \equiv \exp\sum_{i=1}^{M}(\eta^i X_i^\dagger - \text{h.c.}), \tag{12.31}$$

where η^i with $i = 1, \cdots, M$ are complex parameters and Ω is a unitary coset representative of G/H. Here G/H is the basic geometrical manifold of $\mathfrak{H}$. This can be seen from the associated generalized coherent states of G

$$|\Omega\rangle \equiv \Omega|\psi_0\rangle, \tag{12.32}$$

which shows that the manifold induced by the coherent states is isomorphic to the coset space G/H, while the generalized coherent states offer a continuous basis for spanning $\mathfrak{H}$.

However, whether G/H is the desired manifold depends on whether or not it is a complex space with a symplectic structure. As we have emphasized before, the complex structure is the fundamental criterion of quantum mechanics, and the symplectic structure is demanded by the existence of a phase space. First, we shall show that G/H is a complex space. We shall confine our attention to the semisimple Lie groups whose $\mathfrak{g}$ satisfy the Cartan canonical decomposition $\mathfrak{g} = \mathfrak{k} + \mathfrak{p}$:

$$[\mathfrak{k}, \mathfrak{k}] \subset \mathfrak{k},\ [\mathfrak{k}, \mathfrak{p}] \subset \mathfrak{p},\ [\mathfrak{p}, \mathfrak{p}] \subset \mathfrak{k}. \tag{12.33}$$

The dimensions for $\mathfrak{k}$ and $\mathfrak{p}$ are l and m, respectively. When Eq. (12.33) is satisfied, Eq. (12.31) can be expressed in terms of an $l \times m$ block matrix

$$\exp\begin{pmatrix} \mathbf{0} & \boldsymbol{\eta} \\ \pm\boldsymbol{\eta}^\dagger & \mathbf{0} \end{pmatrix} = \begin{pmatrix} \sqrt{\boldsymbol{I} \mp \boldsymbol{X}^\dagger \boldsymbol{X}} & \boldsymbol{X} \\ \mp\boldsymbol{X}^\dagger & \sqrt{\boldsymbol{I} \mp \boldsymbol{X}^\dagger \boldsymbol{X}} \end{pmatrix}, \tag{12.34}$$

where $\boldsymbol{\eta}$ and $\boldsymbol{X}$ are $l \times m$ matrices determined by

$$\boldsymbol{X} = \boldsymbol{\eta}\frac{\sin\sqrt{\boldsymbol{\eta}^\dagger\boldsymbol{\eta}}}{\sqrt{\boldsymbol{\eta}^\dagger\boldsymbol{\eta}}} \quad \text{for compact } G, \tag{12.35a}$$

$$\boldsymbol{X} = \boldsymbol{\eta}\frac{\sinh\sqrt{\boldsymbol{\eta}^\dagger\boldsymbol{\eta}}}{\sqrt{\boldsymbol{\eta}^\dagger\boldsymbol{\eta}}} \quad \text{for non-compact } G, \tag{12.35b}$$

and the plus and minus signs correspond to non-compact and compact G, respectively. It has been shown that a group transformation,

$$u = \begin{pmatrix} \boldsymbol{A} & \boldsymbol{B} \\ \pm\boldsymbol{C} & \boldsymbol{D} \end{pmatrix} \in G \tag{12.36}$$

on G/H is a homomorphic mapping of G/H onto itself:

$$u\Omega : \boldsymbol{Z}' = \frac{\boldsymbol{AZ}+\boldsymbol{B}}{\boldsymbol{CZ}+\boldsymbol{D}}, \boldsymbol{Z} \equiv \boldsymbol{X}\sqrt{\boldsymbol{I} \mp \boldsymbol{X}^\dagger\boldsymbol{X}}. \tag{12.37}$$

The above equation shows that G/H is a complex manifold. A complex manifold can be understood physically, namely, the wave functions should only depend on $\{z\}$ or $\{z^*\}$, which is analogous to the dependence on coordinates or momenta, but it cannot depend on both. Hence, G/H must be a complex space if it is a manifold of the Hilbert space.

Next it can be shown that G/H has a symplectic structure. Since it is a complex space, there exists at least a Hilbert space of functions $\bar{L}^2(G/H)$ on G/H. Let us denote these functions as $f_1, f_2, \cdots$ to be any orthonormal basis of $\bar{L}^2(G/H)$. A kernel $K(z, z^*)$, called the Bergmann kernel, can be defined as follows:

$$K(z, z^*) = \sum_n f_n(z) f_n^*(z). \tag{12.38}$$

The metric of G/H is then defined by

$$d^2s = \sum_j g_{ij} dz^i dz^{j*} = \sum_j \frac{\partial^2 \ln K(z, z^*)}{\partial z^i \partial^{j*}} dz^i dz^{j*}. \tag{12.39}$$

According to differential geometry theory, there exists a closed non-degenerate 2-form ω on G/H whose explicit form in the complex local coordinate system is

$$\omega = i\hbar \sum_{ij} g_{i\bar{j}} dz^i \wedge dz^{j*}, \tag{12.40}$$

and the corresponding Poisson bracket is

$$\{f, g\} \equiv \frac{1}{i\hbar} \sum_{ij} g^{i\bar{j}} \left(\frac{\partial f}{\partial z^i} \frac{\partial g}{\partial z^{j*}} - \frac{\partial g}{\partial z^i} \frac{\partial f}{\partial z^{j*}} \right). \tag{12.41}$$

In the above equation, f and g are two functions defined on G/H and $g_{i\bar{j}} g^{jk} = \delta_{ik}$. Hence, the existence of the symplectic structure is demonstrated. It should be noted that in the expression of the closed non-degenerate 2-form ω, we have added a factor $i\hbar$. This is because we want to embed G/H into a phase space with proper physical dimension.

An explicit construction of the symplectic structure is as follows: By using the coherent state theory, a simple and useful realization of an orthogonal basis $\{f_n(z)\}$ is $f_\gamma(z) = \langle \gamma || z \rangle$, where $\{|\gamma\rangle\}$ is a basis of $\mathfrak{H}$ and $||z\rangle$ is the unnormalized form of the coherent states $|\Omega\rangle$,

$$|\Omega\rangle = N^{-1/2}(z, z^*) \exp\left(\sum_{i=1}^{M} z^i X_i^\dagger \right) |\psi_0\rangle \equiv N^{-1/2}(z, z^*) ||z\rangle. \tag{12.42}$$

It can be shown that the Bergmann kernel is the normalization constant

$$\mathcal{K}(z, z^*) = N(z, z^*) = \langle z || z \rangle = |\langle \psi_0 | \Omega \rangle|^{-2}. \tag{12.43}$$

Hence, the geometrical structure of G/H can be obtained directly from the coherent states. When $\mathfrak{g}$ is a semisimple Lie algebra and satisfies the decomposition of Eq. (12.33), the explicit function of the Bergmann kernel is given as follows:

$$\mathcal{K}(z, z^*) = \det(\boldsymbol{I} \pm \mathbf{Z}^\dagger \mathbf{Z})^{\pm \Xi}, \tag{12.44}$$

where the plus and minus signs correspond to the compact and non-compact group, respectively, and Ξ, referred to as the quenching index hereafter, is defined by $h_i|\psi_0\rangle = \pm\Xi|\psi_0\rangle$ or 0 for $h_i \in \mathfrak{k}$. The index Ξ comes from the mapping $G \to G/H$ induced by the fixed point of $|\psi_0\rangle$. This is the geometrical origin of this topological index. Also, each Ξ specifies an irreducible representation of G and represents a quantum system. In the later sections, we will also see that the quenching index is indeed the inherent quantity to determine quantum-classical correspondence and deduce the classical limit.

By introducing the canonical coordinates $\boldsymbol{q}$ and $\boldsymbol{p}$ of G/H,

$$\frac{1}{\sqrt{2\hbar\Xi}}(\boldsymbol{Q}+i\boldsymbol{P})=\boldsymbol{X}, \tag{12.45}$$

the Poisson bracket can be transformed into the standard form

$$\omega=\sum_{i=1}^{M}dp^i\wedge dq^i,\{f,g\}=\sum_{i=1}^{M}\left(\frac{\partial f}{\partial q^i}\frac{\partial g}{\partial p^i}-\frac{\partial f}{\partial p^i}\frac{\partial g}{\partial q^i}\right), \tag{12.46}$$

which offers the familiar phase space structure. Since G/H has both a complex structure and a symplectic structure, it satisfies the demand of quantum mechanics and allows the development of classical mechanics on it. Hence, one may call the coherent state space G/H a quantum phase space.

Still, at this point there is no dynamical information about the quantum-classical correspondence. The problem of the dynamical correspondence of quantum-classical systems is discussed in the next section.

12.3 Semi-Quantal Dynamics

Based on the properties of the quantum phase space G/H, the phase space representation of a quantum system can be defined if there is an explicit mapping

$$\{\mathfrak{g},\mathfrak{H}\}\rightarrow\{G/H,\mathcal{L}^2(G/H)\}, \tag{12.47}$$

such that $A\rightarrow\mathcal{A}(p,q)$ and $|\psi\rangle\rightarrow f(z)$, where $f(z)$ is a square-integrable function defined on G/H. For a given coset space G/H, this mapping can uniquely be realized by its coherent states, for which the procedure is:

(a) Wave functions. It is well known that coherent states are over-complete, namely,

$$\int_{G/H}|\Omega\rangle d\mu_H(z)\langle\Omega|=I \text{ or } \int_{G/H}||z\rangle d\mu(z)\langle z||=I, \tag{12.48}$$

where $d\mu(z,z^*)=K^{-1}(z,z^*)d\mu_H(z,z^*)$ and $d\mu_H(z,z^*)$ is the Haar measure of G/H. The above equation implies that any quantum state $|\psi\rangle\in\mathfrak{H}$ can be expanded in terms of the coherent state basis as

$$|\psi\rangle=\int_{G/H}||z\rangle f(z^*)d\mu(z) \text{ or } \langle\psi|=\int_{G/H}f(z)d\mu(z)\langle z||. \tag{12.49}$$

However, since $||z\rangle$ is over-complete, this expansion is generally not unique. In other words, for a given $|\psi\rangle$, there exist several square-integrable functions

$\{f_k\}$ which satisfy Eq. (12.49). If $f_1(z)$ and $f_2(z)$ are two functions that satisfy Eq. (12.49) for the same $|\psi\rangle$, their difference must satisfy the following equation:

$$\int_{G/H} K(z, z'^*)(f_1(z') - f_2(z'))d\mu(z') = 0, \tag{12.50}$$

where the kernel $K(z, z'^*) = \langle z' || z \rangle$. Evidently, the uniqueness of the expansion of Eq. (12.49) demands the following identity:

$$f(z) = \int_{G/H} K(z, z'^*) f(z') d\mu(z'). \tag{12.51}$$

In other words, any two functions which represent the same $|\psi\rangle$ must be the same if Eq. (12.51) is satisfied. One can easily verify that the function $f(z) = \langle \psi || z \rangle$ or $f(z^*) = \langle z || \psi \rangle$ is the solution of Eq. (12.49) and satisfies Eq. (12.51). Moreover, it can be shown that all the above solutions $\{f(z)\}$ span the square-integrable Hilbert space $\bar{L}^2(G/H)$ with $\{f_\gamma(z)\}$ as the basis,

$$(f_1, f_2) = \int_{G/H} f_1^*(z) f_2(z) d\mu'(z, z^*) < \infty, \tag{12.52}$$

where

$$f(z) = \sum_\gamma c_\gamma f_\gamma(z), \tag{12.53}$$

and c_γ are the expansion coefficients. Also, according to the above equation, $f_\gamma(z)$ is in one-to-one correspondence with $|\gamma\rangle$, which is the basis of $\mathfrak{H}$. Hence, $\bar{H}^2(G/H)$ is isomorphic with $\mathfrak{H}$, and the quantum dynamics can be described within the Hilbert space $\bar{L}^2(G/H)$. The normalized wave function is $f_N(z) = \langle \psi | \Omega \rangle = K^{-1/2}(z, z^*) f(z)$. With this representation, the orthonormal and completeness relations are standard

$$\int_{G/H} f_{N\gamma}^*(z) f_{N\gamma}(z) d\mu_H(z, z^*) = \delta_{\gamma\gamma'}, \sum_\gamma f_{N\gamma}^*(z) f_{N\gamma}(z) = 1, \tag{12.54}$$

where $f_{N\gamma}(z) = K^{-1/2}(z, z^*) f_\gamma(z)$.

(b) Operators. There are several ways to define the phase space representation of an operator on G/H. We shall begin our discussion with a general description. Since coherent states are over-complete, any arbitrary operator A acting on $\mathfrak{H}$

can be expanded as

$$A = \int ||z\rangle\langle z||A||z'\rangle\langle z'||d\mu(z)d\mu(z') \tag{12.55}$$
$$= \int |\Omega\rangle \mathcal{A}(z, z^*; z', z'^*)\langle\Omega'|d\mu_H(z)d\mu_H(z'),$$

where $\mathcal{A}(z, z^*; z', z'^*) = \langle\Omega|A|\Omega'\rangle = N^{-1/2}(z, z^*)\langle z||A||z'\rangle N^{-1/2}(z', z'^*)$. The above equation can be reduced to three special representations, from which one can study the properties of the operator A in the quantum phase space. Some of these representations have been widely used in the study of the phase space distributions and statistical averages. They also form the basis for our investigation of quantum non-integrability and quantum chaos, as we shall show later.

(I) P-representation. Equation (12.55) can be expressed in a diagonal form as follows:

$$A = \int |\Omega\rangle \mathcal{A}_P(z, z^*)\langle\Omega|d\mu_H(z). \tag{12.56}$$

This is known as the P-representation of the coherent states. However, this expansion is not unique and may not even exist in some cases. It is also, in general, difficult to compute, and thus will not be discussed here.

(II) Q-representation. Every operator A that maps an $\mathfrak{H}$ onto itself in terms of the coherent states $|\Omega\rangle$ is a Q-representation:

$$A \to \mathcal{A}(z, z^*) = \langle\Omega|A|\Omega\rangle = K^{-1}(z, z^*)\langle z||A||z\rangle. \tag{12.57}$$

By definition, the Q-representation of A in $\mathfrak{H}$ is unique.

(III) W-representation. This representation, known as the Wigner distribution, is chronologically the first to describe the quantum distribution function in the phase space. This representation satisfies the following two conditions:

$$A \to \mathcal{A}_W(z, z^*), \tag{12.58}$$

and

$$(A, B) = \mathrm{Tr}_\Lambda(A^\dagger B) = \int d\mu_H(z)\mathcal{A}^*_W(z, z^*)\mathcal{B}_W(z, z^*), \tag{12.59}$$

where A and B are arbitrary two operators acting on the Hilbert space $\mathfrak{H}$.

To discuss integrability of a dynamical system in quantum phase space, an important ingredient is the algebraic structure of the operators in their phase space representation. In the following, we shall discuss the algebraic structure in the Q-representation.

First, we consider the generators T_i of G. The commutation relations in the Q-representation will have the same algebraic structure:

$$i\hbar\{\mathcal{T}_i, \mathcal{T}_j\} = \sum_{k=1}^{n} C_{ij}^k \mathcal{T}_k, \tag{12.60}$$

where $\mathcal{T}_i \equiv \langle \mathbf{\Lambda}, \Omega | T_i | \mathbf{\Lambda}, \Omega\rangle$, $T_i \in \mathfrak{g}$, and $\{\cdot, \cdot\}$ denotes the Poisson bracket on G/H. A global proof of this is as follows: By taking the phase space representation of the commutation relationship, one obtains

$$\langle\Omega|[T_i, T_j]|\Omega\rangle = \sum_{k=1}^{n} C_{ij}^k \mathcal{T}_k. \tag{12.61}$$

Thus, the proof is reduced to verify the following identity:

$$\langle\Omega|[T_i, T_j]|\Omega\rangle = i\hbar\{\mathcal{T}_i, \mathcal{T}_j\}. \tag{12.62}$$

Mathematically, the phase space representation of T, which is a linear combination of T_i, corresponds to a co-adjoint representation of $\mathfrak{g}$

$$\mathcal{T} = \langle \mathbf{\Lambda}, \Omega | T | \mathbf{\Lambda}, \Omega\rangle = \langle\psi_0| Ad_\Omega T |\psi_0\rangle \in \mathfrak{g}^*, T \in \}, \Omega \in G, \tag{12.63}$$

where $\mathfrak{g}^*$ is a space dual to $\mathfrak{g}$. The gradient of $\mathcal{T}$, denoted by grad$\mathcal{T}$, defines a vector field on G/H. By the definition of the symplectic structure on G/H, one obtains

$$\hbar\{\mathcal{T}, \mathcal{T}'\} = \omega(\text{grad}\mathcal{T}, \text{grad}\mathcal{T}') = \frac{1}{i}\langle\Omega|[T, T']|\Omega\rangle. \tag{12.64}$$

Since $T, T' \in \mathfrak{g}$ are linear combinations of the generators T_i, one completes the proof.

However, when A is a nonlinear function of T_i, the quantum fluctuation is manifested in its phase space representation. In this case, the algebraic structure between the operators is usually not preserved. Hence, the preservation of the algebraic structure in the phase space representation can be achieved only through the condition $\Xi \to \infty$. Generally, for any two arbitrary operators A and B,

$$\langle\Omega|[A, B]|\Omega\rangle = i\hbar\{A, B\} + O(1/\Xi), \tag{12.65}$$

where $O(1/\Xi)$ is the second- and higher-order terms of the quenching index, $1/\Xi$. In summary, the above discussions show that based on the quantum phase space G/H, the phase space representation of a quantum system has the following explicit form:

$$A(T_i) \to \mathcal{A}(z, z^*) = \langle \Omega | A | \Omega \rangle, T_i \in \mathfrak{g}, \tag{12.66a}$$

$$|\psi\rangle \to f(z) = \langle \psi | \Omega \rangle, |\psi\rangle \in \mathfrak{H}, \tag{12.66b}$$

where $|\Omega\rangle$ is the coherent state of G/H. This is a realization of the kinematical quantum-classical correspondence.

We are now ready to formulate quantum mechanics in G/H and introduce the idea of semi-quantal description [168]. In order to evaluate quantum dynamics in G/H, it is useful to use the path integrals formalism. The standard Feynman path integral was derived by slicing the time propagator $\exp(-iHt)$ into a product of operators of the form $[\exp(-iHt/N)]^N$ and then inserting the resolution of identity between the operators. The resolution of identity is usually expressed in terms either of the continuous coordinate states $|x\rangle$ or the momentum states $|p\rangle$. Hence, when the system lacks an explicit classical phase space structure, the path integrals formalism appears to be of limited use. This is precisely where the coherent state formulation resolves this limitation.

In the above discussions, we have shown that G/H has a well-defined symplectic structure and the associated coherent state basis is a continuous basis that satisfies a resolution of identity. With these two properties, the path integrals formalism can be established for any quantum system even if it does not have a clear classical limit. The basic procedure is the same. We shall slice the time interval into N equal segments $\epsilon = (t_2 - t_1)/N$ and then insert the resolution of identity at each intermediate point between t_2 and t_1. Finally, we let $N \to \infty$ and obtain

$$\mathcal{U}(z, z_0; t - t_0) = \lim_{N\to\infty} \int \cdots \int \left(\prod_{i=1}^{N-1} d\mu(z_i) \right) \prod_{i=1}^{N} \langle z_i | \exp\left(-\frac{iH\epsilon}{\hbar} \right) | z_{i-1} \rangle. \tag{12.67}$$

When $N \to \infty$ and $\epsilon \to 0$, one obtains up to the first order in ϵ

$$\langle z_i | \exp\left(-\frac{iH\epsilon}{\hbar} \right) | z_{i-1} \rangle \approx \langle z_i | z_{i-1} \rangle \exp\left(-\frac{i\epsilon}{\hbar} \frac{\langle z_i | H | z_{i-1} \rangle}{\langle z_i | z_{i-1} \rangle} \right) \tag{12.68}$$

and

$$\langle z_i | z_{i-1} \rangle \approx \exp\left(-\frac{\langle z_i | \Delta z_i \rangle}{\langle z_i | z_i \rangle} \right), \epsilon \frac{\langle z_i | H | z_{i-1} \rangle}{\langle z_i | z_{i-1} \rangle} \approx \epsilon \frac{\langle z_i | H | z_i \rangle}{\langle z_i | z_i \rangle} = \epsilon \mathcal{H}(z_i, z_i^*). \tag{12.69}$$

Thus, the path integral representation of a propagation kernel is

$$\mathcal{U}(z, z_0; t - t_0) = \int_{z_0}^{z_f} \mathcal{D}\mu(z) \exp\left\{ \frac{i}{\hbar} \mathcal{S}[z(t), z^*(t)] \right\}, \tag{12.70}$$

where

$$\mathcal{S}[z(t), z^*(t)] = \int_{t_0}^{t} d\tau \left\{ \langle\Omega|i\hbar\frac{\partial}{\partial\tau}|\Omega\rangle - \mathcal{H}(z(\tau), z^*(\tau)) \right\} \tag{12.71}$$

is the effective quantum action and $\mathcal{H}(z, z^*)$ the expectation value of the Schrödinger operator H evaluated with the time-dependent coherent states. Furthermore, one finds that

$$\langle\Omega|i\hbar\frac{\partial}{\partial t}|\Omega\rangle dt = \frac{i\hbar}{2}\langle\Omega|\overleftrightarrow{\partial}_t|\Omega\rangle dt = \frac{i\hbar}{2}\sum_{i=1}^{M}\left(\frac{\partial lnK}{\partial z^i}dz^i - \frac{\partial lnK}{\partial z^{i*}dz^{i*}}\right), \tag{12.72}$$

which is a differential 1-form on G/H, and

$$\mathcal{D}\mu(p, q) \equiv \prod_{t_0\leq\tau\leq t} d\mu[z(\tau)] \tag{12.73}$$

is a functional measure of the path integration. In this regard, the equations for quantum dynamics can be formulated in the phase space manifold G/H.

Except for few system, the path integral can only be solved approximately. One well-known approximation is the **stationary phase approximation**. The stationary phase approximation for the present purpose is defined as follows: One may expand the effective quantum action about its stationary path $z(t) = z_c(t)$,

$$\mathcal{S}[z(t), z^*(t)] = \mathcal{S}_0[z(t), z^*(t)] + \cdots, \tag{12.74}$$

where the stationary path is determined by the variation of the action

$$\delta\mathcal{S}[z(t), z^*(t)] = \delta\int_{t_0}^{t} d\tau \left\{ \langle\Omega|i\hbar\frac{\partial}{\partial\tau}|\Omega\rangle - \mathcal{H}(z(\tau), z^*(\tau)) \right\} = 0. \tag{12.75}$$

Using Eq. (12.72) and the definition of g_{ij}, one may derive the following classical-like equations of motion:

$$i g_{ij}\frac{dz^i}{dt} = \frac{\partial\mathcal{H}}{\partial z^{i*}}, -g_{ij}\frac{dz^{j*}}{dt} = \frac{\partial\mathcal{H}}{\partial z^i}, \tag{12.76}$$

as a consequence of the symplectic structure of G/H. If the canonical form of G/H is used, then the equations of motion become the standard Hamilton's equations

$$\frac{dq^i}{dt} = \frac{\partial\mathcal{H}(q, p)}{\partial p^i}, \frac{dp^i}{dt} = -\frac{\partial\mathcal{H}(q, p)}{\partial q^i}. \tag{12.77}$$

It is crucial to recognize that although these equations of motion are classical-like, they should not be misinterpreted as the classical under such an approximation. Mathematically, a stationary phase approximation is an approximate way to evaluate an integral in which the integrand is an exponential phase. How good such an approximation is depends on the details of the phase function. In the original Feynman path integrals formalism, the stationary phase approximation leads to a classical limit because the phase function is entirely constructed from a classical Hamiltonian and thus the corresponding equations under stationary phase approximation are the classical equations of motion.

The present usage of the stationary phase approximation is fundamentally different because the effective quantum action Eq. (12.71) is not classical but quantum mechanical in a coherent state basis. This is why it is referred to as an effective quantum action. In this formulation, the stationary trajectories are not determined by the classical mechanics since the Hamiltonian is different from the classical one. Hence, although the resulting equations are classical-like, the concept of classical mechanics has not been invoked. This means that we can only logically refer to the dynamics determined by the above equations of motion as a semi-quantal dynamics, where $\mathcal{H}$ is the semi-quantal Hamiltonian. The prefix "semi" is used because it is an approximation of quantum mechanics. On the other hand, since the Hamiltonian is not the classical Hamiltonian, it does not, therefore, describe the ordinary classical mechanics. In this regard, the intriguing question is: what physical phenomena does the semi-quantal dynamics describe? We find that since the semi-quantal Hamiltonian function is derived entirely from quantum mechanics, it differs from the classical Hamiltonian function by the leading order quantum fluctuation. The above equations of motion indeed give a description of the phase space structure of quantal dynamics in which some quantum interference is averaged out by the variational procedure.

In the above discussions, we introduced a parameter Ξ, which is called the quenching index. To be more precise, this index appears in the Bergmann kernel of the quantum phase space for different quantum systems. This index plays a crucial role in determining the classical limit of quantum dynamics. Below we shall provide a proof for this statement.

Since the classical limit is a physical concept, one must consider observables which have proper dimensions. The generators of G are dimensionless. Hence, a physical observable with meaningful dimensions should be a function of $\hbar T_i$. To this end, one may consider an explicit operator $A(\hbar T_i)$:

$$A(\hbar T) = \sum_{ij} f_{ij}(\hbar T_i)(\hbar T_j), \tag{12.78}$$

where f_{ij} are some coefficients. Most realistic Hamiltonians have such a canonical form. By using the Baker-Campbell-Hausdorff formula, the transformation of T_i

under G/H is given by

$$\Omega^{-1}T_i\Omega = \sum_l a_{il}K_l + \sum_m b_{im}X_m^\dagger + \sum_n c_{in}X_n, \Omega \in G/H, \tag{12.79}$$

where the transformation coefficients a_{il}, b_{im}, and c_{in} are functions of the phase space coordinates of G/H and $X_m^\dagger, X_m \in \mathfrak{p}$, and $K_l \in \mathfrak{k}$ for the decomposition $\mathfrak{g} = \mathfrak{k} + \mathfrak{p}$. Hence,

$$\langle\Omega|T_i|\Omega\rangle = \sum_l a_{il}\langle\psi_0|K_l|\psi_0\rangle \tag{12.80}$$

and

$$\langle\Omega|T_iT_j|\Omega\rangle = \langle\Omega|T_i|\Omega\rangle\langle\Omega|T_j|\Omega\rangle + \sum_{lm}(c_{il}b_{jm})C_{lm}^k\langle\psi_0|K_k|\psi_0\rangle), \tag{12.81}$$

where we have used $X_i|\psi_0\rangle = \langle\psi_0 X_i^\dagger = 0$. The coefficients C_{lm}^k are the structure constants. For most realistic quantum systems, we have

$$\langle\psi_0|K_l|\psi_0\rangle = \Xi \text{ or } \langle\psi_0|K_l|\psi_0\rangle = 0. \tag{12.82}$$

Thus we can directly verify that

$$\langle\Omega|A(\hbar T)|\Omega\rangle = \sum_{ij} f_{ij}\mathcal{T}_i'(p,q,\Xi\hbar)\mathcal{T}(p,q,\Xi\hbar) + \frac{1}{\Xi}\mathcal{F}(p,q,\Xi\hbar) \tag{12.83}$$

so that the leading order of quantum fluctuation is

$$\delta\mathcal{A}(p,q,\Xi,1/\Xi) = \frac{1}{\Xi}\mathcal{F}(p,q,\Xi\hbar). \tag{12.84}$$

Now, the physical implication is clear. While the coordinates z and z^* are dimensionless, p and q have the dimensions of position and momentum, respectively. As is well known, the reduced Planck constant $\hbar$ can be regarded as a scale of the microscopic world. To examine whether a system is microscopic or macroscopic, one may use the quenching index Ξ as a guide. Meanwhile, the above equation shows that the leading order quantum fluctuation is $1/\Xi$ which will vanish in the classical limit.

For more general operators, e.g., a n-degree polynomial of $\hbar T_i$, a similar proof can be given. It is as follows:

$$\langle\Omega|A_n(\hbar T)|\Omega = A_n(\mathcal{T}_i(p,q,\Xi\hbar)) + \frac{1}{\Xi}\mathcal{F}_1(p,q,\Xi\hbar) + \cdots + \frac{1}{\Xi^{n-1}}\mathcal{F}_{n-1}(p,q,\Xi\hbar). \tag{12.85}$$

Evidently, when $\Xi \to \infty$, the quantum fluctuation must vanish and thus the corresponding dynamics becomes classical. This is precisely the meaning of a classical limit of the quantum theory. As we have emphasized, the classical limit must rely on a quantity which is inherent in the theory and can measure precisely the system's size according to the scale of $\hbar$. This quantity is now revealed in our discussion, namely, the quenching index Ξ. To be more specific, let us compute two simple cases, SU(2) and SU(1, 1). A direct computation yields $\Xi = 2j$ and $2k$, so that the quantum fluctuations vanish in the large j and k limits for the SU(2) and SU(1, 1) dynamical groups, respectively.

The above discussion clearly shows that in the quantum theory, by the use of the quenching index, there is a prescription for going over to the classical limit and for relating the quantum observables to those of the corresponding classical system. The two special limits of the semi-quantal dynamics indicate that the quenching index bridges the quantum and classical phenomena. The limiting process of the quenching index Ξ, from finite to infinite, allows us to systematically study the process of "flow" from quantum to classical mechanics.

For instance, as we have shown, for a spin system with a dynamical group SU(2), the quenching index is proportional to the spin quantum number. This is consistent with Lieb's elegant derivation of the classical limit of quantum spin systems and the underlying quantum phase space is compact. For the radial motion of a particle in a central potential with a dynamical group SU(1, 1), the classical limit corresponds to the orbital angular momentum $\to \infty$. This is the familiar Bohr correspondence principle and the quantum phase space is non-compact. For a N-boson system with a dynamical group U(r), where r is the degree of degeneracy, the quenching index is simply the total boson number N. The large N classical limit has been widely used in various branches of theoretical physics. For a N-fermion system with a dynamical group $U(r)$, the quenching index is $\Xi = 1$. It implies that such a system has no classical limit. What is perhaps most remarkable is that starting entirely from geometrical arguments, one can readily show that deep in the geometries of these systems the lack of classicality is inherent.

12.4 Quantum Non-Integrability

Based on the description of quantum-classical correspondence developed in the last sections, we can now discuss the meaning of non-integrability for quantum systems, the core issue of this chapter. As is well known, the study of non-integrability is the phase in the investigation of classical chaos since integrability is the condition for the absence of disorder. An integrable system can be solved analytically by an appropriate transformation operation, e.g., the canonical transformation in classical phase space. Hence, for such a system, there are suitable local coordinates in terms of which the solutions of the equations of motion are simple functions. In this section, we will provide a criterion for the integrability of a quantum system. The bulk of this section will deal with non-integrable systems, in which the general properties of non-integrability and dynamical effect of quantum fluctuation on nonlinear phenomena will be discussed. Before we embark upon this discussion, however, we shall first present a brief discussion about systems which are integrable.

The goal of the semi-quantum dynamics discussed in the last section is the provision of an underlying framework for the deduction of the classical mechanics from quantum mechanics. In classical Hamiltonian mechanics, the definition of integrability is that for M independent degrees of freedom, a classical system to be integrable must have M integrals of motion. In the previous section, we gave a unique definition of the number of quantum dynamical degrees of freedom: the necessary number of quantum numbers to specify a basis of the Hilbert space. Thus, for a quantum system, if there are M non-fully degenerate commuting operators A_i, $i = 1, \cdots, M$,

$$[A_i, A_j] = 0, \tag{12.86}$$

we can then analytically and completely determine its dynamics. These M commuting operators are called commuting integrals of motion. Correspondingly, there are M good quantum numbers associated with the eigenvalues of $A_1, A_2, \cdots, A_M$. The physical eigenstates $\{|\lambda\rangle\}$ are labeled by these M good quantum numbers,

$$A_i|\lambda\rangle = \lambda_i|\lambda\rangle, \tag{12.87}$$

and the energy spectrum is a function of $\{\lambda_i\}$. Quantum integrability is defined as follows:

A quantum system with M independent dynamical degrees of freedom is integrable if and only if there exist M commuting integrals of motion, i.e., M good quantum numbers. In addition, the existence of these M commuting non-fully degenerate observables A_i must be independent of the Hilbert space; that is, they must be globally defined. According to Dirac, any set of commuting observables can be made into a complete set of commuting operators by adding certain observables to the set. The above definition implies that for an integrable quantum system, one can always find a complete set of commuting operators such that the Hamiltonian is diagonal in a basis labeled by the eigenvalues of the complete set of commuting operators. Conversely, quantum integrability means one can simultaneously measure the M non-fully degenerate observables in the energy eigenstates basis.

Obviously, such a definition is akin to its classical counterpart, and one can prove that quantum integrability defined here is consistent with classical mechanics. The proof follows directly from the fact that the algebraic structure of observables is preserved in the classical limit. Thus, one immediately obtains $\{A_i, A_j\} = 0$.

For a classical system which is integrable, the above definition of quantum integrability can also be realized. This can be seen as follows: For classically integrable systems, the trajectories must lie in an M-dimensional invariant tori with action constants $I_i, i = 1, \cdots, M$. These action constants are related to the M good quantum numbers from the Bohr-Sommerfeld-Wilson quantization conditions, or more precisely the Einstein-Brillouin-Keller rule:

$$I_i = (n_i + \alpha_i)\hbar, (i = 1, \cdots, M), \tag{12.88}$$

where α_i is the number of caustic traversed, usually referred to as the Maslov indices. These M good quantum numbers $\{n_i, i = 1, \cdots, M\}$ will specify a basis for the corresponding quantum system. The associated M operators consist of a set of the commuting non-fully degenerate operators in the complete set of commuting operators. This proves the consistency of the quantum and classical integrability. However, if a classical system is non-integrable, the above justification may not be operative.

Now what are the typical observables for an integrable quantum system with M degrees of freedom? We have pointed out that for the integrable system, there is always a basis which is specified by the M good quantum numbers. The basis is also the set of eigenstates of the system. However, unlike classical trajectories, wave functions are not measurable. The most simple observables in quantum dynamics are the energy spectra, and therefore the integrability condition will immediately lead to what Percival refers to as the "regular" and the "irregular" spectra [171,172].

(i) A regular spectrum corresponds to regimes of the integrable motion in which all the states can be quantized according to the Einstein-Brillouin-Keller rule. (ii) An irregular spectrum corresponds to regimes of predominantly chaotic motion in which Einstein-Brillouin-Keller rule is not applicable. Moreover, Percival [171] conjectured that the regular energy spectra could have the following properties: (i) A quantal state may be labeled by the vector quantum number $n = (n_1, n_2, \cdots, n_M)$. (ii) A state with quantum number n corresponds to those phase space trajectories of the corresponding classical system which lie in an M-dimensional invariant toroid with action constants I_k given by the Bohr-Sommerfeld-Wilson quantum conditions $I_k = (n_k + \alpha_k)\hbar$. (iii) The quantal state must resonate at frequencies close to those of the corresponding classical motion. Given two quantal states with n_k differing only by one unity, the Planck relation for their energy differences is $\Delta E_k = \omega_k \hbar$, where ω_k is a fundamental frequency on the corresponding toroid. (iv) A "neighboring state" to a state n^0 with energy E^0 is a state with the vector quantum number n close to n^0, with energy difference no more than a small multiple of the maximum $|\Delta E_k|$. (v) Under some weak external perturbations, the state n^0 is much more strongly coupled to the neighboring states than to other states, with the coupling strength rapidly decreasing with $|n - n^0|$.

According to the properties of quantum integrability, points (i) and (ii) are obviously satisfied by an integrable quantum system. In fact, if the Hamiltonian operator is not included in the M commuting non-fully degenerate operators, i.e., if the Hamiltonian operator is a polynomial of these M commuting non-fully degenerate operators, then (iii) and (iv) may also be satisfied as a direct consequence of the correspondence of quantum and classical integrability.

On the other hand, the energy spectra could reveal deeply the behaviors of quantum integrability and quantum chaos with the random matrix theory [162–164]. But these studies are far beyond the content of coherent state theory explored in this book. In this book, we will not discuss the irregular spectra. Interested readers should consult the review of O. Bohigas and H. A. Weidenmuller [173].

A more interesting problem is as follows: given the explicit definition of quantum integrability, is there a general algorithm for the determination of the integrability of a given system? We should point out that even for a classical system, this is indeed a difficult question, and so far no generic answer has been found. However, in quantum mechanics, it is found that it is possible to establish such a criterion via group theory. In fact, with the quantum-classical correspondence developed in the last section, this long-standing problem for classical mechanics may also be solved. The answer is obtained from the discovery of a relation between integrability and the elegant concept of dynamical symmetry [174–176]. To this end, let us briefly illustrate here the concept of dynamical symmetry and some of its applications in physics.

As is well known, symmetry is a very profound concept in the development of modern physics. Within the known interaction mechanics, it is not easy to answer directly how a dynamical system evolves in time. From the following discussion, we will see that the global properties of the time evolution of a quantum system are in fact classified by a "generalized symmetry," called as dynamical symmetry. For comparison, we shall first discuss the precise definitions of three kinds of symmetry in physics: symmetry, hidden symmetry, and dynamical symmetry.

A system is said to have a symmetry, denoted by a group S, continuous or discrete, if its Hamiltonian H is invariant under the operation of S: $gHg^{-1} = H$, $g\phi_0 = \phi_0$; $g \in S$, where ϕ_0 is the ground state of the system. A typical example of the fundamental symmetries that play an essential role in the development of modern physics is the gauge symmetry. It dictates that interactions in nature. However, nature does not always have precise symmetries. Quite often, a given system may deviate somewhat from the exact ones. One such deviation is associated with a spontaneous symmetry breaking, namely, the hidden symmetry. This symmetry can be defined as follows:

A system has a hidden symmetry S when its Hamiltonian H is invariant under the operation of the associated symmetry group S, continuous or discrete, but the ground state ϕ_0 is not: $gHg^{-1} = H$, $g\phi_0 \neq \phi_0$; $g \in U$. The associated invariance guarantees the existence of degenerate ground states which means that the system can excite a particle without exhausting any additional energy. This profound concept is indeed one of the most important concepts in the electric-weak interaction theory. Finally, dynamical symmetry has played an important role in physics in quantum theory. It is defined as follows:

A quantum system with a dynamical group G has a dynamical symmetry if and only if the Hamiltonian of the system can be expressed in terms of the Casimir operators, or invariant operators, of any particular subgroup chain G^α of G:

$$H = f(C_i^\alpha), (i = 1, \cdots, l^\alpha), \tag{12.89}$$

where $\alpha = 1, 2, \cdots, \gamma$ is the index of a particular subgroup chain, C_i^α is the ith Casimir operator of the subgroup G^α, and l^α represents the rank of the subgroup

G^α. Clearly, dynamical symmetry is the least restrictive of these three symmetries in that neither the Hamiltonian nor the ground state is necessarily invariant under the transformation of G: $gHg^{-1} \neq H$, $g\phi_0 \neq \phi_0$; $g \in G$. Indeed, from this point of view, the first two symmetries are only special cases of dynamical symmetry.

At this point, a few comments about these symmetries seem to be in order. In the literature, there is quite a bit of discussion to the effect that a system must be integrable if it contains enough symmetries. We want to emphasize that this is not a precise statement! In fact, symmetry does not imply integrability at all. In particular, the symmetries one usually refers to are the fundamental symmetries in the nature, such as translational, rotational, gauge, and internal quantum symmetries, which include discrete symmetries such as time reverse symmetry, space reflection, etc. These symmetries are global and the associated invariances correspond to a small number of constants of motion.

Dynamical symmetry is a widely used concept in quantum systems. It determines the detail of dynamical behaviors and has many robust applications in physics. A fairly complete discussion together with some of the important collections of papers on this subject can be found in the two volumes edited by A. Bohm, Y. Ne'eman, and A. O. Barut [177, 178]. Still, it is worth pointing out that the applications of dynamical symmetry are mostly limited to the purpose of simplifying otherwise exceedingly or maybe hopelessly complicated quantum mechanical calculations and for spectroscopic pattern recognition. For this reason, a dynamical symmetry group is sometimes called spectrum generating algebra. However, the profound significance of the application of the concept of dynamical symmetry, which will be the cardinal issue in the study of quantum non-integrability, is that it indeed provides a classification of the global properties of the dynamics. In the following, we shall discuss how this concept is related to the study of the system's dynamics.

The relationship between the integrability of a quantum system and the dynamical symmetry is given precisely as a theorem:

Theorem. *A quantum system described by dynamical group G is integrable if it possesses a dynamical symmetry of G.*

According to group theory, there are two classes of subgroup chains for G: canonical and non-canonical. A subgroup chain is canonical if the set of Casimir operators in various subgroups of the chain can completely specify states in its irreducible presentation space. Otherwise, it is non-canonical. The simple proof of the above theorem is as follows: First, we consider a canonical subgroup chain G^α *of G. The Casimir operators of G,* $\{C_i\}$ *and all the Casimir operators* $\{C_i^\alpha\}$ *of the subgroups in the chain* G^α *will form a complete set of commuting operators of any irreducible representation carrier space* $\mathfrak{H}$ *of G,* $\{C_i, C_i^\alpha\} \equiv \{Q_j, j = 1, \cdots, M\}$. *When the system has a dynamical symmetry of* G^α, *all the operators* Q_j *are by definition the commuting integrals of motion:* $[H, Q_j] = 0$. *In this case the system always has M commuting integrals of motion and thus is integrable.*

For a non-canonical subgroup chain G^α, the number of Casimir operators $\{G_i\}$ of G and all Casimir operators $\{C_i^\alpha\}$ of G^α is less than the number of the complete set of commuting operators of any irreducible representation carrier space of G. However, according to Dirac's definition of the complete set of commuting operators, there must exist other commuting operators $\{A_i\}$ which will commute with $\{C_i\}$ and $\{C_i^\alpha\}$. The complete set of commuting operators is formed by combining these operators with the Casimir operator, $\{C_i\} \cup \{C_i^\alpha\} \cup \{A_i\} \equiv \{Q_j, j = 1, \cdots, M\}$. When the system has a dynamical symmetry of G^α, the operators Q_j satisfy $[H, Q_j] = 0$. Thus, the system is integrable.

Since realistic quantum systems usually have a dynamical Lie group G, the following generic result is true: suppose G is an l-rank and n-dimensional Lie group, then as we have shown previously, the dimension of the complete set of commuting operators of G is independent of the subgroup chain: $d = l + (n - l)/2$, where the l operators are the Casimir operators of G and are fully degenerate for any given irreducible representation of G. Thus, the number M of the non-fully degenerate operators in the complete set of commuting operators for a given irreducible representation of G exceeds $(n - l)/2$, i.e., $M \leq (n - l)/2$. The equality is only true for the canonical irreducible representation of G. If the system has a dynamical symmetry, there are M commuting integrals of motion. This is a prescription to determine the integrability of quantum systems. Moreover, by the embedding algorithm, one can construct all possible dynamical group chains for a given G. From the above theorem, we see that each subgroup chain corresponds to an integrable system. This provides a general procedure to construct all possible integrable systems for any given dynamical system with dynamical group G.

Although a general approach for the construction of various possible integrable systems exists, this does not mean that all of them can physically be realized. Usually the number of physically admissible subgroup chains for a given G is much less than all the possible subgroup chains. For instance, for most realistic quantum systems, such as atomic and nuclear systems, rotational invariance is obeyed. Only those dynamical symmetries which terminate with rotational symmetry are physically realizable. This greatly reduces the number of possible realistic integrable systems. Thus, the relation of dynamical symmetry and integrability further provides a way to construct all physically realizable integrable systems.

By the use of straightforward methods, it can be shown that the dynamical symmetry is also a criterion to determine a system's integrability in the classical limit. In the classical limit, the group algebraic structure is defined by the Poisson brackets. Since the algebraic structure of G in the phase space representation is preserved, the classical limit of a quantum system, if exists, has the same group structure. The concept of dynamical symmetry can also be defined for classical mechanics based on the Poisson brackets. Thus, the theorem about the relation between integrability and dynamical symmetry is operative in the classical limit as well. This means that the general procedure of using dynamical symmetry to construct various possible integrable systems is available for classical mechanics giving rise to a general algorithm to test the classical systems' integrability which is an unsolved long-standing problem in classical mechanics.

It can further be shown that even in the absence of a classical limit, i.e., in the semi-quantal description of a quantum system, dynamical symmetry still survives. Explicitly, we will first show that if the Hamiltonian has a symmetry R, then its phase space representation has the same symmetry. The reason for this is that if $RHR^{-1} = H$, then in the phase space representation, $\langle\Omega|H|\Omega\rangle = \langle\Omega'|H|\Omega'\rangle$, i.e., $\mathcal{H}(p, q) = \mathcal{H}(p', q')$, where

$$|\Omega'\rangle = R^{-1}\Omega|\psi_0\rangle = \Omega' h|\psi_0\rangle = |\Omega'\rangle e^{i\varphi(h)}. \tag{12.90}$$

Moreover, the phase space representation of an invariant operator $C(T_i)$ is generally given by $\langle\Omega|C(T_i)|\Omega\rangle = s(\Xi)C(\langle\Omega|T_i|\Omega\rangle)$, where $s(\Xi)$ is a function of Ξ. In the classical limit, $s(\Xi)$ approaches unity. Thus, if C is an invariant operator, then its phase space representation is also an invariant observable. Consequently, symmetries including the dynamical symmetries are preserved in the semi-quantal description.

In the above discussions, we have presented a precise definition of integrability for quantum systems and have described their possible generic behaviors of an integrable quantum system. A theorem for determining the integrability of the system via the concept of dynamical symmetry is proven. It leads to a discussion of the consistency between quantum and classical integrability. An important conclusion reached in the above discussions is that non-integrability of a quantum system implies the breaking of its dynamical symmetry. If the semi-quantal dynamics is chaotic, the dynamical symmetries of the system must be broken. From this point on, chaos is strictly defined in the semi-quantal description. The precise relation between symmetry breaking and non-integrability is as follows:

For a given quantum system with a dynamical Lie group G which is l-rank and n-dimensional, if the dynamical symmetry is broken such that any of the M commuting integrals of motion for a nonautonomous system is destroyed, the system becomes non-integrable.

Here, non-integrability is rigorously defined as follows: a system does not satisfy Eq. (12.86). Evidently, non-integrable systems include near-integrable systems. In other words, dynamical symmetry breaking may not be sufficient to alter the dynamics of an integrable system so that it exhibits chaotic behavior. Hence, an interesting question is: to what extent must the dynamical symmetry be broken so that chaos sets in? This is equivalent to a search for a quantum KAM theorem. In the following, we will examine the relationship between semi-quantal chaos and the dynamical symmetry breaking of the "parent" quantum system.

When a quantum system possesses a dynamical symmetry, its dynamics is regular. In the semi-quantal description, this regularity is reflected by a topologically stable phase space structure. Dynamical symmetry breaking results in the irregular semi-quantal phase space structure. Thus, our study of the onset of chaos in quantum systems is based on the semi-quantal dynamics.

According to the stability theory, a stable phase space structure is determined by the stability of the Hamiltonian function. Explicitly, for a given dynamical system, a set of stable points in phase space is given by the following condition:

$$x_c : \left.\frac{\partial \mathcal{H}_0(x)}{\partial x}\right|_{x_c} = 0, \det\left[\frac{\partial^2 \mathcal{H}_0(x)}{\partial x^i \partial x^j}\right]_{x_c} > 0, \quad (12.91)$$

where $x = (p, q) \in G/H$, $\mathcal{H}_0(x)$ is the phase space representation of H_0 which possesses dynamical symmetry.

Dynamical symmetry breaking implies that the system may deviate from regularity. In other words, there is a perturbation, H_1, which does not possess the dynamical symmetry of H_0,

$$H = H_0 + \lambda H_1. \quad (12.92)$$

For certain ranges of λ, λH_1 may not affect the stable structure of H_0, so that the system's dynamics may remain in a phase space structure similar to that of H_0. In the language of the KAM theorem, this means that the topology of phase space structure has not been altered. However, outside of this range of λ, the stable structure of Eq. (12.92) could change. When this happens, the dynamical symmetry is sufficiently broken to force the system to undergo a transition. One can evaluate the critical point of such a transition as follows:

$$\det\left[\frac{\partial^2 \mathcal{H}(x)}{\partial x^i \partial x^j}\right]_{x_c} = f(x_c, \lambda_c) = 0, \quad (12.93)$$

where $\mathcal{H}(x) = \mathcal{H}_0(x) + \lambda \mathcal{H}_1(x)$. When $\lambda > \lambda_c$, the dynamical symmetry is completely broken and the associated stable structure is utterly destroyed. Thus, the dynamics of the system becomes irregular, and the corresponding semi-quantal phase space structure must be chaotic.

The above description of stable structure transition is called a structural phase transition. We postulate that for a non-integrable system, if the dynamical symmetry breaking is accompanied by a structural phase transition, it must result in the system's motion being chaotic.

The above analysis indicates that if a non-integrable system undergoes a structural phase transition, its Hamiltonian must consist of at least two parts, one with a specific dynamical symmetry and the other with a different dynamical symmetry, or no dynamical symmetry at all. Succinctly, if a quantum system possesses a dynamical group G, its various dynamical symmetries are characterized by the various subgroup chains $G \supset G^{\gamma}$. In this case, the Hamiltonian is written as

$$H = H_\alpha + \lambda H_p, \quad (12.94)$$

where H_α has the dynamical symmetry $G \supset G^\alpha$ and H_p has another dynamical symmetry $G^p \neq G^\alpha$. If H_p has no dynamical symmetry, then in the parameter space of $\lambda > \lambda_c$, the system will enter into the realm of chaos. If H_p has another dynamical symmetry, then with increasing $\lambda > \lambda_c$, we will encounter a second set of stable points which are determined by

$$\left.\frac{\partial \mathcal{H}_p(x)}{\partial x^i}\right|_{x'_c} = 0, \det\left[\frac{\partial^2 \mathcal{H}_p(x)}{\partial x^i \partial x^j}\right]_{x'_c} > 0. \tag{12.95}$$

In other words, when $\lambda > \lambda_c$, the system is near another stable phase space structure characterized by the dynamical symmetry of H_p. In this case, the dynamics of such a system will have a "generic" regular-chaos-regular structure, in which when λ is near λ_c, the system is strongly chaotic.

In the above, we discussed the physical criterion of integrability and the origin of irregularity. These are generic discussions at both the quantum level and the classical level. We now turn our attention to examine the quantum behavior of a non-integrable system at the onset of chaos.

Chaos is described by the trajectories in the phase space. Unfortunately due to the uncertainty principle and the consequent quantum fluctuation in quantum mechanics, this concept is non-operative. Indeed, it is well known that quantum fluctuation or quantum correlation is the "wedge" between classical and quantum mechanics. To understand quantum chaos, one needs to understand how the dynamical behavior of a non-integrable system deviates from classical mechanics because of quantum fluctuation. In the semi-quantal description, the concept of "trajectory" is still intact since the inherent quantum fluctuations are built into the equations of motion. Therefore the semi-quantal description provides a natural way to study the effect of quantum fluctuation in regular and irregular phase space structure.

The difference between the semi-quantal and classical mechanics is the leading order of the quantum fluctuation $\delta\mathcal{H}$ contained in the semi-quantal Hamiltonian function $\mathcal{H}$, $\delta\mathcal{H} = \mathcal{H} - \mathcal{H}_c$, which is an explicit function of Ξ and the coordinates of G/H. Therefore, the dependence of δH on Ξ offers a way to control systematically the quantum fluctuation, and the classical limit corresponds to $\Xi \to \infty$, as has been shown before. On the other hand, the dependence of $\delta\mathcal{H}$ on dynamical variables allows one to explore the dynamical effect of quantum fluctuation.

The general Hamiltonian of a non-integrable quantum system is given by Eq. (12.92). The H_1 in Eq. (12.92) may be regarded as a "perturbative" term although it may not be small at all when compared to H_0. Also, this term does not in any way have the same dynamical symmetry as H_0. The crucial point is that when H_0 is perturbed by such a term, the quantum fluctuation can drive the system away from its classical phase space structure. If such derivations change the phase portrait topologically, especially when chaos sets in, the resulting phenomena must provide a hint to the quantum behavior in chaos.

For most realistic systems, H_1 is generally a quadratic, or higher-order, function of generators:

$$H_1 = \sum_{ij} c_{ij}(\hbar T_i)(\hbar T_j) + \cdots . \tag{12.96}$$

The quantum fluctuation can be calculated explicitly. To simplify the discussion, let us consider an H_1 that includes only quadratic terms of the generators. Correspondingly, the semi-quantal and classical Hamiltonian functions are

$$\mathcal{H}_1 = \mathcal{H}_{1cl} + \delta\mathcal{H}, \tag{12.97}$$

where $\delta\mathcal{H}$ is the leading order of quantum correlations

$$\delta\mathcal{H} = \sum_{ij} c_{ij}\hbar^2(\langle T_i T_j\rangle - \langle T_i\rangle\langle T_j\rangle) = \frac{1}{\Xi}\mathcal{H}'(q, p, \Xi\hbar). \tag{12.98}$$

For many realistic systems, the quantum fluctuations may be comparable in magnitude with their classical limits. For non-integrable systems, these fluctuations can alter topologically the classical phase space structure. In summary, in this section, we have presented a detailed discussion of the dynamical behaviors of non-integrable quantum systems. A non-integrable quantum system must have dynamical symmetry breaking. However, the dynamical symmetry breaking does not imply the occurrence of chaotic motion. Indeed, chaos can appear only if the dynamical symmetry breaking is accompanied by a structural phase transition. The chaotic behavior of quantum systems is well defined in the semi-quantal description. Moreover, the semi-quantal description provides explicitly the dynamical effect of quantum fluctuation in the classical trajectories.

Exercises

12.1. For the SU(2) case, show that the transformations of the SU(2) generators under the coset representative Ω are given by

$$\begin{aligned}
\Omega^{-1}\hbar J_0\Omega &= (2j\hbar - p^2 - q^2)J_0/j + (q+ip)\sqrt{4j\hbar - p^2 - q^2}J_+/4j \\
&\quad + (q-ip)\sqrt{4j\hbar - p^2 - q^2}J_-/4j, \\
\Omega^{-1}\hbar J_+\Omega &= -(q-ip)\sqrt{4j\hbar - p^2 - q^2}J_0/2j \\
&\quad + (4j\hbar - p^2 - q^2)J_+/4j - (q-ip)^2 J_-/4j, \\
\Omega^{-1}\hbar J_-\Omega &= (\Omega^{-1}\hbar J_+\Omega)^\dagger .
\end{aligned}$$

12.2. Show that the quantum correlations for the quadratic terms of the SU(2) generators are

$$\delta(\hbar J_0)^2 = \frac{1}{4\Xi}(p^2 + q^2)(4j\hbar - p^2 - q^2),$$

$$\delta(\hbar J_+)^2 = -\frac{1}{4\Xi}(q - ip)^2(4j\hbar - p^2 - q^2),$$

$$\delta(\hbar J_0 \hbar J_+) = \frac{1}{4\Xi}(q - ip)(4j\hbar - p^2 - q^2)\sqrt{4j\hbar - p^2 - q^2},$$

$$\delta(\hbar J_- \hbar J_+) = \frac{1}{4}(4j\hbar - p^2 q^2)^2,$$

where $\Xi = 2j$.

12.3. For the SU(1, 1) case, show that the transformations of the SU(1, 1) generators under the coset representative Ω are given by

$$\Omega^{-1}\hbar K_0 \Omega = (2k\hbar + p^2 + q^2)K_0/2k + (q + ip)\sqrt{4k\hbar + p^2 + q^2 K_+/4k}$$
$$+ (q - ip)\sqrt{4k\hbar + p^2 + q^2 K_-/4k},$$

$$\Omega^{-1}\hbar K_0 \Omega = (q - ip)\sqrt{4k\hbar + p^2 + q^2 K_0/2k}$$
$$+ (4k\hbar + p^2 + q^2)K_+/4k + (q - ip)^2 K_-/4k,$$

$$\Omega^{-1} K_- \Omega = (\Omega^{-1} K_+ \Omega)^\dagger.$$

12.4. Show that the quantum correlations for the quadratic terms of the SU(1, 1) generators are

$$\delta(\hbar K_0)^2 = \frac{1}{4\Xi}(p^2 + q^2)(4k\hbar + p^2 + q^2),$$

$$\delta(\hbar K_+)^2 = \frac{1}{4\Xi}(q - ip)^2(4k\hbar + p^2 + q^2),$$

$$\delta(\hbar K_0 \hbar K_+) = \frac{1}{4\Xi}(q - ip)\sqrt{4k\hbar + p^2 + q^2}(4k\hbar + p^2 + q^2),$$

$$\delta(\hbar K_- \hbar K_+) = \frac{1}{4\Xi}(4k\hbar + p^2 + q^2)^2,$$

where $\Xi = 2k$.

Open Quantum Systems 13

13.1 Overview

Nonrelativistic quantum mechanics deals with closed or isolated physical systems whose dynamics are determined by the Schrödinger equation for a given Hamiltonian. However, this is a somewhat artificial scenario. Indeed, any realistic system will inevitably interact with its environment. When such interactions are not negligible, these systems cannot be treated as closed (or isolated) systems. Consequently, the principle of Schrödinger-based quantum mechanics is no longer applicable. In the literature, these systems are called **open systems**. Understanding the quantum dynamics of open systems is one of the most challenging topics in physics, chemistry, engineering, biology, and even social sciences. In particular, the interactions between the system and its environment can induce various dissipation and noises (fluctuations) such that the physical systems can exhibit disorders. The nature of the emergence of disorders is one of the most difficult problems to solve in sciences.

To understand the quantum dynamics of open systems and the origin of disorders, different theories have been proposed and developed in the past century. In practice, an open system is defined as the principal system consisting of only a few relevant dynamical variables in contact with one or more reservoirs made of a huge (infinite) number of degrees of freedom. Of course, the principal system plus its environments together form a closed system, which can still be governed by the Schrödinger equation in terms of the wave function of the total system [51] or by the von Neumann equation in terms of the total density matrix [179]. Additionally, usually it is also assumed that the environment is initially in a thermal equilibrium state with a given temperature T. In this manner, by definition, it is a mixed state. Thus, the Schrödinger picture is again not applicable. It is for this reason that one needs to begin with the von Neumann equation in terms of the density matrix to address the dynamics of the principal system from the total system. The solution

C.-F. Kam et al., *Coherent States*, Lecture Notes in Physics 1011,
https://doi.org/10.1007/978-3-031-20766-2_13

of the reduced density matrix could predict all physical observables of the principal system, including the origin of disorders, as one would expect.

However, in practice, it is well known that it is very complicated and difficult to solve the von Neumann equation of the total system for arbitrary interactions between the system and the environment. The difficulty stems from the fact that the environment contains an infinite number of degrees of freedom and its dynamics is a priori unknown. Furthermore, one is only interested in the dynamics of the principal system, rather than the dynamics of the environment or the total system. Hence, for a long time, the central issue in the investigation of the quantum dynamics of open system has been focused on finding the equation of motion for the reduced density matrix of the principal system. Such an equation of motion is called the **master equation**.

The master equation for open quantum systems plays the same role as the Newtonian equation for macroscopic objects, the Maxwell equations for electrodynamics, and the Schrödinger equation for isolated quantum systems. However, from a more fundamental point of view, the Newtonian equation can be derived by the Lagrangian formalism or Hamiltonian formalism, the Maxwell equations can be derived from the Lagrangian of the quantum electrodynamics (QED), and the Schrödinger equation is a nonrelativistic approximation of the Dirac equation which can also be derived from the Lagrangian of QED. Thus, all the abovementioned fundamental equations of motions can be obtained from the least action based on the Lagrangian which can be constructed from the space-time symmetry and gauge symmetry. However, to date, neither a fundamental principle has been found to determine the master equation nor a fundamental theory has been developed to derive the master equation of open quantum systems. Due to this obstacle, finding the master equation for open quantum systems becomes the most challenging problem. Certainly, if one were to find the general master equation for arbitrary open systems, then many fundamental and interesting realistic phenomena, in particular the nature of disorders, would become clear. Indeed, under such a scenario, it is not inconceivable that physical principles hitherto unknown may also be discovered.

It is worth underscoring that in as early as the 1960s, the fundamental theory for studying non-equilibrium quantum dynamics of open systems was actually proposed by Schwinger and Feynman independently. They developed this in terms of the Green function technique and path integral approach, respectively [180, 181]. They started with the Brownian motion in attempting to provide a full quantum mechanical description of open systems and meanwhile trying to explore the quantum origins of dissipation and fluctuations induced by the environment. Soon after the pioneering works of Schwinger and Feynman, the quantum transport theory and the non-equilibrium Green function technique of many-body systems were developed by Kadanoff, Baym, and also Keldysh, respectively [182–185]. Later, the quantum dissipative dynamics and decoherence theory of individual open systems were symmetrically explored for the quantum Brownian motion, using the Feynman-Vernon influence functional approach [186–190]. In particular, when the Brownian particle is modeled as a harmonic oscillator and the environment is also modeled by a continuous distribution of infinite numbers of harmonic oscillators, the exact master equation for quantum Brownian motion was derived for the first time during the period of the 1980s–1990s [186, 191–193].

The Feynman-Vernon influence functional approach allows one to exactly integrate out all the degrees of freedom in the environment. This was done when the environment is modeled by a continuous distribution of infinite numbers of harmonic oscillators. The resulting environment effect on the Brownian particle is fully given by an influence functional containing the system degrees of freedom only [181]. The influence functional shows how the dissipation and diffusion emerge in the Brownian motion. Only when the Brownian particle is also modeled as a harmonic oscillator, the dynamics of Brownian particle can be exactly solved, and the exact master equation can be obtained [186, 191–193]. Since then, such a quantum Brownian model becomes the prototype example in one's understanding of the quantum dynamics of open systems [190].

Of course, in reality, however, not all open systems can be modeled as a harmonic oscillator. Indeed, in the last century, no exact master equation has been found when one goes beyond such a quantum Brownian model. Thus, the master equation for the quantum dynamics of open systems in general remains a challenge. This challenge remains for many fundamental problems in the studies of physics, chemistry, engineering, and biology.

Fortunately, since the 2000s, WMZ realized that to address physically the more general essence of the quantum dynamics of open systems, the coherent states path integral [17, 194] could be a very useful tool in deriving the exact master equation for a large class of open systems [195–201]. To this end, it is especially useful for these open systems to couple bilinearly to the environments through the exchanges of particles, energies, and information between systems and their environments. Such interactions between the system and the environments can be described with generalized Fano- and Anderson-type Hamiltonians [200, 202, 203] which have wide range of applications in atomic physics, quantum optics, condensed matter physics, and particle physics [204–206]. In this case, both the system and the environment can either be bosonic [197] or fermionic [195, 196]. Also, it may be extendible to spin-like (or anyon) systems [199, 201]. Using this exact master equation, we were able to obtain for a large class of open systems the general non-Markovian dissipation and fluctuation dynamics of open systems. Also, an extended non-equilibrium fluctuation-dissipation theorem is discovered [198, 200]. The quantum transportation theory for both the bosonic and fermionic junctions and the relation with the non-equilibrium Green function technique are established with the same approach [196, 197, 201, 207]. It also shows that the quantum-to-classical transition, the foundation of statistical mechanics, and quantum thermodynamics can be developed from the quantum dynamics of open systems based on the exact master equation [208–210].

In this chapter, using the coherent states path integral approach, we will focus on the derivation of the exact master equation for a large class of open quantum systems. We will also discuss the general non-Markovian dynamics and their modern applications, particularly the applications to quantum transportation and quantum thermodynamics.

13.2 Influence Functional in the Coherent-State Representation

The quantum dynamics of open systems is determined by the reduced density matrix $\rho_S(t)$ of the total system (system plus environment), which is defined as the partial trace of the total density matrix $\rho_{tot}(t)$ over all the environment states,

$$\rho_S(t) = \mathrm{Tr}_E[\rho_{tot}(t)]. \tag{13.1}$$

For an arbitrary initial state $\rho_{tot}(t_0)$ for the system and the environment, the time evolution of $\rho_{tot}(t)$ will obey the von Neumann equation of the quantum mechanics,

$$\frac{\mathrm{d}}{\mathrm{d}t}\rho_{tot}(t) = \frac{1}{i\hbar}[H_{tot}(t), \rho_{tot}(t)]. \tag{13.2}$$

This is because the system plus the environment together form a closed system. If the initial state of the system plus its environment is a pure state, then the von Neumann equation can be reduced to the Schrödinger equation. But the von Neumann equation is more general because it is also valid to statistically mixed states for which the Schrödinger equation is no longer applicable.

The formal solution of the von Neumann equation can be expressed as

$$\rho_S(t) = \mathrm{Tr}_E[U(t, t_0)\rho_{tot}(t_0)U^{\dagger}(t, t_0)]. \tag{13.3}$$

Here, $U(t, t_0)$ and $U^{\dagger}(t, t_0)$ are, respectively, the forward and backward time evolution operators of the total density matrix. Explicitly,

$$U(t, t_0) = T_{\rightarrow} \exp\left\{-\frac{i}{\hbar}\int_{t_0}^{t} H_{tot}(t')dt'\right\}, \tag{13.4}$$

where $T_{\rightarrow}$ is the time-ordering operator and $H_{tot}(t)$ is the system's total Hamiltonian, the environment plus the interaction between them. In most of the cases, the trace over all the environmental states in Eqs. (13.1) or (13.3) is the most difficult problem for the open quantum systems. In the early 1960s, Feynman and his student Vernon developed an approach, which is now known as the Feynman-Vernon influence functional method in the literature [181]. In this approach, Feynman and Vernon used Feynman's path integral to completely and exactly trace over the environmental state for the system, in particular for the Brownian motion which linearly coupled to the position variables of the environment. The results give rise to an effective action in terms of the system degrees of freedom only but describe all influences of the environment on the system. Such an effective action mixed the forward and backward evolution of the system and is called the influence functional.

After the 2000s, one of us discovered that by using the coherent states path integrals formalism, one can non-perturbatively and exactly trace over the environmental states for a large class of open systems coupled to the environment through quantum tunneling of particles. This includes the Fano resonance of a discrete state in a continuum medium in atomic physics, the spontaneous emissions of two-level atomic systems coupled to radiative field in quantum optics, the Anderson localization model of electron scatterings through lattice potentials in condensed matter physics, the quantum transport incorporating quantum tunneling in mesoscopic physics, the integrated photonic circuits for photonic quantum computers, the superconductor-semiconductor hybrid systems for topological phase of matters, etc. [195–197, 200, 201]. Of course, it also includes the quantum Brownian motion as a special example.

In this chapter we will take the simplest form of the total Hamiltonian for such a class of open systems

$$\begin{aligned} H_{tot}(t) &= H_{\rm S}(t) + H_{\rm E}(t) + H_{\rm SE}(t) \\ &= \sum_i \varepsilon_{{\rm S},i}(t) a_i^\dagger a_i + \sum_k \epsilon_k(t) b_k^\dagger b_k + \sum_{ik} \big(V_{ik}(t) a_i^\dagger b_k + V_{ik}^*(t) b_k^\dagger a_i\big). \end{aligned} \tag{13.5}$$

The operators $a_i^\dagger, a_i$ ($b_k^\dagger, b_k$) are either bosonic or fermionic creation and annihilation operators of the system (environment) that obey the standard commutation or anti-commutation relationships. The single-particle energy spectra of the system and the environment, $\varepsilon_{{\rm S},i}$ and ϵ_k, and the system-environment coupling strength V_{ik} could be time-dependently controlled with the rapidly developed nano- and quantum technologies today.

To trace out explicitly the environment states in Eq. (13.3), it is also convenient to assume that the system and the environment are initially decoupled, and the environment is in a thermal state with initial temperature T_0,

$$\rho_{tot}(t_0) = \rho_{\rm S}(t_0) \otimes \rho_{\rm E}^{th}(t_0), \quad \rho_{\rm E}^{th}(t_0) = \frac{1}{Z_{\rm E}} e^{-(H_{\rm E}(t_0) - \mu_0 N_{\rm E})/k_B T_0}. \tag{13.6}$$

The initial system state $\rho_{\rm S}(t_0)$ can be any arbitrary state. In the initial environmental thermal state $\rho_{\rm E}^{th}(t_0)$, $N_{\rm E} = \sum_k b_k^\dagger b_k$ is the environmental particle number operator, μ_0 is an initial chemical potential, and k_B is the Boltzmann constant. If the environment is made of photons or phonons, then $\mu_0 = 0$. The normalization of the environment density matrix defines the partition function $Z_{\rm E} = {\rm Tr}_{\rm E}[e^{-(H_{\rm E}(t_0) - \mu_0 N_{\rm E})/k_B T_0}]$. In practice, such an initial state can be prepared by turning off the coupling between the system and the environment at initial time t_0. Specifically, we also denote $\epsilon_k(t_0) = \epsilon_k$. Although the initial total state $\rho_{tot}(t_0)$ is a decoupled state between the system and its environment, the system initial state $\rho_{\rm S}(t_0)$ can usually be non-Gaussian. It is certainly not an easy task to partially trace over the environment state in Eq. (13.3). In order to complete this partial trace, we shall use the coherent states path integral approach [17, 194] (also see Chap. 4). The result can be expressed only as an influence functional in terms of the coherent state variables of the system only. It corresponds to an effective action induced by the environment through the system-environment interaction that modifies significantly the original quantum dynamics of open systems.

For convenience in the present discussion, here we shall reformulate the coherent states path integral here. We use the unnormalized bosonic or fermionic coherent states,

$$|\boldsymbol{\xi}\rangle = \prod_i \exp(\xi_i a_i^\dagger)|0\rangle, \qquad \int d\mu(\boldsymbol{\xi})|\boldsymbol{\xi}\rangle\langle\boldsymbol{\xi}| = I \tag{13.7a}$$

and

$$|z\rangle = \prod_k \exp(z_k b_k^\dagger)|0\rangle, \qquad \int d\mu(z)|z\rangle\langle z| = I, \tag{13.7b}$$

for the system and the environment, respectively. The Haar measures in the resolution of the identity, $d\mu(\boldsymbol{\xi})$ and $d\mu(z)$, are defined in the coherent state parameter space $\boldsymbol{\xi} \equiv (\xi_1, \xi_2, ...)$, and $z \equiv (z_{k_1}, z_{k_2}, ...)$,

$$d\mu\,(\boldsymbol{\xi}) = \prod_i g_i d\xi_i^* d\xi_i e^{-|\xi_i|^2}, \quad d\mu\,(z) = \prod_k g_k dz_k^* dz_k e^{-|z_k|^2}, \tag{13.8}$$

where ξ_i and z_{k_i} are complex variables for bosons and Grassmann variables for fermions with $g_i(g_k) = 1/2\pi i$ and 1, respectively.

In the coherent state representation, the reduced density matrix of Eq. (13.3) with the initial decoupled state Eq. (13.6) can be expressed as

$$\langle\boldsymbol{\xi}_f|\rho_{\rm S}(t)|\boldsymbol{\xi}'_f\rangle = \int d\mu(z_f)\langle\boldsymbol{\xi}_f z_f|U(t,t_0)[\rho_{\rm S}(t_0)\otimes\rho_{\rm E}^{th}]U^\dagger(t,t_0)|\pm z_f\boldsymbol{\xi}'_f\rangle. \tag{13.9}$$

The partial trace over the environmental state in the coherent state representation is given by $\mathrm{Tr_E}[\cdots] = \int d\mu(z_f)\langle z_f|\cdots|\pm z_f\rangle$ where the minus sign corresponds to the trace over fermionic coherent states. Inserting the resolutions of the identity between the products of the time evolution operator and the initial density matrix operator, we arrive at

$$\begin{aligned}\langle\boldsymbol{\xi}_f|\rho_{\rm S}(t)|\boldsymbol{\xi}'_f\rangle = &\int d\mu(z_f)d\mu(\boldsymbol{\xi}_0)d\mu(z_0)d\mu(\boldsymbol{\xi}'_0)d\mu(z'_0)\langle\boldsymbol{\xi}_0|\rho_{\rm S}(t_0)|\boldsymbol{\xi}'_0\rangle \\ &\times\langle\boldsymbol{\xi}_f z_f|U(t,t_0)|z_0\boldsymbol{\xi}_0\rangle\langle z_0|\rho_{\rm E}^{th}|z'_0\rangle\langle\boldsymbol{\xi}'_0 z'_0|U^\dagger(t,t_0)|\pm z_f\boldsymbol{\xi}'_f\rangle.\end{aligned} \tag{13.10}$$

Here $\langle\boldsymbol{\xi}_f z_f|U(t,t_0)|z_0\boldsymbol{\xi}_0\rangle$ is the forward time evolution matrix element and $\langle\boldsymbol{\xi}'_0 z'_0|U^\dagger(t,t_0)|\pm z_f\boldsymbol{\xi}'_f\rangle$ the backward evolution matrix element with respect to the initial decoupled state $\langle\boldsymbol{\xi}_0 z_0|\rho_{tot}(t_0)|z'_0\boldsymbol{\xi}'_0\rangle = \langle\boldsymbol{\xi}_0|\rho_{\rm S}(t_0)|\boldsymbol{\xi}'_0\rangle\langle z_0|\rho_{\rm E}^{th}|z'_0\rangle$.

In terms of the coherent states path integral discussed in Chap. 4, the forward time evolution matrix element can now be expressed as

$$\langle \boldsymbol{\xi}_f z_f | U(t, t_0) | z_0 \boldsymbol{\xi}_0 \rangle = \exp\Big\{\Phi_{\rm G}(\boldsymbol{\xi}) + \Phi_{\rm G}(z)\Big\} \times \int \mathcal{D}[\boldsymbol{\xi}]\mathcal{D}[z] \exp\Big\{\frac{i}{\hbar}\int_{t_0}^{t} d\tau \big(\mathcal{L}_{\rm S}[\boldsymbol{\xi}] + \mathcal{L}_{\rm E}[z] - \mathcal{H}_{\rm SE}[\boldsymbol{\xi}, z]\big)\Big\}. \tag{13.11}$$

It is worth noting that the coherent states path integrals are very different from Feynman's conventional path integrals formalism. They are defined in terms of the complex or Grassmann variables so that the path integral end-point fixed values are given by the both sides of the path integrals $z_k(t_0) = z_{k0}$, $\xi_i(0) = \xi_{i0}$ and $z^*_k(t) = z^*_{kf}$, $\xi^*_i(t) = \xi_{if}$. The first exponential function in Eq. (13.11) is a boundary factor associated with the end-point values of the path integrals in the coherent state representation

$$\Phi_{\rm G}(\boldsymbol{\xi}) = \frac{1}{2}\sum_i \big[\xi^*_{if}\xi_i(t) + \xi^*_i(t_0)\xi_{i0}\big], \tag{13.12a}$$

$$\Phi_{\rm G}(z) = \frac{1}{2}\sum_k \big[z^*_{kf} z_k(t) + z^*_k(t_0) z_{k0}\big], \tag{13.12b}$$

where $(\xi^*_i(t_0), z^*_k(t_0))$ and $(\xi_i(t), z_k(t))$ are not the complex conjugate pairs of the fixed end points (ξ_{i0}, z_{k0}) and (ξ_{if}, z^*_{kf}). They are independent and are determined by solving explicitly the path integral, as we shall see later. This manifests the difference from the conventional Feynman's path integrals in terms of the position representation in quantum mechanics. The system and the environment Lagrangians and the interaction Hamiltonian between the system and the environment in Eq. (13.11) are given by

$$\mathcal{L}_{\rm S}[\boldsymbol{\xi}] = \frac{i\hbar}{2}\sum_i \big[\xi^*_i(\tau)\dot{\xi}_i(\tau) - \dot{\xi}^*_i(\tau)\xi_i(\tau)\big] - \sum_i \varepsilon_{{\rm S},i}(\tau)\xi^*_i(\tau)\xi_i(\tau), \tag{13.13a}$$

$$\mathcal{L}_{\rm E}[\boldsymbol{\xi}] = \frac{i\hbar}{2}\sum_k \big[z^*_k(\tau)\dot{z}_k(\tau) - \dot{z}^*_k(\tau)z_k(\tau)\big] - \sum_k \epsilon_k(\tau)z^*_k(\tau)z_k(\tau), \tag{13.13b}$$

$$\mathcal{H}_{\rm SE}[\boldsymbol{\xi}, z] = \sum_{ik} \big[V_{ik}(\tau)\xi^*_i(\tau)z_k(\tau) + V^*_{ik}(\tau)z^*_k(\tau)\xi_i(\tau)\big]. \tag{13.13c}$$

The path integral measures over the coherent state parameter space are defined by

$$\mathcal{D}[\boldsymbol{\xi}] = \prod_{i, t_0<\tau<t} g_i d\xi^*_i(\tau) d\xi_i(\tau), \tag{13.14a}$$

$$\mathcal{D}[z] = \prod_{k, t_0<\tau<t} g_k dz^*_k(\tau) dz_k(\tau). \tag{13.14b}$$

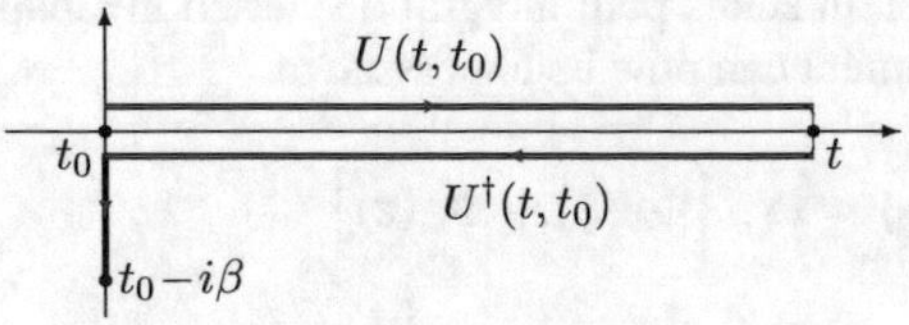

Fig. 13.1 (Color online) A schematic plot of the forward and backward evolutions of the total density matrix. The black and green lines with arrows represent the evolution paths of the environment and the system, respectively

The backward evolution matrix element is the complex conjugate of the forward evolution matrix element,

$$\langle \boldsymbol{\xi}'_0 z'_0 | U^\dagger(t,t_0) | z_f \boldsymbol{\xi}'_f \rangle = \exp\Big\{\Phi^*_{\rm G}(\boldsymbol{\xi}') + \Phi^*_{\rm G}(z')\Big\} \\ \times \int \mathcal{D}[\boldsymbol{\xi}']\mathcal{D}[z'] \exp\Big\{\frac{i}{\hbar}\int_t^{t_0} d\tau \big(\mathcal{L}^*_{\rm S}[\boldsymbol{\xi}'] + \mathcal{L}^*_{\rm E}[z'] - \mathcal{H}^*_{\rm SE}[\boldsymbol{\xi}', z']\big)\Big\}, \tag{13.15}$$

with the end-point boundary conditions: $z'_k(t) = z'_{kf}$, $\xi'_i(t) = \xi'_{if}$, and $z'^*_k(t_0) = z'^*_{k0}$, $\xi'^*_i(t_0) = \xi'_{i0}$. Equation (13.13) shows that $\mathcal{L}^*_{\rm S}[\boldsymbol{\xi}'] = \mathcal{L}_{\rm S}[\boldsymbol{\xi}']$, $\mathcal{L}^*_{\rm E}[z'] = \mathcal{L}_{\rm E}[z']$, and $\mathcal{H}^*_{\rm SE}[\boldsymbol{\xi}', z'] = \mathcal{H}_{\rm SE}[\boldsymbol{\xi}', z']$, i.e., they are all real functions, but the corresponding boundary conditions are exchanged as the result of the backward time evolution. The evolutions paths of Eq. (13.15) are depicted by Fig. 13.1.

Thus, the reduced density matrix of Eq. (13.10) can simply be written in terms of the coherent states path integrals as follows:

$$\langle \boldsymbol{\xi}_f | \rho_{\rm S}(t) | \boldsymbol{\xi}'_f \rangle = \int d\mu(\boldsymbol{\xi}_0) d\mu(\boldsymbol{\xi}'_0) \langle \boldsymbol{\xi}_0 | \rho_{\rm S}(t_0) | \boldsymbol{\xi}'_0 \rangle \mathcal{J}\Big(\boldsymbol{\xi}_f, \boldsymbol{\xi}'_f, t; \boldsymbol{\xi}_0, \boldsymbol{\xi}'_0, t_0\Big), \tag{13.16}$$

where $\mathcal{J}(\boldsymbol{\xi}_f, \boldsymbol{\xi}'_f, t; \boldsymbol{\xi}_0, \boldsymbol{\xi}'_0, t_0)$ is defined as the propagating function of the reduced density matrix in the coherent state representation. It is explicitly given by

$$\mathcal{J}(\boldsymbol{\xi}_f, \boldsymbol{\xi}'_f, t; \boldsymbol{\xi}_0, \boldsymbol{\xi}'_0, t_0) = \exp\Big\{\Phi_{\rm G}(\boldsymbol{\xi}) + \Phi^*_{\rm G}(\boldsymbol{\xi}')\Big\} \\ \times \int \mathcal{D}[\boldsymbol{\xi}]\mathcal{D}[\boldsymbol{\xi}'] \exp\Big\{\frac{i}{\hbar}\int_{t_0}^{t} d\tau \big(\mathcal{L}_{\rm S}[\boldsymbol{\xi}] - \mathcal{L}^*_{\rm S}[\boldsymbol{\xi}']\big)\Big\} \mathcal{F}[\boldsymbol{\xi}, \boldsymbol{\xi}']. \tag{13.17}$$

Here $\mathcal{F}[\boldsymbol{\xi}, \boldsymbol{\xi}']$ is the influence functional in the coherent state representation. It contains all the influence of the environment dynamics on the system after one traced over all the environmental states. Explicitly,

$$\mathcal{F}[\boldsymbol{\xi},\boldsymbol{\xi}'] = \int d\mu(z_f)d\mu(z_0)d\mu(z_0')\langle z_0|\rho_{\rm E}^{th}|z_0'\rangle \exp\Big\{\Phi_{\rm G}(z)+\Phi_{\rm G}^*(z')\Big\} \times\int \mathcal{D}[z]\mathcal{D}[z']\exp\Big\{\frac{i}{\hbar}\int_{t_0}^{t} d\tau\big(\mathcal{L}_{\rm ES}[\boldsymbol{\xi},z]-\mathcal{L}_{\rm ES}^*[\boldsymbol{\xi}',z']\big)\Big\}, \tag{13.18}$$

and

$$\begin{aligned}\mathcal{L}_{\rm ES}[\boldsymbol{\xi},z] &= \mathcal{L}_{\rm E}[z]-\mathcal{H}_{\rm SE}[\boldsymbol{\xi},z] \\ &= \sum_k\Big\{\frac{i\hbar}{2}\big[z_k^*(\tau)\dot{z}_k(\tau)-\dot{z}_k^*(\tau)z_k(\tau)\big]-\epsilon_k(\tau)z_k^*(\tau)z_k(\tau)\Big\} \\ &\quad -\sum_{ik}\big[V_{ik}(\tau)\xi_i^*(\tau)z_k(\tau)+V_{ik}^*(\tau)z_k^*(\tau)\xi_i(\tau)\big].\end{aligned} \tag{13.19}$$

is the generalized Lagrangian of the environment plus the system-environment interaction in the coherent states path integrals.

The trace over the environmental degrees of freedom can be computed with the coherent states path integrals. Because the generalized Lagrangian (13.19) is a quadratic function of environmental variables z_k and z_k^*, the path integrals in Eq. (13.18) can be exactly calculated using either the Gaussian integral or the stationary path approximation. The latter can also produce the exact solution when the Lagrangian is only a quadratic function of path integral variables. In this case, only the stationary paths have the contribution to the path integrals, i.e.,

$$\int\mathcal{D}[z]\exp\Big\{\frac{i}{\hbar}\int_{t_0}^{t}d\tau\mathcal{L}_{\rm ES}[\boldsymbol{\xi},z]\Big\} = \exp\Big\{\frac{i}{\hbar}\int_{t_0}^{t}d\tau\mathcal{L}_{\rm ES}^{\rm SP}(\boldsymbol{\xi},z)\Big\}, \tag{13.20}$$

where $\mathcal{L}_{\rm ES}^{\rm SP}(\boldsymbol{\xi},z)$ denotes the Lagrangian of the stationary paths only. The stationary paths are determined by the last action principle, i.e., the variation of the action functional with respect to the environment variables is zero:

$$\delta S_{\rm ES}[\boldsymbol{\xi},z] = \delta\int_{t_0}^{t}d\tau\mathcal{L}_{\rm ES}[\boldsymbol{\xi},z]=0. \tag{13.21}$$

It results in the classical-like Euler-Lagrangian equations:

$$\frac{d}{d\tau}\frac{\partial\mathcal{L}_{\rm ES}}{\partial\dot{z}_k}-\frac{\partial\mathcal{L}_{\rm ES}}{\partial z_k}=0,\quad \frac{d}{d\tau}\frac{\partial\mathcal{L}_{\rm ES}}{\partial\dot{z}_k^*}-\frac{\partial\mathcal{L}_{\rm ES}}{\partial z_k^*}=0. \tag{13.22}$$

From the above equations, one can find the equations of motion for the stationary paths of every environmental mode:

$$\frac{d}{d\tau}z_k(\tau) + \frac{i}{\hbar}\epsilon_k(\tau)z_k(\tau) = -\frac{i}{\hbar}\sum_i V_{ik}^*(\tau)\xi_i(\tau), \tag{13.23a}$$

$$\frac{d}{d\tau}z_k^*(\tau) - \frac{i}{\hbar}\epsilon_k(\tau)z_k^*(\tau) = \frac{i}{\hbar}\sum_i V_{ik}(\tau)\xi_i^*(\tau), \tag{13.23b}$$

subjected to the boundary condition $z_k(t_0) = z_{k0}$ and $z_k^*(t) = z_{kf}$. As it was shown, $z_k(\tau)$ and $z_k^*(\tau)$ obey the different boundary condition. That is, the stationary paths $z_k(\tau)$ and $z_k^*(\tau)$ $(t_0 \le \tau \le t)$ are two independent dynamical variables that are not complex conjugate to each other. This is indeed the general property of quantum mechanics presented in terms of the complex space, as a manifestation of quantum mechanics complex structure. Ignoring such a complex structure breaks the fundamental principle of quantum mechanics, namely, the unitarity. Furthermore, Eq. (13.23) can be solved analytically because it is linear. The result is

$$z_k(\tau) = u_{k0}(\tau, t_0)z_{k0} - \frac{i}{\hbar}\sum_i\int_{t_0}^{\tau} d\tau' u_{k0}(\tau, \tau')V_{ik}^*(\tau')\xi_i(\tau), \tag{13.24a}$$

$$z_k^*(\tau) = u_{k0}(t, \tau)z_{kf}^* - \frac{i}{\hbar}\sum_i\int_{\tau}^{t} d\tau' V_{ik}(\tau')\xi_i^*(\tau')u_{k0}(\tau', \tau), \tag{13.24b}$$

where

$$u_{k0}(t, t') = \exp\left\{-\frac{i}{\hbar}\int_{t'}^{t} d\tau\epsilon_k(\tau)\right\} \tag{13.25}$$

is the free particle propagating function in the environment.

Similarly, the path integrals of the environment part in the backward time evolution are only contributed by the backward stationary paths as well:

$$\begin{aligned}\int\mathcal{D}[z']\exp\left\{-\frac{i}{\hbar}\int_{t_0}^{t} d\tau\mathcal{L}_{\rm ES}^*[\boldsymbol{\xi}', z']\right\} \\ = \mathcal{N}(t)\exp\left\{-\frac{i}{\hbar}\int_{t_0}^{t} d\tau\mathcal{L}_{\rm ES}^{\rm SP}(\boldsymbol{\xi}', z')\right\},\end{aligned} \tag{13.26}$$

where $\mathcal{N}(t)$ is a pure time-dependent factor arising from the quantum fluctuations around the stationary paths. Because the generalized Lagrangian is real, $\mathcal{L}_{\rm ES}^*[\boldsymbol{\xi}', z'] = \mathcal{L}_{\rm ES}[\boldsymbol{\xi}', z']$, and it has the same form as that of the forward evolution, the backward stationary paths obey the same equations of the motion:

$$\frac{d}{d\tau}z'_k(\tau)+\frac{i}{\hbar}\epsilon_k(\tau)z'_k(\tau)=-\frac{i}{\hbar}\sum_i V^*_{ik}(\tau)\xi'_i(\tau), \tag{13.27a}$$

$$\frac{d}{d\tau}z'^*_k(\tau)-\frac{i}{\hbar}\epsilon_k(\tau)z'^*_k(\tau)=\frac{i}{\hbar}\sum_i V_{ik}(\tau)\xi'^*_i(\tau). \tag{13.27b}$$

But they are subjected to the different boundary conditions $z'_k(t) = \pm z_{kf}$ and $z'^*_k(t_0) = z'^*_{k0}$. The minus sign in the boundary condition comes from the trace with the fermionic coherent state integral; see Eq. (13.9). It is interesting to see that the backward stationary paths are the complex conjugate to the forward stationary paths:

$$z'_k(\tau)=\pm u_{k0}(\tau,t)z_{kf}+\frac{i}{\hbar}\int_\tau^t d\tau'\sum_i u_{k0}(\tau,\tau')V^*_{ik}(\tau)\xi'_i(\tau'), \tag{13.28a}$$

$$z'^*_k(\tau)=u_{k0}(t_0,\tau)z'^*_{k0}+\frac{i}{\hbar}\int_{t_0}^\tau d\tau'\sum_i V_{ik}(\tau')\xi'^*_i(\tau')u_{k0}(\tau',\tau). \tag{13.28b}$$

Using the equations of motion (13.23) and (13.27), the influence functional of Eq. (13.18) is reduced to

$$\begin{aligned}\mathcal{F}[\xi,\xi'] = \mathcal{N}(t)\int d\mu(z_f)d\mu(z_0)d\mu(z'_0)\langle z_0|\rho_{\rm E}^{th}|z'_0\rangle e^{\{\Phi_{\rm G}^{\rm SP}(z)+\Phi_{\rm G}^{*\rm SP}(z')\}} \\ \times \exp\left\{\frac{i}{2\hbar}\int_{t_0}^t d\tau\big(\mathcal{H}_{\rm ES}^{\rm SP}(\xi,z)-\mathcal{H}_{\rm ES}^{\rm SP}(\xi',z')\big)\right\}.\end{aligned} \tag{13.29}$$

where z, z' are the solution of the forward and backward stationary paths given by Eqs. (13.24) and (13.28).

Note that the matrix element of the initial environment state in the coherent state representation is

$$\langle z_0|\rho_{\rm E}^{th}|z'_0\rangle=\frac{1}{Z_{\rm E}}\exp\left\{\sum_k z^*_{k0}e^{-(\epsilon_k-\mu_0 n_k)/k_BT_0}z'_{k0}\right\}. \tag{13.30}$$

The partition function $Z_{\rm E}$ for noninteracting bosonic or fermionic environment is given by

$$Z_{\rm E}=\prod_k\Big(1\mp\exp\left\{-(\epsilon_k-\mu_0)/k_BT_0\right\}\Big)^{\mp}. \tag{13.31}$$

The left integrals over the end-point environmental variables in Eq. (13.29) are z_{k0}, z'_{k0}, z_{kf}, and their complex conjugates, which can be easily calculated with the

Gaussian integrals

$$\prod_k \int g_i dz_k^* dz_k \exp\{-z^\dagger A z + z^\dagger \alpha + \alpha^\dagger z\} = (\det A)^{\mp} e^{\alpha^\dagger A^{-1} \alpha}. \tag{13.32}$$

Here we have used the matrix notation: $z^\dagger A z \equiv \sum_{kk'} z_k^* A_{kk'} z_{k'}$, $z^\dagger \alpha \equiv \sum_k z_k^* \alpha_k$. The resulting influence functional can be expressed in terms of the system variables only:

$$\mathcal{F}[\xi, \xi'] = \mathcal{N}(t) \exp\left\{\frac{i}{\hbar}\int_{t_0}^{t} d\tau \mathcal{L}_{\rm IF}[\xi, \xi']\right\} \tag{13.33}$$

where $\mathcal{L}_{\rm IF}[\xi, \xi']$ is the environment-induced influence Lagrangian acting on the system:

$$\begin{aligned}\mathcal{L}_{\rm IF}[\xi, \xi'] =& i\hbar \sum_{ij}\left\{\int_{t_0}^{\tau} d\tau' \big(\xi_i^*(\tau) \mp \xi_i'^*(\tau)\big) g_{ij}(\tau, \tau') \xi_j(\tau')\right. \\ &\mp \int_{\tau}^{t} d\tau' \xi_i'^*(\tau) g_{ji}(\tau, \tau') \big(\xi_j(\tau') \mp \xi_j'(\tau')\big) \\ &\left.\pm \int_{t_0}^{t} d\tau' \big(\xi_i^*(\tau) \mp \xi_i'^*(\tau)\big) \widetilde{g}_{ij}(\tau, \tau') \big(\xi_j(\tau') \mp \xi_j'(\tau')\big)\right\}.\end{aligned} \tag{13.34}$$

In this solution, the up and down signs of $\mp$ correspond to the system being bosonic and fermionic, respectively. The first two terms in Eq. (13.34) come from the system-environment interaction through the forward and backward evolution of the environment dynamics, respectively. The last term is the mixed effects of forward and backward evolution through the initial environment state, also due to the system-environment interaction. The system-environment two-time correlation functions $g_{ij}(\tau, \tau')$ and $\widetilde{g}_{ij}(\tau, \tau')$ characterize the non-Markovian memory effect, i.e., the back-reactions between the system and the environment:

$$g_{ij}(\tau, \tau') = \left(\frac{1}{\hbar}\right)^2 \sum_k V_{ik}(\tau) u_{k0}(\tau, \tau') V_{jk}^*(\tau'), \tag{13.35a}$$

$$\widetilde{g}_{ij}(\tau, \tau') = \left(\frac{1}{\hbar}\right)^2 \sum_k V_{ik}(\tau) u_{k0}(\tau, \tau') V_{jk}^*(\tau') \langle b_k^\dagger b_k \rangle, \tag{13.35b}$$

and

$$\langle b_k^\dagger b_k \rangle = {\rm Tr}_{\rm E}[b_k^\dagger b_k \rho_{\rm E}^{\rm th}(t_0)] = \frac{1}{e^{(\epsilon_k - \mu_0)/k_B T_0} \mp 1} = f(\epsilon_k, T_0) \tag{13.36}$$

is the initial mode k particle occupation, given by the Bose-Einstein or the Fermi-Dirac distributions, in the environment at initial temperature T_0 with initial chemical potential μ_0. The influence Lagrangian Eq. (13.34) encompasses all the environmental effects on the system dynamics, including the renormalization of the system Hamiltonian, the dissipation, and fluctuation dynamics induced by the environment.

13.3 Exact Master Equation of the Reduced Density Matrix

The environment-induced influence Lagrangian significantly modifies the dynamics of the system. To find such significant changes of the system dynamics, we shall rewrite the propagating function of Eq. (13.17) in terms of an effective system Lagrangian:

$$\mathcal{J}(\boldsymbol{\xi}_f, \boldsymbol{\xi}'_f, t; \boldsymbol{\xi}, \boldsymbol{\xi}'_0, t_0) = \mathcal{N}(t) \exp\Big\{\Phi_{\rm G}(\boldsymbol{\xi}) + \Phi^*_{\rm G}(\boldsymbol{\xi}')\Big\} \times \int \mathcal{D}[\boldsymbol{\xi}]\mathcal{D}[\boldsymbol{\xi}'] \exp\Big\{\frac{i}{\hbar}\int_{t_0}^{t} d\tau \mathcal{L}_{\rm eff}[\boldsymbol{\xi}, \boldsymbol{\xi}']\Big\}. \tag{13.37}$$

The effective system Lagrangian $\mathcal{L}_{\rm eff}[\boldsymbol{\xi}, \boldsymbol{\xi}']$ mixes the forward and backward system Lagrangian through the interaction with the environment, and it is given by

$$\begin{aligned}
\mathcal{L}_{\rm eff}[\boldsymbol{\xi}, \boldsymbol{\xi}'] &= \mathcal{L}_{\rm S}[\boldsymbol{\xi}] - \mathcal{L}^*_{\rm S}[\boldsymbol{\xi}'] + \mathcal{L}_{\rm IF}[\boldsymbol{\xi}, \boldsymbol{\xi}'] \\
&= \sum_i \Big\{\frac{i\hbar}{2}[\xi_i^*(\tau)\dot{\xi}_i(\tau) - \dot{\xi}_i^*(\tau)\xi_i(\tau)] - \varepsilon_{{\rm S},i}(\tau)\xi_i^*(\tau)\xi_i(\tau)\Big\} \\
&\quad - \sum_i \Big\{\frac{i\hbar}{2}[\xi_i'^*(\tau)\dot{\xi}_i'(\tau) - \dot{\xi}_i'^*(\tau)\xi_i'(\tau)] - \varepsilon_{{\rm S},i}(\tau)\xi_i'^*(\tau)\xi_i'(\tau)\Big\} \\
&\quad + i\hbar \sum_{ij} \Big\{\int_{t_0}^{\tau} d\tau' \big(\xi_i^*(\tau) \mp \xi_i'^*(\tau)\big) \boldsymbol{g}_{ij}(\tau, \tau')\xi_j(\tau') \\
&\quad \mp \int_{\tau}^{t} d\tau' \xi_i'^*(\tau) \boldsymbol{g}_{ji}(\tau, \tau')\big(\xi_j(\tau') \mp \xi_j'(\tau')\big) \\
&\quad \pm \int_{t_0}^{t} d\tau' \big(\xi_i^*(\tau) \mp \xi_i'^*(\tau)\big)\widetilde{\boldsymbol{g}}_{ij}(\tau, \tau')\big(\xi_j(\tau') \mp \xi_j'(\tau')\big)\Big]\Big\}.
\end{aligned} \tag{13.38}$$

The last three terms are the environment-induced influence Lagrangian that would take into account all the back-reactions between the system and the environment. It results in quantum memory on its historical dynamical evolution.

To explore the memory dynamics of the system under the influence of the environment, we need to compute explicitly the path integrals in the propagating function of Eq. (13.37) to determine the time evolution of the reduced density matrix Eq. (13.16). Again, because the effective Lagrangian (13.38) is a quadratic function of the system variables, we can solve exactly the path integrals with the stationary path method. Similar to solving the stationary paths of all the modes of the environment, the stationary paths of the particles in the i-state of the system are determined by the classical-like Euler-Lagrangian equations of motion. This leads to the following equations of motion for the stationary paths:

$$\frac{d}{d\tau}\xi_i(\tau) + \frac{i}{\hbar}\varepsilon_{S,i}(\tau)\xi_i(\tau) + \sum_j \int_{t_0}^{\tau} d\tau' \boldsymbol{g}_{ij}(\tau,\tau')\xi_j(\tau') \tag{13.39a}$$

$$= \mp \sum_j \int_{t_0}^{t} d\tau' \widetilde{\boldsymbol{g}}_{ij}(\tau,\tau')[\xi_j(\tau') \mp \xi'_j(\tau)],$$

$$\frac{d}{d\tau}\xi'_i(\tau) + \frac{i}{\hbar}\varepsilon_{S,i}(\tau)\xi'_i(\tau) \pm \sum_j \int_{t_0}^{\tau} d\tau' \boldsymbol{g}_{ij}(\tau,\tau')\xi_j(\tau') \tag{13.39b}$$

$$= \mp \sum_j \int_{\tau}^{t} d\tau' \boldsymbol{g}_{ij}(\tau,\tau')[\xi_j(\tau') \mp \xi'_j(\tau')]$$

$$- \sum_j \int_{t_0}^{t} d\tau' \widetilde{\boldsymbol{g}}_{ij}(\tau,\tau')[\xi_j(\tau') \mp \xi'_j(\tau)],$$

$$\frac{d}{d\tau}\xi^*_i(\tau) - \frac{i}{\hbar}\varepsilon_{S,i}(\tau)\xi^*_i(\tau) - \sum_j \int_{t_0}^{\tau} d\tau' \boldsymbol{g}^*_{ij}(\tau,\tau')[\xi^*_j(\tau') \mp \xi'^*_j(\tau)] \tag{13.39c}$$

$$= \mp \sum_j \int_{\tau}^{t} d\tau' \boldsymbol{g}^*_{ij}(\tau,\tau')\xi'^*_j(\tau')$$

$$\pm \sum_j \int_{t_0}^{t} d\tau' \widetilde{\boldsymbol{g}}^*_{ij}(\tau,\tau')[\xi^*_j(\tau') \mp \xi'^*_j(\tau')],$$

$$\frac{d}{d\tau}\xi'^*_i(\tau) - \frac{i}{\hbar}\varepsilon_{S,i}(\tau)\xi'^*_i(\tau) \mp \sum_j \int_{\tau}^{t} d\tau' \boldsymbol{g}^*_{ij}(\tau,\tau')\xi'^*_j(\tau') \tag{13.39d}$$

$$= \sum_j \int_{t_0}^{t} d\tau' \widetilde{\boldsymbol{g}}^*_{ij}(\tau,\tau')[\xi^*_j(\tau') \mp \xi'^*_j(\tau')],$$

subjected to the boundary conditions $\xi_i(0) = \xi_{i0}$, $\xi'_i(t) = \xi'_{if}$, $\xi^*_i(t) = \xi^*_{if}$, and $\xi'^*_i(t_0) = \xi'^*_{i0}$, where $t_0 \le \tau \le t$. Note that the above equations of motion give rise to a number of interesting properties for open system dynamics. First of all, ξ_i and ξ^*_i obey very different equations of motion. They are no longer complex conjugate to each other, which breaks the complex structure of quantum mechanics.

This indicates that the dynamics of open systems no longer follow the unitary evolution of quantum mechanics. The breakdown of the unitary evolution or the complex structure of quantum mechanics arises from the mixture of the forward and backward propagating paths through the matter, energy, and information exchanges between the system and the environment. One must note that the above equations of motion also mix the past and future events together, which will give rise to the breakdown of causality. This is precisely the origin of the entanglement between the system and the environment which cannot happen in the classical mechanics.

Equations (13.39) are not directly solvable because of the lack of the causality, namely, the path $\xi_i(\tau)$ at time τ depends on both the past and the further states. To find the solution of the above equations of motion for the stationary paths, one may introduce the new variables $\chi_i(\tau) = \xi_i(\tau) \mp \xi_i'(\tau)$. Then Eq. (13.39) can be reduced to

$$\frac{d}{d\tau}\chi_i(\tau) + \frac{i}{\hbar}\varepsilon_{S,i}(\tau)\chi_i(\tau) + \int_{t_0}^{\tau} d\tau_1 \sum_j \boldsymbol{g}_{ij}(\tau,\tau_1)\chi_j(\tau_1) = 0, \tag{13.40a}$$

$$\frac{d}{d\tau}\chi_i^*(\tau) - \frac{i}{\hbar}\varepsilon_{S,i}(\tau)\chi_i^*(\tau) - \int_{\tau}^{t} d\tau_1 \sum_j \boldsymbol{g}_{ij}^*(\tau,\tau_1)\chi_j^*(\tau_1) = 0. \tag{13.40b}$$

These equations are apparently solvable because there is no obvious violation of causality. However, there is a new difficulty that now appears, namely, $\chi_i(\tau) = \xi_i(\tau) - \xi_i'(\tau)$ has no unique boundary conditions because $\xi_i(\tau)$ and $\xi_i'(\tau)$ obey the different boundary conditions: $\xi_i(t_0) = \xi_{i0}$ and $\xi_i'(t) = \xi_{if}'$. In other words, the breakdown of causality in quantum mechanics is now manifested in the boundary conditions of the path integrals in terms of the classical-like stationary paths. To solve this problem, we can further factorize the boundary values of the stationary paths by introducing the following transformation:

$$\xi_i(\tau) \mp \xi_i'(\tau) = \sum_j \boldsymbol{u}_{ij}(\tau,t)[\xi_j(t) \mp \xi_f'], \tag{13.41a}$$

$$\xi_i(\tau) = \sum_j \boldsymbol{u}_{ij}(\tau,t_0)\xi_{j0} \pm \sum_j \boldsymbol{v}_{ij}(\tau,t)[\xi_j(t) \mp \xi_{jf}'] \tag{13.41b}$$

and a similar transformation for their conjugate variables (by the replacement of $\xi_i \to \xi_i'^*$ with $\xi_i' \to \xi_i^*$ for the boundary conditions $\xi_i'^*(t_0) = \xi_{i0}'^*$ and $\xi_i^*(t) = \xi_{if}^*$). Here the bold notation $\boldsymbol{u}(t,t_0)$ and $\boldsymbol{v}(t,t)$ represent $N \times N$ non-equilibrium Green function matrices, where N is the total number of energy levels in the system. Then the stationary path equations of motion (13.39) are transformed to the following equations of motion for the non-equilibrium Green functions $\boldsymbol{u}_{ij}(\tau, t)$ and $\boldsymbol{v}_{ij}(\tau, t)$:

$$\frac{d}{dt}\boldsymbol{u}_{ij}(\tau, t_0) + \frac{i}{\hbar}\varepsilon_{S,i}(\tau)\boldsymbol{u}_{ij}(\tau, t_0) + \sum_{j'}\int_{t_0}^{\tau} d\tau' \boldsymbol{g}_{ij'}(\tau, \tau')\boldsymbol{u}_{j'j}(\tau', t_0) = 0. \tag{13.42a}$$

$$\frac{d}{dt}\boldsymbol{v}_{ij}(\tau, t) + \frac{i}{\hbar}\varepsilon_{S,i}(\tau)\boldsymbol{v}_{ij}(\tau, t) + \sum_{j'}\int_{t_0}^{\tau} d\tau' \boldsymbol{g}_{ij'}(\tau, \tau')\boldsymbol{v}_{j'j}(\tau', t) \tag{13.42b}$$

$$= + \sum_{j'}\int_{t_0}^{t} d\tau' \widetilde{\boldsymbol{g}}_{ij'}(\tau, \tau')\boldsymbol{u}_{j'j}(\tau', t).$$

subjected to the unique boundary conditions: $\boldsymbol{u}_{ij}(t_0, t_0) = \delta_{ij}$ and $\boldsymbol{v}_{ij}(t_0, t) = 0$. The boundary condition $\boldsymbol{v}_{ij}(t_0, t) = 0$ immediately leads to the following general solution of Eq. (13.42b):

$$\boldsymbol{v}_{ij}(\tau, t) = \int_{t_0}^{\tau} dt_1 \int_{t_0}^{t} dt_2 \sum_{j'j''} \boldsymbol{u}_{ij'}(\tau, t_1)\widetilde{\boldsymbol{g}}_{j'j''}(t_1, t_2)\boldsymbol{u}^*_{jj''}(t, t_2). \tag{13.43}$$

This solution is the generalized non-equilibrium fluctuation-dissipation relation in the time domain [196–198]. As we will show later, these two non-equilibrium Green functions fully capture the dissipation and fluctuation dynamics of open systems.

With the equations of motion (13.39) for the stationary paths, it is easy to find that the effective Lagrangian vanishes for the stationary paths: $\mathcal{L}^{\text{SP}}_{\text{eff}}[\boldsymbol{\xi}, \boldsymbol{\xi}'] = 0$. Then the propagating function (13.37) is simply contributed by the boundary terms in the coherent states path integrals:

$$\begin{aligned}\mathcal{J}(\boldsymbol{\xi}_f, \boldsymbol{\xi}'_f, t; \boldsymbol{\xi}, \boldsymbol{\xi}'_0, t_0) &= A(t) \exp\left\{\Phi_{\text{G}}(\boldsymbol{\xi}) + \Phi^*_{\text{G}}(\boldsymbol{\xi}')\right\} \\ &= A(t) \exp\left\{\frac{1}{2}\sum_i \left[\xi^*_{if}\xi_i(t) + \xi^*_i(t_0)\xi_{i0} + \xi'^*_{i0}\xi'_i(t_0) + \xi'^*_i(t)\xi'_{if}\right]\right\} \\ &= A(t) \exp\left\{\frac{1}{2}\left[\boldsymbol{\xi}^\dagger_f\boldsymbol{\xi}(t) + \boldsymbol{\xi}^\dagger(t_0)\boldsymbol{\xi}_0 + \boldsymbol{\xi}'^\dagger_0\boldsymbol{\xi}'(t_0) + \boldsymbol{\xi}'^\dagger(t)\boldsymbol{\xi}'_f\right]\right\},\end{aligned} \tag{13.44}$$

where $A(t)$ counts the fluctuations around the stationary paths from the path integrals of Eq. (13.37), which can be determined later by the normalization condition of the propagating function. Using the transformation Eq. (13.41), we can express the end points of the path integrals: $\xi_i(t)$, $\xi^*_i(t_0)$ and $\xi'_i(t_0)$, $\xi'^*_i(t)$, in terms of the non-equilibrium functions $u_{ij}(t, t_0)$ and $v_{ij}(t, t)$ and the boundary conditions of the fixed end points in the path integrals,

$$\boldsymbol{\xi}(t) = [\mathbf{1} \pm \boldsymbol{v}(t,t)]^{-1}[\boldsymbol{u}(t,t_0)\boldsymbol{\xi}_0 + \boldsymbol{v}(t,t)\boldsymbol{\xi}'_f] \tag{13.45a}$$

$$\begin{aligned}\boldsymbol{\xi}^\dagger(t_0) = \boldsymbol{\xi}^\dagger_f[\mathbf{1} \pm \boldsymbol{v}(t,t)]^{-1}\boldsymbol{u}(t,t_0) \\ \pm \boldsymbol{\xi}'^\dagger_0[\mathbf{1} - \boldsymbol{u}^\dagger(t,t_0)[\mathbf{1} \pm \boldsymbol{v}(t,t)]^{-1}\boldsymbol{u}(t,t_0)].\end{aligned} \tag{13.45b}$$

$$\begin{aligned}\boldsymbol{\xi}'(t_0) = \boldsymbol{u}^\dagger(t,t_0)[\mathbf{1} \pm \boldsymbol{v}(t,t)]^{-1}\boldsymbol{\xi}'_f \\ \pm [\mathbf{1} - \boldsymbol{u}^\dagger(t,t_0)[\mathbf{1} \pm \boldsymbol{v}(t,t)]^{-1}\boldsymbol{u}(t,t_0)]\boldsymbol{\xi}_0.\end{aligned} \tag{13.45c}$$

$$\boldsymbol{\xi}'^\dagger(t) = [\boldsymbol{\xi}'^\dagger_0\boldsymbol{u}^\dagger(t,t_0) + \boldsymbol{\xi}^\dagger_f\boldsymbol{v}(t,t)][\mathbf{1} \pm \boldsymbol{v}(t,t)]^{-1}. \tag{13.45d}$$

Thus, the propagating function is analytically solved

$$\begin{aligned}\mathcal{J}(\boldsymbol{\xi}_f, \boldsymbol{\xi}'_f, t; \boldsymbol{\xi}, \boldsymbol{\xi}'_0, t_0) = A(t)\exp\Big\{\boldsymbol{\xi}^\dagger_f\boldsymbol{K}_1(t,t_0)\boldsymbol{\xi}_0 + \boldsymbol{\xi}'^\dagger_0\boldsymbol{K}^\dagger_1(t,t_0)\boldsymbol{\xi}'_f \\ \pm \boldsymbol{\xi}'^\dagger_0\boldsymbol{K}_3(t,t_0)\boldsymbol{\xi}_0 + \boldsymbol{\xi}^\dagger_f\boldsymbol{K}_2(t)\boldsymbol{\xi}'_f\Big\},\end{aligned} \tag{13.46}$$

where

$$\boldsymbol{K}_1(t,t_0) = [\mathbf{1} \pm \boldsymbol{v}(t,t)]^{-1}\boldsymbol{u}(t,t_0), \tag{13.47a}$$

$$\boldsymbol{K}_2(t) = \boldsymbol{v}(t,t)/[\mathbf{1} \pm \boldsymbol{v}(t,t)], \tag{13.47b}$$

$$\boldsymbol{K}_3(t,t_0) = \mathbf{1} - \boldsymbol{u}^\dagger(t,t_0)[\mathbf{1} \pm \boldsymbol{v}(t,t)]^{-1}\boldsymbol{u}(t,t_0). \tag{13.47c}$$

The up and down signs of $\pm$ correspond to the system being bosonic or fermionic. The functions $\boldsymbol{u}(t,t_0)$ and $\boldsymbol{v}(t,t)$ are the non-equilibrium Green's functions of the open system we introduced and can be dynamically determined by Eqs. (13.42) and (13.43) [195–197].

With the propagating function solution (13.46), the reduced density matrix (13.16) is also solved exactly:

$$\begin{aligned}\langle\boldsymbol{\xi}_f|\rho_S(t)|\boldsymbol{\xi}'_f\rangle = A(t)\int d\mu(\boldsymbol{\xi}_0)d\mu(\boldsymbol{\xi}'_0)\exp\Big\{\boldsymbol{\xi}^\dagger_f\boldsymbol{K}_1(t,t_0)\boldsymbol{\xi}_0 + \boldsymbol{\xi}'^\dagger_0\boldsymbol{K}^\dagger_1(t,t_0)\boldsymbol{\xi}'_f \\ \pm \boldsymbol{\xi}'^\dagger_0\boldsymbol{K}_3(t,t_0)\boldsymbol{\xi}_0 + \boldsymbol{\xi}^\dagger_f\boldsymbol{K}_2(t)\boldsymbol{\xi}'_f\Big\}\langle\boldsymbol{\xi}_0|\rho_S(t_0)|\boldsymbol{\xi}'_0\rangle.\end{aligned} \tag{13.48}$$

Furthermore, using the probability normalization condition that the trace of the reduced density matrix equals to one,

$$\int d\mu(\boldsymbol{\xi}_f)\langle\boldsymbol{\xi}_f|\rho_S(t)| \pm \boldsymbol{\xi}_f\rangle = 1, \tag{13.49}$$

we find the normalized constant $A(t)$ in the propagating function (13.46):

$$A(t) = \big(\det[\mathbf{1} \pm \boldsymbol{v}(t)]\big)^{\mp 1}. \tag{13.50}$$

Thus, for any given initial state of the system, one can obtain the time evolution of its reduced density matrix from the above solution (13.48).

The above solution of the reduced density matrix is given in the coherent state representation. For the study of open system dynamics, it is more convenient if we could find the equation of motion for the reduced density matrix in the operator form. This can be done by taking a time derivative to the solution (13.48) of the reduced density matrix, and using the D-algebra of particle creation and annihilation operators acting on the coherent state [17]

$$a_i|\boldsymbol{\xi}\rangle = \xi_i|\boldsymbol{\xi}\rangle, \qquad a_i^\dagger|\boldsymbol{\xi}\rangle = \pm\frac{\partial}{\partial\xi_i}|\boldsymbol{\xi}\rangle, \tag{13.51a}$$

$$\langle\boldsymbol{\xi}|a_i^\dagger = \langle\boldsymbol{\xi}|\xi_i^*, \qquad \langle\boldsymbol{\xi}|a_i = \frac{\partial}{\partial\xi_i^*}\langle\boldsymbol{\xi}|, \tag{13.51b}$$

it is not difficult to find the following exact master equation:

$$\begin{aligned}\frac{d}{dt}\rho_{\rm S}(t) = \frac{1}{i\hbar}\big[H_{\rm S}^r(t), \rho_{\rm S}(t)\big] \\ +\sum\nolimits_{ij}\big\{\boldsymbol{\gamma}_{ij}(t,t_0)\big[2a_j\rho_{\rm S}(t)a_i^\dagger - a_i^\dagger a_j\rho_{\rm S}(t) - \rho_{\rm S}(t)a_i^\dagger a_j\big] \\ +\widetilde{\boldsymbol{\gamma}}_{ij}(t,t_0)\big[a_i^\dagger\rho_{\rm S}(t)a_j \pm a_j\rho_{\rm S}(t)a_i^\dagger \mp a_i^\dagger a_j\rho_{\rm S}(t) - \rho_{\rm S}(t)a_j a_i^\dagger\big]\big\}.\end{aligned} \tag{13.52}$$

Here, again the upper and lower signs of $\pm$ correspond, respectively, to the bosonic and fermionic systems. In this exact master equation, all the renormalization effects arise from the system-reservoir interactions that have been taken into account when the environmental degrees of freedoms are integrated out non-perturbatively and exactly. These renormalization effects are manifested by the renormalized system Hamiltonian,

$$H_{\rm S}^r(t) = \sum\nolimits_{ij}\boldsymbol{\varepsilon}_{{\rm S},ij}^r(t,t_0)a_i^\dagger a_j \tag{13.53}$$

and the dissipation and fluctuation coefficients $\boldsymbol{\gamma}_{ij}(t,t_0)$ and $\widetilde{\boldsymbol{\gamma}}_{ij}(t,t_0)$ in Eq. (13.52). These time-dependent coefficients are determined non-perturbatively and exactly by the following relations:

$$\boldsymbol{\varepsilon}_{{\rm S},ij}^r(t,t_0) = -\hbar\,{\rm Im}\big[\dot{\boldsymbol{u}}(t,t_0)\boldsymbol{u}^{-1}(t,t_0)\big]_{ij}, \tag{13.54a}$$

$$\boldsymbol{\gamma}_{ij}(t,t_0) = -{\rm Re}\big[\dot{\boldsymbol{u}}(t,t_0)\boldsymbol{u}^{-1}(t,t_0)\big]_{ij}, \tag{13.54b}$$

$$\widetilde{\boldsymbol{\gamma}}_{ij}(t,t_0) = \dot{\boldsymbol{v}}_{ij}(t,t) - \big[\dot{\boldsymbol{u}}(t,t_0)\boldsymbol{u}^{-1}(t,t_0)\boldsymbol{v}(t,t) + {\rm h.c.}\big]_{ij}. \tag{13.54c}$$

The master equation shows that the environment induced the dissipation and fluctuations, described by the terms proportional to the dissipation coefficient $\gamma_{ij}(t, t_0)$ and the fluctuation coefficient $\widetilde{\gamma}_{ij}(t, t_0)$. The former determines the relaxation of the system dynamics, while the latter characterize the dephasing of the system that makes the system approach to thermal equilibrium in the steady-state limit. It also shows how disorders of quantum dynamics emerge and how open systems are eventually thermalized, as we will discuss later. In other words, the randomness in nature is originated microscopically from the interaction between the system and its environment.

The exact master equation (13.52) is the most general and exact one discovered thus far for open quantum systems [195–197]. It has been extended to Majorana states for dissipative topological systems [199, 201], and it has also been applied to the generalized quantum Brownian motion with momentum-dependent couplings between the system and the environment [210]. The conventional quantum Brownian motion that was originally proposed by Feynman and Vernon and by Caldeira and Leggett is a special case of the generalized quantum Brownian motion, as we have shown very recently [210]. The exact master equation for the conventional quantum Brownian motion is the only exact master equation one obtained in the last century and is only valid for a harmonic oscillator coupled to the environment consisting of many harmonic oscillators with an initially decoupled system-environment state. Now one can see that the exact master equation (13.52) is valid for a large class of bosonic and fermionic open systems, and it has also been extended for the initially system-environment entangled states [201]. As we shall discuss in the next section, other approximated master equations, such as the Redfield master equation and the Markov master equation that are often used [211, 212], can easily be reduced from the exact master equation (13.52). From the exact master equation (13.52), we can also easily derive the quantum transport theory for the non-equilibrium open systems [196, 197]; it can reproduce the Meir-Wingreen formula describing the electric current in mesoscopic system [213, 214] and the Landauer-Büttiker formula [215, 216] for the equilibrium limit; see Sect. 13.6.

13.4 Master Equation in the Weak-Coupling Limit

Before the 2000s, except for the quantum Brownian model, one does not know how to discover the exact master equation for other open systems. Therefore, in studying the open quantum systems, one often deploys various approximations and assumptions. Typical assumptions were the Born approximation and Markov approximation. The Born approximation assumes that the environment remains invariant, and the Markov approximation assumes that all the memory effect can be ignored during the non-equilibrium evolution of the system. Such approximations allow one to make a perturbation reduction of the von Neumann equation with respect to the system-environment interaction Hamiltonian. In the literature, the resulting equation of motion is known as the Redfield master equation or the Born-

Markov approximation in the literature [211, 212]. However, rigorously speaking, the Born and Markov approximations are illogical. If the environment remains unchanged, namely, the environment keeps in the same initial thermal state, then the state of the system is always decoupled from the environment state. This simply cannot be true because the coupling between the system and the environment makes them entangled together so that the environment state must necessarily be altered. Here, we shall use the exact master equation and explicitly carry out a perturbation expansion to the second order with respect to the system-environment couplings. In this manner, we can naturally produce the Redfield master equation and the Born-Markov (BM) master equation without the need to resort to the Born and Markov assumptions.

As we have seen from the exact master equation (13.52), the dynamics of open systems is embedded into the time-dependent coefficients of the energy renormalization, the dissipation, and fluctuations in the master equation. These time-dependent coefficients are determined non-perturbatively by integrodifferential equations of motion for the non-equilibrium Green functions, Eqs. (13.72) and (13.43). These integrodifferential equations of motion carry the typical form of the Dyson equation in quantum field theory which possess a natural perturbation expansion with respect to the interaction Hamiltonian. Here the interaction Hamiltonian is the system-environment coupling Hamiltonian H_{ES} which is proportional to the system-environment coupling V_{ik}. We can find the coefficients of the energy renormalization, the dissipation, and fluctuations in Redfield/Born-Markov master equation from the exact results of Eq. (13.54) by simply taking a perturbation expansion up to the second order in terms of the system environment coupling V_{ik}. In this respect, there is no need for the additional Born and Markov approximations.

To be explicit, the two-time system-environment correlations $\boldsymbol{g}_{ij}(\tau, \tau')$ and $\widetilde{\boldsymbol{g}}_{ij}(\tau, \tau')$ are proportional to the product of two system-environment couplings $\sim \sum_k V_{ik}(\tau)V^*_{jk}(\tau')$. For convenience and without any loss of generality, we may let the system and environment energy spectra as well as the system-environment couplings in the total Hamiltonian (13.5) to be time-independent. Then the system-environment correlation functions $\boldsymbol{g}_{ij}(\tau, \tau')$ and $\widetilde{\boldsymbol{g}}_{ij}(\tau, \tau')$ of Eq. (13.35) are reduced to

$$\boldsymbol{g}_{ij}(\tau, \tau') = \frac{1}{\hbar^2}\int \frac{d\epsilon}{2\pi} \boldsymbol{J}_{ij}(\epsilon) e^{-\frac{i}{\hbar}\epsilon(\tau-\tau')}, \tag{13.55a}$$

$$\widetilde{\boldsymbol{g}}_{ij}(\tau, \tau') = \frac{1}{\hbar^2}\int \frac{d\epsilon}{2\pi} \boldsymbol{J}_{ij}(\epsilon) f(\epsilon, T_0) e^{-\frac{i}{\hbar}\epsilon(\tau-\tau')}, \tag{13.55b}$$

where

$$\boldsymbol{J}_{ij}(\epsilon) = 2\pi \sum_k V_{ik}V^*_{jk}\delta(\epsilon - \epsilon_k) = 2\pi \varrho(\epsilon) V_i(\epsilon) V^*_j(\epsilon) \tag{13.56}$$

is the spectral density matrix of the environment and the $\varrho(\epsilon)$ is the environmental density of states. When the system-environment couplings are weak, we can take the perturbation expansion of the time-dependent coefficients in Eq. (13.54) up to the second-order of the coupling strength $V_i(\epsilon)$, namely, up to the terms proportional to the spectral density $\boldsymbol{J}_{ij}(\epsilon)$. This implies that the non-equilibrium Green functions $\boldsymbol{u}(\tau, t_0)$ and $\boldsymbol{u}^{-1}(t, t_0)$ in the right-hand side of Eq. (13.72) need only to keep to the zeroth order: $\boldsymbol{u}_0(\tau, t_0) = e^{-\frac{i}{\hbar}\boldsymbol{\varepsilon}_S(\tau-t_0)}$ and $\boldsymbol{u}_0^{-1}(t, t_0) = e^{\frac{i}{\hbar}\boldsymbol{\varepsilon}_S(t-t_0)}$. Thus, up to the second-order perturbation expansion, we can obtain directly from Eq. (13.42a) that

$$\dot{\boldsymbol{u}}(t)\boldsymbol{u}^{-1}(t) \simeq -\frac{i}{\hbar}\boldsymbol{\varepsilon}_S - \frac{1}{\hbar^2}\int_{t_0}^{t} d\tau \int \frac{d\epsilon}{2\pi} \boldsymbol{J}(\epsilon) e^{-\frac{i}{\hbar}(\epsilon-\boldsymbol{\varepsilon}_S)(t-\tau)}. \tag{13.57}$$

Similarly, up to the second-order perturbation expansion, Eq. (13.43 can simply be taken as

$$\dot{\boldsymbol{v}}(t, t) \simeq \frac{2}{\hbar^2}\int_{t_0}^{t} d\tau \int \frac{d\epsilon}{2\pi} \boldsymbol{J}(\epsilon) f(\epsilon, T) \cos\Big[\frac{1}{\hbar}(\epsilon - \boldsymbol{\varepsilon}_S)(t-\tau)\Big]. \tag{13.58}$$

The term $\boldsymbol{v}(t, t)\dot{\boldsymbol{u}}(t, t_0)\boldsymbol{u}^{-1}(t, t_0)$+H.c in Eq. (13.54c) is proportional to $|V(\epsilon)|^4$ and therefore can be neglected. As a result, in the second-order perturbative expansion, the coefficients of Eq. (13.54) are reduced to

$$\boldsymbol{\varepsilon}^{\mathrm{BM}}(t, t_0) \simeq \boldsymbol{\varepsilon}_S - \frac{1}{\hbar}\int_{t_0}^{t} d\tau \int \frac{d\epsilon}{2\pi} \boldsymbol{J}(\epsilon) \sin\Big[\frac{1}{\hbar}(\epsilon - \boldsymbol{\varepsilon}_S)(t-\tau)\Big], \tag{13.59a}$$

$$\boldsymbol{\gamma}^{\mathrm{BM}}(t, t_0) \simeq \frac{1}{\hbar^2}\int_{t_0}^{t} d\tau \int \frac{d\epsilon}{2\pi} \boldsymbol{J}(\epsilon) \cos\Big[\frac{1}{\hbar}(\epsilon - \boldsymbol{\varepsilon}_S)(t-\tau)\Big], \tag{13.59b}$$

$$\widetilde{\boldsymbol{\gamma}}^{\mathrm{BM}}(t, t_0) \simeq \frac{2}{\hbar^2}\int_{t_0}^{t} d\tau \int \frac{d\epsilon}{2\pi} \boldsymbol{J}(\epsilon) f(\epsilon, T) \cos\Big[\frac{1}{\hbar}(\epsilon - \boldsymbol{\varepsilon}_S)(t-\tau)\Big]. \tag{13.59c}$$

Substituting these coefficients into Eq. (13.52), we obtain the Redfield/Born-Markov master equation

$$\begin{aligned} \frac{d}{dt}\rho_S(t) = & \frac{1}{i\hbar}\big[H_S^{\mathrm{BM}}(t, t_0), \rho_S(t)\big] \\ & + \sum_{ij}\Big\{\boldsymbol{\gamma}_{ij}^{\mathrm{BM}}(t, t_0)\big[2a_j\rho_S(t)a_i^\dagger - a_i^\dagger a_j\rho_S(t) - \rho_S(t)a_i^\dagger a_j\big] \\ & + \widetilde{\boldsymbol{\gamma}}_{ij}^{\mathrm{BM}}(t, t_0)\big[a_i^\dagger\rho_S(t)a_j \pm a_j\rho_S(t)a_i^\dagger \mp a_i^\dagger a_j\rho_S(t) - \rho_S(t)a_j a_i^\dagger\big]\Big\}. \end{aligned} \tag{13.60}$$

Here $H_S^{\mathrm{BM}}(t, t_0) = \sum_{ij} \boldsymbol{\varepsilon}_{S,ij}^{\mathrm{BM}}(t, t_0)a_i^\dagger a_j$ the second-order renormalized Hamiltonian of the system. Obviously, this master equation is an approximative one that is valid only when the system-environment couplings are very weak, although formally it has the same form as the exact master equation (13.52). In the literature, one usually obtains such approximated master equation from the von Neumann equation under the Born and Markov assumptions. In our derivation, only the perturbation is used from the exact solution. Therefore there is no need to invoke the Born and Markov assumptions.

Besides the weak couplings, for some open quantum systems, the dynamical time scale of the system is much longer than the time scale of the environment. That is, the environment modes are dominated by the rapid oscillations compared to the decay time of the system. In such a situation, the τ integration in Eq. (13.59) is dominated by much shorter time of the decay time of reservoir correlations. In other words, one can effectively take the τ integration to infinity in Eq. (13.59):

$$\lim_{t\to\infty}\int_0^{t-t_0} dt' e^{\pm\frac{i}{\hbar}(\epsilon-\varepsilon_{\rm S})t'} = \pi\hbar\delta(\epsilon-\varepsilon_{\rm S}) \mp i\hbar\frac{\mathcal{P}}{\epsilon-\varepsilon_{\rm S}}. \tag{13.61}$$

This is called the Markov limit or the secular approximation, where $\mathcal{P}$ denotes the Cauchy principal value of the integral. As a result, all the coefficients in the Born-Markov master equation, given by Eq. (13.59), can be further reduced to the time-independent constants:

$$\boldsymbol{\varepsilon}^{\rm BM}_{{\rm S},ij}t, t_0) \to \varepsilon_{{\rm S},i}\delta_{ij} + \delta\boldsymbol{\varepsilon}_{{\rm S},ij} \tag{13.62a}$$

$$\boldsymbol{\gamma}^{\rm BM}_{ij}(t, t_0) \to \frac{1}{2\hbar}\boldsymbol{J}_{ij}(\varepsilon_{{\rm S},i}) = \boldsymbol{\gamma}_{ij} \tag{13.62b}$$

$$\widetilde{\boldsymbol{\gamma}}^{\rm BM}_{ij}(t, t_0) \to \frac{1}{\hbar}\boldsymbol{J}_{ij}(\varepsilon_{{\rm S},i})f(\varepsilon_{{\rm S},i}, T) = 2\boldsymbol{\gamma}_{ij}f(\varepsilon_{{\rm S},i}, T), \tag{13.62c}$$

and the energy shift is given by $\delta\boldsymbol{\varepsilon}_{{\rm S},ij} = \mathcal{P}\big[\int\frac{d\epsilon}{2\pi}\frac{\boldsymbol{J}_{ij}(\epsilon)}{\epsilon-\varepsilon_{{\rm S},i}}\big]$. The Born-Markov master equation (13.60) is also further simplified to

$$\begin{aligned}\frac{d}{dt}\rho_{\rm S}(t) = \sum_{ij}\Big\{&\frac{1}{i\hbar}(\varepsilon_{{\rm S},i}\delta_{ij} + \delta\boldsymbol{\varepsilon}_{{\rm S},ij})[a^\dagger_i a_j, \rho_{\rm S}(t)] \\ &+ \boldsymbol{\gamma}_{ij}\big[2a_j\rho_{\rm S}(t)a^\dagger_i - a^\dagger_i a_j\rho_{\rm S}(t) - \rho_{\rm S}(t)a^\dagger_j a_i\big] \\ &+ 2\boldsymbol{\gamma}_{ij}f(\varepsilon_{{\rm S},i}, T)\big[a^\dagger_i\rho_{\rm S}(t)a_j + a_j\rho_{\rm S}(t)a^\dagger_i - a^\dagger_i a_j\rho_{\rm S}(t) - \rho_{\rm S}(t)a_j a^\dagger_i\big]\Big\}.\end{aligned} \tag{13.63}$$

This master equation is known as the Markov-limit master equation because it is obtained under the long-time Markov limit. Nevertheless, neither the Redfield/Born-Markov master equation nor the Markov master equation is incapable exhibiting the non-Markovian effect. This is because the memory effect has been entirely ignored either by the Markovian assumption or by the long-time Markov limit. This conclusion will become more transparent in the next section where we will show that in the weak coupling regime, the exact master equation and the BM master equation give qualitatively the same dynamics, namely, the damping dynamics is a simple exponential decay, and non-Markovian effect no longer can exist.

It is worth mentioning that Eq. (13.64) can be further written in the following form:

$$\frac{d}{dt}\rho_S(t) = \sum_{ij}\Big\{\frac{1}{i\hbar}(\varepsilon_{S,i}\delta_{ij} + \delta\boldsymbol{\varepsilon}_{S,ij})[a_i^\dagger a_j, \rho_S(t)] \tag{13.64}$$

$$+ 2\gamma_{ij}\big[1 + f(\varepsilon_{S,i}, T)\big]\big[a_j\rho_S(t)a_i^\dagger - \frac{1}{2}a_i^\dagger a_j\rho_S(t) - \frac{1}{2}\rho_S(t)a_i^\dagger a_j\big]$$

$$+ 2\gamma_{ij} f(\varepsilon_{S,i}, T)\big[a_i^\dagger\rho_S(t)a_j - \frac{1}{2}a_j a_i^\dagger\rho_S(t) - \frac{1}{2}\rho_S(t)a_j a_i^\dagger\big]\Big\}.$$

In the literature, one often refers to the above form of the master equation as the Lindblad master equation. This is because it apparently has the same form as the Lindblad-GKS master equation. But, unfortunately, this is a misunderstanding. The Lindblad-GKS master equation was derived mathematically by Lindblad and by Gorini, Kossakowski, and Sudarshan independently through the quantum dynamical semigroup which is defined by the completely positive maps for the irreversible processes [217,218]. It requires N^2-1 independent Lindblad generators to construct the completely positive maps for a N-dimensional Hilbert space of the concerned system. Otherwise, the Lindblad-GKS master equation cannot be formulated without the completeness of the Lindblad generators. On the other hand, the number of the generators a_i and $a_i^\dagger$ in Eq. (13.64) is at most linearly proportional to the dimension of the Hilbert space of the open system. Therefore, the number of the generators in the exact master equation (13.64) is much less than the independent generators of the dynamical semigroup in the Lindblad-GKS master equation for the same system. Except for the spin-1/2 system (which accidentally has the same number of the Lindblad generators with the independent spin operators), there is not any master equation derived from the von Neumann equation (either exactly or approximately) that can be identified with the Lindblad-GKS master equation for the completely positive maps. In other words, the pure mathematically derived Lindblad-GKS master equation has no connection with the fundamental evolution of open systems in terms of the fundamental degrees of freedom. In the literature, there is much confusion regarding the pictures of completely positive maps with the physical master equation of open systems derived from the von Neumann equation. In fact, the exact master equation (13.52) and the second-order perturbation master equation (13.60) all have the same operator form as Eq. (13.64), but they have nothing to do with Lindblad's dynamical semigroups for the completely positive maps of the density matrix.

13.5 General Non-Markovian Dynamics of Open Systems

From the exact master equation in the last section, we can define more physically the concept of non-Markovian dynamics and discuss the general non-Markovian properties in open systems. Non-Markovian dynamics represents memory processes of the dynamical evolution of open systems. That is, the present state of a physical system is determined not only by its previous states in an infinitesimal time interval but also explicitly by all the past historical states. In other words, the equation of the motion for the state evolution, given by the master equation for the reduced density matrix, is usually a time-convolution equation. However, the exact master equation derived in the last section, Eq. (13.52), is a time-convolutionless equation. This is because the time convolution that characterizes non-Markovian memory processes is converted into the equations of motion for the non-equilibrium Green functions $\boldsymbol{u}(t, t_0)$ and $\boldsymbol{v}(t, t)$, as shown by Eq. (13.42). These non-equilibrium Green functions determine all the non-Markovian memory dynamics through the time-dependent coefficients in the master equations, including the energy renormalization, the dissipation, and fluctuation coefficients. Dynamical equations of motion without involving time convolution cannot manifest the memory effect.

Furthermore, the correlation Green function $\boldsymbol{v}(t, t)$ is determined by the retarded Green function $\boldsymbol{u}(t, t_0)$ through the fluctuation-dissipation theorem; see Eq. (13.43). Then the general properties of non-Markovian dynamics can be extracted from the propagating Green function $\boldsymbol{u}(t, t_0)$. From Eqs. (13.54), the formal solution of propagating Green function $\boldsymbol{u}(t, t_0)$ can be expressed in terms of the renormalized system energy spectrum and the dissipative damping coefficient $\boldsymbol{\gamma}(t)$ as

$$\boldsymbol{u}(t, t_0) = \mathcal{T} \exp\left\{-\int_{t_0}^{t} d\tau \left[\frac{i}{\hbar}\widetilde{\boldsymbol{\varepsilon}}^{r}_{s}(\tau) + \boldsymbol{\gamma}(\tau)\right]\right\}, \tag{13.65}$$

where $\mathcal{T}$ is the time-ordering operator. This solution shows clearly that $\boldsymbol{u}(t, t_0)$ contains explicitly a damping factor given by dissipation coefficient in the exact master equation, which is induced by the coupling to the environment. However, due to the time dependence of the dissipation coefficients, the detailed dissipation dynamics can vary significantly for different environments, as the manifestation of different non-Markovian memory effects.

Without any loss of generality, we simplify again the system and environment energy spectra as well as the system-environment couplings to be time-independent. Then, we can rewrite the system-environment two-time correlation functions $\boldsymbol{g}(\tau, \tau') = \boldsymbol{g}(\tau - \tau')$ and $\widetilde{\boldsymbol{g}}(\tau, \tau') = \widetilde{\boldsymbol{g}}(\tau - \tau')$, see Eq. (13.55). As a result, the non-equilibrium propagating Green function also has the property $\boldsymbol{u}(t, t_0) = \boldsymbol{u}(t - t_0)$. Applying the modified Laplace transform,

$$\boldsymbol{U}(z) = \int_{t_0}^{\infty} dt\, \boldsymbol{u}(t - t_0) e^{\frac{i}{\hbar} z (t - t_0)}, \tag{13.66}$$

to Eq. (13.42a), we obtain the Laplace transform of $\boldsymbol{u}(t-t_0)$,

$$\boldsymbol{U}(z) = \frac{i\hbar}{z\mathbf{I} - \boldsymbol{\varepsilon}_S - \boldsymbol{\Sigma}(z)}, \tag{13.67}$$

where $\boldsymbol{I}$ is the identity matrix. The function $\boldsymbol{\Sigma}(z)$ is the self-energy correction, i.e., the Laplace transform of the integral part in Eq. (13.42a),

$$\boldsymbol{\Sigma}(z) = \int \frac{d\epsilon}{2\pi}\frac{\boldsymbol{J}(\epsilon)}{z-\epsilon} \xrightarrow{z=\epsilon\pm i0^+} \boldsymbol{\Delta}(\epsilon) \mp i\frac{\boldsymbol{J}(\epsilon)}{2}, \tag{13.68}$$

and $\boldsymbol{\Delta}(\epsilon) = \mathcal{P}\left[\int \frac{d\epsilon'}{2\pi}\frac{\boldsymbol{J}(\epsilon')}{\epsilon-\epsilon'}\right]$ induces the energy shift of the system. It can be shown that the general solution of $\boldsymbol{u}(t,t_0)$ is given by

$$\begin{aligned}\boldsymbol{u}(t-t_0) = &\sum_l \boldsymbol{\mathcal{Z}}_l e^{-\frac{i}{\hbar}\varepsilon_l(t-t_0)} \\ &+ \sum_b \int_{B_b} \frac{d\epsilon}{2\pi\hbar}\Big[\boldsymbol{U}(\epsilon+i0^+) - \boldsymbol{U}(\epsilon-i0^+)\Big]e^{-\frac{i}{\hbar}\epsilon(t-t_0)} \\ = &\sum_l \boldsymbol{\mathcal{Z}}_l e^{-\frac{i}{\hbar}\varepsilon_l(t-t_0)} + \int \frac{d\epsilon}{2\pi}\boldsymbol{D}(\epsilon)e^{-\frac{i}{\hbar}\epsilon(t-t_0)},\end{aligned} \tag{13.69}$$

where

$$\begin{aligned}\boldsymbol{D}(\epsilon) &= \boldsymbol{U}(\epsilon+i0^+) - \boldsymbol{U}(\epsilon-i0^+) \\ &= \frac{1}{\epsilon-\boldsymbol{\varepsilon}_S-\boldsymbol{\Delta}(\epsilon)+i\frac{\boldsymbol{J}(\epsilon)}{2}}\boldsymbol{J}(\epsilon)\frac{1}{\epsilon-\boldsymbol{\varepsilon}_S-\boldsymbol{\Delta}(\epsilon)-i\frac{\boldsymbol{J}(\epsilon)}{2}}.\end{aligned} \tag{13.70}$$

The first term in (13.69) is the contribution of localized bounded states with poles $\{\varepsilon_l\}$ located at the real z axis with $\boldsymbol{J}(\varepsilon_l) = 0$, i.e.,

$$\varepsilon_l \boldsymbol{I} - \boldsymbol{\varepsilon}_S - \boldsymbol{\Delta}(\varepsilon_l) = 0 \quad \text{and} \quad \boldsymbol{J}(\varepsilon_l) = 0. \tag{13.71}$$

The coefficients $\{\boldsymbol{\mathcal{Z}}_l\}$ are the corresponding residues of the poles. It shows that the localized bound states exist only when the environmental spectral density has band gaps or a finite band structure in the corresponding energy spectrum regions; see Fig. 13.2. These localized modes do not decay, and give the dissipationless long-time non-Markovian dynamics. The second term in (13.69) is the contribution from the branch cuts $\{B_b\}$, due to the discontinuity of $\boldsymbol{\Sigma}(z)$, so does $\boldsymbol{U}(z)$, across the real axis on the complex space z; see Eq. (13.68). The branch cuts usually generate non-exponential decays, which is a short-time character of the non-Markovian dynamics. When the system is weakly coupled to the environment, the non-exponential decays are reduced to exponential-like decays.

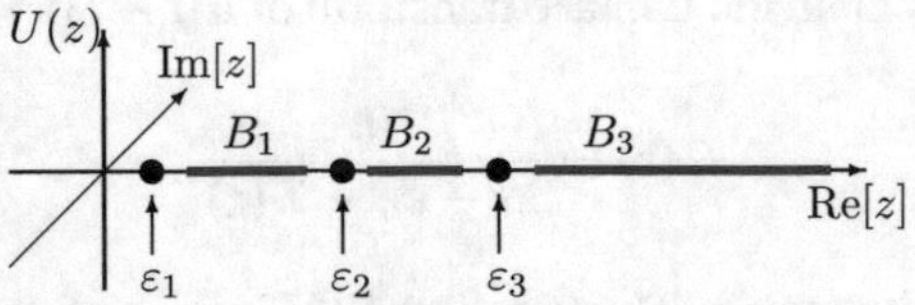

Fig. 13.2 (Color online) A schematic pole structure of the retarded Green function $\boldsymbol{U}(z)$. The thick red lines on the real z axis correspond to $\boldsymbol{J}(z) \neq 0$ [198]

Equation (13.69) is the general solution of the non-Markovian dissipation dynamics. It shows that the non-Markovian dissipation dynamics consists of non-exponential decays plus dissipationless localized bound states. This is the general dynamic feature of any microscopic particle (considered as the principal system) moving in many-body systems (treated as the environment). Such a solution of the propagating Green function $\boldsymbol{u}(t, t_0)$ is generic and can be proven from the quantum field theory, even if particle-particle interactions are included.

As an example, we consider the open system of a single particle in the state with energy ε_s, in contact with the surrounding as an arbitrary environment at finite initial temperature T_0. Then the solution (13.69) can be written further as

$$u(t-t_0) = \sum_l \mathcal{Z}_l(\varepsilon_l) e^{-\frac{i}{\hbar}\varepsilon_l(t-t_0)} + \int \frac{d\epsilon}{2\pi} \mathcal{D}_d(\epsilon) e^{-\frac{i}{\hbar}\epsilon(t-t_0)}. \tag{13.72}$$

The localized bound state energy (frequency) ε_l and its amplitude $\mathcal{Z}_l(\varepsilon_l)$ are simply determined by

$$\varepsilon_l - \varepsilon_s - \Delta(\varepsilon_l) = 0 \quad \text{and} \quad \mathcal{Z}_l(\varepsilon_l) = \frac{1}{1 - \partial_\epsilon \Sigma(\epsilon)|_{\epsilon=\varepsilon_l}}, \tag{13.73}$$

where

$$\Sigma(\epsilon) = \int \frac{d\epsilon'}{2\pi} \frac{J(\epsilon')}{\epsilon - \epsilon'} \quad \text{and} \quad \Delta(\epsilon) = \mathcal{P}\Big[\int \frac{d\epsilon'}{2\pi} \frac{J(\epsilon')}{\epsilon - \epsilon'}\Big] \tag{13.74}$$

are the environment-induced self-energy correction and energy shift, respectively. The dissipation spectrum $\mathcal{D}_d(\epsilon)$ is given by

$$\begin{aligned} \mathcal{D}_d(\epsilon) &= U(\epsilon + i0^+) - U(\epsilon - i0^+) \\ &= \frac{J(\epsilon)}{[\epsilon - \varepsilon_s - \Delta(\epsilon)]^2 + J^2(\epsilon)/4}. \end{aligned} \tag{13.75}$$

It is particularly important to note that the localized bound states (the first term in Eq. (13.72)) only exist if the spectral density $J(\epsilon)$ contains band gap(s) or zero energy points with sharp slopes, while the dissipation spectrum $\mathcal{D}_d(\epsilon)$ (proportional

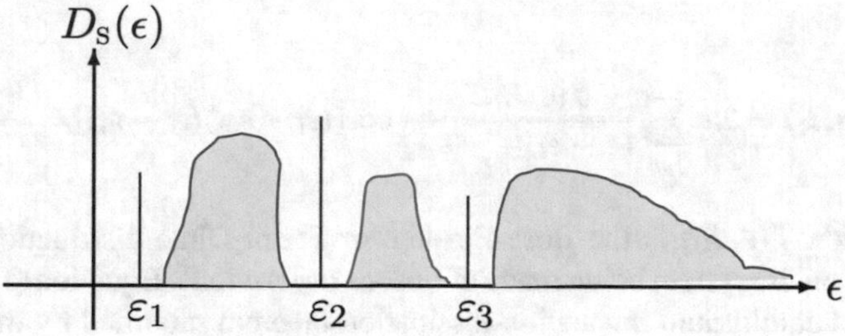

Fig. 13.3 (Color online) A schematic plot of the energy spectrum of a single-particle open system modified by the interaction with the environment

to $J(\epsilon)$; see Eq. (13.151a)) is crucially determined by the spectral density profile; see Fig. 13.3. In general, the solution (13.72) can be written as

$$u(t-t_0) = \int \frac{d\epsilon}{2\pi} \mathcal{D}_S(\epsilon) e^{-\frac{i}{\hbar}\epsilon(t-t_0)}, \tag{13.76}$$

where

$$\mathcal{D}_S(\epsilon) = 2\pi \sum_l \mathcal{Z}_l(\varepsilon_l)\delta(\epsilon - \varepsilon_l) + \mathcal{D}_d(\epsilon). \tag{13.77}$$

This shows that the original energy level of the system is modified as a combination of localized bound state and a continuous spectrum due to the influence of the environment. In the steady-state limit $t = t_s \to \infty$, only the dissipationless oscillation terms remain:

$$u(t_s - t_0) = \sum_l \mathcal{Z}_l e^{-i\varepsilon_l(t_s - t_0)}. \tag{13.78}$$

If the localized bound state does not exist, $u(t_s - t_0) \to 0$ in the steady-state limit, resulting in a complete relaxation process.

On the other hand, the general solution of the correlation (fluctuating) Green function $\boldsymbol{v}(\tau, t)$ is given by Eq. (13.43). For an open system of a single particle in the state with energy ε_s coupled to a thermal environment with initial temperature T_0, the steady-state ($t \to \infty$) solution of Eq. (13.43) is reduced to

$$v(t_s, t_s \to \infty) = \int_{t_0}^{\infty} dt_1 \int_{t_0}^{\infty} dt_2 u(\tau, t_1)\widetilde{g}(t_1, t_2)u^*(t, t_2) = \int \frac{d\epsilon}{2\pi}\chi(\epsilon) \tag{13.79}$$

where

$$\chi(\epsilon) = [\mathcal{D}_b(\epsilon, t_s) + \mathcal{D}_d(\epsilon)] f(\epsilon, T_0), \tag{13.80}$$

and

$$\mathcal{D}_b(\epsilon, t_s) = 2\pi \sum_{j,m} \frac{J(\epsilon)\mathcal{Z}_l \mathcal{Z}_m}{(\epsilon - \varepsilon_l)(\epsilon - \varepsilon_m)} \cos[(\varepsilon_l - \varepsilon_m)(t_s - t_0)]. \tag{13.81}$$

The function $f(\epsilon, T_0)$ is the Bose-Einstein or Fermi-Dirac distribution function, depending on the system being made of bosons or fermions. Equation (13.80) is the generalized equilibrium fluctuation-dissipation theorem modified by the localized bound states of open systems. If there is no localized bound state, $\mathcal{Z}_l = 0$, then

$$\chi(\epsilon) = \mathcal{D}_d(\epsilon) f(\epsilon, T_0). \tag{13.82}$$

This is the equilibrium fluctuation-dissipation theorem at arbitrary temperature.

With the analytical solutions (13.72) and (13.79), we can depict the dissipation and fluctuation dynamics through the time-dependent dissipation and fluctuation coefficients, $\gamma(t, t_0)$ and $\widetilde{\gamma}(t, t_0)$, in the exact master equation (13.52). To be more specific, we consider a single-mode bosonic nanosystem, such as a nanophotonic or optomechanical resonator, coupled to a general non-Markovian environment with spectral density

$$J(\epsilon) = 2\pi\eta\epsilon\Big(\frac{\epsilon}{\epsilon_c}\Big)^{s-1} \exp\Big(-\frac{\epsilon}{\epsilon_c}\Big), \tag{13.83}$$

where η is a dimensionless parameter characterizing the coupling strength between the system and the environment and ϵ_c is the frequency cutoff. When $s = 1, < 1$, and > 1, the corresponding environments are called as the Ohmic, sub-Ohmic, and super-Ohmic, respectively. For simplicity, we also set $t_0 = 0$. The analytical solution of the retarded Green function is analytically given by

$$u(t) = \mathcal{Z}e^{-\frac{i}{\hbar}\epsilon' t} + \int_0^\infty \frac{d\epsilon}{2\pi} \frac{J(\epsilon)e^{-\frac{i}{\hbar}\epsilon t}}{[\epsilon - \varepsilon_s - \Delta(\epsilon)]^2 + J^2(\epsilon)/4}, \tag{13.84}$$

where $\Delta(\epsilon) = \frac{1}{2}[\Sigma(\epsilon + i0^+) + \Sigma(\epsilon - i0^+)]$ and the self-energy correction

$$\Sigma(\epsilon) = -\eta\epsilon_c\Gamma(s+1)(-\widetilde{\epsilon})^s e^{-\widetilde{\epsilon}}\Gamma(-s, -\widetilde{\epsilon}), \tag{13.85}$$

where $\Gamma(x) = \int_0^\infty dt\, t^{x-1}e^{-t}$ is the gamma function, $\Gamma(\alpha, z) = \int_z^\infty dt\, t^{\alpha-1}/e^t$ is the complementary incomplete Gamma function, and $\widetilde{\epsilon} = \epsilon/\epsilon_c$. Due to the vanishing spectral density for $\epsilon < 0$, a localized mode occurs at

$$\epsilon' = \varepsilon_s - \Sigma(\epsilon') < 0 \quad \text{for} \quad \eta\epsilon_c\Gamma(s) > \varepsilon_s. \tag{13.86}$$

The corresponding residue is $\mathcal{Z} = 1/[1 - \Sigma'(\epsilon')]$. Figure 13.4 shows that the dissipation and fluctuation of a single-mode nanocavity coupled to a sub-Ohmic

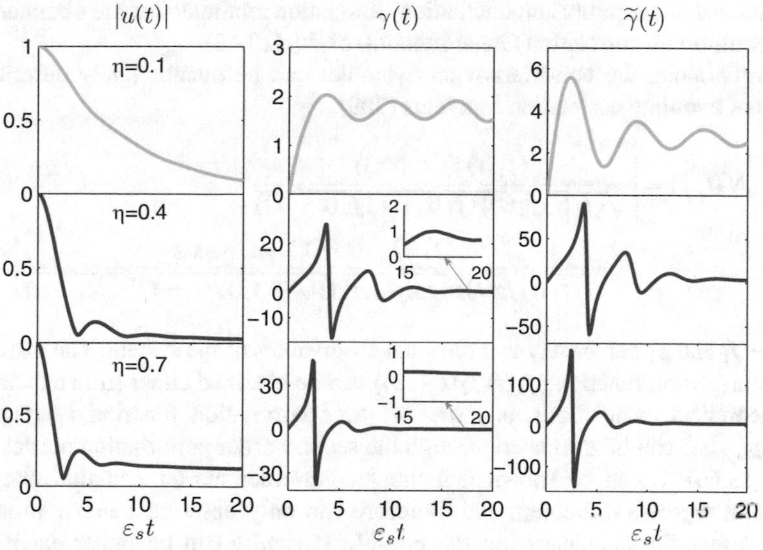

Fig. 13.4 The time evolution of the retarded Green function $|u(t)|$, the dissipation and the fluctuation coefficients, $\gamma(t)$ and $\widetilde{\gamma}(t)$ of a single-mode nanocavity coupled to sub-Ohmic thermal bath ($s = 1/2$), for different values of the coupling constant η. Other parameters are taken as $\epsilon_c = \varepsilon_s$ and also $k_B T = \varepsilon_s$ [198]

thermal bath with $s = 1/2$. One can see that for a small η, the dissipation dynamics is an exponential-like decay. The corresponding $\gamma(t)$ and $\widetilde{\gamma}(t)$ are time-dependent but always positive (corresponding to Markovian dynamics). When $\eta > 0.3$, the non-exponential decay dominates, and $\gamma(t)$ and $\widetilde{\gamma}(t)$ oscillate in positive and negative values with nonzero asymptotical values. When $\eta > 0.6$, the localized state occurs, and $u(t)$ does not decay to zero. Correspondingly, $\gamma(t)$ and $\widetilde{\gamma}(t)$ asymptotically approach to zero.

The above solution gives the general answer to the non-Markovian memory dynamics in open systems [198]: The non-exponential decays, the second terms in Eq. (13.72), are induced by the discontinuity in the imaginary part of the environmental-induced self-energy correction to the system Hamiltonian, $\Sigma(\epsilon \pm i0^+) = \Delta(\epsilon) \mp i J(\epsilon)/2$. Depending on the detailed spectral density structure of $J(\omega)$, it could result in damping coefficients oscillating between positive and negative values in short times. As a short-time non-Markovian memory effect, such oscillations correspond to the forward and backward information flow between the system and the environment. The dissipationless oscillations, characterized by the localized bound states which arise from band gaps or a finite band structure of environment spectral densities, provide a long-time non-Markovian memory effect. This is because information flowing between the system and environment could rest forever, due to the existence of the dissipationless localized bound state. Fluctuation dynamics induces similar non-Markovian dynamics as dissipation through the generalized non-equilibrium fluctuation-dissipation relations that are obtained from non-equilibrium correlation Green function of Eq. (13.43).

Furthermore, the non-Markovian dynamics can be quantitatively described in terms of two-time correlation functions [209],

$$\mathcal{N}(t,\tau) = \Bigg| \frac{\langle f_1(t) f_2(t+\tau)\rangle}{\sqrt{\langle f_1(t) f_2(t)\rangle \langle f_1(t+\tau) f_2(t+\tau)\rangle}} - \frac{\langle f_1(t) f_2(t+\tau)\rangle_{boldsymbol}}{\sqrt{\langle f_1(t) f_2(t)\rangle_{boldsymbol} \langle f_1(t+\tau) f_2(t+\tau)\rangle_{boldsymbol}}} \Bigg|, \tag{13.87}$$

where f_1 and f_2 can be any two physical observables of the system. The exact two-time correlation function $\langle f_1(t) f_2(t+\tau)\rangle$ can be obtained either from experiments or theoretical calculations, and the two-time correlation function $\langle f_1(t) f_2(t+\tau)\rangle_{boldsymbol}$ can be evaluated through the second-order perturbation master equation (13.60). It can be shown that the second-order master equation obeys the quantum regression theorem and therefore can only describe Markov processes. Such a quantity characterizing the non-Markovianity can be rather easily measured experimentally. For example, the exact two-time current-current correlation $\langle I(t) I(t+\tau)\rangle$ in nanoelectronic systems has been studied experimentally in nanoelectronics. The two-time photon number correlation $\langle n(t) n(t+\tau)\rangle$ is also experimentally measured extensively through photon bunching and anti-bunching experiments. Detailed discussions can be found from [209]. Thus, a quantitative measurement of non-Markovian dynamics can be obtained through two-time correlation functions. This gives the general physical picture of the quantum memory in open systems.

13.6 Quantum Transport Theory of Mesoscopic Nanosystems

Mesoscopic systems refer to systems measuring in the nanometer (10^{-9} m) scale of nanostructures which is in size intermediate between molecule and bacterium. The characteristic dimension of a mesoscopic nanodevice is smaller than one or more of these length scales: the de Broglie wavelength of the electrons, their mean free path, and the phase coherence length (distance over which an electron can interfere with itself). Such devices usually do not follow the Ohmic law because of the quantum mechanical wave nature of electrons. Understanding how electrons behave over such tiny distant scales is therefore of great importance to the quantum electronics, quantum communication, and quantum computations. The enormous mobilities of electrons in nanodevices enable one to explore fundamental physics of quantum nature, because except for confinement and effective mass, the electrons do not interact with each other very often, so that they can travel several micrometers before colliding. As a result, the quantum coherence of electron wave may play an important role in its transport processes. Theoretically, electrons transport in mesoscopic systems described as physical systems consisting of a nanoscale active region (the device system) attached to multiple reservoirs of charges. The device system exchanges the particles, energy, and information with the surroundings; it is thus a typical open system described by Eq. (13.5) with an extension of the system coupled to multiple reservoirs. The issues of open quantum systems, such as dissipation, fluctuation, and decoherence, will inevitably arise. Furthermore, when a bias is present, the transient transport properties have to be examined within a non-equilibrium framework.

Preliminary quantum transport theories were based on the Landauer-Büttiker approach [215, 216]. However, in this approach, systems are usually treated as perfect conductors without any dissipation. Hence, such an approach can only be applied to ballistic systems near thermal equilibrium. In order for nanodevices to functionally operate in present logic gate-based computational dynamics, it must be subjected to high source-drain voltages and high-frequency bandwidths. When it is far from equilibrium, for highly transient and highly nonlinear regimes, a more microscopic theory has been developed for quantum transport in terms of non-equilibrium Green functions [213, 214]. Furthermore, the device system exchanges the particles, energy, and information with the leads and is a typical open system in which the dissipation, fluctuation, and decoherence all are heavily involved. The master equation approach which fully addresses the dissipation, fluctuation, and decoherence dynamics could provide a more microscopic description for the transient quantum transport [196, 197].

We shall extend the Hamiltonian (13.5) to the case with multiple reservoirs, such as quantum electronic devices coupled to several electrodes (leads) in nano- and quantum electronics and also nano- or micro-cavities coupled with several waveguides in integrated photonic circuits. The corresponding Hamiltonian is

$$
\begin{aligned}
H_{\rm tot}(t) &= \sum_i \varepsilon_{{\rm S},i}(t) a_i^\dagger a_i + \sum_{\alpha k} \epsilon_{\alpha k}(t) b_{\alpha k}^\dagger b_{\alpha k} \\
&\quad + \sum_{i\alpha k} \left(V_{i\alpha k}(t) a_i^\dagger b_{\alpha k} + V_{i\alpha k}^*(t) b_{\alpha k}^\dagger a_i\right),
\end{aligned} \tag{13.88}
$$

where the index α denotes different reservoirs. Following the same procedure given in Sec. II, we can find the exact master equation of (13.88). The exact master equation remains the same form, but we can also rewrite it in terms of suppercurrent form [196, 197], in order to find the connection with the transient current in quantum transport

$$
\begin{aligned}
\frac{d\rho(t)}{dt} &= \frac{1}{i\hbar}[H_{\rm S}^r(t,t_0), \rho(t)] \\
&\quad + \sum_{ij} \gamma_{ij}(t,t_0)[2a_j\rho(t)a_i^\dagger - \rho(t)a_i^\dagger a_j - a_i^\dagger a_j \rho(t)]
\end{aligned} \tag{13.89}
$$

$$+\sum_{ij}\widetilde{\boldsymbol{\gamma}}_{ij}(t,t_0)[a_i^\dagger\rho(t)a_j \pm a_j\rho(t)a_i^\dagger \mp a_i^\dagger a_j\rho(t) - \rho(t)a_ja_i^\dagger],$$

$$=\frac{1}{i\hbar}[H_S(t),\rho(t)] + \sum_\alpha[\mathcal{L}_\alpha^+(t,t_0) + \mathcal{L}_\alpha^-(t,t_0)]\rho(t).$$

In the second equality of the above equation, $\mathcal{L}_\alpha^+(t,t_0)$ and $\mathcal{L}_\alpha^-(t,t_0)$ are the suppercurrent operators:

$$\mathcal{L}_\alpha^+(t,t_0)\rho(t) = \sum_{ij}\{\boldsymbol{\lambda}_{\alpha ij}(t,t_0)[a_j\rho(t)a_i^\dagger - \rho(t)a_ja_i^\dagger] \tag{13.90a}$$
$$-\boldsymbol{\kappa}_{\alpha ij}(t,t_0)a_i^\dagger a_j\rho(t) + \text{H.c.}\},$$

$$\mathcal{L}_\alpha^-(t,t_0)\rho(t) = \sum_{ij}\{\boldsymbol{\lambda}_{\alpha ij}(t,t_0)[a_i^\dagger\rho(t)a_j - a_i^\dagger a_j\rho(t)] \tag{13.90b}$$
$$+\boldsymbol{\kappa}_{\alpha ij}(t,t_0)a_j\rho(t)a_i^\dagger + \text{H.c.}\}.$$

The superoperators $\mathcal{L}_\alpha^+(t,t_0)$ and $\mathcal{L}_\alpha^-(t,t_0)$ are intimately related to the transport current through the reservoir α, as we will see next. The renormalized system Hamiltonian $H_S^r(t,t_0)$ with the associated renormalized energy $\omega'_{ij}(t,t_0)$ and the dissipation coefficient $\boldsymbol{\gamma}_{ij}(t,t_0)$ as well as the fluctuation coefficients $\widetilde{\boldsymbol{\gamma}}_{ij}(t,t_0)$ in the master equation (13.89) remain the same forms as that given by Eq. (13.54). But they can also be expressed in terms of new coefficients $\boldsymbol{\kappa}_\alpha(t,t_0)$ and $\boldsymbol{\lambda}_\alpha(t,t_0)$ given in the suppercurrent operators (Eq. (13.90)):

$$H_S^r(t,t_0) = \sum_{ij}\varepsilon_{S,ij}^r(t,t_0)a_i^\dagger a_j, \tag{13.91a}$$

$$\boldsymbol{\varepsilon}_{S,ij}^r(t,t_0) = -\hbar\,\text{Im}\big[\dot{\boldsymbol{u}}(t,t_0)\boldsymbol{u}^{-1}(t,t_0)\big]_{ij} \tag{13.91b}$$
$$= \varepsilon_{S,i}(t)\delta_{ij} - \frac{i\hbar}{2}\sum_\alpha[\boldsymbol{\kappa}_\alpha(t,t_0) - \boldsymbol{\kappa}_\alpha^\dagger(t,t_0)]_{ij},$$

$$\boldsymbol{\gamma}_{ij}(t,t_0) = -\text{Re}\big[\dot{\boldsymbol{u}}(t,t_0)\boldsymbol{u}^{-1}(t,t_0)\big]_{ij} \tag{13.91c}$$
$$= \frac{1}{2}\sum_\alpha[\boldsymbol{\kappa}_\alpha(t,t_0) + \boldsymbol{\kappa}_\alpha^\dagger(t,t_0)]_{ij},$$

$$\widetilde{\boldsymbol{\gamma}}_{ij}(t,t_0) = \dot{\boldsymbol{v}}_{ij}(t,t) - \big[\dot{\boldsymbol{u}}(t,t_0)\boldsymbol{u}^{-1}(t,t_0)\boldsymbol{v}(t,t) + \text{h.c.}\big]_{ij} \tag{13.91d}$$
$$= \sum_\alpha[\boldsymbol{\lambda}_\alpha(t,t_0) + \boldsymbol{\lambda}_\alpha^\dagger(t,t_0)]_{ij}$$

They are determined by the non-equilibrium Green function $\boldsymbol{u}(t, t_0)$ and $\boldsymbol{v}(t, t_0)$,

$$\boldsymbol{\kappa}_\alpha(t, t_0) = \int_{t_0}^{t} d\tau \boldsymbol{g}_\alpha(t, \tau)\boldsymbol{u}(\tau, t_0)\boldsymbol{u}^{-1}(t, t_0)\,, \tag{13.92a}$$

$$\boldsymbol{\lambda}_\alpha(t, t_0) = -\boldsymbol{\kappa}_\alpha(t)\boldsymbol{v}(t, t) + \int_{t_0}^{t} d\tau[\boldsymbol{g}_\alpha(t, \tau)\boldsymbol{v}(\tau, t) - \widetilde{\boldsymbol{g}}_\alpha(t, \tau)\bar{\boldsymbol{u}}(\tau, t)], \tag{13.92b}$$

where $\boldsymbol{g}_\alpha(t, \tau)$ and $\widetilde{\boldsymbol{g}}_\alpha(t, \tau)$ are the two-time correlation functions between the system and the reservoir α,

$$g_{\alpha,ij}(\tau, \tau') = \left(\frac{1}{\hbar}\right)^2 \sum_k V_{i\alpha k}(\tau)u_{\alpha k0}(\tau, \tau')V^*_{j\alpha k}(\tau'), \tag{13.93a}$$

$$\widetilde{g}_{\alpha ij}(\tau, \tau') = \left(\frac{1}{\hbar}\right)^2 \sum_k V_{i\alpha k}(\tau)u_{\alpha k0}(\tau, \tau')V^*_{j\alpha k}(\tau')f(\epsilon_{\alpha k}, T_{\alpha,0}), \tag{13.93b}$$

and $f(\epsilon_{\alpha k}, T_{\alpha,0}) = 1/[e^{(\epsilon_{\alpha k}-\mu_{\alpha,0})/k_B T_{\alpha,0}} \mp 1]$ is the initial electron distribution of the reservoir α at initial temperature $T_{\alpha,0}$ with initial chemical potential $\mu_{\alpha,0}$. The non-equilibrium Green functions $\boldsymbol{u}(t, t_0)$, $\boldsymbol{v}(t, t)$ obey the same time-convolution equation of motion (13.42) with the extension to multiple reservoirs.

To be more specific, we shall now consider the electron current flowing from the system into the reservoir α. In the Heisenberg picture, such a current is defined by

$$\begin{aligned} I_\alpha(t) &\equiv -e\frac{d\langle N_\alpha(t)\rangle}{dt} = \frac{ie}{\hbar}\langle [N_\alpha(t), H(t)]\rangle \\ &= -\frac{ie}{\hbar}\sum_{ki}[V_{i\alpha k}(t)\langle a_i^\dagger(t)c_{\alpha k}(t)\rangle - V^*_{i\alpha k}(t)\langle c^\dagger_{\alpha k}(t)a_i(t)\rangle], \end{aligned} \tag{13.94}$$

where e is the electron charge and $N_\alpha = \sum_k c^\dagger_{\alpha k}c_{\alpha k}$ is the particle number operator in reservoir α. To find the transport current from the exact master equation (13.89), we will introduce the one-particle reduced density matrix:

$$\boldsymbol{\rho}^{(1)}_{ij}(t) \equiv \mathrm{tr}_s[a_j^\dagger a_i \rho(t)] = \langle a_j^\dagger(t)a_i(t)\rangle. \tag{13.95}$$

With the abovementioned conditions, from the Heisenberg equation of motion, it is easy to find the equation of motion as follows:

$$\frac{d\boldsymbol{\rho}^{(1)}(t)}{dt} = -\frac{i}{\hbar}[\boldsymbol{\varepsilon}_S(t), \boldsymbol{\rho}^{(1)}(t)] - \sum_\alpha \boldsymbol{\mathcal{I}}_\alpha(t)\,, \tag{13.96}$$

where $\mathcal{I}_\alpha(t)$ is the current matrix flowing into the reservoir α:

$$\mathcal{I}_{\alpha ij}(t) = \frac{i}{\hbar}\sum_k [V_{i\alpha k}(t)\langle a_j^\dagger(t)c_{\alpha k}(t)\rangle - V_{j\alpha k}^*(t)\langle c_{\alpha k}^\dagger a_i(t)\rangle]. \tag{13.97a}$$

Comparing Eqs. (13.94) and (13.97a), one can see that the trace of the current matrix is just the transport current flowing from the system into the reservoir α:

$$I_\alpha(t) = -e\,\mathrm{Tr}[\mathcal{I}_\alpha(t)]. \tag{13.98}$$

where Tr denotes the trace over the $N \times N$ matrix of the N energy levels in the system.

On the other hand, one can also obtain Eq. (13.96) directly from the exact master equation (13.89). The result is

$$\frac{d}{dt}\boldsymbol{\rho}_{ij}^{(1)}(t) = -\frac{i}{\hbar}[\boldsymbol{\varepsilon}_S(t), \boldsymbol{\rho}^{(1)}(t)]_{ij} + \sum_\alpha \mathrm{tr}_s[a_j^\dagger a_i[\mathcal{L}_\alpha^+(t) + \mathcal{L}_\alpha^-(t)]\rho(t)]\,. \tag{13.99}$$

Comparing Eqs. (13.96) and (13.99) for the single-particle reduced density matrix, with the help of Eqs. (13.90) and (13.92), we obtain the explicit formula of the current matrix:

$$\mathcal{I}_\alpha(t) = \int_{t_0}^t d\tau\{\boldsymbol{g}_\alpha(t,\tau)\boldsymbol{n}(\tau,t) - \widetilde{\boldsymbol{g}}_\alpha(t,\tau)\boldsymbol{u}^\dagger(\tau,t) + \mathrm{H.c.}\}. \tag{13.100}$$

Here we have introduced the one-particle correlation function $\boldsymbol{n}_{ij}(\tau,t) = \langle a_j^\dagger(t)a_i(\tau)\rangle$ which can also be calculated directly from the Heisenberg equation of motion,

$$\boldsymbol{n}(\tau,t) = \boldsymbol{u}(\tau,t_0)\boldsymbol{n}(t_0)\boldsymbol{u}^\dagger(t,t_0) + \boldsymbol{v}(\tau,t), \tag{13.101}$$

and $\boldsymbol{\rho}^{(1)}(t) = \boldsymbol{n}(t,t)$. Thus, the current flowing into the reservoir α is simply given by

$$\begin{aligned} I_\alpha(t) &= -2e\mathrm{Re}\int_{t_0}^t d\tau \mathrm{Tr}[\boldsymbol{g}_\alpha(t,\tau)\boldsymbol{n}(\tau,t) - \widetilde{\boldsymbol{g}}_\alpha(t,\tau)\boldsymbol{u}^\dagger(\tau,t)] \qquad (13.102) \\ &= -2e\mathrm{Re}\int_{t_0}^t d\tau \mathrm{Tr}[\boldsymbol{g}_\alpha(t,\tau)\boldsymbol{v}(\tau,t) - \widetilde{\boldsymbol{g}}_\alpha(t,\tau)\boldsymbol{u}^\dagger(\tau,t) \\ &\quad + \boldsymbol{g}_\alpha(t,\tau)\boldsymbol{u}(\tau,t_0)\boldsymbol{n}(t_0)\boldsymbol{u}^\dagger(t,t_0)]. \end{aligned}$$

It shows that the transient transport current flowing into the reservoir α from the system is fully determined by the time-correlation functions $\boldsymbol{g}_\alpha(t,\tau)$ and $\widetilde{\boldsymbol{g}}_\alpha(t,\tau)$

between the system and the reservoir and the non-equilibrium Green functions $\boldsymbol{u}(\tau, t)$ and $\boldsymbol{v}(\tau, t)$ of the system. The dissipation, fluctuation, and decoherence dynamics during the non-equilibrium evolution are fully taken into account. The last term in the second equality depends on the initial state which is important for the transient transport.

We shall now apply the above discussed theory for a practical application. The simplest application is the noninteracting resonant-level model. The resonant-level model describes a single-level quantum dot or molecular with energy ε_S that is in contact with two leads, the source and drain. The leads are treated as the free electron gas. The Hamiltonian is Eq. (13.88) with $\alpha = L, R$ labeling the left and the right leads. Also, for simplicity, we consider the wideband limit (WBL), namely, the spectral density coupled to each lead is a constant, given by

$$J_\alpha(\epsilon) = 2\pi \sum_k V_{\alpha k} V^*_{\alpha k} \delta(\epsilon - \epsilon_{\alpha k}) = \Gamma_\alpha, \quad \alpha = L, R. \tag{13.103}$$

Also, we have made the assumption that two leads are in the same initial temperature $T_{L,0} = T_{R,0} = T_0$. (With different initial temperatures for the source and the drain, one can also study the heating transport with this formulation.) Then the system-lead corrections function is reduced to

$$g_{\alpha,ij}(\tau, \tau') = \frac{1}{\hbar^2} \int_{-\infty}^{\infty} \frac{d\epsilon}{2\pi} J_\alpha(\epsilon) e^{-\frac{i}{\hbar}\epsilon(t-t')} = \frac{\Gamma_\alpha}{\hbar} \delta(t - t'), \tag{13.104a}$$

$$\begin{aligned} \widetilde{g}_{\alpha ij}(\tau, \tau') &= \frac{1}{\hbar^2} \int_{-\infty}^{\infty} \frac{d\epsilon}{2\pi} J_\alpha(\epsilon) f(\epsilon, T_0) e^{-\frac{i}{\hbar}\epsilon(t-t')} \\ &= \frac{\Gamma_\alpha}{\hbar^2} \int_{-\infty}^{\infty} \frac{d\epsilon}{2\pi} \frac{e^{-\frac{i}{\hbar}\epsilon(t-t')}}{e^{(\epsilon - \mu_{\alpha,0})/k_B T_0} + 1}, \end{aligned} \tag{13.104b}$$

where μ_L and μ_R are the chemical potentials of the leads. In the WBL, the equations of motion for the non-equilibrium Green functions $u(\tau, t)$ and $v(\tau, t)$ are reduced to

$$\frac{d}{dt} u(\tau, t_0) + \frac{i}{\hbar} \varepsilon_0(\tau) u(\tau, t_0) + \frac{\Gamma}{2\hbar} u(\tau, t_0) = 0. \tag{13.105a}$$

$$v(\tau, t) = \sum_\alpha \frac{\Gamma_\alpha}{\hbar^2} \int_{t_0}^{\tau} dt_1 \int_{t_0}^{t} dt_2 \int_{-\infty}^{\infty} \frac{d\epsilon}{2\pi} \frac{u(\tau, t_1) e^{-\frac{i}{\hbar}\epsilon(t_1 - t_2)} u^*(t_2, t)}{e^{(\epsilon - \mu_\alpha)/k_B T_0} + 1}, \tag{13.105b}$$

where $\Gamma = \Gamma_L + \Gamma_R$. The solution of the non-equilibrium Green functions can be solved analytically (set $t_0 = 0$)

$$u(t) = e^{-(i\varepsilon_0 + \frac{\Gamma}{2})t/\hbar}, \tag{13.106a}$$

$$v(t,t) = v_{\rm st} + \int \frac{d\epsilon}{2\pi} \frac{\Gamma_L f_L(\epsilon) + \Gamma_R f_R(\epsilon)}{(\varepsilon_0-\epsilon)^2 + (\Gamma/2)^2} \times \left\{ e^{-\Gamma t/\hbar} - 2e^{-\Gamma t/2\hbar} \cos[(\varepsilon_0-\epsilon)t/\hbar] \right\}, \tag{13.106b}$$

where $\Gamma = \Gamma_L + \Gamma_R$, $f_{L,R}(\epsilon) = 1/[e^{(\epsilon-\mu_{L,R})/k_B T_0} + 1]$, and $v_{\rm st}$ is the solution of $v(t)$ at the steady-state limit:

$$v_{\rm st} = \int \frac{d\omega}{2\pi} \frac{\Gamma_L f_L(\epsilon) + \Gamma_R f_R(\epsilon)}{(\varepsilon_0-\epsilon)^2 + (\Gamma/2)^2}. \tag{13.107}$$

The electron occupation in the dot can be calculated by Eq. (13.95) as

$$N(t) = \rho^{(1)}(t) = e^{-\Gamma t} \rho^{(1)}(0) + v(t), \tag{13.108}$$

where $N(0) = \rho^{(1)}(0)$ is the initial electron occupation in the dot. In the steady-state limit, $N_{\rm st} = v_{\rm st}$.

The electron transient current can also be analytically computed:

$$\begin{aligned} I_\alpha(t) = {} & I_{\alpha,\rm st} - \frac{e}{\hbar}\Gamma_\alpha[N(t) - N_{\rm st}] \\ & - \frac{e}{\hbar} e^{-\frac{\Gamma t}{2\hbar}} \int \frac{d\epsilon}{2\pi} \frac{\Gamma_\alpha f_\alpha(\epsilon)}{(\varepsilon_0-\epsilon)^2 + (\Gamma/2)^2} \Big\{ \Gamma \cos[(\varepsilon_0-\epsilon)t/\hbar] \\ & - 2(\varepsilon_0-\epsilon)\sin[(\varepsilon_0-\epsilon)t/\hbar] \Big\}, \end{aligned} \tag{13.109}$$

where the steady-state current is

$$I_{\alpha,\rm st} = \frac{e}{\hbar}\Gamma_\alpha \int \frac{d\epsilon}{2\pi} \frac{\Gamma_L[f_\alpha(\epsilon) - f_L(\epsilon)] + \Gamma_R[f_\alpha(\epsilon) - f_R(\epsilon)]}{(\varepsilon_0-\epsilon)^2 + (\Gamma/2)^2}. \tag{13.110}$$

It is also straightforward to calculate the net current:

$$\begin{aligned} I_{\rm net}(t) &= I_L(t) - I_R(t) \\ &= I_{\rm st} - \frac{e}{\hbar}(\Gamma_L - \Gamma_R)[N(t) - N_{\rm st}] \\ &\quad - \frac{e}{\hbar} e^{-\frac{\Gamma t}{2\hbar}} \int \frac{d\epsilon}{2\pi} \frac{\Gamma_L f_L(\epsilon) - \Gamma_R f_R(\epsilon)}{(\varepsilon_0-\epsilon)^2 + (\Gamma/2)^2} \\ &\quad \times \Big\{ \Gamma\cos[(\varepsilon_0-\epsilon)t/\hbar] - 2(\varepsilon_0-\epsilon)\sin[(\varepsilon_0-\epsilon)t/\hbar] \Big\} \end{aligned} \tag{13.111}$$

where the stationary net current is

$$I_{\rm st} = \frac{2e}{\hbar}\int \frac{d\epsilon}{2\pi}\frac{\Gamma_L\Gamma_R}{(\varepsilon_0-\epsilon)^2+(\Gamma/2)^2}[f_L(\epsilon)-f_R(\epsilon)]. \tag{13.112a}$$

From the above, we see that once one solves the non-equilibrium Green functions from Eq. (13.42), the time dependence of all the physical quantities, such as the electron occupations in the central system and the currents flowing from each lead to the central region, can be obtained with the explicit dependence on the time. From Eq. (13.109), we also see that the initial current $I_\alpha(0) = -\Gamma_\alpha N(0)$ which depends on the initial occupation of the dot. This result is consistent with the electron occupation in the dot, Eq. (13.108). For zero initial occupation, the initial current is zero.

It is worth noting that some of the above results can also be obtained using the non-equilibrium Green function technique. In fact, we can easily find the connection with Keldysh's non-equilibrium Green function technique which has been widely used in quantum transport theory and many-body dynamics. From Eq. (13.102), one sees that the transient current is completely determined by the time-correlation functions $\boldsymbol{g}_\alpha(t,\tau)$ and $\widetilde{\boldsymbol{g}}_\alpha(t,\tau)$ of the waveguides plus the non-equilibrium Green function $\boldsymbol{u}(\tau,t)$ and $\boldsymbol{n}(\tau,t)$ of the central system. In fact, we have shown [196] that the functions $\boldsymbol{u}(\tau,t_0)$, $\boldsymbol{u}^\dagger(\tau,t)$, and $\boldsymbol{n}(\tau,t)$ are related to the retarded, advanced, and lesser Green functions of the system in Keldysh's non-equilibrium formalism [180, 182, 183]:

$$\boldsymbol{u}_{ij}(t_1,t_2) = \theta(t_1-t_2)\langle[a_i(t_1),a_j^\dagger(t_2)]\rangle \equiv iG^r_{ij}(t_1,t_2), \tag{13.113a}$$

$$\boldsymbol{u}^\dagger_{ij}(t_1,t_2) = \theta(t_2-t_1)\langle[a_i(t_1),a_j^\dagger(t_2)]\rangle \equiv -iG^a_{ij}(t_1,t_2), \tag{13.113b}$$

$$\boldsymbol{n}_{ij}(t_1,t_2) = \langle a_j^\dagger(t_2)a_i(t_1)\rangle \equiv -iG^<_{ij}(t_1,t_2). \tag{13.113c}$$

The time-correlation functions $\boldsymbol{g}_\alpha(t,\tau)$ and $\widetilde{\boldsymbol{g}}_\alpha(t,\tau)$ correspond to the retarded and lesser self-energy functions that arose from the couplings between the system and the reservoirs:

$$g_{\alpha ij}(t_1,t_2) = i\Sigma^r_{\alpha ij}(t_1,t_2), \tag{13.114a}$$

$$\widetilde{g}_{\alpha ij}(t_1,t_2) = -i\Sigma^<_{\alpha ij}(t_1,t_2). \tag{13.114b}$$

The explicit form of these self-energy functions is simply given by Eq. (13.93).

Explicitly, the equation of motion (13.42a) for $\boldsymbol{u}(\tau,t_0)$ obtained in the master equation formulation can be rewritten as

$$\Big\{\frac{d}{d\tau}-\frac{i}{\hbar}\boldsymbol{\varepsilon}_S(t)\Big\}\boldsymbol{G}^r(\tau,t_0)+\int_{t_0}^{\tau}\boldsymbol{\Sigma}^r(\tau,\tau')\boldsymbol{G}^r(\tau',t_0)d\tau' = \delta(\tau-t_0). \tag{13.115}$$

This is just the standard Kadanoff-Baym equation for the retarded Green function. The advanced Green function obeys the relation: $\boldsymbol{G}^a(t_1, t_2) = [\boldsymbol{G}^r(t_2, t_1)]^\dagger$ by the definition. The central and also the most difficult part in the non-equilibrium Green function technique is the lesser Green function $\boldsymbol{G}^<(\tau, t)$. The lesser Green function $\boldsymbol{G}^<(t_1, t_2)$ fully determines the quantum kinetic theory of non-equilibrium system. From Eq. (13.101), we have solved already the exact analytical solution of the lesser Green function:

$$\begin{aligned}\boldsymbol{G}^<(\tau, t) = & \; \boldsymbol{G}^r(\tau, t_0)\boldsymbol{G}^<(t_0, t_0)\boldsymbol{G}^a(t_0, t) \\ & + \int_{t_0}^{\tau} d\tau_1 \int_{t_0}^{t} d\tau_2 \, \boldsymbol{G}^r(\tau, \tau_1)\boldsymbol{\Sigma}^<(\tau_1, \tau_2)\boldsymbol{G}^a(\tau_2, t),\end{aligned} \tag{13.116}$$

where $\boldsymbol{G}^<(t_0, t_0) = \boldsymbol{n}(t_0, t_0)$ is the initial particle distribution in the system.

In the standard Green function formulation, one usually takes the initial time $t_0 \to -\infty$ such that the first term will disappear. This ignores the information of the initial state dependence in quantum transport, an important effect on non-Markovian memory dynamics. Thus, the resulting lesser Green function obtained in the mesoscopic electron transport contains only the last term in Eq. (13.116) [214]. In other words, Eq. (13.116) gives the exact and general solution for the lesser Green function in mesoscopic systems. With the above relations and solutions, the transient transport current, Eq. (13.102), can be re-expressed as

$$I_\alpha(t) = 2e\mathrm{Re}\int_{t_0}^{t} d\tau \mathrm{Tr}[\boldsymbol{\Sigma}^r_\alpha(t, \tau)\boldsymbol{G}^<(\tau, t) + \boldsymbol{\Sigma}^<_\alpha(t, \tau)\boldsymbol{G}^a(\tau, t)]. \tag{13.117}$$

This reproduces the Meir-Wingreen formula for transport current in the non-equilibrium Green function technique, the latter has been widely used in the investigation of various electron transport phenomena in mesoscopic systems [214] in which the extension that the initial state effect is explicitly included [196].

Furthermore, if we were to take the steady-state limit ($t \to \infty$), we can reproduce the generalized Landauer-Büttiker formula of the transport current. Using the Laplace transformation, $f(z) = \int_{t_0}^{\infty} dt e^{\frac{i}{\hbar}z(t-t_0)} f(t)$, the system-environment correlation function $\boldsymbol{g}_\alpha(t - t')$ and the retarded Green function $\boldsymbol{u}(t, t')$ give

$$\boldsymbol{g}_\alpha(\epsilon) = \frac{i}{\hbar}\int \frac{d\epsilon'}{2\pi} \frac{\boldsymbol{\Gamma}_\alpha(\epsilon')}{\epsilon - \epsilon' + i0^+} = i\boldsymbol{\Sigma}^r_\alpha(\epsilon), \tag{13.118a}$$

$$\boldsymbol{u}(\epsilon) = \frac{i\hbar}{\epsilon - \boldsymbol{\varepsilon} - \boldsymbol{\Sigma}^r(\epsilon)} = i\boldsymbol{G}^r(\epsilon) \tag{13.118b}$$

The advanced Green function is simply given by $\bar{\boldsymbol{u}}(\omega) = -i\boldsymbol{G}^a(\omega) = \boldsymbol{u}^\dagger(\omega)$. Furthermore, for the correlation Green function $\boldsymbol{v}(t, t)$ at $t \to \infty$, its Laplace transformation gives

$$v(\epsilon) = u(\epsilon)\widetilde{g}(\epsilon)u^{\dagger}(\epsilon) = -iG^{r}(\epsilon)\Sigma^{<}(\epsilon)G^{a}(\epsilon) = -iG^{<}(\epsilon). \tag{13.119}$$

Here $\widetilde{g}(\epsilon) = -i\Sigma^{<}(\epsilon)$. Thus, in the steady-state limit $t \to \infty$, substituting the above results into Eq. (13.102), we obtain the steady-state transport current:

$$I_{\alpha,\mathrm{st}} = \frac{ie}{\hbar}\int \frac{d\epsilon}{2\pi}\,\mathrm{Tr}\Big(\Gamma_{\alpha}(\epsilon)\big[G^{<}(\epsilon) + f_{\alpha}(\epsilon)\{G^{r}(\epsilon) - G^{a}(\epsilon)\}\big]\Big). \tag{13.120}$$

This reproduces the steady-state current in terms of the non-equilibrium Green functions in the frequency domain that has been widely used. When we consider specifically a system coupled with left (source) and right (drain) electrodes, i.e., $\alpha = L$ and R, respectively, and also assume that the spectral densities for the left and right leads have the same energy dependence, $\Gamma_L(\omega) = \lambda\Gamma_R(\omega)$, where λ is a constant, then the steady-state net current flowing from the left to the right lead is given by

$$I_{\mathrm{st}} = \frac{2e}{\hbar}\int \frac{d\epsilon}{2\pi}\big[f_L(\epsilon) - f_R(\epsilon)\big]\mathcal{T}(\epsilon), \tag{13.121}$$

where

$$\mathcal{T}(\epsilon) = \mathrm{Tr}\Big\{\frac{\Gamma_L(\epsilon)\Gamma_R(\epsilon)}{\Gamma_L(\epsilon) + \Gamma_R(\epsilon)}\mathrm{Im}[G^{a}(\epsilon)]\Big\}, \tag{13.122}$$

is the transmission coefficient. This is the generalized Landauer-Büttiker formula.

13.7 Quantum Thermodynamics

The classical thermodynamics is built on the hypothesis of equilibrium. Specifically, a macroscopic system at equilibrium is completely described by relation between the internal energy U and a set of other extensive parameters: the entropy S, the volume V, the particle number N, the magnetic moment $\boldsymbol{M}$, etc.

$$U = U(S, V, N, \boldsymbol{M}, \cdots). \tag{13.123}$$

This relation obeys the extremum principle of either maximizing the entropy or minimizing the internal energy (the second law of thermodynamics) and is known as the fundamental equation of thermodynamics. The thermodynamic temperature T, the pressure P, the chemical potential μ, and the magnetic field $\boldsymbol{B}$ are defined by the first derivative of the internal energy with respect to the entropy, the volume, the particle number, and the magnetic moment, respectively, from this relation:

$$T = \left.\frac{\partial U}{\partial S}\right|_{V,N,\boldsymbol{M},\cdots}, \quad P = \left.\frac{\partial U}{\partial V}\right|_{S,N,\boldsymbol{M},\cdots}, \tag{13.124}$$
$$\mu = \left.\frac{\partial U}{\partial N}\right|_{S,V,\boldsymbol{M},\cdots}, \quad \boldsymbol{B} = \left.\frac{\partial U}{\partial \boldsymbol{M}}\right|_{S.V,N}.$$

As a result, the relation (13.123) directly leads to the first law of thermodynamics:

$$dE = TdS + PdV + \mu dN + \boldsymbol{B}d\boldsymbol{M} + \cdots. \tag{13.125}$$

And the first derivatives in Eq. (13.124) give a complete set of equations of state in classical thermodynamics. Furthermore, the second derivatives characterize various intrinsic properties of individual macroscopic systems, such as the specific heat, compressibility, magnetic susceptibility, etc.

Microscopically, all thermodynamic parameters at equilibrium can be obtained as an average over all possible microstates with equal probability (known as the ergodic hypothesis) at fixed energy and particle number (known as the microcanonical ensemble). Alternatively, they can be alternatively determined from the probability distribution $\{p_i\}$ for each microstate which is determined by maximizing Shannon entropy,

$$S = -k \sum_i p_i \ln p_i \tag{13.126}$$

under the condition of fixed average energy (the canonical ensemble) or both the fixed average energy and the average particle number (the grand canonical ensembles). This is also a natural result of the second law of thermodynamics. Furthermore, maximizing the Shannon entropy under the condition of fixed average energy, and maybe also the average particle number, will determine the equilibrium probability distribution in terms of the Gibbs state,

$$\rho_{\text{th}} = \begin{cases} \frac{1}{Z} e^{-\beta H}, & \text{canonical ensemble;} \\ \frac{1}{Z} e^{-\beta(H - \sum_i \mu_i N_i)}, & \text{grand canonical ensemble,} \end{cases} \tag{13.127}$$

where Z is the partition function coming arising from the normalized condition $\text{tr}\rho_{\text{th}} = 1$, $\beta = 1/kT$ is the inverse temperature of the reservoir, and H is the Hamiltonian of the system. Thus, the internal energy, the entropy, the particle numbers, the magnetic moment, etc. can systematically be calculated from the thermal equilibrium state ρ_{th}:

$$U = \text{Tr}[H\rho_{\text{th}}], \quad S = -k\,\text{Tr}[\rho_{\text{th}} \ln \rho_{\text{th}}]. \tag{13.128a}$$

$$N = \text{Tr}[\hat{N}\rho_{\text{th}}], \quad \boldsymbol{M} = \text{Tr}[\hat{\boldsymbol{M}}\rho_{\text{th}}], \quad \cdots. \tag{13.128b}$$

The fundamental equation of thermodynamics is microscopically manifested from these thermodynamic quantities. In the energy basis, the entropy in Eq. (13.128a) is identical to the Shannon entropy.

Equations (13.123)–(13.128a) is the axiomatic description of thermodynamics and statistical mechanics, from which all the thermodynamical phenomena can be determined naturally [219]. For more than a century, it has been a significant challenge as to whether and how the thermodynamics and statistical mechanics can be deduced from the dynamics of the quantum systems. The equilibrium hypothesis states that over a sufficiently long time period, a given macroscopic system can always reach thermal equilibrium with its environment (reservoir), and the corresponding equilibrium statistical distribution does not depend on its initial state. This process is known as **thermalization**. Solving the problem of thermalization within the framework of quantum mechanics has been a dream for many physicists and is also the foundation in the investigation of quantum thermodynamics and non-equilibrium statistical mechanics [220].

Clearly, thermalization relies on a profound understanding of the quantum dynamics of systems under the interactions with their environments. In the last couple of decades, there has been an attempt to develop a thermodynamics formulation which is far from equilibrium for the nanoscale or atomic-scale quantum systems, in which the particle number is much less than the order of 10^{23}. If one could develop such a thermodynamics for arbitrary small quantum systems, and prove the consistency with equilibrium thermodynamics, then the foundation of thermodynamics can be established. This is a newly emerged research field now in the literature which is known as quantum thermodynamics. The aim of quantum thermodynamics is also to develop the thermodynamics within the framework of quantum principle that is generally valid for arbitrary dynamics of quantum systems interacting with their arbitrary environments. Recently, this arduous task has been completed by one of the authors [210].

Specifically, based on the exact master equation and the exact solution of the reduced density matrix given in the previous sections, we can show how the system is thermalized under the time evolution of quantum mechanics. The general solution of the non-equilibrium Green functions presented in Sect. 13.5 shows that, if there are no localized bound states (modes), the non-equilibrium Green functions in the steady-state limit become

$$\lim_{t\to\infty} \boldsymbol{u}(t, t_0) = 0\,, \tag{13.129a}$$

$$\lim_{t\to\infty} \boldsymbol{v}(t, t) = \int \frac{d\epsilon}{2\pi} f(\epsilon, T_0)\boldsymbol{D}(\epsilon), \tag{13.129b}$$

where T_0 is the initial temperature of the environment at time t_0. This solution is valid for arbitrary continuous spectral density matrix of the reservoir $\boldsymbol{J}_{ij}(\epsilon)$ that covers every point of the whole energy frequency domain. Then the coefficients in the propagating function of the reduced density matrix, Eq. (13.46), are largely simplified at the steady-state limit $t \to \infty$,

$$\boldsymbol{K}_1(t, t_0) = 0, \quad \boldsymbol{K}_2(t) = \frac{\boldsymbol{v}(t,t)}{\mathbf{1} \pm \boldsymbol{v}(t,t)}, \quad \boldsymbol{K}_3(t, t_0) = \mathbf{1}. \tag{13.130}$$

As a result, the propagating function of the reduced density matrix, Eq. (13.46), is reduced to

$$\lim_{t\to\infty} \mathcal{J}(\boldsymbol{\xi}_f, \boldsymbol{\xi}'_f, t; \boldsymbol{\xi}_0, \boldsymbol{\xi}'_0, t_0) = \lim_{t\to\infty} \left(\det[\mathbf{1} \pm \boldsymbol{v}(t,t)]\right)^{\mp 1} \times \exp\left\{ \pm \boldsymbol{\xi}'^{\dagger}_0 \boldsymbol{\xi}_0 + \boldsymbol{\xi}^{\dagger}_f \frac{\boldsymbol{v}(t,t)}{\mathbf{1} \pm \boldsymbol{v}(t,t)} \boldsymbol{\xi}'_f \right\}. \tag{13.131}$$

Substituting this result into Eq. (13.16), we obtain the exact steady-state reduced density matrix:

$$\lim_{t\to\infty} \langle \boldsymbol{\xi}_f | \rho_S(t) | \boldsymbol{\xi}'_f \rangle = \lim_{t\to\infty} \int d\mu(\boldsymbol{\xi}_0) d\mu(\boldsymbol{\xi}'_0) \rho_S(\boldsymbol{\xi}_0, \boldsymbol{\xi}'_0, t_0) \times \left(\det[\mathbf{1} \pm \boldsymbol{v}(t,t)]\right)^{\mp 1} \exp\left\{ \pm \boldsymbol{\xi}'^{\dagger}_0 \boldsymbol{\xi}_0 + \boldsymbol{\xi}^{\dagger}_f \frac{\boldsymbol{v}(t,t)}{\mathbf{1} \pm \boldsymbol{v}(t,t)} \boldsymbol{\xi}'_f \right\}. \tag{13.132}$$

Notice the normalization condition

$$\int d\mu(\boldsymbol{\xi}_0) d\mu(\boldsymbol{\xi}'_0) \rho_S(\boldsymbol{\xi}_0, \boldsymbol{\xi}'_0, t_0) \exp\left\{ \pm \boldsymbol{\xi}'^{\dagger}_0 \boldsymbol{\xi}_0 \right\} = 1, \tag{13.133}$$

we have

$$\lim_{t\to\infty} \langle \boldsymbol{\xi}_f | \rho_S(t) | \boldsymbol{\xi}'_f \rangle = \lim_{t\to\infty} \left(\det[\mathbf{1} \pm \boldsymbol{v}(t,t)]\right)^{\mp 1} \langle \boldsymbol{\xi}_f | \frac{\boldsymbol{v}(t,t)}{\mathbf{1} \pm \boldsymbol{v}(t,t)} \boldsymbol{\xi}'_f \rangle. \tag{13.134}$$

This shows that as a consequence of thermalization, the steady-state reduced density matrix is independent of its initial states. Equation (13.134) can be rewritten in an operator form,

$$\lim_{t\to\infty} \rho_S(t) = \lim_{t\to\infty} \left(\frac{1}{\det[\mathbf{1} \pm \boldsymbol{v}(t,t)]} \right)^{\pm 1} \exp\left\{ \boldsymbol{a}^{\dagger} \left(\ln \frac{\boldsymbol{v}(t,t)}{\mathbf{1} \pm \boldsymbol{v}(t,t)} \right) \boldsymbol{a} \right\}, \tag{13.135}$$

where $\boldsymbol{a} \equiv (a_1, a_2, \cdots, a_N)^T$ is a one-column matrix operator. This is the exact steady-state solution of the exact master equation (13.52).

As one can see, the exact steady-state solution (13.135) is indeed a generalized Gibbs-type state. Here $\lim_{t\to\infty} \boldsymbol{v}(t,t)$ is the steady-state one-particle density matrix of Eq. (13.95). Its diagonal elements are the particle occupations (the particle statistical distributions) at each corresponding energy level:

$$\mathbf{lim}_{t\to\infty}\boldsymbol{v}_{ij}(t,t) = \boldsymbol{n}_{ij}(t) \equiv \mathrm{Tr_S}[a_i^\dagger a_j \rho_S(t)] = \boldsymbol{\rho}_{ij}^{(1)}(t). \tag{13.136}$$

Thus, the non-equilibrium internal energy, entropy, and particle number can be defined by

$$U_S(t) \equiv \mathrm{Tr_S}[H_S^r(t)\rho_S(t)] = \sum_{ij} \boldsymbol{\varepsilon}_{ij}^r(t)\boldsymbol{n}_{ij}(t), \tag{13.137a}$$

$$N_S(t) \equiv \mathrm{Tr_S} \sum_i [a_i^\dagger a_i \rho_S(t)] = \sum_i \boldsymbol{n}_{ii}(t), \tag{13.137b}$$

$$S_S(t) \equiv -k\,\mathrm{Tr_S}[\rho_S(t)\ln\rho_S(t)]. \tag{13.137c}$$

Here $H_S^r(t,t_0)$ is the renormalized system Hamiltonian obtained in the exact master equation (13.52). The reduced density matrix $\rho_S(t)$ of the system is determined by the master equation (13.52). Both of them contain all the information of the system under the influence of the environment. This provides the dynamical description of thermodynamics under the fundamental principle of quantum mechanics. Meanwhile, the quantum thermodynamic quantities defined by Eq. (13.137) may be related to each other and form the fundamental equation for quantum thermodynamics: $U_S(t) = U_S(\varepsilon_s^r(t), S_S(t), N_S(t))$. Physically, as we see, energy levels play a similar role as the volume in quantum mechanics [221]. Thus,

$$dU_S(t) = dW_S(t) + T^r(t)dS_S(t) + \mu^r(t)dN_S(t). \tag{13.138}$$

This is the first law of non-equilibrium quantum thermodynamics.

Explicitly, the quantum work $dW_S(t)$ done on the system has arisen from the changes of energy levels:

$$\frac{dW_S(t)}{dt} = \mathrm{Tr_S}\left[\rho_S(t)\frac{dH_S^r(t)}{dt}\right] = \sum_{ij} \boldsymbol{n}_{ij}(t)\frac{d\boldsymbol{\varepsilon}_{ij}^r(t)}{dt}. \tag{13.139}$$

The quantum heat $dQ_S(t)$ (also including the chemical work $dW_S^c(t)$) comes from the changes of particle distributions and transitions (the one-particle density matrix; see Eq. (13.136)):

$$dQ_S(t) + dW_S^c(t) = \sum_{ij} \boldsymbol{\varepsilon}_{ij}^r(t)d\boldsymbol{n}_{ij}(t) = T^r(t)dS_S(t) + \mu^r(t)dN_S(t). \tag{13.140}$$

It indicates that $d\boldsymbol{n}_{ij}(t)$ characterizes both the state information exchanges (entropy production) and the matter exchanges (chemical process for massive particles) between the systems and the reservoir. For photon or phonon systems, particle number is the number of energy quanta $\hbar\omega$. In this case, both are not a matter, and therefore their respective chemical potential as expected vanishes; $\mu^r(t) = 0$.

Based on the above formulation, we can now define the renormalized temperature and renormalized chemical potential by

$$T^r(t) = \frac{\partial U_{\rm S}(t)}{\partial S_{\rm S}(t)}\bigg|_{\boldsymbol{\varepsilon}^r(t), N_{\rm S}(t)}, \quad \mu^r(t) = \frac{\partial U_{\rm S}(t)}{\partial N_{\rm S}(t)}\bigg|_{\boldsymbol{\varepsilon}^r(t), S_{\rm S}(t)}. \tag{13.141}$$

The renormalization means that all the thermal quantities are renormalized under the influence of environment when the system and the environment strongly coupled to each other [210]. As a result, Eq. (13.135) can be also written as the standard Gibbs state:

$$\lim_{t\to\infty} \rho_{\rm S}^{\rm exact}(t) = \frac{1}{Z^r} \exp\left\{-\beta^r (H_{\rm S}^r - \mu^r \hat{N}_{\rm S})\right\}. \tag{13.142}$$

Here $\hat{N}_{\rm S} = \sum_i a_i^\dagger a_i$ is the particle number operator of the system. It shows that, by introducing the renormalized Hamiltonian $H_{\rm S}^r(t)$, the renormalized temperature $T^r(t)$, and the renormalized chemical potential $\mu^r(t)$ at steady state, we derive the quantum thermodynamic Gibbs state, Eq. (13.142), for all the coupling strengths between the system and the environment. Because the exact solution of the steady state is the standard Gibbs state, classical thermodynamic laws are all preserved in quantum thermodynamics. In this manner, the theory of quantum thermodynamics is complete.

The above formulation of quantum thermodynamics has gone far beyond the classical thermodynamics based on equilibrium hypothesis, namely, the system is very weakly coupled with its environment. We can reproduce the classical thermodynamics under the weak coupling limit. Explicitly, when the coupling strength between the system and the environment is *very weak*, then the spectral density $\boldsymbol{J}(\epsilon)$ and the energy shift $\boldsymbol{\Delta}(\epsilon)$ both tend to vanish, i.e.,

$$\boldsymbol{J}(\epsilon) \to \boldsymbol{0}, \qquad \boldsymbol{\Delta}(\epsilon) \to \boldsymbol{0}. \tag{13.143}$$

Following, the spectrum of the retarded Green function is given by

$$\boldsymbol{D}(\epsilon) = \frac{1}{\epsilon - \boldsymbol{\epsilon}_{\rm S} - \boldsymbol{\Delta}(\epsilon) + i\boldsymbol{J}(\epsilon)/2}\boldsymbol{J}(\epsilon)\frac{1}{\epsilon - \boldsymbol{\epsilon}_{\rm S} - \boldsymbol{\Delta}(\epsilon) - i\boldsymbol{J}(\epsilon)/2}, \tag{13.144}$$

One can find that under the condition (13.143),

$$\boldsymbol{D}(\epsilon) \to 2\pi\delta(\epsilon \boldsymbol{I} - \boldsymbol{\epsilon}_{\rm S}). \tag{13.145}$$

That is, when the system-environment coupling becomes very weak, the spectrum broadening and energy shift of the system energy levels can be negligible, making the spectrum of the system converging to the original one. As a consequence of Eqs. (13.145) and (13.129b), $\boldsymbol{v}(t,t)$ approaches the Bose-Einstein/Fermi-Dirac distribution, i.e.,

$$\lim_{t\to\infty} \boldsymbol{v}(t,t) \to f(\boldsymbol{\epsilon}_S, T_0) = \frac{1}{e^{(\boldsymbol{\epsilon}_S-\mu)/k_B T_0} \mp 1}. \tag{13.146}$$

Thus, Eq. (13.142) converges to

$$\lim_{t\to\infty} \rho_S(t) = \frac{1}{\mathrm{Tr}[e^{-\beta_0 \boldsymbol{a}^\dagger(\boldsymbol{\epsilon}_S-\mu)\boldsymbol{a}}]} \exp\big\{-\beta_0 \boldsymbol{a}^\dagger(\boldsymbol{\epsilon}_S-\mu)\boldsymbol{a}\big\}. \tag{13.147}$$

This is exactly the Gibbs state in the grand canonical ensemble of classical thermodynamics with the original system Hamiltonian $H_S = \boldsymbol{a}^\dagger \boldsymbol{\epsilon}_S \boldsymbol{a}$ at the initial inverse temperature $\beta_0 = 1/k_B T_0$ and initial chemical potential μ_0 of the environment. This is a rigorous proof that in the weak coupling limit, the exact evolution of an open quantum system would reproduce the Gibbs thermal state of the equilibrium statistical mechanics in the steady-state limit. All the classical thermodynamic quantities can be determined from the Gibbs state.

Finally, we will apply the above general theory of quantum thermodynamics to a bosonic system, such as a nanocavity or a nanomechanical resonator, $H_S = \hbar\omega_S a^\dagger a$, which can strongly couple to a thermal bath. The bath is described by the Ohmic spectral density $J(\omega) = 2\pi\eta\omega\exp(-\omega/\omega_c)$. The exact master equation (13.52) is simply reduced to

$$\frac{d}{dt}\rho_S(t) = \frac{1}{i\hbar}\big[H_S^r(t), \rho_S(t)\big] + \gamma(t,t_0)\big\{2a\rho_S(t)a^\dagger - a^\dagger a\rho_S(t) - \rho_S(t)a^\dagger a\big\}$$

$$+\widetilde{\gamma}(t,t_0)\big\{a^\dagger\rho_S(t)a + a\rho_S(t)a^\dagger - a^\dagger a\rho_S(t) - \rho_S(t)aa^\dagger\big\}. \tag{13.148}$$

with the renormalized Hamiltonian

$$H_S^r(t) = \hbar\omega_S^r(t,t_0)a^\dagger a. \tag{13.149}$$

The reduced density matrix can also be exactly solved from Eq. (13.48) in the coherent state representation which is reduced to

$$\langle\xi_f|\rho_S(t)|\xi_f'\rangle = \big(\det[1 \pm v(t,t)]\big)^{\mp 1}\int d\mu(\xi_0)d\mu(\xi_0')\langle\xi_0|\rho_S(t_0)|\xi_0'\rangle \tag{13.150}$$

$$\exp\big\{\xi_f^* K_1(t,t_0)\xi_0 + \xi_0'^* K_1^*(t,t_0)\xi_f' \pm \xi_0'^* K_3(t,t_0)\xi_0 + \xi_f^* K_2(t)\xi_f'\big\}.$$

for a single-mode bosonic system. All the coefficients in the master equation (13.148) and in the reduced density matrix (13.150) are determined by the non-equilibrium Green functions $u(t, t_0)$ and $v(t, t)$:

$$\omega_S^r(t, t_0) = -\mathrm{Im}\Big[\frac{\dot{u}(t, t_0)}{u(t, t_0)}\Big], \ \gamma(t, t_0) = -\mathrm{Re}\Big[\frac{\dot{u}(t, t_0)}{u(t, t_0)}\Big], \tag{13.151a}$$

$$\widetilde{\gamma}(t, t_0) = \dot{v}(t, t) - 2v(t, t)\mathrm{Re}\Big[\frac{\dot{u}(t, t_0)}{u(t, t_0)}\Big]. \tag{13.151b}$$

$$K_1(t) = \frac{u(t, t_0)}{1 + v(t, t)}, \ K_2(t) = \frac{v(t, t)}{1 + v(t, t)}, \ K_3(t) = 1 - \frac{|u(t, t_0)|^2}{1 + v(t, t)}. \tag{13.151c}$$

And $u(t, t_0)$ and $v(t, t)$ have been solved analytically in Sect. 13.5.

Assume that the system is initially in an arbitrary state

$$\rho_S(t_0) = \sum_{l,m=0}^{\infty} \rho_{lm} |l\rangle\langle m| \tag{13.152}$$

which can be either a pure state if $\rho_{lm} = c_l c_m^*$ or a mixed state if $\rho_{lm} \neq c_l c_m^*$, where c_l is a complex number. The time evolution of the reduced density matrix can be solved either from the exact master equation (13.148) or from the exact solution (13.150) in the coherent state representation [208]. The result is

$$\rho_S(t) = \sum_{l,m=0}^{\infty} \rho_{lm} \sum_{k=0}^{\min\{l,m\}} d_k A_{lk}^{+}(t)\widetilde{\rho}[v(t, t)]A_{mk}(t) \tag{13.153}$$

where

$$\widetilde{\rho}[v(t, t)] = \sum_{n=0}^{\infty} \frac{[v(t, t)]^n}{[1 + v(t, t)]^n} |n\rangle\langle n|, \tag{13.154a}$$

$$A_{lk}^{\dagger}(t) = \frac{\sqrt{l!}}{(l-k)!\sqrt{k!}}\Big[\frac{u(t, t_0)}{1 + v(t, t)} a^{\dagger}\Big]^{l-k}, \tag{13.154b}$$

$$d_k = \Big[1 - \frac{|u(t, t_0)|^2}{1 + v(t, t)}\Big]^k. \tag{13.154c}$$

It can be shown that when $\eta < \eta_c = \omega_S/\omega_c$, the localized mode does not exist [198]. When the system reaches the steady state, we have $\lim_{t\to\infty} u(t, t_0) \to 0$ (see Fig. 13.4). As a result,

$$\lim_{t\to\infty} \rho_S(t) = \lim_{t\to\infty} \sum_{n=0}^{\infty} \frac{[v(t, t)]^n}{[1 + v(t, t)]^{n+1}} |n\rangle\langle n| \tag{13.155}$$

$$= \lim_{t\to\infty} \frac{1}{1 + v(t, t)} e^{\ln\left[\frac{v(t,t)}{1+v(t,t)}\right] a^{\dagger} a},$$

Equation (13.155) is the exact solution at steady state for arbitrary system-reservoir coupling strengths $\eta < \eta_c$.

To seek the new physics in the quantum thermodynamics, we shall look at the average particle occupation in the system which can be calculated exactly either from the master equation (13.148), $\overline{n}(t) \equiv \mathrm{Tr}_{\mathrm{S}}[a^\dagger a \rho_{\mathrm{S}}(t)]$, or from the Heisenberg equation of motion directly: $\overline{n}(t) \equiv \mathrm{Tr}_{\mathrm{SE}}[a^\dagger(t) a(t) \rho_{\mathrm{tot}}(t_0)]$. Both calculations give the same result:

$$\overline{n}(t) = u^*(t, t_0)\overline{n}(t_0)u(t.t_0) + v(t, t). \tag{13.156}$$

Its steady-state solution is given by

$$\lim_{t\to\infty} \overline{n}(t) = \lim_{t\to\infty} v(t, t) = \int \frac{d\omega}{2\pi} D(\omega)\overline{n}(\omega, T_0), \tag{13.157}$$

where $D(\omega) = \frac{J(\omega)}{[\omega-\omega_s-\Delta(\omega)]^2+J^2(\omega)/4}$ describing the system spectrum broadening and $\Delta(\omega) = \mathcal{P}\big[\int d\omega' \frac{J(\omega')}{\omega-\omega'}\big]$ gives the system frequency shift.

Figure 13.5a (the red dashed line) is the exact solution $\overline{n}(t \to \infty)$ of Eq. (13.157) as a function of the coupling strength η for the Ohmic spectral density. As one can see, except for the very weak coupling regime $\eta \ll \eta_c$, $\overline{n}(t \to \infty)$ deviates significantly from the Bose-Einstein distribution $\overline{n}(\omega_{\mathrm{S}}, T_0)$ (the green dot line) as η increases. Physically, this deviation comes from the strong system-reservoir coupling which changes the intrinsic thermal property of the system when we go beyond the classical thermodynamics. Note that the classical thermodynamics is valid when the system is very weakly coupled to the bath, as we have discussed in the beginning of this section.

The first change we observed is induced by the strong system-reservoir coupling which is the system energy, which is shifted as we have shown explicitly in the exact master equation (13.148). In other words, the Hamiltonian of the system is renormalized from H_{S} to H_{S}^r with the energy $\hbar\omega_{\mathrm{S}}$ being shifted to $\hbar\omega_{\mathrm{S}}^r$ when we trace over all the environmental state. Due to the non-negligible system-reservoir coupling effect, the renormalized frequency $\omega_{\mathrm{S}}^r = \omega_{\mathrm{S}}^r(t \to \infty)$ can be calculated exactly from Eq. (13.151), and the result is presented in Fig. 13.5b. The blue dashed-dot line in Fig. 13.5a is the particle occupation with the renormalized energy: $\overline{n}(\omega_{\mathrm{S}}^r, T_0) = 1/[e^{\hbar\omega_{\mathrm{S}}^r/k_B T_0} - 1]$. It shows that $\overline{n}(\omega_{\mathrm{S}}^r, T_0)$ changes with increasing η, similar to the exact solution $\overline{n}(t \to \infty)$. Of course, it must be noted that there is still obvious difference between them.

Secondly, the non-negligible system-reservoir coupling effect also changes the solution of the reduced density matrix. This is shown by Eqs. (13.155) and (13.157) which are all related to the spectral broadening $D(\omega)$ and the coupling strength η explicitly. The change of the energy implies the change of the system temperature. In other words, the system and the reservoir in the steady state have a different temperature other than the initial temperature of the reservoir T_0. This new temperature characterizes the equilibrium of the system and the reservoir at the steady state. We

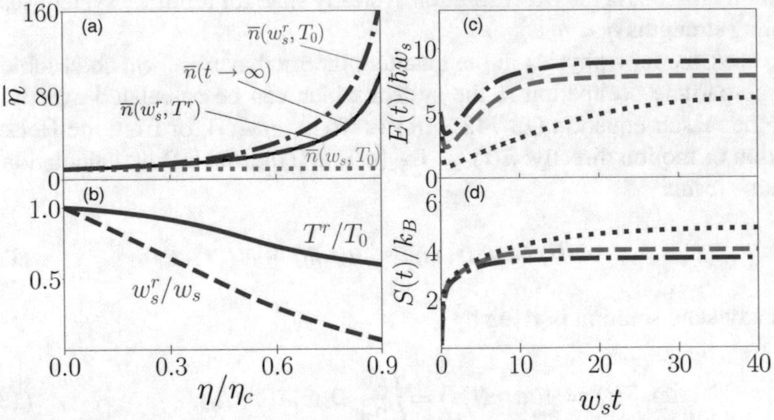

Fig. 13.5 (**a**) The steady-state particle distribution as a function of coupling strength η/η_c. The exact solution $\overline{n}(t \to \infty)$ of Eq. (13.157) (red dashed line) is identical to $\overline{n}(\omega_s^r, T^r)$ (black dot line). The green dot line and blue dashed-dot line are $\overline{n}(\omega_s, T_0)$ and $\overline{n}(\omega_s^r, T_0)$, respectively. (**b**) The steady-state values of the renormalized frequency and renormalized temperature as a function of the coupling. (**c**)–(**d**) The internal energy and entropy production in time for different coupling strengths $\eta/\eta_c = 0.3, 0.5, 0.8$ (the blue solid, green dashed, red dot lines). Other parameters: $\omega_c = 5\omega_s$, $T_0 = 10\hbar\omega_s$, $\eta_c = \omega_s/\omega_c$ [210]

shall refer to it as the renormalized temperature T^r. It can be determined by the change of the internal energy with respect to the thermal entropy of the system. The internal energy and the entropy can both be calculated from the renormalized Hamiltonian in the exact master equation (13.148), and its solution (13.153), as defined by Eq. (13.137) is

$$U_S(t) \equiv \mathrm{Tr}_S[H_S^r(t)\rho_S(t)] = \omega_S^r(t)\overline{n}(t), \quad S_S(t) = -k_B \,\mathrm{Tr}_S[\rho_S(t)\ln\rho_S(t)]. \tag{13.158}$$

Thus, a renormalized temperature is given by Eq. (13.141):

$$T^r(t) = \left.\frac{\partial U_S(t)}{\partial S_S(t)}\right|_{\omega_S^r}. \tag{13.159}$$

The chemical potential is zero for the massless bosons, as expected.

Furthermore, the change of the internal energy in time has two parts, corresponding to quantum work done on the system and the quantum heat associated with the entropy production:

$$\begin{aligned} dU_S(t) &= \mathrm{Tr}_S[\rho_S(t)dH_S^r(t)] + \mathrm{Tr}_S[H_S^r(t)d\rho_S(t)] \\ &= dW(t) + dQ(t) = dW(t) + T^r(t)dS_S(t). \end{aligned} \tag{13.160}$$

Because the exact steady-state solution Eq. (13.155) is a Gibbs state, it can be also expressed as

$$\rho_S = \sum_{n=0}^{\infty} \frac{[\overline{n}(\omega_S^r, T^r)]^n}{[1+\overline{n}(\omega_S^r, T^r)]^{n+1}} |n\rangle\langle n| = \frac{1}{Z^r} e^{-\beta^r H_S^r}, \tag{13.161}$$

where $\overline{n}(\omega_S^r, T^r) = \frac{1}{e^{\hbar\omega_S^r/kT^r}-1}$ is the Bose-Einstein distribution and $Z^r = \mathrm{Tr}_S[e^{-\beta^r H_S^r}]$ with $\beta^r = 1/kT^r$ and $T^r = T^r(t \to \infty)$ is the renormalized equilibrium temperature at steady state (see Fig. 13.5b). We plot $\overline{n}(\omega_S^r, T^r)$ with the renormalized energy and temperature (the black dot line) in Fig. 13.5a. Remarkably, it precisely reproduces the exact solution Eq. (13.157), i.e., $\overline{n}(t \to \infty) = \overline{n}(\omega_S^r, T^r)$. This is a most rigorous proof that strong coupling quantum thermodynamics must be defined in terms of the renormalized Hamiltonian and temperature given above.

Furthermore, in the very weak coupling regime $\eta \ll \eta_c$, we have $\Delta(\omega) \to 0$ and $D(\omega) \to \delta(\omega - \omega_S)$ so that in the steady state, Eq. (13.157) is directly reduced to $\overline{n} \to \overline{n}(\omega_S, T_0)$, and

$$\rho_S = \sum_{n=0}^{\infty} \frac{[\overline{n}(\omega_S, T_0)]^n}{[1+\overline{n}(\omega_S, T_0)]^{n+1}} |n\rangle\langle n| = \frac{1}{Z} e^{-\beta_0 H_S}. \tag{13.162}$$

It recovers the expected solution in the weak coupling regime. Figure 13.5 also shows that $\hbar\omega_S^r \to \hbar\omega_S$ and $T^r \to T_0$ at very weak coupling. Thus, the equilibrium hypothesis of thermodynamics and statistical mechanics is proven rigorously from the dynamics of quantum systems, which solves the long-standing problem lasted for more than hundred years for the foundation of statistical mechanics and thermodynamics [220]. On the other hand, for the very strong coupling $\eta > \eta_c$, the system exists a dissipationless localized bound state (localized mode) at frequency $\omega_b = \omega_s + \Delta(\omega_b)$ with $J(\omega_b) = 0$. Then, the asymptotic value of the Green function $u(t \to \infty, t_0)$ never vanishes. As a result, the steady state of the reduced density matrix Eq. (13.153) cannot be reduced to Eq. (13.155). It always depends on the initial state distribution ρ_{lm} of Eq. (13.152). In other words, the system cannot be thermalized with the reservoir [210], [?]. This corresponds to a realization of the thermalization to many-body localization transition that is currently a hot topic in research.

The theory of open quantum systems and quantum thermodynamics is an ongoing research field. The results presented in this chapter are solved from the fundamental principle of quantum mechanics by extending the system to include environment as a closed system. When the system involves many-body interaction, its dynamics should become much more complicated. With the rapid development of quantum science and technology which mainly deals with the dynamics of open quantum systems, let us witness the new progress in this emerging research field in the near future.

Exercises

13.1. Apply the coherent states path integral (13.11) to the single-mode harmonic oscillator to show that $\langle z'|e^{-\frac{i}{\hbar}H(t-t_0)}|z\rangle$ is fully determined by the end-point boundary term in Eq. (13.11).

13.2. Derive Eq. (13.34) from Eq. (13.29).

13.3. Taking a time derivative to Eq. (13.48) to derive the exact master equation (13.52). (hint: use the D-algebra of the creation and annihilation operators in the coherent state representation).

13.4. Calculate the dissipation and fluctuation coefficients in the second-order perturbation expansion for the sub-Ohmic spectral density $s = 1/2$, and make a comparison with the exact solution given in Fig. 13.4.

13.5. Consider a quantum dot with two energy levels (serves as a charge qubit) coupled to two leads, the source and the drain at initial temperatures $T_{L,R}$ and chemical potential $\mu_{L,R}$, and apply a bias voltage $V(t)$ to the two leads to solve the exact master equation (13.89) to see how to use the bias $V(t)$ to control the qubit states. Also explore the heating transfer in this quantum device.

13.6. Consider a nano-resonator interacting with a thermal bath and initially in a coherent state $|\alpha\rangle$, solve the master equation (13.148) to find the exact solution of the reduced density matrix, and study its thermalization (the steady-state limit).

Introduction of Lie Group

A

A.1 Lie Group and Lie Algebra

What are Lie groups and Lie algebras? To illustrate them, one needs to start by introducing the concept of a group. A **group** is a set G equipped with a binary operation $f : G \times G \to G$ which obeys the four axioms: (1) If $a, b \in G$, $f(a, b) \in G$ (closure); (2) If a, b and $c \in G$, $f(f(a, b), c) = f(a, f(b, c))$ (associativity); (3) there is an identity element e, such that for all $a \in G$, $f(e, a) = f(a, e) = g$ (identity); (4) every element $a \in G$ has an inverse $a^{-1} \in G$ such that $f(a, a^{-1}) = f(a^{-1}, a) = e$ (invertibility).

As an example, one may consider the symmetries of a square, which is a set with four elements: $\{R_0, R_{\pi/2}, R_\pi, R_{3\pi/2}\}$, where R_θ is an anticlockwise rotation of the square by θ degree. Let the binary operation $f(a, b)$ be first applying the anticlockwise rotation a and then applying the anticlockwise rotation b; one can immediately show that the set $\{R_0, R_{\pi/2}, R_\pi, R_{3\pi/2}\}$ is a group with the following group Table A.1:

In linear algebra, one represents linear transformations in vector spaces via matrices. Similarly, one may represent symmetrical transformations of a group as linear operators in vector spaces. Here, a **representation** of a group G is mapping $g : G \to V$ onto a vector space V, which obeys the following properties: (1) $g(e) = I$, with I being the identity operator of the vector space; (2) $g(a)g(b) = g(f(a, b))$ for $a, b \in G$ and f being the binary operation for G.

As an example, the representation of the group of rotation symmetries of a square onto the vector space $\mathbb{R}^2$ has the form

$$g(R_0) = \begin{pmatrix} 1 & 0 \\ 0 & 1 \end{pmatrix}, g(R_{\pi/2}) = \begin{pmatrix} 0 & -1 \\ 1 & 0 \end{pmatrix}, \tag{A.1}$$

$$g(R_\pi) = \begin{pmatrix} -1 & 0 \\ 0 & -1 \end{pmatrix}, g(R_{3\pi/2}) = \begin{pmatrix} 0 & 1 \\ -1 & 0 \end{pmatrix}.$$

C.-F. Kam et al., *Coherent States*, Lecture Notes in Physics 1011,
https://doi.org/10.1007/978-3-031-20766-2

Table A.1 Group table of the rotation symmetries of a square

	R_0	$R_{\pi/2}$	R_π	$R_{3\pi/2}$
R_0	R_0	$R_{\pi/2}$	R_π	$R_{3\pi/2}$
$R_{\pi/2}$	$R_{\pi/2}$	R_π	$R_{3\pi/2}$	R_0
R_π	R_π	$R_{3\pi/2}$	R_0	$R_{\pi/2}$
$R_{3\pi/2}$	$R_{3\pi/2}$	R_0	$R_{\pi/2}$	R_π

We now introduce the concept of algebra. An **algebra** V over a field is a vector space equipped with a bilinear product $f : V \times V \to V$, where a **field** is a set equipped with the operations of addition, subtraction, multiplication, and division. The most commonly used fields are the fields of real numbers and complex numbers. Notice that the field of complex numbers can also be regarded as an algebra over the vector space $\mathbb{R}^2$, where the bilinear product is given by the product of complex numbers $(a + ib) \cdot (c + id)$.

In our previous example, the group of rotation symmetries of a square contains finite elements. However, as most important symmetries in physics are continuous symmetries, one needs to consider groups of continuous symmetries, whose elements are continuous or smooth functions of some variables.

A **Lie group** is set G with two structures: $G = (G, \cdot)$ is a group and G is a smooth manifold, such that the multiplication $\cdot : G \times G \to G, (a, b) \mapsto ab$ and inversion $()^{-1} : G \to G, a \mapsto a^{-1}$ are smooth maps. If one replaces the word smooth manifold with topological space equipped with continuous multiplication and inversion in the definition of a Lie group, one obtains a topological group. Remarkably, the requirement of smoothness in the definition of a Lie group is redundant. A highly nontrivial result (Gleason-Yanabe theorem [222], a positive resolution to Hilbert's fifth problem) asserts that if G is a topological group and is locally Euclidean, then G is isomorphic to a Lie group.

As an example, the symmetry of a unit circle is that its shape is invariant under rotations by any angle θ. One may represent these rotations in $\mathbb{R}^2$ by the following 2×2 matrix:

$$R(\theta) \equiv \begin{pmatrix} \cos\theta & -\sin\theta \\ \sin\theta & \cos\theta \end{pmatrix}, \tag{A.2}$$

where $\theta \in \mathbb{R}/2\pi\mathbb{Z}$. It describes a compact connected Lie group, the special orthogonal group SO(2,$\mathbb{R}$), which is diffeomorphic to a unit circle. The multiplication of elements of SO(2,$\mathbb{R}$) corresponds to addition of the angles, i.e., $R(\theta_1)R(\theta_2) = R(\theta_1 + \theta_2)$, and the inverse of an element corresponds to taking the opposite angle, i.e., $R^{-1}(\theta) = R^{\mathsf{T}}(\theta) = R(-\theta)$. As both the multiplication and inversion are smooth maps, SO(2,$\mathbb{R}$) is a Lie group, containing both the group and the smooth manifold structures.

As another example, the isospin symmetry in particle physics is described by the special unitary group of order 2, SU(2, $\mathbb{C}$), which is the group of 2×2 unitary matrices with complex entries. Here, a unitary matrix means a matrix with determinant 1 in which its conjugate transpose equals its inverse. The SU(2, ℂ) group can be written explicitly as

$$\mathrm{SU}(2,\mathbb{C}) \equiv \left\{ \begin{pmatrix} \alpha & -\beta^* \\ \beta & \alpha^* \end{pmatrix} \,|\, \alpha, \beta \in \mathbb{C}, |\alpha|^2 + |\beta|^2 = 1 \right\}. \tag{A.3}$$

The verification of SU(2, ℂ) being a group is left to readers. Let $(\alpha, \beta) \mapsto (a + bi, -c + di)$ be a bijection from $\mathbb{C}^2$ to $\mathbb{R}^4$. Then the equation $|\alpha|^2 + |\beta|^2 = 1$ becomes $a^2+b^2+c^2+d^2 = 1$. This is exactly the equation of the three-dimensional unit sphere S^3. As SU(2, ℂ) has both the group structure and the smooth manifold structure (the unit sphere S^3), it is a Lie group.

In fact, the **general linear group** $\mathrm{GL}(n, \mathbb{R})$, the group of all invertible $n \times n$ matrices with real entries, forms a Lie group. It forms a group, as the product of two invertible matrices is invertible, the inverse of an invertible matrix is invertible, and the identity matrix represents the identity element of the group. To show that the general linear group is a smooth manifold, one needs a fact: the space of $n \times n$ matrices $M_n(\mathbb{R})$ is isomorphic to $\mathbb{R}^{n^2}$. It can be directly verified by assigning a bijection from a given $n \times n$ matrix $A = (a_{ij})$ to the n^2-tuple

$$(a_{11}, a_{12}, \ldots, a_{1n}, a_{21}, a_{22}, \ldots, a_{2n}, \ldots, a_{n1}, a_{n2}, \ldots, a_{nn}). \tag{A.4}$$

This bijection naturally induces a metric $d(A, B) \equiv \sqrt{\sum_{i,j}(a_{ij} - b_{ij})^2}$ on $M_n(\mathbb{R})$, which makes $M_n(\mathbb{R})$ a topological manifold, a topological space which locally resembles the Euclidean space. By definition, the general linear group $\mathrm{GL}(n, \mathbb{R}) \equiv \{A \in M_n(\mathbb{R}) | \det A \neq 0\}$ consists of all the $n \times n$ matrices with nonzero determinants. The determinant function $\det A \equiv \sum_{\sigma \in S_n} \mathrm{sgn}(\sigma) \prod_{i=1}^{n} a_{i,\sigma(i)}$ is a sum involving all permutations of the set $\{1, 2, \ldots, n\}$ multiplied by the signature of the permutation, which is a homogeneous polynomial in the entries of the matrix. Hence, the determinant function $\det : M_n(\mathbb{R}) \to \mathbb{R}$ is a smooth function. Since $\mathbb{R}\backslash\{0\}$ is an open subset of $\mathbb{R}$, the general linear group $\mathrm{GL}(n, \mathbb{R}) \equiv \det^{-1}(\mathbb{R}\backslash\{0\})$ is also an open subset of $M_n(\mathbb{R})$, which makes it a smooth manifold, as any open subset of a manifold is a manifold. Finally, since $\mathrm{GL}(n, \mathbb{R})$ has both the group structure and smooth manifold structure, it is by definition a Lie group.

Using the same argument, it can be shown that both the general linear group over the complex numbers $\mathrm{GL}(n, \mathbb{C})$ and the general linear group over the quaternions $\mathrm{GL}(n, \mathbb{H})$ are Lie groups. As real Lie groups, the dimensions of $\mathrm{GL}(n, \mathbb{R})$, $\mathrm{GL}(n, \mathbb{C})$, and $\mathrm{GL}(n, \mathbb{H})$ are n^2, $2n^2$, and $4n^2$, respectively. In particular, $\mathrm{GL}(n, \mathbb{R})$ as a real manifold is not connected but rather consists of two connected components: the matrices with positive determinants and the matrices with negative determinants. One may denote the identity component as $\mathrm{GL}^+(n, \mathbb{R})$ and the other component as $\mathrm{GL}^-(n, \mathbb{R})$, respectively. $\mathrm{GL}^+(n, \mathbb{R})$ is also a Lie group of dimension n^2.

We now focus on the **orthogonal group** in an arbitrary dimension n, denoted as $\mathrm{O}(n, \mathbb{R})$, which is the group of distance preserving transformations of a n-

dimensional Euclidean space described by

$$\mathrm{O}(n, \mathbb{R}) \equiv \{R \in \mathrm{GL}(n, \mathbb{R}) \mid R^{\mathsf{T}} R = I\}. \tag{A.5}$$

It can be verified that the set $\mathrm{O}(n, \mathbb{R})$ forms a group under matrix multiplication. In order to prove that $\mathrm{O}(n, \mathbb{R})$ has a smooth manifold structure, one may construct a continuous map $f : M_n(\mathbb{R}) \to S_n(\mathbb{R})$, $R \mapsto R^{\mathsf{T}} R$, from the space of $n \times n$ matrices to its symmetric subspace. By construction, the orthogonal group is the preimage of a single element $\{I\}$ in the symmetric subspace $S_n(\mathbb{R})$, $\mathrm{O}(n, \mathbb{R}) = f^{-1}(I)$. As f is continuous, the inverse of a single element $\{I\}$, which is obviously closed in $S_n(\mathbb{R})$, is a closed subset in $M_n(\mathbb{R})$. According to the preimage theorem, it suffices to prove that $\{I\}$ is a regular value of f, i.e., the differential map $df_R : M_n(\mathbb{R}) \to S_n(\mathbb{R})$ is surjective for all orthogonal matrices $R \in f^{-1}(I)$. Let $X \in M_n(\mathbb{R})$ be an arbitrary $n \times n$ matrix; one obtains

$$\begin{aligned} df_R(X) &= \lim_{s \to 0} \frac{f(R + sX) - f(R)}{s} \\ &= \lim_{s \to 0} \frac{(R + sX)^{\mathsf{T}}(R + sX) - R^{\mathsf{T}} R}{s} = X^{\mathsf{T}} R + R^{\mathsf{T}} X. \end{aligned} \tag{A.6}$$

This is a surjective map. Since if $Y \in S_n(\mathbb{R})$ is an arbitrary symmetric $n \times n$ matrix, one may always choose $X = \frac{1}{2} RY$, so that

$$df_R(\frac{1}{2} RY) = \frac{1}{2}(Y^{\mathsf{T}} R^{\mathsf{T}} R + R^{\mathsf{T}} RY) = Y. \tag{A.7}$$

Hence, according to the preimage theorem, the orthogonal group $\mathrm{O}(n, \mathbb{R})$ is a real sub-manifold of $M_n(\mathbb{R})$ of dimension $N = \dim M_n(\mathbb{R}) - \dim S_n(\mathbb{R})$. As $\dim M_n(\mathbb{R}) = n^2$ and $\dim S_n(\mathbb{R}) = n(n+1)/2$, the orthogonal group $\mathrm{O}(n, \mathbb{R})$ is a real sub-manifold of $M_n(\mathbb{R})$ of dimension $n(n-1)/2$. Since the orthogonal group $\mathrm{O}(n, \mathbb{R})$ has both the group structure and the smooth manifold structure, it is a Lie group.

Similarly, one may consider the **unitary group** in an arbitrary dimension n, denoted as $\mathrm{U}(n, \mathbb{C})$, which is the group of transformations preserving the length of complex vectors described by

$$\mathrm{U}(n, \mathbb{C}) \equiv \{U \in \mathrm{GL}(n, \mathbb{C}) | U^{\dagger} U = I\}. \tag{A.8}$$

As the product of two unitary matrices is a unitary matrix, the inverse of a unitary matrix is another unitary matrix, and the identity matrix is unitary, the set of $n \times n$ unitary matrices forms a group. By taking the continuous map $f : M_n(\mathbb{C}) \to H_n(\mathbb{C})$, $U \mapsto U^{\dagger} U$, from the space of $n \times n$ matrices with complex entries to its Hermitian subspace, one may show that $\{I\}$ is a regular value of f. Hence, the unitary group $\mathrm{U}(n, \mathbb{C})$, as a preimage of $\{I\}$, is a real sub-manifold of $M_n(\mathbb{C})$ of dimension $N = \dim M_n(\mathbb{C}) - \dim H_n(\mathbb{C})$. As the space of $n \times n$ matrices with complex entries has real dimension $2n^2$, and the space of $n \times n$ Hermitian matrices has real dimension n^2, the unitary group $\mathrm{U}(n, \mathbb{C})$ is a Lie group of dimension n^2.

It is understood that a Lie group is a group which has a smooth manifold structure. In order to study the Lie group in more details, we turn our attention to the concept of a **Lie algebra** and their associated structures. Historically, the Lie algebra was introduced by the Norwegian mathematician Marius Sophus Lie in the 1870s to study the infinitesimal generators of the continuous transformation groups, in which the original group structure is encoded in the commutator bracket.

Being a vector space rather than a manifold, Lie algebra is a simpler object than Lie group: the former one has a linear structure, which admits a global coordinate chart, while the latter cannot be covered by a single coordinate chart, and does not admit a linear structure. Nevertheless, as the Lie algebra still encodes the infinitesimal structure of a Lie group, one may reconstruct the Lie group at the local level by using the exponential map and the Baker-Campbell-Hausdorff formula. As such, the local structure of a Lie group is completely described by its Lie algebra.

To be more precise, a Lie algebra $\mathfrak{g}$ is the tangent space of a Lie group G at the identity, which includes an operation $[,] : \mathfrak{g} \times \mathfrak{g} \to \mathfrak{g} : (X, Y) \mapsto [X, Y]$, called the **Lie bracket**, an alternating bilinear map which satisfies the **Jacobi identity**. In other words, the Lie bracket obeys the following conditions:

$$[\lambda_1 X_1 + \lambda_2 X_2, Y] = \lambda_1 [X_1, Y] + \lambda_2 [X_2, Y], \tag{A.9a}$$

$$[X, Y] = -[Y, X], \tag{A.9b}$$

$$[X, [Y, Z]] + [Y, [Z, X]] + [Z, [X, Y]] = 0, \tag{A.9c}$$

for arbitrary $X, Y, Z, X_1, X_2 \in \mathfrak{g}$ and $\lambda_1, \lambda_2 \in \mathbb{C}$.

As a first example of Lie algebra, one may consider the tangent space of the orthogonal group $\mathrm{SO}(3, \mathbb{R})$ at the identity element, denoted as $\mathfrak{so}(3, \mathbb{R})$, which is a vector space spanned by the basis

$$L_x = \begin{pmatrix} 0 & 0 & 0 \\ 0 & 0 & -1 \\ 0 & 1 & 0 \end{pmatrix}, L_y = \begin{pmatrix} 0 & 0 & 1 \\ 0 & 0 & 0 \\ -1 & 0 & 0 \end{pmatrix}, L_z = \begin{pmatrix} 0 & -1 & 0 \\ 1 & 0 & 0 \\ 0 & 0 & 0 \end{pmatrix}. \tag{A.10}$$

Then, $\mathfrak{so}(3, \mathbb{R})$ forms a Lie algebra, where the Lie bracket of arbitrary two elements $X, Y \in \mathfrak{so}(3, \mathbb{R})$ is given by the **commutator** $[X, Y] \equiv XY - YX$. The commutation relations of the basis elements are

$$[L_x, L_y] = L_z, [L_z, L_x] = L_y, [L_y, L_z] = L_x. \tag{A.11}$$

As arbitrary two elements $X, Y \in \mathfrak{so}(3, \mathbb{R})$ can be written as

$$X = \begin{pmatrix} 0 & -x_3 & x_2 \\ x_3 & 0 & -x_1 \\ -x_2 & x_1 & 0 \end{pmatrix}, Y = \begin{pmatrix} 0 & -y_3 & y_2 \\ y_3 & 0 & -y_1 \\ -y_2 & y_1 & 0 \end{pmatrix}, \tag{A.12}$$

a direction computation yields

$$[X, Y] = (x_2y_3 - x_3y_2)L_x + (x_3y_1 - x_1y_3)L_y + (x_1y_2 - x_2y_1)L_z. \tag{A.13}$$

It shows that the Lie bracket of $\mathfrak{so}(3, \mathbb{R})$ corresponds to the cross product in $\mathbb{R}^3$. In other words, the Lie algebra $\mathfrak{so}(3, \mathbb{R})$ along with the matrix commutator is isomorphic to the Lie algebra $\mathbb{R}^3$ along with the cross product.

As another example, one may consider the **general linear Lie algebra** $\mathfrak{gl}(n, \mathbb{R})$ of the general linear group $\mathrm{GL}(n, \mathbb{R})$. As the general linear group $\mathrm{GL}(n, \mathbb{R})$ is an open sub-manifold of the space of $n \times n$ matrices $M_n(\mathbb{R})$, one may identify the tangent space of $\mathrm{GL}(n, \mathbb{R})$ at any invertible matrix A with the tangent space of $M_n(\mathbb{R})$ at the matrix A. Moreover, since $M_n(\mathbb{R})$ is isomorphic to the Euclidean space $\mathbb{R}^{n^2}$, one may identify $M_n(\mathbb{R})$ with its tangent space at the matrix A. In particular, one may identify $M_n(\mathbb{R})$ with the tangent space of $\mathrm{GL}(n, \mathbb{R})$ at the identity matrix I and thus with the Lie algebra $\mathfrak{gl}(n, \mathbb{R})$ of $\mathrm{GL}(n, \mathbb{R})$: $\mathfrak{gl}(n, \mathbb{R}) \cong M_n(\mathbb{R})$.

In general, let $\mathfrak{g}$ be a Lie algebra in dimension n, and let $X_1, \cdots, X_n$ be a basis of $\mathfrak{g}$. One may assume that

$$[X_i, X_j] = \sum_{k=1}^{n} c_{ij}^k X_k, \ 1 \le i, j \le n, \tag{A.14}$$

where c_{ij}^k $(i, j, k = 1, 2, \cdots, n)$ are called the **structure constants** of the Lie algebra $\mathfrak{g}$. Then the Lie bracket between two arbitrary elements in $\mathfrak{g}$, e.g., $X = \sum_{i=1}^{n} \lambda_i X_i$ and $Y = \sum_{j=1}^{n} \mu_j X_j$, is completely determined by the structure constants as

$$[X, Y] = \sum_{i,j,k=1}^{n} \lambda_i \mu_j c_{ij}^k X_k, \ 1 \le i, j \le n. \tag{A.15}$$

It is not difficult to verify that the structure constants of a Lie algebra satisfy the following relations:

$$c_{ij}^k = -c_{ji}^k, \ 1 \le i, j, k \le n, \tag{A.16a}$$

$$\sum_{s=1}^{n} (c_{ij}^s c_{sk}^l + c_{jk}^s c_{si}^l + c_{ki}^s c_{sj}^l) = 0 \tag{A.16b}$$

for $1 \le i, j, k, l \le n$. As an example, the structure constants of the Lie algebra $\mathfrak{so}(3, \mathbb{R})$ are given by $c_{ij}^k = \epsilon_{ijk}$, where ϵ_{ijk} is the Levi-Civita symbol in three dimensions defined by $\epsilon_{ijk} = 1$, if (i, j, k) is an even permutation of (x, y, z); $\epsilon_{ijk} = -1$, if (i, j, k) is an odd permutation of (x, y, z), and $\epsilon_{ijk} = 0$ for other cases.

A.2 Special Orthogonal Group: SO(N)

The special orthogonal group, denoted as SO(N), is the group of rotations in a Euclidean space of dimension N, which is formed by all $N \times N$ special orthogonal matrices. A square matrix R is a special orthogonal matrix if $RR^\mathsf{T} = R^\mathsf{T}R = I$, where I is the identity matrix and its determinant satisfies $\det R = I$. Near the identity, the first condition is solved by writing $R \approx I + M$ and requiring $M = -M^\mathsf{T}$, i.e., M is a skew-symmetric matrix. Hence, the SO(N) group has $\frac{1}{2}(N^2 - N)$ generators M_{ab} which satisfy $M_{ab} = -M_{ba}$. For physicists, as the generators are required to be Hermitian, i.e., $M^\dagger = M$, a convenient choice would be

$$(M_{ab})_{kl} = -i(\delta_{ak}\delta_{bl} - \delta_{al}\delta_{bk}). \tag{A.17}$$

To compute the Lie algebra for SO(N), one may take SO(4) as an inspiring example. For those generators whose index sets have no integer in common, e.g., M_{12} and M_{34}, the commutator vanishes, i.e., $[M_{12}, M_{34}] = 0$. For those generators whose index sets have one integer in common, e.g., M_{23} and M_{31}, a direct computation yields $[M_{23}, M_{31}] \equiv [M_x, M_y] = iM_z = iM_{12}$, where we have used the fact that M_{12}, M_{23}, and M_{31} form a SO(3) subgroup of SO(4). It implies that when the indices b and c are equal, the commutator $[M_{ab}, M_{cd}]$ equals iM_{da}. Hence, taking into account the fact that the generators M_{ab} are skew-symmetric, and the commutator vanishes identically when the index sets for the generators have two integers in common, one immediately obtains the Lie algebra for SO(N)

$$[M_{ab}, M_{cd}] = -i(\delta_{bc}M_{ad} + \delta_{ad}M_{bc} - \delta_{ac}M_{bd} - \delta_{bd}M_{ac}). \tag{A.18}$$

In order to find the root vectors of SO(N), one may start with the first nontrivial case of SO(4) beyond the almost trivial case of SO(3). Of the six generators of SO(4), M_{12} and M_{34} form a maximal subset of mutually commuting generators. Hence, one may diagonalize it simultaneously and denote them by H_1 and H_2 in the new basis, respectively:

$$H_1 = \mathrm{diag}(1, -1, 0, 0), \tag{A.19a}$$

$$H_2 = \mathrm{diag}(0, 0, 1, -1). \tag{A.19b}$$

As the rank of SO(4) is $r = 2$, all the weights and roots are two-dimensional vectors. One can directly read off the four weights of the defining representation from the entries of H_1 and H_2, scanning Eq. (A.19) vertically

$$\boldsymbol{w}_1 = (1, 0), \boldsymbol{w}_2 = (-1, 0), \boldsymbol{w}_3 = (0, 1), \boldsymbol{w}_4 = (0, -1). \tag{A.20}$$

Hence, the four root vectors which connect the weights are given by

$$\boldsymbol{\alpha}_1 \equiv \boldsymbol{w}_1 - \boldsymbol{w}_4 = (1, 1), \tag{A.21a}$$

$$\boldsymbol{\alpha}_2 \equiv \boldsymbol{w}_1 - \boldsymbol{w}_3 = (1, -1), \tag{A.21b}$$

$$\boldsymbol{\alpha}_3 \equiv \boldsymbol{w}_4 - \boldsymbol{w}_1 = (-1, -1), \tag{A.21c}$$

$$\boldsymbol{\alpha}_4 \equiv \boldsymbol{w}_3 - \boldsymbol{w}_1 = (-1, 1). \tag{A.21d}$$

If one denotes the basis vectors of a two-dimensional Euclidean space as $\boldsymbol{e}_1$ and $\boldsymbol{e}_2$, then the four roots of SO(4) can be expressed as

$$\pm\, \mathbf{e}_1 \pm \mathbf{e}_2. \tag{A.22}$$

One may choose $\boldsymbol{\alpha}_1$ and $\boldsymbol{\alpha}_2$ as positive roots, which can be expressed as

$$\mathbf{e}_1 \pm \mathbf{e}_2, \tag{A.23}$$

where both of them are simple roots.

Before we directly jump onto SO($2r$), it is helpful to discuss two more examples in low dimensions, for which new features emerge. For SO(6), the maximal subgroup of mutually commuting generators consists of the following three traceless matrices:

$$H_1 = \mathrm{diag}(1, -1, 0, 0, 0, 0), \tag{A.24a}$$

$$H_2 = \mathrm{diag}(0, 0, 1, -1, 0, 0), \tag{A.24b}$$

$$H_3 = \mathrm{diag}(0, 0, 0, 0, 1, -1). \tag{A.24c}$$

As the rank of SO(6) is $r = 3$, all the weights and roots are three-dimensional vectors. One can directly read off the six weights of the defining representation by scanning Eq. (A.24) vertically

$$\boldsymbol{w}_1 = (1, 0, 0), \boldsymbol{w}_2 = (-1, 0, 0), \boldsymbol{w}_3 = (0, 1, 0), \tag{A.25a}$$

$$\boldsymbol{w}_4 = (0, -1, 0), \boldsymbol{w}_5 = (0, 0, 1), \boldsymbol{w}_6 = (0, 0, -1). \tag{A.25b}$$

The 12 roots of SO(6) are then given by

$$\pm\, \boldsymbol{e}_1 \pm \boldsymbol{e}_2, \pm \boldsymbol{e}_1 \pm \boldsymbol{e}_3, \pm \boldsymbol{e}_2 \pm \boldsymbol{e}_3. \tag{A.26}$$

One may choose the six positive roots to be

$$\boldsymbol{e}_1 \pm \boldsymbol{e}_2, \boldsymbol{e}_1 \pm \boldsymbol{e}_3, \boldsymbol{e}_2 \pm \boldsymbol{e}_3. \tag{A.27}$$

A convenient choice of the simple roots would be

$$\boldsymbol{\alpha}_1 \equiv \boldsymbol{e}_1 - \boldsymbol{e}_2, \boldsymbol{\alpha}_2 \equiv \boldsymbol{e}_2 - \boldsymbol{e}_3, \boldsymbol{\alpha}_3 \equiv \boldsymbol{e}_2 + \boldsymbol{e}_3. \tag{A.28}$$

Hence, the remaining positive roots in terms of the simple roots are

$$\boldsymbol{\alpha}_1 + \boldsymbol{\alpha}_2, \boldsymbol{\alpha}_1 + \boldsymbol{\alpha}_3, \boldsymbol{\alpha}_1 + \boldsymbol{\alpha}_2 + \boldsymbol{\alpha}_3, \tag{A.29}$$

For SO(8), the maximal subgroup of mutually commuting generators consists of the following four traceless matrices:

$$H_1 = \mathrm{diag}(1, -1, 0, 0, 0, 0, 0, 0), \tag{A.30a}$$

$$H_2 = \mathrm{diag}(0, 0, 1, -1, 0, 0, 0, 0), \tag{A.30b}$$

$$H_3 = \mathrm{diag}(0, 0, 0, 0, 1, -1, 0, 0), \tag{A.30c}$$

$$H_4 = \mathrm{diag}(0, 0, 0, 0, 0, 0, 1, -1). \tag{A.30d}$$

As the rank of SO(8) is $r = 4$, all the weights and roots are four-dimensional vectors. One can directly read off the eight weights of the defining representation by scanning Eq. (A.30) vertically

$$\boldsymbol{w}_1 = (1, 0, 0, 0), \boldsymbol{w}_2 = (-1, 0, 0, 0), \boldsymbol{w}_3 = (0, 1, 0, 0), \tag{A.31a}$$

$$\boldsymbol{w}_4 = (0, -1, 0, 0), \boldsymbol{w}_5 = (0, 0, 1, 0), \boldsymbol{w}_6 = (0, 0, -1, 0), \tag{A.31b}$$

$$\boldsymbol{w}_7 = (0, 0, 0, 1), \boldsymbol{w}_8 = (0, 0, 0, -1). \tag{A.31c}$$

The 24 roots of SO(8) which connect the weights are then given by

$$\pm \mathbf{e}_1 \pm \mathbf{e}_2, \pm\mathbf{e}_1 \pm \mathbf{e}_3, \pm\mathbf{e}_1 \pm \mathbf{e}_4, \pm\mathbf{e}_2 \pm \mathbf{e}_3, \pm\mathbf{e}_2 \pm \mathbf{e}_4, \pm\mathbf{e}_3 \pm \mathbf{e}_4. \tag{A.32}$$

One may choose the 12 positive roots of SO(8) to be

$$\mathbf{e}_1 \pm \mathbf{e}_2, \mathbf{e}_1 \pm \mathbf{e}_3, \mathbf{e}_1 \pm \mathbf{e}_4, \mathbf{e}_2 \pm \mathbf{e}_3, \mathbf{e}_2 \pm \mathbf{e}_4, \mathbf{e}_3 \pm \mathbf{e}_4. \tag{A.33}$$

Among them, a convenient choice of the four simple roots is

$$\boldsymbol{\alpha}_1 \equiv \mathbf{e}_1 - \mathbf{e}_2, \boldsymbol{\alpha}_2 \equiv \mathbf{e}_2 - \mathbf{e}_3, \boldsymbol{\alpha}_3 \equiv \mathbf{e}_3 - \mathbf{e}_4, \boldsymbol{\alpha}_4 \equiv \mathbf{e}_3 + \mathbf{e}_4. \tag{A.34}$$

Hence, the remaining eight positive roots in terms of the simple roots are

$$\boldsymbol{\alpha}_1 + \boldsymbol{\alpha}_2, \boldsymbol{\alpha}_2 + \boldsymbol{\alpha}_3, \boldsymbol{\alpha}_2 + \boldsymbol{\alpha}_4, \tag{A.35a}$$

$$\boldsymbol{\alpha}_1 + \boldsymbol{\alpha}_2 + \boldsymbol{\alpha}_3, \boldsymbol{\alpha}_1 + \boldsymbol{\alpha}_2 + \boldsymbol{\alpha}_4, \boldsymbol{\alpha}_2 + \boldsymbol{\alpha}_3 + \boldsymbol{\alpha}_4, \tag{A.35b}$$

$$\boldsymbol{\alpha}_1 + \boldsymbol{\alpha}_2 + \boldsymbol{\alpha}_3 + \boldsymbol{\alpha}_4, \boldsymbol{\alpha}_1 + 2\boldsymbol{\alpha}_2 + \boldsymbol{\alpha}_3 + \boldsymbol{\alpha}_4. \tag{A.35c}$$

In general, for SO$(2r)$, the maximal subgroup of mutually commuting generators are

$$\begin{aligned} H_1 &= \mathrm{diag}(1, -1, 0, 0, \cdots, 0, 0), \\ H_2 &= \mathrm{diag}(0, 0, 1, -1, \cdots, 0, 0), \\ &\vdots \\ H_r &= \mathrm{diag}(0, 0, 0, 0, 0, 0, 1, -1). \end{aligned} \tag{A.36}$$

from which one can read off the $2r$ weights for the defining representation

$$\begin{aligned} \boldsymbol{w}_1 &= (1, 0, \cdots, 0), \\ \boldsymbol{w}_2 &= (-1, 0, \cdots, 0), \\ &\vdots \\ \boldsymbol{w}_{2r-1} &= (0, 0, \cdots, 1), \\ \boldsymbol{w}_{2r} &= (0, 0, \cdots, -1). \end{aligned} \tag{A.37}$$

Exercises

A.1. Prove that every 2×2 unitary matrix U with determinant 1 can be written in the form

$$\begin{aligned} U &= \begin{pmatrix} e^{i\phi_1}\cos\theta & -e^{i\phi_2}\sin\theta \\ e^{-i\phi_2}\sin\theta & e^{-i\phi_1}\cos\theta \end{pmatrix} \\ &= \begin{pmatrix} e^{i\Psi} & 0 \\ 0 & e^{-i\Psi} \end{pmatrix} \begin{pmatrix} \cos\theta & -\sin\theta \\ \sin\theta & \cos\theta \end{pmatrix} \begin{pmatrix} e^{i\Delta} & 0 \\ 0 & e^{-i\Delta} \end{pmatrix}, \end{aligned}$$

where $\phi_1 \equiv \Psi + \Delta$ and $\phi_2 \equiv \Psi - \Delta$.

A.2. Verify that the set $\mathrm{O}(n, \mathbb{C})$ under matrix multiplication forms a group.

A.3. Prove that the orthogonal group $O(n, \mathbb{R})$ is compact. (Hint: Prove that $O(n, \mathbb{R})$ is both closed and bounded. Then, use the Heine-Borel theorem to prove that $O(n, \mathbb{R})$ is compact.)

A.4. Verify that the real dimension of the space of $n \times n$ Hermitian matrices is n^2.

References

1. E. Schrödinger, Der stetige übergang von der Mikro-zur Makromechanik. Naturwissenschaften **14**(28), 664–666 (1926)
2. R.J. Glauber, Photon correlations. Phys. Rev. Lett. **10**(3), 84 (1963)
3. R.J. Glauber, The quantum theory of optical coherence. Phys. Rev. **130**(6), 2529 (1963)
4. R.J. Glauber, Coherent and incoherent states of the radiation field. Phys. Rev. **131**(6), 2766 (1963)
5. E.C.G. Sudarshan, Equivalence of semiclassical and quantum mechanical descriptions of statistical light beams. Phys. Rev. Lett. **10**(7), 277 (1963)
6. J.R. Klauder, Continuous-representation theory. I. postulates of continuous-representation theory. J. Math. Phys. **4**(8), 1055–1058 (1963)
7. J.R. Klauder, Continuous-representation theory. II. generalized relation between quantum and classical dynamics. J. Math. Phys. **4**(8), 1058–1073 (1963)
8. A.M. Perelomov, Coherent states for arbitrary lie group. Commun. Math. Phys. **26**(3), 222–236 (1972)
9. R. Gilmore, Geometry of symmetrized states. Ann. Phys. **74**(2), 391–463 (1972)
10. R. Gilmore, On the properties of coherent states. Revista Mexicana de Fisica **23**, 143–187 (1974)
11. F.A. Berezin, Quantization in complex bounded domains. Doklady Akademii Nauk SSSR **211**(6), 1263–1266 (1973)
12. F.A. Berezin, Quantization. Math. USSR-Izvestiya **8**(5), 1109 (1974)
13. F.A. Berezin, Quantization in complex symmetric spaces. Math. USSR-Izvestiya **9**(2), 341 (1975)
14. F.A. Berezin, General concept of quantization. Commun. Math. Phys. **40**(2), 153–174 (1975)
15. A.M. Perelomov, Generalized coherent states and some of their applications. Soviet Phys. Uspekhi **20**(9), 703 (1977)
16. W.M. Zhang, Integrability and chaos in quantum systems (as viewed from geometry and dynamical symmetry). Doctoral Thesis. Drexel University, 1989
17. W.M. Zhang, D.H. Feng, R. Gilmore, Coherent states: theory and some applications. Rev. Mod. Phys. **62**(4), 867 (1990)
18. R.P. Feynman, Space-time approach to non-relativistic quantum mechanics. Rev. Mod. Phys. **20**(2), 367 (1948)
19. R.P. Feynman, The principle of least action in quantum mechanics, in *Feynman's Thesis–A New Approach To Quantum Theory* (World Scientific, Singapore, 2005), pp. 1–69
20. R.P. Feynman, A.R. Hibbs, D.F. Styer, *Quantum Mechanics and Path Integrals* (Courier Corporation, North Chelmsford, 2010)
21. J.R. Klauder, The action option and a Feynman quantization of spinor fields in terms of ordinary c-numbers. Ann. Phys. **11**(2), 123 (1960)
22. J.R. Klauder, Path integrals and stationary-phase approximations. Phys. Rev. D **19**(8), 2349 (1979)

C.-F. Kam et al., *Coherent States*, Lecture Notes in Physics 1011,
https://doi.org/10.1007/978-3-031-20766-2

23. J.P. Blaizot, H. Orland, Path integrals for the nuclear many-body problem. Phys. Rev. C **24**(4), 1740 (1981)
24. H. Kuratsuji, T. Suzuki, Path integral approach to many-nucleon systems and time-dependent Hartree-Fock. Phys. Lett. B **92**(1–2), 19–22 (1980)
25. H. Kuratsuji, Geometric canonical phase factors and path integrals. Phys. Rev. Lett. **61**(15), 1687 (1988)
26. G.H. Lang, C.W. Johnson, S.E. Koonin, W.E. Ormand, Monte Carlo evaluation of path integrals for the nuclear shell model. Phys. Rev. C **48**(4), 1518 (1993)
27. J.K. Freericks, V. Zlatić, Exact dynamical mean-field theory of the Falicov-Kimball model. Rev. Modern Phys. **75**(4), 1333 (2003)
28. A. Polkovnikov, Phase space representation of quantum dynamics. Ann. Phys. **325**(8), 1790–1852 (2010)
29. A. Altland, B.D. Simons, *Condensed Matter Field Theory* (Cambridge University Press, Cambridge, 2010)
30. S. Tomsovic, Complex saddle trajectories for multidimensional quantum wave packet and coherent state propagation: Application to a many-body system. Phys. Rev. E **98**(2), 023301 (2018)
31. E. Keçecioğlu and A. Garg. SU(2) instantons with boundary jumps and spin tunneling in magnetic molecules. Phys. Rev. Lett. **88**(23), 237205 (2002)
32. A.A. Kovalev, L.X. Hayden, G.E.W. Bauer, Y. Tserkovnyak, Macrospin tunneling and magnetopolaritons with nanomechanical interference. Phys. Rev. Lett. **106**(14), 147203 (2011)
33. J.R. Klauder, Noncanonical quantization of gravity. I. foundations of affine quantum gravity. J. Math. Phys. **40**(11), 5860–5882 (1999)
34. L. Qin, Y.G. Ma, Coherent state functional integrals in quantum cosmology. Phys. Rev. D **85**(6), 063515 (2012)
35. A. Perez, The spin-foam approach to quantum gravity. Living Rev. Relativity **16**(1), 1–128 (2013)
36. A. Haldar, S. Bera, S. Banerjee, Rényi entanglement entropy of Fermi and non-Fermi liquids: Sachdev-Ye-Kitaev model and dynamical mean field theories. Phys. Rev. Res. **2**(3), 033505 (2020)
37. R. Ghosh, N. Dupuis, A. Sen, K. Sengupta, Entanglement measures and nonequilibrium dynamics of quantum many-body systems: a path integral approach. Phys. Rev. B **101**(24), 245130 (2020)
38. E.H. Lieb, The classical limit of quantum spin systems. Commun. Math. Phys. **31**(4), 327–340 (1973)
39. F.A. Berezin, Covariant and contravariant symbols of operators. Math. USSR-Izvestiya **6**(5), 1117 (1972)
40. D.H. Feng, R. Gilmore, S.R. Deans, Phase transitions and the geometric properties of the interacting boson model. Phys. Rev. C **23**(3), 1254 (1981)
41. A.M. Perelomov, *Generalized Coherent States and Their Applications* (Springer, Berlin, 1986)
42. T.D. Lee, F.E. Low, D. Pines, The motion of slow electrons in a polar crystal. Phys. Rev. **90**(2), 297 (1953)
43. P.W. Anderson, Coherent excited states in the theory of superconductivity: Gauge invariance and the Meissner effect. Phys. Rev. **110**(4), 827 (1958)
44. J.G. Valatin, D. Butler, On the collective properties of a Boson system. Il Nuovo Cimento **10**(1), 37–54 (1958)
45. J. Schwinger, On the Green's functions of quantized fields. I. Proc. Nat. Acad. Sci **37**(7), 452–455 (1951)
46. J. Schwinger, The theory of quantized fields. VI. Phys. Rev. **94**(5), 1362 (1954)
47. E. Schrödinger, *Collected Papers on Wave Mechanics*, vol. 302 (American Mathematical Society, Providence, 2003)

48. A. Omran, H. Levine, A. Keesling, G. Semeghini, T.T. Wang, S. Ebadi, H. Bernien, A.S. Zibrov, H. Pichler, S. Choi, et al., Generation and manipulation of Schrödinger cat states in Rydberg atom arrays. Science **365**(6453), 570–574 (2019)
49. C. Song, K. Xu, H. Li, Y.R. Zhang, X. Zhang, W. Liu, Q.J. Guo, Z. Wang, W. Ren, J. Hao, et al., Generation of multicomponent atomic Schrödinger cat states of up to 20 qubits. Science **365**(6453), 574–577 (2019)
50. H. Weyl, *The Theory of Groups and Quantum Mechanics* (Dover Publications, Mineola, 1950)
51. P.A.M. Dirac, *The Principles of Quantum Mechanics* (Oxford University Press, Oxford, 1930)
52. R. Shankar, *Principles of Quantum Mechanics* (Springer Science & Business Media, Chem, 2012)
53. R. Hanbury Brown, R.Q. Twiss, Correlation between photons in two coherent beams of light. Nature **177**(4497), 27–29 (1956)
54. L. Mandel, E. Wolf, Coherence properties of optical fields. Rev. Modern Phys. **37**(2), 231 (1965)
55. E.P. Wigner, On the quantum correction for thermodynamic equilibrium. Phys. Rev. **40**(5), 749 (1932)
56. A. Einstein, B. Podolsky, N. Rosen, Can quantum-mechanical description of physical reality be considered complete? Phys. Rev. **47**(10), 777 (1935)
57. E. Schrödinger, Die gegenwärtige situation in der quantenmechanik. Naturwissenschaften **23**(49), 823–828 (1935)
58. A.O. Caldeira, A.J. Leggett, Influence of damping on quantum interference: an exactly soluble model. Phys. Rev. A **31**(2), 1059 (1985)
59. D.F. Walls, G.J. Milburn, Effect of dissipation on quantum coherence. Phys. Rev. A **31**(4), 2403 (1985)
60. W.H. Zurek, Decoherence and the transition from quantum to classical. Phys. Today **44**(10), 36 (1991)
61. B. Yurke, D. Stoler, Generating quantum mechanical superpositions of macroscopically distinguishable states via amplitude dispersion. Phys. Rev. Lett **57**(1), 13 (1986)
62. C.H. Bennett, G. Brassard, C. Crépeau, R. Jozsa, A. Peres, W.K. Wootters, Teleporting an unknown quantum state via dual classical and Einstein-Podolsky-Rosen channels. Phy. Rev. Lett. **70**(13), 1895 (1993)
63. S.J. van Enk, O. Hirota, Entangled coherent states: Teleportation and decoherence. Phys. Rev. A **64**(2), 022313 (2001)
64. M.A. Nielsen, I. Chuang, *Quantum Computation and Quantum Information* (Cambridge University Press, Cambridge, 2000)
65. T.C. Ralph, A. Gilchrist, G.J. Milburn, W.J. Munro, S. Glancy, Quantum computation with optical coherent states. Phys. Rev. A **68**(4), 042319 (2003)
66. J. Feinberg, Self-pumped, continuous-wave phase conjugator using internal reflection. Opt. Lett. **7**(10), 486–488 (1982)
67. P. Marek, J. Fiurášek, Elementary gates for quantum information with superposed coherent states. Phys. Rev. A **82**(1), 014304 (2010)
68. J. Wenger, R. Tualle-Brouri, P. Grangier, Non-Gaussian statistics from individual pulses of squeezed light. Phys. Rev. Lett. **92**(15), 153601 (2004)
69. J.L. Martin, Generalized classical dynamics, and the 'classical analogue' of a Fermi oscillator. Proc. R. Soc. London A: Math. Phys. Sci. **251**(1267), 536–542 (1959)
70. J.L. Martin, The Feynman principle for a Fermi system. Proc. R. Soc. London A: Math. Phys. Sci. **251**(1267), 543–549 (1959)
71. Y. Ohnuki, T. Kashiwa, Coherent states of Fermi operators and the path integral. Prog. Theor. Phys. **60**(2), 548 (1978)

72. H. Grassmann, *Die lineale Ausdehnungslehre ein neuer Zweig der Mathematik: dargestellt und durch Anwendungen auf die übrigen Zweige der Mathematik, wie auch auf die Statik, Mechanik, die Lehre vom Magnetismus und die Krystallonomie erläutert*, vol. 1 (O. Wigand, Germany, 1844)
73. J. Schwinger, The theory of quantized fields. vi. Phys. Rev. **92**(5), 1283 (1953)
74. F.A. Berezin, *The Method of Second Quantization* (Academic, New York, 1966)
75. K.E. Cahill, R.J. Glauber, Density operators for Fermions. Phys. Rev. A **59**(2), 1538 (1999)
76. J.R. Klauder, B.S. Skagerstam, *Coherent States: Applications in Physics and Mathematical Physics* (World Scientific, Singapore, 1985)
77. M Baranger, M.A.M. de Aguiar, F. Keck, H.J. Korsch, B. Schellhaaß, Semiclassical approximations in phase space with coherent states. J. Phys. A Gen. Phys. **34**(36), 7227 (2001)
78. J.M. Radcliffe, Some properties of coherent spin states. J. Phys. A Gen. Phys. **4**(3), 313 (1971)
79. F.T. Arecchi, E. Courtens, R. Gilmore, H. Thomas, Atomic coherent states in quantum optics. Phys. Rev. A **6**(6), 2211 (1972)
80. M.V. Berry, Quantal phase factors accompanying adiabatic changes. Proc. Roy. Soc. A **392**(1802), 45 (1984)
81. Y. Aharonov, J. Anandan, Phase change during a cyclic quantum evolution. Phys. Rev. Lett. **58**(16), 1593 (1987)
82. J.I. Cirac, P. Zoller, Preparation of macroscopic superpositions in many-atom systems. Phys. Rev. A **50**(4), R2799 (1994)
83. C.C. Gerry, R. Grobe, Generation and properties of collective atomic Schrödinger-cat states. Phys. Rev. A **56**(3), 2390 (1997)
84. C.C. Gerry, R. Grobe, Cavity-QED state reduction method to produce atomic Schrödinger-cat states. Phys. Rev. A **57**(3), 2247 (1998)
85. G.S. Agarwal, R.R. Puri, R.P. Singh, Atomic Schrödinger cat states. Phys. Rev. A **56**(3), 2249 (1997)
86. E.T. Jaynes, F.W. Cummings, Comparison of quantum and semiclassical radiation theories with application to the beam maser. Proc. IEEE **51**(1), 89–109 (1963)
87. S. Haroche, J.M. Raimond, *Exploring the Quantum: Atoms, Cavities, and Photons* (Oxford University Press, Oxford, 2006)
88. D.F. Walls, G.J. Milburn, *Quantum Optics* (Springer Science & Business Media, Berlin, 2007)
89. M.J. Holland, D.F. Walls, P. Zoller, Quantum nondemolition measurements of photon number by atomic beam deflection. Phys. Rev. Lett. **67**(13), 1716 (1991)
90. A. Blais, R.S. Huang, A. Wallraff, S.M. Girvin, R.J. Schoelkopf, Cavity quantum electrodynamics for superconducting electrical circuits: an architecture for quantum computation. Phys. Rev. A **69**(6), 062320 (2004)
91. M. Brune, S. Haroche, J.M. Raimond, L. Davidovich, N. Zagury, Manipulation of photons in a cavity by dispersive atom-field coupling: quantum-nondemolition measurements and generation of "Schrödinger cat"states. Phys. Rev. A **45**(7), 5193 (1992)
92. G.S. Agarwal, Relation between atomic coherent-state representation, state multipoles, and generalized phase-space distributions. Phys. Rev. A **24**(6), 2889 (1981)
93. G.S. Agarwal, Perspective of Einstein-Podolsky-Rosen spin correlations in the phase-space formulation for arbitrary values of the spin. Phys. Rev. A **47**(6), 4608 (1993)
94. E. Shojaee, C.S. Jackson, C.A. Riofrío, A. Kalev, I.H. Deutsch, Optimal pure-state qubit tomography via sequential weak measurements. Phys. Rev. Lett. **121**(13), 130404 (2018)
95. L.C. Biedenharn, J.D. Louck, *Angular Momentum in Quantum Physics: Theory and Application* (Cambridge University Press, Cambridge, 1984)
96. D. Stoler, Equivalence classes of minimum uncertainty packets. Phys. Rev. D **1**(12), 3217 (1970)
97. D. Stoler, Equivalence classes of minimum-uncertainty packets. ii. Phys. Rev. D **4**(6), 1925 (1971)

98. H.P. Yuen, Two-photon coherent states of the radiation field. Phys. Rev. A **13**(6), 2226 (1976)
99. R.E. Slusher, L.W. Hollberg, B. Yurke, J.C. Mertz, J.F. Valley, Observation of squeezed states generated by four-wave mixing in an optical cavity. Phys. Rev. Lett. **55**(22), 2409 (1985)
100. R.M. Shelby, M.D. Levenson, S.H. Perlmutter, R.G. DeVoe, D.F. Walls, Broad-band parametric deamplification of quantum noise in an optical fiber. Phys. Rev. Lett. **57**(6), 691 (1986)
101. L.A. Wu, H.J. Kimble, J.L. Hall, H. Wu, Generation of squeezed states by parametric down conversion. Phys. Rev. Lett. **57**(20), 2520 (1986)
102. B.P. Abbott, R. Abbott, R. Adhikari, P. Ajith, B. Allen, G. Allen, R.S. Amin, S.B. Anderson, W.G. Anderson, M.A. Arain, et al., Ligo: the laser interferometer gravitational-wave observatory. Rep. Progr. Phys. **72**(7), 076901 (2009)
103. T. Accadia, F. Acernese, M. Alshourbagy, P. Amico, F. Antonucci, S. Aoudia, N. Arnaud, C. Arnault, K.G. Arun, P. Astone, et al., Virgo: a laser interferometer to detect gravitational waves. J. Instr. **7**(03), P03012 (2012)
104. J. Aasi, J. Abadie, B.P. Abbott, R. Abbott, T.D. Abbott, M.R. Abernathy, C. Adams, T. Adams, P. Addesso, R.X. Adhikari, et al., Enhanced sensitivity of the LIGO gravitational wave detector by using squeezed states of light. Nat. Photonics **7**(8), 613 (2013)
105. J.N. Hollenhorst, Quantum limits on resonant-mass gravitational-radiation detectors. Phys. Rev. D **19**(6), 1669 (1979)
106. C.M. Caves, Quantum-mechanical radiation-pressure fluctuations in an interferometer. Phys. Rev. Lett. **45**(2), 75 (1980)
107. C.M. Caves, Quantum-mechanical noise in an interferometer. Phys. Rev. D **23**(8), 1693 (1981)
108. K. McKenzie, N. Grosse, W.P. Bowen, S.E. Whitcomb, M.B. Gray, D.E. McClelland, P.K. Lam, Squeezing in the audio gravitational-wave detection band. Phys. Rev. Lett. **93**(16), 161105 (2004)
109. H. Vahlbruch, S. Chelkowski, B. Hage, A. Franzen, K. Danzmann, R. Schnabel, Coherent control of vacuum squeezing in the gravitational-wave detection band. Phys. Rev. Lett. **97**(1), 011101 (2006)
110. H. Vahlbruch, S. Chelkowski, K. Danzmann, R. Schnabel, Quantum engineering of squeezed states for quantum communication and metrology. New J. Phys. **9**(10), 371 (2007)
111. H. Grote, K. Danzmann, K.L. Dooley, R. Schnabel, J. Slutsky, H. Vahlbruch, First long-term application of squeezed states of light in a gravitational-wave observatory. Phys. Rev. Lett. **110**(18), 181101 (2013)
112. L. Mandel, Sub-Poissonian photon statistics in resonance fluorescence. Opt. Lett. **4**(7), 205–207 (1979)
113. R. Demkowicz-Dobrzański, K. Banaszek, R. Schnabel, Fundamental quantum interferometry bound for the squeezed-light-enhanced gravitational wave detector GEO 600. Phys. Rev. A **88**(4), 041802 (2013)
114. M. Xiao, L.A. Wu, H.J. Kimble, Precision measurement beyond the shot-noise limit. Phys. Rev. Lett. **59**(3), 278 (1987)
115. P. Grangier, R.E. Slusher, B. Yurke, A. LaPorta, Squeezed-light–enhanced polarization interferometer. Phys. Rev. Lett. **59**(19), 2153 (1987)
116. K. McKenzie, D.A. Shaddock, D.E. McClelland, B.C. Buchler, P.K. Lam, Experimental demonstration of a squeezing-enhanced power-recycled Michelson interferometer for gravitational wave detection. Phys. Rev. Lett. **88**(23), 231102 (2002)
117. H. Vahlbruch, S. Chelkowski, B. Hage, A. Franzen, K. Danzmann, R. Schnabel, Demonstration of a squeezed-light-enhanced power-and signal-recycled Michelson interferometer. Phys. Rev. Lett. **95**(21), 211102 (2005)
118. H. Vahlbruch, M. Mehmet, S. Chelkowski, B. Hage, A. Franzen, N. Lastzka, S. Gossler, K. Danzmann, R. Schnabel, Observation of squeezed light with 10-dB quantum-noise reduction. Phys. Rev. Lett. **100**(3), 033602 (2008)

119. K. Goda, O. Miyakawa, E.E. Mikhailov, S. Saraf, R. Adhikari, K. McKenzie, R. Ward, S. Vass, A.J. Weinstein, N. Mavalvala, A quantum-enhanced prototype gravitational-wave detector. Nat. Phys. **4**(6), 472–476 (2008)
120. J. Abadie, B.P. Abbott, R. Abbott, T.D. Abbott, M. Abernathy, C. Adams, R. Adhikari, C. Affeldt, B. Allen, G.S. Allen, et al., A gravitational wave observatory operating beyond the quantum shot-noise limit. Nat. Phys. **7**(12), 962 (2011)
121. J. Gea-Banacloche, G. Leuchs, Squeezed states for interferometric gravitational-wave detectors. J. Modern Opt. **34**(6–7), 793–811 (1987)
122. S. Lloyd, S.L. Braunstein, Quantum computation over continuous variables. Phys. Rev. Lett. **82**(8), 1784–1787 (1999)
123. S.L. Braunstein, Error correction for continuous quantum variables. Phys. Rev. Lett. **80**(18), 4084–4087 (1998)
124. M. Kitagawa, M. Ueda, Squeezed spin states. Phys. Rev. A **47**(6), 5138 (1993)
125. C.C. Gerry, Dynamics of SU(1,1) coherent states. Phys. Rev. A **31**(4), 2721 (1985)
126. V. Bužek, Jaynes-Cummings model with intensity-dependent coupling interacting with Holstein-Primakoff su(1,1) coherent state. Phys. Rev. A **39**(6), 3196 (1989)
127. A. Zee, *Quantum Field Theory in a Nutshell*, vol. 7 (Princeton University Press, Princeton, 2010)
128. G. Weinreich, *Solids: Elementary Theory for Advanced Students* (Wiley, Hoboken, 1965)
129. P. Ring, P. Schuck, *The Nuclear Many-Body Problem* (Springer Science & Business Media, Berlin, 2004)
130. J. Bardeen, L.N. Cooper, J.R. Schrieffer, Theory of superconductivity. Phys. Rev. **108**(5), 1175 (1957)
131. B. Simon, The classical limit of quantum partition functions. Commun. Math. Phys. **71**(3), 247–276 (1980)
132. M. Rasetti, Coherent states and partition function. Int. J. Theoret. Phys. **14**(1), 1–21 (1975)
133. M. Vojta, Quantum phase transitions. Rep. Progr. Phys. **66**(12), 2069 (2003)
134. H.v. Löhneysen, A. Rosch, M. Vojta, P. Wölfle, Fermi-liquid instabilities at magnetic quantum phase transitions. Rev. Modern Phys. **79**(3), 1015 (2007)
135. J.A. Hertz, Quantum critical phenomena. Phys. Rev. B **14**(3), 1165 (1976)
136. R. Gilmore, D.H. Feng, Phase transitions in nuclear matter described by pseudospin hamiltonians. Nuclear Phys. A **301**(2), 189–204 (1978)
137. R.P. Feynman, *Statistical Mechanics: A Set of Lectures* (CRC Press, Boca Raton, 1998)
138. J. Pecaric, T. Furuta, J.M. Hot, Y. Seo, *Mond-Pecaric Method in Operator Inequalities*. (Element Zagreb, Zagreb, 2005)
139. L.G. Yaffe, Large N limits as classical mechanics. Rev. Modern Phys. **54**(2), 407 (1982)
140. W. Rudin, *Principles of Mathematical Analysis*, 3rd edn. (McGraw-hill, New York, 1976)
141. W. Rudin, *Real and Complex Analysis*, 3rd edn. (McGraw-Hill, New York, 1986)
142. N. Dunford, J.T. Schwartz, *Linear Operators, Part 1: General Theory*, vol. 10 (Wiley, Hoboken, 1988)
143. J.R. Munkres, *Topology*, 2nd edn. (Prentice Hall, Hoboken, 2000)
144. R. Gilmore, The classical limit of quantum nonspin systems. J. Math. Phys. **20**(5), 891–893 (1979)
145. M.A. Caprio, P. Cejnar, F. Iachello, Excited state quantum phase transitions in many-body systems. Ann. Phys. **323**(5), 1106–1135 (2008)
146. D. Petrellis, A. Leviatan, F. Iachello, Quantum phase transitions in Bose–Fermi systems. Ann. Phys. **326**(4), 926–957 (2011)
147. H.J. Lipkin, N. Meshkov, A.J. Glick, Validity of many-body approximation methods for a solvable model:(I). exact solutions and perturbation theory. Nuclear Phys. **62**(2), 188–198 (1965)
148. A.E.L. Dieperink, O. Scholten, F. Iachello, Classical limit of the interacting-boson model. Phys. Rev. Lett. **44**(26), 1747 (1980)
149. O. Scholten, F. Iachello, A. Arima, Interacting boson model of collective nuclear states III. the transition from SU(5) to SU(3). Ann. Phys. **115**(2), 325–366 (1978)

150. F. Iachello, A. Arima, *The Interacting Boson Model* (Cambridge University Press, Cambridge, 1987)
151. W.M. Zhang, D.H. Feng, J.N. Ginocchio, Geometrical interpretation of SO(7): A critical dynamical symmetry. Phys. Rev. Lett. **59**(18), 2032 (1987)
152. W.M. Zhang, D.H. Feng, J.N. Ginocchio, Geometrical structure and critical phenomena in the fermion dynamical symmetry model: SO(8). Phys. Rev. C **37**(3), 1281 (1988)
153. J. Vidal, G. Palacios, R. Mosseri, Entanglement in a second-order quantum phase transition. Phys. Rev. A **69**(2), 022107 (2004)
154. J.I. Latorre, R. Orús, E. Rico, J. Vidal, Entanglement entropy in the lipkin-meshkov-glick model. Phys. Rev. A **71**(6), 064101 (2005)
155. S. Dusuel, J. Vidal, Finite-size scaling exponents of the Lipkin-Meshkov-Glick model. Phys. Rev. Lett. **93**(23), 237204 (2004)
156. T. Barthel, S. Dusuel, J. Vidal, Entanglement entropy beyond the free case. Phys. Rev. Lett. **97**(22), 220402 (2006)
157. D.A. Varshalovich, A.N. Moskalev, V.K. Khersonskii, *Quantum Theory of Angular Momentum* (World Scientific, Singapore, 1988)
158. T.Y. Li, J.A. Yorke, Period three implies chaos. Amer. Math. Monthly **82**(10), 985–992 (1975)
159. R.M. May, Simple mathematical models with very complicated dynamics. Nature **261**(5560), 459–467 (1976)
160. M.J. Feigenbaum, Quantitative universality for a class of nonlinear transformations. J. Statist. Phys. **19**(1), 25–52 (1978)
161. O. Bohigas, M.J. Giannoni, C. Schmit, Characterization of chaotic quantum spectra and universality of level fluctuation laws. Phys. Rev. Lett. **52**(1), 1 (1984)
162. E.P. Wigner, Characteristic vectors of bordered matrices with infinite dimensions. Ann. Math. **62**(3), 548–564 (1955)
163. F.J. Dyson, Statistical theory of the energy levels of complex systems. i. J. Math. Phys. **3**(1), 140–156 (1962)
164. M.L. Mehta, *Random Matrices*, 3rd edn. (Elsevier, Amsterdam, 2004)
165. E.J. Heller, Bound-state eigenfunctions of classically chaotic hamiltonian systems: scars of periodic orbits. Phys. Rev. Lett. **53**(16), 1515 (1984)
166. H. Bernien, S. Schwartz, A. Keesling, H. Levine, A. Omran, H. Pichler, S. Choi, A.S. Zibrov, M. Endres, M. Greiner, V. Vuletić, M.D. Lukin, Probing many-body dynamics on a 51-atom quantum simulator. Nature **551**(7682), 579–584 (2017)
167. C.J. Turner, A.A. Michailidis, D.A. Abanin, M. Serbyn, Z. Papić, Quantum scarred eigenstates in a rydberg atom chain: entanglement, breakdown of thermalization, and stability to perturbations. Phys. Rev. B **98**(15), 155134 (2018)
168. W.M. Zhang, D.H. Feng, Quantum nonintegrability in finite systems. Phys. Rep. **252**(1-2), 1–100 (1995)
169. F. Strocchi, Complex coordinates and quantum mechanics. Rev. Modern Phys. **38**(1), 36 (1966)
170. J. Ping, F. Wang, J.Q. Chen, *Group Representation Theory for Physicists* (World Scientific Publishing Company, Singapore, 2002)
171. I.C. Percival, Regular and irregular spectra. J. Phys. B: Atomic and Molecular Physics **6**(9), L229 (1968-1987)
172. I.C. Percival, Semiclassical theory of bound states. Adv. Chem. Phys, **36**, 1–61 (1977)
173. O. Bohigas, H.A. Weidenmueller, Aspects of chaos in nuclear physics. Ann. Rev. Nuclear Part. Sci. **38**, 421–453 (1988)
174. W.M. Zhang, C.C. Martens, D.H. Feng, J.M. Yuan, Dynamical symmetry breaking and quantum nonintegrability. Phys. Rev. Lett. **61**(19), 2167 (1988)
175. W.M. Zhang, D.H. Feng, J.M. Yuan, S.J. Wang, Integrability and nonintegrability of quantum systems: quantum integrability and dynamical symmetry. Phys. Rev. A **40**(1), 438 (1989)
176. W.M. Zhang, D.H. Feng, J.M. Yuan, Integrability and nonintegrability of quantum systems. ii. dynamics in quantum phase space. Phys. Rev. A **42**(12), 7125 (1990)

177. A. Bohm, Y. Ne'eman, A.O. Barut, et al., *Dynamical Groups and Spectrum Generating Algebras*, vol. 1 (World Scientific, Singapore, 1988)
178. A. Bohm, Y. Ne'eman , A.O. Barut, et al., *Dynamical Groups and Spectrum Generating Algebras*, vol. 2 (World Scientific, Singapore, 1988)
179. J. von Neumann, *Mathematische Grundlagen der Quantenmechanik* (Springer, Berlin, 1932)
180. J. Schwinger, Brownian motion of a quantum oscillator. J. Math. Phys. **2**(3), 407–432 (1961)
181. R.P. Feynman, F.L. Vernon Jr., The theory of a general quantum system interacting with a linear dissipative system. Ann. Phys. **24**, 118–173 (1963)
182. L.P. Kadanoff, G. Baym, *Quantum Statistical Mechanics: Green's Function Methods in Equilibrium and Nonequilibrium Problems* (Benjamin, New York, 1962)
183. L.V. Keldysh, Diagram technique for nonequilibrium processes. Sov. Phys. JETP **20**(4), 1018–1026 (1965)
184. K.C. Chou, Z.B. Su, B.L. Hao, L. Yu, Equilibrium and nonequilibrium formalisms made unified. Phys. Rep. **118**(1–2), 1–131 (1985)
185. J. Rammer, H. Smith, Quantum field-theoretical methods in transport theory of metals. Rev. Modern Phys. **58**(2), 323 (1986)
186. A.O. Caldeira, A.J. Leggett, Path integral approach to quantum Brownian motion. Phys. A Statist. Mech. Appl. **121**(3), 587–616 (1983)
187. A.O. Caldeira, A.J. Leggett, Quantum tunnelling in a dissipative system. Ann. Phys. **149**(2), 374–456 (1983)
188. A.J. Leggett, S. Chakravarty, A.T. Dorsey, M.P.A. Fisher, A. Garg, W. Zwerger, Dynamics of the dissipative two-state system. Rev. Modern Phys. **59**(1), 1 (1987)
189. H. Grabert, P. Schramm, G.L. Ingold, Quantum Brownian motion: the functional integral approach. Phys. Rep. **168**(3), 115–207 (1988)
190. U. Weiss, *Quantum Dissipative Systems* (World Scientific, Singapore, 2021)
191. F. Haake, R. Reibold, Strong damping and low-temperature anomalies for the harmonic oscillator. Phys. Rev. A **32**(4), 2462 (1985)
192. B.L. Hu, J.P. Paz, Y.H. Zhang, Quantum Brownian motion in a general environment: exact master equation with nonlocal dissipation and colored noise. Phys. Rev. D **45**(8), 2843 (1992)
193. R. Karrlein, H. Grabert, Exact time evolution and master equations for the damped harmonic oscillator. Phys. Rev. E **55**(1), 153 (1997)
194. L.D. Faddeev, A.A. Slavnov, *Gauge Fields: Introduction to Quantum Theory* (Benjamin-Cummings, Reading, 1980)
195. M.W.Y. Tu, W.M. Zhang, non-Markovian decoherence theory for a double-dot charge qubit. Phys. Rev. B **78**(23), 235311 (2008)
196. J.S. Jin, M.W.Y. Tu, W.M. Zhang, Y.J. Yan, Non-equilibrium quantum theory for nanodevices based on the Feynman–Vernon influence functional. New J. Phys. **12**(8), 083013 (2010)
197. C.U. Lei, W.M. Zhang, A quantum photonic dissipative transport theory. Ann. Phys. **327**(5), 1408–1433 (2012)
198. W.M. Zhang, P.Y. Lo, H.N. Xiong, M.W.Y. Tu, F. Nori, General non-Markovian dynamics of open quantum systems. Phys. Rev. Lett. **109**(17), 170402 (2012)
199. H.L. Lai, P.Y. Yang, Y.W. Huang, W.M. Zhang, Exact master equation and non-Markovian decoherence dynamics of majorana zero modes under gate-induced charge fluctuations. Phys. Rev. B **97**(5), 054508 (2018)
200. W.M. Zhang, Exact master equation and general non-Markovian dynamics in open quantum systems. Europ. Phys. J. Spec. Topics **227**(15), 1849–1867 (2019)
201. Y.W. Huang, P.Y. Yang, W.M. Zhang, Quantum theory of dissipative topological systems. Phys. Rev. B **102**(16), 165116 (2020)
202. P.W. Anderson, Absence of diffusion in certain random lattices. Phys. Rev. **109**(5), 1492 (1958)
203. U. Fano, Effects of configuration interaction on intensities and phase shifts. Phys. Rev. **124**(6), 1866 (1961)
204. A.E. Miroshnichenko, S. Flach, Y.S. Kivshar, Fano resonances in nanoscale structures. Rev. Modern Phys. **82**(3), 2257 (2010)

205. P. Lambropoulos, G.M. Nikolopoulos, T.R. Nielsen, S. Bay, Fundamental quantum optics in structured reservoirs. Rep. Progr. Phys. **63**(4), 455 (2000)
206. G.D. Mahan, *Many-Particle Physics* (Springer Science & Business Media, Berlin, 2000)
207. P.-Y. Yang, W.-M. Zhang, Master equation approach to transient quantum transport in nanostructures. Front. Phys. **12**(4), 127204 (2017)
208. H.N. Xiong, P.Y. Lo, W.M. Zhang, D.H. Feng, F. Nori, Non-Markovian complexity in the quantum-to-classical transition. Sci. Rep. **5**(1), 13353 (2015)
209. M.M. Ali, P.Y. Lo, M.W.Y. Tu, W.M. Zhang, Non-Markovianity measure using two-time correlation functions. Phys. Rev. A **92**(6), 062306 (2015)
210. W.M. Huang, W.M. Zhang, Nonperturbative renormalization of quantum thermodynamics from weak to strong couplings. Phys. Rev. Res. **4**(2), 023141 (2022)
211. A.G. Redfield, On the theory of relaxation processes. IBM J. Res. Develop. **1**(1), 19–31 (1957)
212. H.P. Breuer, F. Petruccione, *The Theory of Open Quantum Systems* (Oxford University Press, Oxford, 2002)
213. Y. Meir, N.S. Wingreen, Landauer formula for the current through an interacting electron region. Phys. Rev. Lett. **68**(16), 2512 (1992)
214. H. Haug, A.P. Jauho, *Quantum Kinetics in Transport and Optics of Semiconductors*, 2nd edn. (Springer, Berlin, 2008)
215. M. Büttiker, Scattering theory of current and intensity noise correlations in conductors and wave guides. Phys. Rev. B **46**(19), 12485 (1992)
216. Y. Imry, *Introduction to Mesoscopic Physics*, 2nd edn. (Oxford University Press, Oxford, 2002)
217. V. Gorini, A. Kossakowski, E.C.G. Sudarshan, Completely positive dynamical semigroups of N-level systems. J. Math. Phys. **17**(5), 821–825 (1976)
218. G. Lindblad, On the generators of quantum dynamical semigroups. Commun. Math. Phys. **48**(2), 119–130 (1976)
219. H.B. Callen, *Thermodynamics and an Introduction to Thermostatistics* (Wiley, Hoboken, 1985)
220. K. Huang, *Statistical Mechanics* (Elsevier, Amsterdam, 1987)
221. M.W. Zemansky, R. Dittman, *Heat and Thermodynamics: An Intermediate Textbook* (McGraw-Hill, New York, 1997)
222. T. Tao, *Hilbert's Fifth Problem and Related Topics*, vol. 153 (American Mathematical Society, Providence, 2014)

Printed in the United States
by Baker & Taylor Publisher Services